W9-BHA-323

Professional Engineers' Examination Questions and Answers

Professional Engineers' Examination Questions and Answers

WILLIAM S. LA LONDE, JR.

Licensed Professional Engineer, New Jersey, New York
Chairman Emeritus, Department of Civil and
Environmental Engineering
New Jersey Institute of Technology, Newark, N.J.

WILLIAM J. STACK-STAIKIDIS

Licensed Professional Engineer, New Jersey, New York
Dean, School of Engineering
Pratt Institute, Brooklyn, N.Y.

FOURTH EDITION

McGRAW-HILL BOOK COMPANY

New York   St. Louis   San Francisco   Auckland
Bogotá   Hamburg   Johannesburg   London
Madrid   Mexico   Montreal   New Delhi
Panama   Paris   São Paulo   Singapore
Sydney   Tokyo   Toronto

Library of Congress Cataloging in Publication Data

La Londe, William S.
 Professional engineers' questions and answers

 1. Engineering—Examinations, questions, etc.
 2. Engineers—Licenses—United States. I. Stack-
Staikidis, William J., date. II. Title
TA159.L3   1983   620'.0076   82-23393
ISBN 0-07-036099-5

Copyright © 1984, 1976, 1966, 1960, 1956 by McGraw-Hill, Inc.
All rights reserved. Printed in the United States of America.
Except as permitted under the Copyright Act of 1976, no part
of this publication may be reproduced or distributed in any
form or by any means, or stored in a data base or retrieval
system, without the prior written permission of the publisher.

1234567890   BKP/BKP   89876543

ISBN 0-07-036099-5

*The editors for this book were Diane Heiberg and Alice Goehring,
the designer was Naomi Auerbach, and the production
supervisor was Teresa F. Leaden. It was set in Century Schoolbook
by Techna-Type.*

Printed and bound by The Book Press.

CONTENTS

PREFACE

This book has been prepared to help both the newly graduated engineer and the older practicing engineer in their review of engineering fundamentals in preparation for the professional engineer examinations as part requirement for the engineer-in-training certificate or the professional engineer license.

About 80 percent of the problems appearing in this book were taken directly from sample question sheets or past professional engineer examinations as given by the many state boards of professional engineer examiners throughout the United States. These problems may be considered a composite of the many examinations which have been given in the past, but they purposely do not represent the examination questions of any state. The problems that have been added were included so as to give better coverage of some of the subject matter for those preparing to take future professional engineer examinations.

Recent professional engineer license examinations indicated a need to extend the scope of this book to provide added coverage in some parts of the several professional fields. The author also received letters from users of this book suggesting more practice questions and answers in certain sections. Further, some state boards of professional engineers have adopted a multiple-choice format of questions for the general engineering examination area. This second edition, through the addition of one hundred questions and answers, attempts to meet these needs.

Grateful acknowledgment is given to the many state boards of professional engineers or engineering examiners and to the National Council of Engineering Examiners for contributing specimen or sample questions or information on the scope of examinations and of registration; and to professional engineers R. E. Anderson, S. Fishman, J. Joffe, W. Jordan, G. C. Keeffe, F. G. Lehman, R. D. Mangasarian, C. L. Mantell, E. Miller,

A. A. Nims, J. L. Polaner, J. M. Robbins, F. A. Russell, and L. Shapiro, all of the faculty of the Newark College of Engineering, for the services which they contributed toward the solutions of the problems in their fields of specialization, and to the past users of this book who spotted and took the time to report discrepancies in the earlier printings.

William S. La Londe, Jr.

ACKNOWLEDGMENTS FOR THE THIRD EDITION

Since the First Edition was published in 1956, engineering knowledge has expanded and with it the codes of practice. The questions asked in the professional engineering license examinations have accordingly kept pace with these changes. This Third Edition brings this book in step with these changes. About 20 percent of the questions and answers in the professional engineering areas were either modified or new material substituted to reflect new engineering knowledge and codes of practice.

Of particular importance to me was my success in enlisting professional engineer William J. Stack-Staikidis to serve as co-author. He showed his ability in the work he did for this new edition. We both are grateful to professional engineers J. Carluccio, R. D. Mangasarian, E. Miller, and D. P. Tassios, all of the faculty of the New Jersey Institute of Technology for their contributions in the fields of their specialization to ensure that the questions and answers here presented represent the latest to be found in the professional engineering license examinations.

A WORD ABOUT THE FOURTH EDITION

Some 80 percent of the Professional Engineer Licensing Agencies use the services of the National Council of Engineering Examiners (NCEE) to supply questions for part or all of their examinations and to score the answers to the questions. Accordingly, as many of the questions in this book as it was considered practicable to convert have been placed in a multiple-choice format as used by NCEE. By giving the examinee a choice of answers that are reasonably close together, blunders can be avoided, but it is imperative that the examinee be accurate in making the calculations. The choice of answer is not to be guessed at.

Questions and answers have been dualized in part with the metric system. For those using this book outside the United States, it is hoped that this will be helpful. Questions dealing with costs have been changed in part to reasonably reflect present-day costs. Questions have been modified or new ones added to satisfy code changes or new material.

William S. La Londe, Jr.
William J. Stack-Staikidis

HOW TO USE THIS BOOK

It is suggested that the General Information in Part One be reviewed to estimate your eligibility and qualifications for making application for the engineer-in-training certificate or professional engineer's license.

The definitions set forth in the Model Law for the Registration of Professional Engineers and Land Surveyors should be helpful. Your professional experience should be carefully scrutinized and compared with the Experience Qualification Guide prepared by the Committee on Qualifying Experience of the National Council of Engineering Examiners. After you have done this and you feel that you have had the prerequisite training and years of professional experience to qualify you to apply for the professional engineer's license, then write to the secretary of your state board of registration for professional engineers and land surveyors and request an application blank and information with respect to the laws covering the requirements. After you have filled out and submitted your application blank, you will be informed by the license board whether there will be any written examinations or not. Some license boards will give bona fide applicants sample or specimen questions, and all will at least inform the applicant as to the scope of the examinations.

Having at least learned the scope of your examination, you will find many problems worked out in this book that will be helpful to you in reviewing the engineering fundamentals with which you should be familiar. It is the intent here only to help you review what you know. Space limitations prevent the derivation of formulas or explanations concerning the subject material which may be found in standard engineering texts. Formulas and symbols will be identified as well as source material taken from special tables. The solutions to these problems should be studied and used as a guide for minimum clarity and explanation when later working out the problems of your particular examination. So often engineers, when taking these examinations, present their work so nonunderstandingly as to make it diffcult for the examiner to follow the specific solutions.

In studying for the examinations, you should have a sufficient duplication of problems to do a certain number with the solutions as practice and a certain number without the solutions to test yourself for competency and speed. There is no equal in the preparation for these license examinations to the working of a sufficiently large number of these problems.

In practically all the state board examinations, the applicant will be permitted some choice. It is desirable at the beginning of each examination to calmly and thoughtfully read over all the questions and to pick out those you wish to do; and it might be a good idea to do first those problems which you can do readily. If eight problems are to be done in four hours, try first to budget your time with not over 15 or 20 minutes per problem. This will allow an opportunity to go back and review the problems afterwards. It is also better to answer six problems correctly than to try to get all eight problems answered with only part scores. Sometimes if one has difficulty with a problem, it is well not to work on it too long as a suggestion in another problem or the feeling of confidence gained from the solution of a number of problems will stimulate the thinking or relax the pressure, and the solution may be obtained when one returns to the problem. Neatness and carefulness in the arithmetical work and in making tabulations will ensure the best results. Never omit a check if it is possible to make one.

Sometimes needed information is purposely omitted from the examination question to test the examinee's ability to size up the situation and to supply reasonable data. In answering such a question, merely indicate your assumptions and proceed from that point. It is well to be careful, however, not to make assumptions where none are needed.

A satisfactory speed for the solution of these problems, indicative of reasonable familiarity with the fundamentals of any given section, is to be able to work two or three problems per hour. As much work as possible should be done without the use of texts or notes, as in many cases their use or the use of books such as this one are not permitted at the examinations. It is best not to bank on the use of more than a handbook of tables and formulas, and where the necessary information is supplied to the questions, perhaps not more than a slide rule or nonprogrammed computer will be permitted. Where books are permitted, the questions may be more searching, and unless you have a knowledge of your subject, much time can be lost browsing in books looking for the answer. The solutions to the problems in this book are purposely separated from the questions to encourage you as far as it may be possible to do your own thinking and work.

Those who are preparing for the engineer-in-training examinations would do well to review all questions and solutions 1.101 *to* 1.825, 5.01 *to* 5.25, *and* 7.01 *to* 7.75.

Professional
Engineers'
Examination Questions
and Answers

PART ONE

General Information

LEGAL REGISTRATION

The office of the National Council of Engineering Examiners (NCEE) will furnish upon request any specific information available regarding the legal registration or licensing of professional engineers that may be required in addition to the following.

Engineers who wish to secure legal registration or professional license in any particular state must communicate directly with the state board or department in that state and request its application form and regulations. All 50 states, Puerto Rico, and the provinces of Canada have registration laws regulating the practice of engineering.

The engineering registration laws are not uniform because of local conditions, date of enactment, etc. Most state laws, however, require graduation in an accredited engineering curriculum, a written engineering examination, adequate and satisfactory engineering experience, or a combination of these standards.

INTERSTATE REGISTRATION

Legal registration or professional license in any one state or certification by the NCEE does not permit or authorize the practice of engineering in other states. Most states, however, grant registration to registrants of other states or temporary permits to practice, provided they have the required qualifications and comply with the regulations of those states.

NATIONAL ENGINEERING CERTIFICATION

The National Engineering Certification is a service of the NCEE. It was established primarily to minimize the effort and expense of registered engineers desiring to secure registration or license in more than one state. The registration laws or regulations of most states specifically provide for acceptance of the Certificates of Qualification as competent evidence. However, such certification is not binding upon the State Boards and is not a form of registration or license. The "Information Booklet" regarding NCEE Services, Council Records, Certification, and the Procedures, Requirements, and Schedule of Fees may be secured from the Executive Director, NCEE, P.O. Box 1686, Clemson, South Carolina 29633-1686.

A MODEL LAW FOR THE REGISTRATION OF PROFESSIONAL ENGINEERS AND LAND SURVEYORS AND PROVIDING FOR THE CERTIFICATION OF ENGINEERS-IN-TRAINING

DEFINITIONS

Engineer: The term Engineer as used in this Act should mean a Professional Engineer as hereinafter defined.

Professional Engineer: The term Professional Engineer within the meaning and intent of this Act shall mean a person who, by reason of special knowledge of the mathematical and physical sciences and the principles and methods of engineering analysis and design, acquired by professional education and practical experience, is qualified to practice engineering as hereinafter defined, as attested by that person's legal registration as a Professional Engineer.

Engineer-in-Training: The term Engineer-in-Training as used in this Act shall mean a candidate for registration as Professional Engineer who is a graduate in an approved engineering curriculum of four years or more from a school or college approved by the Board as of satisfactory standing, or who has had four years or more of experience in engineering work of a character satisfactory to the Board; in addition, the person shall have successfully passed the examination in the fundamental engineering subjects prior to completion of the requisite years of experience in engineering work, as provided in Section 14 of this Act, and shall have received from the Board, as hereinafter defined, a certificate stating that he or she has successfully passed this portion of the professional examinations.

Practice of Engineering: The term Practice of Engineering within the meaning and intent of this Act shall mean any professional service or creative work requiring engineering education, training, and experience and the application of special knowledge of the mathematical, physical, and engineering sciences to such professional services or creative work as consultation, investigation, evaluation, planning, design, and supervision of construction for the purpose of assuring compliance with specifications and design, in connection with any public or private utilities, structures, buildings, machines, equipment, processes, works, or projects.

Persons shall be construed to practice or offer to practice engineering, within the meaning and intent of this Act, who practice any branch of the profession of engineering, or who, by verbal claim, sign, advertisement, letterhead, card, or in any other way represent themselves as professional engineers, or through the use of some other title imply that they are professional engineers, or who hold themselves out as able to perform, or who do perform any engineering service or work or any other professional service designated by the practitioner or recognized by educational authorities as engineering.

Land Surveyor: The term Land Surveyor as used in this Act shall mean a person who engages in the practice of land surveying as hereinafter defined.

Land Surveying: The practice of land surveying within the meaning and intent of this Act includes surveying of areas for their correct determination and description and for conveyancing, or for the establishment or reestablishment of land boundaries and the plotting of lands and subdivisions thereof.

SUGGESTED EXPERIENCE QUALIFICATION GUIDE

CHEMICAL ENGINEERING

Professional Experience

Design, Specifications, and/or Construction: Equipment for chemical enterprises; processing plants; plant layouts; equipment; economic balances; production planning; pilot plants; heat-transmission apparatus.

Development: Processes; pilot plants; refrigeration systems for food processing.

Research: Laboratory; pilot plants; director; markets; unit operations; kinetics; processes; calculation and correlation of physical properties.

Responsible charge of broader fields of chemical engineering; executive.

Consultation; appraisals; evaluations; reports; economics; chemical patent laws.

Operation of pilot plants; product testing; technical service; technical sales.

Teaching full time, at college level; editing and writing.

Subprofessional Experience

Construction: Process equipment; pilot plants; piping systems.

Operation: Shift operator; special chemicals manufacture (solutions); pilot plants; troubleshooting; glassblower.

Drafting: Flow-sheet layout.

Instrument making and servicing; routine analyses; routine sampling and tests; lathe operator.

Analyst: Routine, under direction; laboratory assistant (commercial, college); computations, under direction; data taking.

Sales: Routine, of standard equipment.

CIVIL ENGINEERING

Professional Experience

Design and/or Construction: Highways; streets; subways; tunnels; drainage; drainage structures; sewerage; sewage-disposal structures; railroad spurs and turnouts; railroads; rivers, harbors, and waterfront; waterworks; water supply; water power; airports and/or airways; bridges; dams; irrigation; irrigation structures; water-purification systems; incinerators; storage elevators and structures; flood control.

Development: Structures for regions; ports; wharves and docks; irrigation works; flood control; drainage.

Investigations; appraisals and evaluations; cost analysis; consultations; testing; research and economic studies; reports.

Teaching full time, at college level, in an accredited college of engineering.

Editing and writing.

Surveys: Topographic; hydrographic; earthwork; triangulation; railroad; irrigation; drainage; water power.

Independent responsibility and supervision of the work of others.

Subprofessional Experience

Fieldwork: Instrument operator; chain operator; rod person; chief of party; recorder; surveyor; level operator; staking out foundations; staking out dredging ranges; railroad lines and turnouts; sounding; instrumental observations.

Drafting: Plotting field notes; map drafting and tracing; tracing designs; structural detailing; plotting soundings.

Construction: Inspection; timekeeping; supervision, without authority to change designs.

Computations: Routine, under direction; taking off quantities; monthly estimates.

Cost Data: Routine assembling and computing; estimating, under direction.

Testing: Routine testing of constuction materials; water.

Teaching: As an associate without full responsibility in an engineering college.

ELECTRICAL ENGINEERING

Professional Experience

Design, Specifications, and/or Construction: Power plants; electronic and transmission equipment; application of electric equipment to industry; electric service and supply; communication systems; control systems and servomechanisms; manufacturing plants; operating procedures; distribution networks; transportation; radio circuits; major electric installations.

Development and Production: New techniques and devices; electric and electronic devices.

Supervision and/or Management: Power plants; manufacturing plants.

Research and investigations in electric fields; water, steam, and electric power; electronics; new techniques and devices; high-voltage cables.

Consultation; appraisals; valuations; estimates; cost analysis; reports; determination of rates; economics and cost figures; calculations of system and device performance.

Sales and application of equipment to industry; service engineering; testing, when not of a routine character.

Editing and writing on electrical-engineering subjects.

Teaching full time, at college level, in an accredited school.

Subprofessional Experience

Construction and Installation: Construction of electric installations under competent direction; electronic and radio equipment; wiring, lineperson; switchboard wiring; electrician; meterperson; armature winding.

Inspection: Routine, under direction.

Installation: Heating, ventilating, and air conditioning; radio equipment.

Drafting: Elementary layout; circuit drafting; detailing.

Testing: Routine; performance of electric machinery.

Operation: Radio; power stations; substations; switchboards.

Repairs: Motors; generators; electric devices; electronic and radio devices.

Maintenance: Electronic and radio devices; radio and television technician.

Apprentice: Work as junior in fields of illumination, communication, and transportation.

Sales: Routine; almost all customer service; inventory.

Teaching: Manual skills; instructor in accredited school.

MECHANICAL ENGINEERING

Professional Experience

Design and/or Construction: Machines; machinery and mill layouts; heating, ventilating, and air conditioning; power plants; power-plant equipment; industrial layouts; refrigeration; tools and processes; designs involving application of existing equipment and principles; internal-combustion engines.

Development of industrial plants and processes; consultation.

Appraisals; investigations; research and economic studies; cost analyses; reports.

Operation: Major mechanical installations; manufacturing plants; power plants; testing; application of instruments and control to manufacturing processes.

Editing and writing on mechanical-engineering subjects.

Teaching full time, at college level, in accredited schools.

Subprofessional Experience

Construction and Installation: Machinery; heating, ventilating, and air conditioning; mechanical structures.

Operation: Heating, ventilating, and air-conditioning equipment; power plants; mechanical manufacturing plant; stationary machinery (small operations); foundry and machine shops.

Drafting: Detailing; shop drawings; checking shop drawings; tracing; layout work; designing tools, jigs, and fixtures.

Recording; routine computing under competent direction; laboratory assistant; routine testing under competent direction; inspecting materials, etc.

Maintenance; repairs; welding.

Teaching as an assistant without full responsibility, in an accredited college.

SANITARY ENGINEERING

Professional Experience

Design, Specifications, and/or Construction: Water-treatment plants; sewage-disposal plants; waterworks; sewers; incinerators; distribution systems; sewage- and industrial-waste-treatment plants.

Consultations; reports for proposed works; original research for processes and equipment; reviewing plans for works for state and federal authorities; planning and analyzing surveys and pollution data; advising on water-purification plants; counseling municipal officials concerning needed sanitary facilities; planning and supervising environmental sanitation work in sanitation of food, milk, etc.; investigations; studies; reports and evaluation of proposed or existing water systems or sewerage; industrial-hygiene surveys and reports; research in sanitary engineering or public health.

Abatement procedures on stream pollution; supervision of works for large cities; supervision of pollution surveys; control of rodents and insects.

Teaching full time, at college level, in an accredited school.

Editing and writing.

Subprofessional Experience

Routine laboratory tests; installing machinery; gathering records for reports; stream gaging; surveying; plant operation and maintenance; record keeping; sanitarian's inspection; mapping; inspection; routine chemical and bacteriological tests; pest control; collecting data; take off; sampling; establishing lines and grades; inspecting eating and drinking establishments; routine vessel-sanitation inspections; routine watering-point inspections; orientation periods in public health agencies; inspecting construction.

Maintenance and operation of filter and sewage plant; routine laboratory examinations.

Same as Civil Engineering.

STRUCTURAL ENGINEERING

Professional Experience

Design, specifications, and/or construction on structures; hydraulic design; stress analysis; soils analysis; mechanical analysis.

Research on structures; testing; layout studies and comparisons; new design procedures and methods of analysis; reports; consultations; investigation evaluations.

Project supervision; field execution of projects; senior structural draftsperson; senior stress analyst; senior engineer on materials testing; senior concrete laboratory technician; resident (field) engineer on bridges, dams, and buildings; research engineer.

Same as Civil Engineering.

Teaching full time, at college level, in an accredited school.

Subprofessional Experience

Structural drafting, detailing; drafting layout and details; junior stress analyst; computer; checking detail drawings; checking design computations; collecting data for reports; inspector; junior laboratory engineer; research assistant; junior concrete technician.

TRAFFIC ENGINEERING

Professional Experience

Design of traffic islands; design of new and augmented control systems; design of relief systems; design of highway lighting; traffic-engineering planning and roadway design; design of mechanical traffic control.

Planning and analyzing surveys and traffic counts; interpreting field data; computing earning power of proposed routes; establishing traffic-count stations; transportation studies (major or complete); traffic-education programs.

Traffic-engineering organization and administration. Teaching, at college level.

Subprofessional Experience

Traffic counts; compiling data; timing of automatic signals; plotting traffic-flow charts; plotting traffic-dispersion charts; origin and destination data; weighting stations; accident study and tabulation; sign and signal checking.

ADDRESSES OF REGISTRATION BOARDS

UNITED STATES

Alabama State Board of Registration for Professional Engineers and Land Surveyors, Suite 212, 750 Washington Ave., Montgomery, Alabama 36130

Alaska State Board of Registration for Architects, Engineers, and Land Surveyors, Pouch D (State Office Bldg., 9th Floor), Juneau, Alaska 99811

Arizona State Board of Technical Registration, 1645 W. Jefferson St., Suite 140, Phoenix, Arizona 85007

Arkansas State Board of Registration for Professional Engineers and Land Surveyors, P.O. Box 2541, 1818 W. Capitol, Little Rock, Arkansas 72203

California Board of Registration for Professional Engineers, 1006 Fourth St., Sixth Floor, Sacramento, California 95814

Colorado State Board of Registration for Professional Engineers and Land Surveyors, 600-B State Service Bldg., 1525 Sherman St., Denver, Colorado 80203

Connecticut State Board of Registration for Professional Engineers and Land Surveyors, The State Office Bldg., Room G-3A, 165 Capitol Ave., Hartford, Connecticut 06106

Delaware Association of Professional Engineers, 2005 Concord Pike, Wilmington, Delaware 19803

Delaware State Board of Registration for Professional Land Surveyors, 820 N. French St., 3rd Level, Wilmington, Delaware 19801

District of Columbia Board of Registration for Professional Engineers, 614 H St., N.W., Room 910, Washington, D.C. 20001

Florida State Board of Professional Engineers and Board of Registration for Land Surveyors, 130 N. Monroe St., Tallahassee, Florida 32301

Georgia State Board of Registration for Professional Engineers and Land Surveyors, 166 Pryor St., S.W., Atlanta, Georgia 30309

Guam Territorial Board of Registration for Professional Engineers, Architects, and Land Surveyors, Dept. of Public Works, Govt. of Guam, P.O. Box 2950, Agana, Guam 96910

Hawaii State Board of Registration for Professional Engineers, Architects, Land Surveyors, and Landscape Architects, P.O. Box 3469 (1010 Richards St.), Honolulu, Hawaii 96801

Idaho Board of Professional Engineers and Land Surveyors, 842 La Cassia Dr., Boise, Idaho 83705

Illinois Department of Registration & Education, Professional Engineers' Examining Committee, 320 W. Washington St., Springfield, Illinois 62786

Indiana State Board of Registration for Professional Engineers and Land Surveyors, 1021 State Office Bldg., 100 N. Senate Ave., Indianapolis, Indiana 46204

Iowa State Board of Engineering Examiners, Capitol Complex, 1209 East Court Ave., West Des Moines, Iowa 50319

Kansas State Board of Technical Professions, 535 Kansas Ave., Suite 1105, Topeka, Kansas 66603

Kentucky State Board of Registration for Professional Engineers and Land Surveyors, P.O. Box 612 (Rt. 3, Millville Rd.), Frankfort, Kentucky 40602

Louisiana State Board of Registration for Professional Engineers and Surveyors, 1055 St. Charles Ave., Suite 415, New Orleans, Louisiana 70130

Maine State Board of Registration for Professional Engineers, State House, Augusta, Maine 04333

Maine State Board of Registration for Land Surveyors, State House, Station 98, Augusta, Maine 04333

Maryland State Board of Registration for Professional Engineers and Board of Registration for Professional Land Surveyors, 501 St. Paul Place, Room 902, Baltimore, Maryland 21202

Massachusetts State Board of Registration of Professional Engineers and of Land Surveyors, Room 1512, Leverett Saltonstall Bldg., 100 Cambridge St., Boston, Massachusetts 02202

Michigan Board of Registration for Professional Engineers and Board of Registration for Land Surveyors, P.O. Box 30018 (808 Southland), Lansing, Michigan 48909

Minnesota State Board of Registration for Architects, Engineers, Land Surveyors, and Landscape Architects, 5th Floor, Metro Square, 7th and Roberts Sts., St. Paul, Minnesota 55101

Mississippi State Board of Registration for Professional Engineers and Land Surveyors, P.O. Box 3 (200 S. President St., Suite 516), Jackson, Mississippi 39205

Missouri Board for Architects, Professional Engineers, and Land Surveyors, P.O. Box 184 (3523 N. Ten Mile Dr.), Jefferson City, Missouri 65102

Montana State Board of Professional Engineers and Land Surveyors, Dept. of Commerce, 1424 9th Ave., Helena, Montana 59620

Nebraska State Board of Examiners for Professional Engineers and Architects, P.O. Box 94751 (301 Centennial Mall, South), Lincoln, Nebraska 68509

Nevada State Board of Registered Professional Engineers and Land Surveyors, 1755 East Plum Lane, Suite 102, Mezzanine Floor, Reno, Nevada 89502

New Hampshire State Board for Professional Engineers, 77 N. Main St., Room 214, Concord, New Hampshire 03301

New Jersey State Board of Professional Engineers and Land Surveyors, 1100 Raymond Blvd., Newark, New Jersey 07102

New Mexico State Board of Registration for Professional Engineers and Land Surveyors, P.O. Box 4847 (1505 Llano St.), Santa Fe, New Mexico 87501

New York State Board for Engineering and Land Surveying, The State Education Department, Cultural Education Center, Madison Ave., Albany, New York 12230

North Carolina Board of Registration for Professional Engineers and Land Surveyors, 3620 Six Forks Rd., Raleigh, North Carolina 27609

North Dakota State Board of Registration for Professional Engineers and Land Surveyors, P.O. Box 1264 (609 N. Broadway), Minot, North Dakota 58701

Ohio State Board of Registration for Professional Engineers and Surveyors, 65 S. Front St., Room 302, Columbus, Ohio 43215

Oklahoma State Board of Registration for Professional Engineers and Land Surveyors, Oklahoma Engineering Center, Room 120, 201 N.E. 27th St., Oklahoma City, Oklahoma 73105

Oregon State Board of Engineering Examiners, Dept. of Commerce, 4th Floor, Labor and Industries Bldg., Salem, Oregon 97310

Pennsylvania State Registration Board for Professional Engineers, P.O. Box 2649 (Transportation & Safety Bldg., 6th Floor, Commonwealth Ave. & Forester St.), Harrisburg, Pennsylvania 17105-2649

Puerto Rico Board of Examiners of Engineers, Architects, and Surveyors, P.O. Box 3271 (Tanca Street, 261, Comer Tetuan), San Juan, Puerto Rico 00904

Rhode Island State Board of Registration for Professional Engineers and Land Surveyors, 308 State Office Bldg., Providence, Rhode Island 02903

South Carolina State Board of Engineering Examiners, 2221 Devine St., Suite 404, Columbia, South Carolina 29205

South Dakota State Commission of Engineering and Architectural Examiners, 2040 W. Main St., Suite 212, Rapid City, South Dakota 57701

Tennessee State Board of Architectural and Engineering Examiners, 546 Doctor's Bldg., 706 Church St., North, Nashville, Tennessee 37219

Texas State Board of Registration for Professional Engineers, P.O. Drawer 18329 (1917 1H 35 South), Austin, Texas 78760

Utah Representative Committee of Professional Engineers and Land Surveyors, Div. of Registration, P.O. Box 5802 (160 East 300 South), Salt Lake City, Utah 84110

Vermont State Board of Registration for Professional Engineers, Div. of Licensing and Registration, Pavilion Bldg., Montpelier, Vermont 05602

Vermont State Board of Registration for Land Surveyors, P.O. Box 38, Marshfield, Vermont 05658

Virginia State Board of Architects, Professional Engineers, and Land Surveyors and Certified Landscape Architects, 2 S. Ninth St., Richmond, Virginia 23219

Virgin Islands Board for Architects, Engineers, and Land Surveyors, Submarine Base, P.O. Box 476, St. Thomas, Virgin Islands 00801

Washington State Board of Registration for Professional Engineers and Land Surveyors, P.O. Box 9649 (3rd Floor, 9th and Columbia), Olympia, Washington 98504

West Virginia State Board of Examiners of Land Surveyors, 34 W. Main St., Romney, West Virginia 26757

West Virginia State Board of Registration for Professional Engineers, 608 Union Bldg., Charleston, West Virginia 25301

Wisconsin State Examining Board of Architects, Professional Engineers, Designers, and Land Surveyors, P.O. Box 8936 (1400 E. Washington Ave.), Madison, Wisconsin 53708

Wyoming State Board of Examining Engineers, Barrett Bldg., Cheyenne, Wyoming 82002

CANADA

Canadian Council of Professional Engineers, Suite 401, 116 Albert St., Ottawa, Ontario KIP 5G3

Association of Professional Engineers, Geologists, and Geophysicists of Alberta (Apegga), 1010, One Thornton Ct., Edmonton, Alberta T5J 2E7

Association of Professional Engineers of British Columbia (APEBC), 2210 W. 12th Ave., Vancouver, British Columbia V6K 2N6

Association of Professional Engineers of Manitoba (APEM), 177 Lombard Ave., Room 710, Winnipeg, Manitoba R3B 0W9

Association of Professional Engineers of New Brunswick (APENB), 123 York St., Fredericton, New Brunswick E3B 3N6

Association of Professional Engineers of Newfoundland (APEN), P.O. Box 8414, Postal Station A, St. John's, Newfoundland A1B 3N7

Association of Professional Engineers, Geologists, and Geophysicists of the Northwest Territories (APEGGNWT), Box 1962, Yellowknife, Northwest Territory X1A 2P5

Association of Professional Engineers of Nova Scotia (APENS), P.O. Box 129, Halifax, Nova Scotia B3J 2M4

Association of Professional Engineers of Ontario (APEO), 1027 Yonge St., Toronto, Ontario M4W 3E5

Association of Professional Engineers of Prince Edward Island (APEPEI), P.O. Box 278, Charlottetown, Prince Edward Island CIA 4B1

Order of Engineers of Quebec/Ordre des Ingenieurs du Quebec (OEQ/OIQ), 2075 Rue University, Suite 1100, Montreal, Quebec H3A 1K8

Association of Professional Engineers of Saskatchewan (APES), 220-2220 Twelfth Ave., Regina, Saskatchewan S4P OMB

Association of Professional Engineers of Yukon Territory (APEYT), P.O. Box 4125, Whitehorse, Yukon Territory Y1A 3S9

PART TWO

QUESTIONS

from
Professional
Engineers'
Examinations

Basic Fundamentals

CHEMISTRY, COMPUTER AND MATERIALS SCIENCE, AND NUCLEONICS

1.101 The weight of sulfuric acid and sodium chloride that would be required to prepare 1 liter of hydrochloric acid, density 1.201, is: (a) 595.3, (b) 646.8, (c) 673.2, or (d) 692.4 g? HCl of density 1.201 contains 40.09 percent by weight HCl. Chemical equation and atomic weights are:

$$2NaCl + H_2SO_4 \rightarrow Na_2SO_4 + 2HCl$$

Atomic weights: Na, 23; Cl, 35.46; H, 1.008; O, 16; S, 32.

1.102 Explain:
  a. Oxidation and reduction
  b. The periodic law of the chemical elements
  c. Radioactivity
  d. Acidity

1.103 Cupric sulfide (CuS) ore may be roasted in air, the sulfur driven off as sulfur dioxide and later converted to sulfuric acid. The amount of acid (93 percent H_2SO_4) that can be made from 1 ton of ore that shows 60 percent copper, allowing for 10 percent loss of sulfur during the process, is: (a) 1528.5, (b) 1645.8, (c) 1775.0, or (d) 1893.8 lb?
  Atomic weights: Cu, 63.6; H, 1.00; S, 32; O, 16.

1.104 Describe the commercial method for the production of sodium hydroxide (caustic soda) from soda ash and indicate the type of equipment used for each step in the process. What engineering problems are involved?

1.105 A 4.00-g solution of an unknown substance in 200 g benzene is found to freeze at 4.82°C; the benzene freezes at 5.49°C. The molecular weight of the substance is: (a) 135.7, (b) 141.4, (c) 147.1, or (d) 152.8?

1.106 An aqueous solution of sodium sulfate is saturated at 32°C. The temperature to which the solution must be cooled to crystallize 60 percent of the solution as $Na_2SO \cdot 10H_2O$ is: (a) 18.7, (b) 16.6, (c) 15.4, or (d) 14.7°C?

1.107 The number of cubic feet (cubic meters) of a gas composed of 28 percent CO and 72 percent N_2 (by volume) and measured at 85°F (29.4°C) and 100 psig (689.5 kPa gage) required to reduce 1 ton (907.2 kg) of ore that is 84 percent Fe_3O_4 to metallic iron is: (a) 4975.3 (140.9), (b) 5060.3 (143.3), (c) 5145.1 (145.7), or (d) 5230.0 ft³ (148.1 m³)? The reactions are:

$$Fe_3O_4 + CO \rightarrow 3FeO + CO_2$$
$$FeO + CO \rightarrow Fe + CO_2$$

Atomic weights: C, 12.00; Fe, 55.84; O, 16.00; N, 14.008.

1.108 Balance the following chemical equation. Atomic weights: Al, 26.97; Fe, 55.84; S, 32.06; O, 16.00.

$$?Al + ?Fe_2(SO_4)_3 \rightarrow Al_2(SO_4)_3 + ?Fe\ SO_4$$

In questions 1.109 to 1.115, select only one correct answer:

1.109 The chemical most commonly used to speed sedimentation of sewage is: (a) sulfuric acid, (b) copper sulfate, (c) lime, (d) methylene blue, (e) sodium permanganate.

1.110 The gas from sludge-digestion tanks is mainly composed of: (a) nitrogen, (b) hydrogen sulfide, (c) carbon dioxide, (d) carbon monoxide, (e) methane.

1.111 The quantity of chlorine required to satisfactorily chlorinate sewage is usually: (a) 0–25, (b) 30–60, (c) 65–90, (d) 95–120, (e) 125–150 ppm?

1.112 The ratio of oxygen available to the oxygen required for stabilization of sewage is called: (a) biochemical oxygen demand, (b) oxygen-ion concentration, (c) relative stability, (d) bacterial-stability factor, (e) concentration factor.

1.113 Most of the bacteria in sewage are: (*a*) parasitic, (*b*) saprophytic, (*c*) pathogenic, (*d*) anaerobic, (*e*) dangerous?

1.114 In the design of grit chambers: (*a*) temperature is an important factor; (*b*) baffles are essential; (*c*) there should be a 5:1 ratio of length to depth; (*d*) the detention period should be at least 30 min; (*e*) the maximum velocity of flow is 1 ft/s?

1.115 In an Imhoff tank: (*a*) the effluent contains very little dissolved oxygen; (*b*) there are no settling compartments; (*c*) the sludge and fresh sewage are well mixed to give complete digestion; (*d*) the sludge and raw sewage are not mixed; (*e*) much larger settling compartments are required than for an equivalent septic tank?

1.116 The vapor pressure of benzophenone at several temperatures is tabulated as:

T, °C	108.2	141.7	157.6	175.8	195.7	208.2	224.4
P, mmHg	1	5	10	20	40	60	100

The following program is a computer subroutine that estimates the vapor pressure of benzophenone for any temperature between 108.2 and 224.4°C, using two-point interpolation:

Question line	Computer line	
1		`DIMENSION T(10), P(10)`
2		`READ (5,1) (T(I), P(I), I=1,7)`
3	1	`FORMAT (2E14.7)`
4		`READ (5,2) TX`
5	2	`FORMAT (F10.1)`
6		`DO 50 J = 2,7`
7		`IF (TX − T(J))60, 60, 50`
8	50	`CONTINUE`
9	60	`PX = P(J−1) + (TX−T(J−1)`
		`* (P(J)−P(J−1))/(T(J)−T(J−1))`
10		`WRITE(6,2) TX, PX`
11	2	`FORMAT (1H0, 'TB =', F6.1, 5x, 'P_b =',`
		`F7.2)`
12		`STOP`
13		`END`

1. If the given corresponding tabulated points between temperature and pressure were 12 instead of 7, you would have to change lines: (1*a*) 2 and 6, (1*b*) 3 and 7, (1*c*) 1 only, or (1*d*) 8 only?

2. The data are fed into the computer by lines: (2*a*) 2 and 4, (2*b*) 3 and 5, (2*c*) 6 only, or (2*d*) 1 only?

1.117 The rate at which a substance passes through a semipermeable membrane is determined by the diffusivity $D(\text{cm}^2/\text{s}) \times 10^6$ of the gas. D varies with the membrane temperature $T(\text{K})$ according to Arrhenius' law: $D = D_0 \exp(-E/RT)$, where D_0 = exponential factor, E = activation energy, and $R = 1.987$ cal/(g·mol)(K).

Diffusivities of SO_2 in a fluorosilicone rubber tube are measured at several temperatures, with the following results:

T, K	D, $(\text{cm}^2/\text{s}) \times 10^6$
347.0	1.34
374.2	2.50
396.2	4.55
420.7	8.52
447.7	14.07
471.2	19.99

The following computer program reads the (T, D) data and calculates D_o and E by using the method of least squares.

Question line	Computer line	
1		DIMENSION T(6), D(6), TINV(6), DLN(6)
2		READ (5,1) T
3	1	FORMAT (6F 10.1)
4		READ (5,2) D
5	2	FORMAT (6E 10.2)
6		DO 20 J = 1,6
7		TINV (J) = 1.0/T(J)
8	20	DLN(J) = ALOGR (D(J))
9		SX = 0.0
10		SY = 0.0
11		SXX = 0.0
12		SXY = 0.0
13		DO 100 J = 1,6
14		SX = SX + TINV(J)
15		SY = SY .+ DLN (J)
16		SXX = SXX + TINV(J) ** 2
17	100	SXY = SXY + TINV(J) * DLN(J)
18		SX = SX/6.0
19		SY = SY/6.0
20		SXX = SXX/6.0
21		SXY = SXY/6.0

```
22                     SLOPE = (SXY - SX * SY)/
                       (SXX - SX ** 2)
23                     DZLN = SY - SLOPE * SX
24                     SLOPE = -E/R, DZLN = LN(DO)
25                     DZERO = EXP (DZLN)
26                     E = -1.987 * SLOPE
27                     WRITE (6,3) DZERO, E
28            3        FORMAT (2E14.7)
29                     DO 200 J = 1,6
30                     DC = DZERO * EXP (SLOPE * TINV (J))
31                     WRITE (6,5) T(J), D(J), DC
32            5        FORMAT (3E14.7)
33           200       CONTINUE
34                     STOP
35                     END
```

1. The values of T(K) are fed into the computer by line(s): (1*a*) 6, 16, (1*b*) 2, (1*c*) 8, or (1*d*) 31?

2. The values of D are fed into the computer by line: (2*a*) 4, (2*b*) 7, (2*c*) 13, or (2*d*) 23?

3. The line that expresses Arrhenius' law is: (3*a*) 22, (3*b*) 30, (3*c*) 31, or (3*d*) 32?

1.118 The van der Waals' equation of state is $(P + a/\hat{V}^2) \times (\hat{V} - b) = RT$, where a and b are constants.

The following program solves van der Waals' equation for $\hat{V}$ from given values of P and T, using Newton's iteration method, stopping when the fractional change in $\hat{V}$ from one iteration to the next is E. The program reads in and prints out values of a, b, E and the number of (P, T) pairs for each $\hat{V}$ to be calculated; then for each pair it calculates $\hat{V}$ and prints out P, T, $\hat{V}$. The upper limit of iteration is 10.

Question line	Computer line	
1		READ (5,1) A, B, EPS, NP
2	1	FORMAT (3F10.2, I 10)
3		WRITE (6,2) A, B, EPS
4	2	FORMAT (3F10.2)
5		DO 500 N = 1, NP
6		READ (5,3) P, T
7	3	FORMAT (2F10.2)
8		CALCULATE FIRST ESTIMATE FROM IDEAL GAS LAW
9		TK = T + 273.2
10		V = 0.08206 * TK/P

```
11                      CON = P * B + 0.08206 * TK
12                      DO 50 ITER = 1, 10
13                      CON 1 = A/V
14                      CON 2 = A * B/V ** 2
15                      F = P * V + CON 1 - CON 2 - CON
16                      FPR = P - CON 1/V + 2.0 * CON 2/V
17                      DELV = F/FPR
18                      VNEW = V - DELV
19                      IF (ABS (DELV/VNEW), LT, EPS) GO TO 55
20                      V = VNEW
21            50        CONTINUE
22                      WRITE (6,4) P, T
23            4         FORMAT (2F10.2, 3X 'NO CONVERGENCE')
24                      GO TO 500
25            55        WRITE (6,5) P,T, VNEW, ITER
26            5         FORMAT (3F10.2, I 10)
27            500       CONTINUE
28                      STOP
29                      END
```

1. The line that gives the answers to the values of P, T, and $\hat{V}$ is: (1a) 3, (1b) 7, (1c) 15, or (1d) 25?

2. When the condition of the if statement is satisfied, the computer goes to the statement of line: (2a) 23, (2b) 20, (2c) 25, or (2d) 27?

3. In line 4, the 2 in the symbol F10.2 indicates that the output: (3a) is a two-digit even integer, (3b) is a two-digit odd integer, (3c) has two digits before the decimal point, or (3d) has two digits after the decimal point?

1.119 The deflection of a simple beam with a uniform cross section, loaded with a single concentrated force, is given by

$$
Y = \begin{cases} \dfrac{-Pb}{6EIL} [2L(L - x) - b^2 - (L - x)^2] & \text{for } 0 \leq x \leq a \\[2ex] \dfrac{-Pa(L - x)}{6EIL} [2Lb - b^2 - (L - x)^2] & \text{for } a \leq x \leq L \end{cases}
$$

where Y = deflection at location x

　　　P = applied load

　　　L = length of beam

　　　a = distance to load application point from left end of beam

　　　b = load application point from right end of beam

　　　E = modulus of elasticity of beam material

　　　I = moment of inertia of cross section of beam

The following computer program calculates the deflection at several equally spaced points along the beam:

Question line	Computer line	
1		READ (5,50) P,A, XL, E, XI, N
2	50	FORMAT (6E14.7)
3		B = XL − A
4		FAC = −P/(6. * E * XI * XL)
5		SECTS = N + 1
6		DO 3 I = 1,N
7		POINT = 1
8		X = POINT * XL/SECTS
		IF (A − X) 1,2,2
9	1	Y = FAC * A * (XL − X) * ((2. * XL − B) * B − (XL − X) ** 2)
10	2	Y = FAC * B * X * (2. * XL * (XL − X)−B * B − (XL − X) ** 2)
11	3	WRITE (6,51) X,Y
12	51	FORMAT (6H AT X = F8.3, 3X, 3HY = F8.4)
13		STOP
14		END

1. The data are fed into the computer by the statement on line: (1a) 2, (1b) 1, (1c) 11, or (1d) 12?

2. The deflection at a point x when $0 \leqslant x \leqslant a$ is given by the statement on line: (2a) 4, (2b) 7, (2c) 9, or (2d) 10?

3. From line 12, the deflection Y will be given with: (3a) 2, (3b) 3, (3c) 4, or (3d) 5 decimal points?

1.120 Organic matter in sewage decomposes through chemical and bacterial action. In this process, free oxygen is consumed and the sewage is deoxygenated. A standard procedure for determining the rate of deoxygenation of sewage involves diluting a sewage sample with water containing a known amount of dissolved oxygen and determining the loss in oxygen after the mixture has been maintained at a temperature of 20°C for 5 days. This loss is called the biochemical oxygen demand (BOD). The BOD for a period of 20 days at a temperature of 20°C is called the first-stage demand and denoted λ20. For any temperature T, the first-stage demand is given by

$$\lambda = \lambda_{20}(0.02T + 0.6) \tag{1}$$

The oxygen is also absorbed from the air. However, the deoxygenation and reoxygenation rates are different. The oxygen deficit at any time t is given by

$$D_t = \frac{K_d \lambda_m}{K_r - K_d} [(10^{-K_d t} - 10^{K_r t}) + D_0(10^{-K_r t})] \qquad (2)$$

where K_d and K_r are coefficients of deoxygenation and reoxygenation.

If Tm = TMIX, K_d = CD, $K_d 20$ = CD20, D_0 = DO, D_t = DT, Q_s = QSEW, Q_r = QSTR, (BOD)s = BODSEW, (BOD)r = BODSTR, (BOD)m = BODMIX, and $\lambda 20$ = FBOD, then the following program calculates the maximum deficit in the stream:

Question line	Computer line	
1		READ CR, QSTR, QSEW, CD20, TMIX, BODSTR, BODSEW, DO
2		CD = CD20 * 1.047 ** (TMIX - 20.)
3		BODMIX = (BODSTR * QSTR + BODSEW * QSEW)/(QSTR + QSEW)
4		FBOD = BODMIX * (0.02 * TMIX + 0.6)
5		DTOLD = DO
6		DO 1 I = 2700,3000
7		Z = I
8		T = Z/1000
9		A = 10. ** (-CR * T)
10		DT = ((CD * FBOD)/(CR - CD)) * (10. ** (-CD * T) - A) + DO * A
11		IF (DT - DTOLD) 2,2,1
12	1	DTOLD = DT
13		PRINT 'THE TEST FAILED'
14		GO TO 7
15	2	PRINT 55,T,DT
16	55	FORMAT ('THE MAXIMUM DEFICIT AT T =' F9.4, 'is' ,F9.4)
17	7	STOP
18		END

1. The oxygen deficit is given by line: (1a) 11, (1b) 10, (1c) 4, or (1d) 3?

2. The first-day demand is given by line: (2a) 4, (2b) 8, (2c) 11, or (2d) 14?

1.121 The most important factor in determining high-temperature behavior of an alloy is: (*a*) dispersion, (*b*) ionization, (*c*) crystallization, (*d*) composition, or (*e*) endurance?

1.122 At relatively high temperatures and low rates of strains, structures will perform better if their material is: (*a*) fine-grained, (*b*) coursegrained, or (*c*) their behavior is independent of the grain?

1.123 With regard to corrosion of metals, passivation is the process that: (*a*) intensifies deterioration, (*b*) intensifies deterioration temporarily, (*c*) changes the composition of the metal, (*d*) inhibits further deterioration, or (*e*) alters the grain size of the metal?

1.124 In the process of pair formation, a pair cannot be formed unless the quantum has an energy greater than: (*a*) 0.5 MeV, (*b*) $2m_0C^2$, (*c*) $\frac{1}{2}mV^2$, (*d*) $h\nu/C$, or (*e*) 3.5 MeV?

1.125 One of the two types of nonmaterial nuclear radiation is: (*a*) transmutation radiation, (*b*) Walton radiation, (*c*) gamma radiation, (*d*) betatron radiation, or (*e*) Rutherford radiation?

ELECTRICITY

1.201 An electric truck is propelled by a dc motor driven by a 110-V storage battery. The truck is required to exert a tractive effort of 200 lb at a speed of 5 mph. The overall efficiency of the motor and drive is 70 percent. The current taken from the battery is: (*a*) 25.8, (*b*) 28.0, (*c*) 30.2, or (*d*) 32.4 A?

1.202 In the electric circuit in Fig. 1.202, if the total resistance from *A* to *B* is 16 ohms, will the resistance R_2 be: (*a*) 26.2, (*b*) 26.8, (*c*) 27.2, or (*d*) 27.4 ohms?

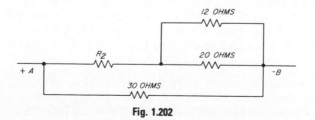

Fig. 1.202

1.203 The resistances in ohms are as indicated in Fig. 1.203. Will the current supplied by the battery in the network be: (*a*) 3.4, (*b*) 2.8, (*c*) 2.4, or (*d*) 2.2 A?

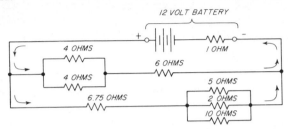

Fig. 1.203

1.204 Two coils of wire have resistance of 5 and 17 ohms, respectively.

1. If the two coils are connected in parallel to the terminals of a 6-V battery, will the current in coil 1 be: (1*a*) 1.0, (1*b*) 1.2, (1*c*) 1.3, or (1*d*) 1.35 A?

2. Will the current in coil 2 be: (2*a*) 0.295, (2*b*) 0.320, (2*c*) 0.340, or (2*d*) 0.353 A?

1.205 In the electrolytic refining of copper, the original weight of the cathode is 15.0 lb (6.80 kg). After the cathode has been in a bath of $CuSO_4$ and free H_2SO_4 for 15 days, the weight of the electrode is 300 lb (136.08 kg). Assume 100 percent efficiency.

1. Will the total ampere-hours required be: (1*a*) 109,000, (1*b*) 110,000, (1*c*) 112,000, or (1*d*) 115,000 Ah?

2. Will the current flow be: (2*a*) 290, (2*b*) 297, (2*c*) 303, or (2*d*) 307 A?

3. If the cathode is 40 × 36 in (1.094 × 0.984 m), will the current density be: (3*a*) 30.0 (322.9), (3*b*) 30.3 (326.2), (3*c*) 30.7 (330.5), or (3*d*) 31.2 A/ft² (335.8 A/m²)?

1.206 A 50-kVA 2300/230-V 60-cycle single-phase transformer has a core loss of 400 W and a full-load copper loss of 600 W.

1. Will the efficiency at 100 percent load and 90 percent power factor be: (1*a*) 96.25, (1*b*) 97.00, (1*c*) 97.57, or (1*d*) 97.83 percent?

2. Will the maximum efficiency occur at: (2*a*) 80.2, (2*b*) 81.0, (2*c*) 81.6, or (2*d*) 82.0 percent of full load?

1.207 Two groups of electric lights are connected in series across a 220-V line. The first group has 25 lights in parallel; the second group has 15 lights in parallel. Each light will carry 0.9 A at 110 V. Assume the resistance of the lights constant at varying voltages.

1. Will the potential difference across the 25-lamp group be: (1a) 82.5, (1b) 83.5, (1c) 85.0, or (1d) 87.0 V?

2. Will the potential difference across the 15-lamp group be: (2a) 141.0, (2b) 139.0, (2c) 137.5, or (2d) 136.5 V?

3. Will the current through each lamp of the 25-lamp group be: (3a) 0.675, (3b) 0.700, (3c) 0.750, or (3d) 0.850 A?

4. Will the current through each lamp of the 15-lamp group be: (4a) 0.67, (4b) 0.85, (4c) 1.00, or (4d) 1.12 A?

1.208   A dc load of 10 kW at 120 V is to be supplied from a copper line that has a diameter of 0.204 in. The distance from generator to load is 500 ft.

1. Is the required voltage of the generator: (1a) 141.1, (1b) 147.3, (1c) 153.3, or (1d) 159.7 V?

2. Is the power loss in the line: (2a) 1665, (2b) 1710, (2c) 1755, or (2d) 1800 W?

1.209   Two condensers of 2- and 10-μF capacitance are connected in parallel, and this combination is connected in series with a third condenser of 8-μF capacitance across a 120-V dc supply circuit.

1. Will the voltage across the 8-μF condenser be: (1a) 36, (1b) 54, (1c) 66, or (1d) 72 V?

2. Will the voltage across the 2- and 10-μF condensers be: (2a) 36, (2b) 48, (2c) 54, or (2d) 57 V?

3. Will the total energy stored be: (3a) 0.0345, (3b) 0.0375, (3c) 0.0420, or (3d) 0.0480 J?

1.210   *a.* Two storage cells connected in parallel supply jointly a 20-A current to an external circuit. One cell has 2.1-V emf and the other 2.0-V emf, and each cell has an internal resistance of 0.05 ohm.

       1. Will the current I_1 be: (1a) 10.0, (1b) 11.0, (1c) 12.5, or (1d) 15.5 A?

       2. Will the current I_2 be: (2a) 11.1, (2b) 10.2, (2c) 9.5, or (2d) 9.0 A?

b. Explain why the specific gravity of the electrolyte of a lead storage cell decreases as it discharges.

1.211   A dc generator has four poles with faces 6 in square and a flux density in the air gap of 40,000 lines/in^2. The machine has 360 conductors arranged in four parallel paths through the armature. The emf of the generator when driven at 1200 rpm is: (a) 97.1, (b) 99.3, (c) 101.5, or (d) 103.7 V?

1.212 Two storage batteries of 10 and 15 V and an internal resistance of 1 and 2 ohms, respectively, have their positive terminals attached to one end of a variable resistance and their negative terminals to the other end. So that the same current will flow from each battery, will this variable resistance be: (*a*) 0.2, (*b*) 0.3, (*c*) 0.5, or (*d*) 0.8 ohm?

1.213 Could two 30-W and one 60-W 120-V lamps be connected across a 240-V circuit in such a way as to give normal illumination? If so, draw a circuit diagram and explain.

1.214 Electric resistances of 7 and 11 ohms are connected in parallel and the combination is then placed in series with a resistance of 15 ohms. The whole combination is connected across 110-V dc mains.
   1. Will the total current be: (1*a*) 5.7, (1*b*) 6.0, (1*c*) 6.6, or (1*d*) 7.8 A?
   2. Will the current through the 7-ohm branch be: (2*a*) 3.40, (2*b*) 3.48, (2*c*) 3.54, or (2*d*) 3.58 A?
   3. Will the current through the 11-ohm branch be: (3*a*) 2.46, (3*b*) 2.34, (3*c*) 2.26, or (3*d*) 2.22 A?

1.215 A dc motor having an 85 percent efficiency is rated 5 hp and 220 V. It is located 150 ft from the supply lines. A maximum voltage drop of 5 percent based on rated voltage is permitted. Is the recommended wire size for overhead wiring: (*a*) #10 (25.9), (*b*) #11 (23.0), (*c*) #12 (20.5), or (*d*) #13 AWG (18.3 mm in diameter)?

1.216 Can the number of 100-W lamps be: (*a*) 18, (*b*) 17, (*c*) 16, or (*d*) 15 if placed in a 115-V circuit, as shown in Fig. 1.216, which is protected by a 15-A fuse? (The fuse "burns out" and opens the circuit when the current through it rises to 15 A.)

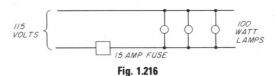

Fig. 1.216

1.217 If two voltmeters are connected in series to a power line and the first voltmeter, which has a resistance of 16,000 ohms, reads 60 V, will the voltage between the wires of the power lines, provided the second voltmeter's resistance is 20,000 ohms, be: (*a*) 125, (*b*) 130, (*c*) 135, or (*d*) 140 V?

1.218 A circuit consists of three devices and a 30-V battery having an internal resistance of 1 ohm. One device, which has a 7-ohm resistance, is connected in series with the other two, and these devices, one with 3 ohms and the other with 6 ohms of resistance, are connected in parallel.

1. Will the voltage drop on the 7-ohm device be: (1a) 21, (1b) 24, (1c) 27, or (1d) 30 V?

2. Will the power expended in the 3-ohm device be: (2a) 9, (2b) 12, (2c) 15, or (2d) 18 W?

3. Will the current in the 6-ohm device be: (3a) 2.0, (3b) 1.5, (3c) 1.0, or (3d) 0.5 A?

1.219   A 10-hp 220-V dc shunt motor has armature and field resistances of 0.25 and 100 ohms, respectively. The full-load efficiency is 83 percent. The value of the starting resistance, so that the starting current will not exceed 200 percent of the full-load value, is: (a) 2.2, (b) 2.5, (c) 2.8, or (d) 3.1 ohms?

1.220   A circuit consisting of a 63-ohm resistance, a 0.15-H inductance, and a 15-μF capacitor connected in series is placed on a 110-V 60-cycle supply line. Will the value of the current in the circuit be: (a) 0.700, (b) 0.740, (c) 0.780, or (d) $0.815\underline{/62.2°}$ A?

1.221   A three-phase Y-connected induction motor operating on 220 V has a power factor of 70 percent and an efficiency of 82 percent when delivering 8 hp (5.97 kW).

1. The voltage across each phase is: (1a) 127.0, (1b) 133.5, (1c) 140.0, or (1d) 146.5 V?

2. The current in each phase is: (2a) 18.9, (2b) 21.7, (2c) 24.5, or (2d) 27.3 A?

1.222   Explain how the plate resistance, amplification factor, and mutual conductance may be calculated for a three-element electron tube from a family of plate-characteristic curves (Fig. 1.222, p. 32) for the tube.

1.223   1. If the current from a short-circuited 1.5-V dry cell is 25 A, will the internal resistance of the cell be: (1a) 0.06, (1b) 0.10, (1c) 0.14, or (1d) 0.18 ohm?

2. The resistance of the short circuit is negligible. If this cell is connected through a 1-ohm external circuit, will the current be: (2a) 1.000, (2b) 1.200, (2c) 1.414, or (2d) 1.664 A?

3. Will the voltage across the battery terminals, when connected to the 1-ohm load, be: (3a) 0.707, (3b) 1.000, (3c) 1.200, or (3d) 1.414 V?

1.224   A series *LCR* circuit with inductance of 100 H has a transient resonant frequency of 5 cps.

1. Will the period be: (1a) 0.2, (1b) 0.3, (1c) 0.5, or (1d) 0.7 s?

2. Will the capacitance be: (2a) 9.0, (2b) 9.5, (2c) 10.1, or (2d) 10.8 μF? Neglect effect of *R* on frequency.

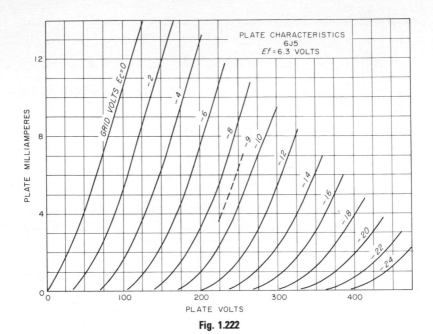

Fig. 1.222

1.225 Same circuit as in question 1.224. The condenser is given an initial charge Q and at time $T = 0$ is switched to discharge through L and R in series.

    a. Write the differential equation.

    b. What is the physical meaning of each term?

FLUID MECHANICS

1.301 A cylindrical water tower 40 ft (12.19 m) in diameter contains water to a depth of 60 ft (18.29 m).

    1. Will the total pressure on the base be: (1*a*) 2355 (20.95), (1*b*) 2501 (22.25), (1*c*) 2573 (22.89), or (1*d*) 2609 tons (23.21 MN)?

    2. Will the total pressure on the side be: (2*a*) 6077 (54.06), (2*b*) 6218 (55.32), (2*c*) 6500 (57.82), or (2*d*) 7065 tons (62.85 MN)?

1.302 A vertical gate 4 ft (1.22 m) wide and 6 ft (1.83 m) high hinged at the upper edge is kept closed by the pressure of water standing 8 ft (2.44 m) deep over its top edge. Will the force needed to open the gate, when applied normally at the bottom of the gate, be: (*a*) 8000 (35.6), (*b*) 9000 (40.0), (*c*) 10,000 (44.5), or (*d*) 11,000 lb (48.9 kN)?

1.303 A 3-ft (0.91-m) square plate with two of its edges horizontal has its center of pressure 1 in (2.54 cm) from its center of gravity. Will the top of the plate be submerged: (a) 6.0 (1.83), (b) 7.0 (2.13), (c) 7.5 (2.29), or (d) 8.0 ft (2.44 m)?

1.304 Water at a temperature of 80°F (26.67°C) flows through two connecting pipes 10 (25.4) and 12 in (30.48 cm) in diameter. If the mean velocity of flow in the 12-in (30.48-cm) pipe is 6 fps (1.83 m/s),

1. Will the velocity of flow in the 10-in (25.4-cm) pipe be: (1a) 6.6 (2.01), (1b) 6.8 (2.07), (1c) 7.43 (2.26), or (1d) 8.64 fps (2.63 m/s)?

2. Will the Reynolds number for the 12-in (30.48-cm) pipe be: (2a) 624,000, (2b) 636,000, (2c) 645,000, or (2d) 660,000?

1.305 If the water in the 10-in (25.4-cm) pipe in question 1.304 is replaced by oil having a specific gravity of 0.80 and a μ value of 0.000042, will the oil's velocity for similarity in the two flows be: (a) 21.0 (6.40), (b) 21.5 (6.55), (c) 22.0 (6.71), or (d) 22.5 fps (6.86 m/s)?

1.306 A flat-bottomed scow is built with vertical sides and straight sloping ends. Its length on deck is 80 ft (24.38 m), and on the bottom, 65 ft (19.81 m); its width is 20 ft (6.10 m) and its vertical depth 12 ft (3.66 m). The scow weighs 250 tons (226.8 Mg). Will the scow draw: (a) 4.0 (1.22), (b) 5.0 (1.52), (c) 5.84 (1.78), or (d) 6.55 ft (2.00 m)?

1.307 A tank with vertical sides is 4 ft (1.22 m) square, 10 ft (3.05 m) deep, and filled to a depth of 9 ft (2.74 m) with water. A cube of wood, specific gravity 0.5, measuring 2 ft (0.61 m) on the edge is placed in the water so as to float with one face horizontal. Will the pressure on one side of the tank increase: (a) 569 (2.53), (b) 618 (2.75), (c) 674 (3.00), or (d) 741 lb (3.30 kN)?

1.308 Water from a reservoir is pumped over a hill through a pipe 3 ft (0.914 m) in diameter, and a pressure of 30 psi (206.8 kPa) is maintained at the summit, where the pipe is 300 ft (91.44 m) above the reservoir. The quantity pumped is 49.5 ft³/s (1.40m³/s), and because of friction in the pump and pipe 10 ft (3.05 m) of head is lost between reservoir and summit. The energy to be furnished by the pump each second to the water must be: (a) 2000 (1491), (b) 2130 (1588), (c) 2250 (1678), or (d) 2360 hp (1760 kW)?

1.309 A centrifugal pump draws water from a pit through a vertical 12-in (30.48-cm) pipe that extends below the surface. It discharges into a 6-in (15.24-cm) horizontal pipe 15 ft (4.57 m) above the water surface. While the pump pumps 2 ft³/s (0.0566 m³/s), a pressure gage on the discharge pipe reads 24 psi (165.5 kPa) and a gage on the suction pipe registers 5 psi (34.48 kPa) below atmosphere. Both gages are close to the pump and separated by a vertical distance of 5 ft (1.52 m). Will the output of the pump be: (a) 12.0 (8.95), (b) 14.0 (10.44), (c) 16.7 (12.45), or (d) 20.1 hp (14.99 kW)?

1.310 A rectangular steel box floats with a draft of 4 ft (1.22 m). If the box is 20 ft (6.10 m) long, 10 ft (3.05 m) wide, and 6 ft (1.83) deep, will the time necessary to sink its top edge by opening a standard orifice, 6 in (1.22 m) in diameter, in its bottom be: (a) 167, (b) 185, (c) 200, or (d) 213 s?

1.311 A rectangular weir with end contractions has a crest 10 (3.048) and 1.00 ft (0.305 m) above the bottom of the channel approach. If the channel width is to be 16 ft (4.88 m), will the amount of water discharged under a head of 0.875 ft (0.267 m) be: (a) 26.0 (0.74), (b) 26.8 (0.76), (c) 27.4 (0.78), or (d) 27.8 ft³ (0.79 m³)/s?

1.312 A suppressed weir having a crest length of 7.00 ft (2.13 m), a height of 1.00 ft (0.305 m) above the bottom of the channel discharges under a head of 0.875 ft (0.267 m). Will the rate of discharge be: (a) 19.1 (0.54), (b) 19.4 (0.55), (c) 19.9 (0.56), or (d) 20.6 ft³ (0.58 m³)/s?

1.313 Two reservoirs with a difference in level of 100 ft (30.48 m) are connected by 12,000 ft (3658 m) of 12-in (30.48-cm) cast-iron pipe. Will the rate of discharge be: (a) 3.5 (0.10), (b) 3.75 (0.11), (c) 4.25 (0.12), or (d) 5.0 ft³ (0.14 m³)/s?

1.314 An 18-in (45.72-cm) cast-iron pipe is discharging 7500 gpm (28.32 m³/min). At point A, 4000 ft (1219 m) from the supplying reservoir (measured along the pipe), the center of the pipe is 180 ft (54.86 m) below the reservoir surface. Will the pressure expected at point A be: (a) 51.4 (354.4), (b) 53.0 (365.4), (c) 55.4 (382.0), or (d) 58.6 psi (404.0 kPa)?

1.315 A pump at elevation 1000 ft (304.8 m) is pumping 2.00 ft³/s (0.0566 m³/s) through 8000 ft (2438.4 m) of 6-in (15.24-cm) pipe to a reservoir whose level is at elevation 1250 ft (381.0 m). Assume $f = 0.0225$. At a point 100 ft (30.48 m) above and 3200 ft (975.4 m) from the pump, will the pressure in the pipe be: (a) 200 (1379), (b) 217 (1496), (c) 251 (1731), or (d) 302 psi (2082 kPa)?

1.316 From reservoir A, whose surface is at elevation 800 ft (243.8 m), water is pumped through 4000 ft (1219.2 m) of 12-in (30.48-cm) pipe across a valley to a second reservoir B, whose level is at elevation 900 ft (274.3 m). During pumping, the pressure is 100 psi (689.5 kPa) at a point M on the pipe, midway of its length and at elevation 700 ft (213.4 m). Assume $f = 0.02$.

1. Will the rate of discharge be: (1a) 4.75 (0.13), (1b) 4.90 (0.14), (1c) 5.15 (0.15), or (1d) 5.55 ft³ (0.16 m³)/s?

2. Will the power output of the pump be: (2a) 89 (65.5), (2b) 96 (70.6), (2c) 101 (74.3), or (2d) 104 hp (76.5 kW)?

1.317 Reservoir A is at elevation 1000 ft (304.8 m) above datum. From reservoir A, an 8-in (20.32-cm) pipeline leads 3000 ft (914.4 m) to point Y at elevation 800 ft (243.8 m), at which point it branches into two lines: a 6-in (15.24-cm) line running 2000 ft (609.6 m) to reservoir B at elevation 850 ft (259.1 m) and a 6-in (15.24-cm) line running 1000 ft (304.8 m) to reservoir C at elevation 875 ft (266.7 m). Assume $f = 0.02$ in all cases. Will water be delivered to reservoir C at the rate of: (a) 1.20 (0.034), (b) 1.45 (0.041), (c) 1.60 (0.045), or (d) 1.70 ft³ (0.048 m³)/s?

1.318 Reservoir R is at grade 400 ft (121.9 m). From reservoir R an 18-in (45.72-cm) pipe, which is to carry 7 ft³ (0.20 m³)/s of water, leads 2000 ft (609.6 m) to grade 300 ft (91.44 m) at point Y. It there divides, and branch A, 12 in (30.48 cm) in diameter, leads 13,000 ft (3962.4 m) to reservoir A, which is at grade 250 ft (76.2 m). Branch B leads 4000 ft (1219.2 m) to reservoir B which is at grade 50 ft (15.24 m). Neglect all losses except from friction and assume $f = 0.02$ in each case. Will the diameter of the pipe of branch B be: (a) 6 (15.24), (b) 8 (20.32), (c) 10 (25.40), or (d) 12 in (30.48 cm)?

1.319 A 48-in (121.92-cm) main, carrying 75.4 ft³/s (2.14 m³/s), branches at a point A into two pipes: one 2000 ft (609.6 m) long and 36 in (91.44 cm) in diameter and the other 6000 ft (1828.8 m) long and 24 in (60.96 cm) in diameter. Both pipes come together at point B and continue as a single 48-in (121.92-cm) pipe. Assume $f = 0.021$, 0.022, and 0.023, respectively, for the 48-, 36-, and 24-in pipes. Will the rate of flow in the 24-in pipe be: (a) 12.0 (0.340), (b) 12.5 (0.354), (c) 12.9 (0.365), or (d) 13.2 ft³ (0.373 m³)/s?

1.320 A pipeline 25,000 ft (7620 m) long and 5 ft (152.4 cm) in diameter supplies eight nozzles with water from a reservoir whose level is 600 ft (182.88 m) above the nozzles. Each nozzle has an opening 3 in (7.62 cm) in diameter and discharge and velocity coefficients of 0.95. Assuming $f = 0.017$, the aggregate power available in the jets will be: (a) 4200 (3089), (b) 4312 (3171), (c) 4400 (3236), or (d) 4464 hp (3283 kW)?

1.321 A Venturi meter has an area ratio 9:1, the larger diameter being 24 in (60.96 cm). During flow, the recorded pressure head in the large section is 36 ft (10.97 m) and that at the throat 9 ft (2.74 m). If c is 0.99, will the rate of discharge through the meter be: (a) 13.54 (0.383), (b) 13.77 (0.390), (c) 14.10 (0.399), or (d) 14.53 ft³ (0.411 m³)/s?

1.322 A canal has a bottom width of 30 ft (9.14 m) and side slopes of 3 horizontal to 1 vertical. If the water depth is 4 ft (1.22 m) and the slope 1 in 1000, will the probable velocity of discharge, using Kutter's C and $n = 0.02$, be: (a) 826 (23.4), (b) 800 (22.7), (c) 750 (21.2), or (d) 650 ft³ (18.4 m³)/s?

1.323 A cylindrical vessel open at the top is 4 ft (1.22 m) high and 4 ft (1.22 m) in diameter. It is revolved about its center vertical axis at the rate of 56 rpm. A piezometer tube is attached to the side of the vessel and stands vertically 3 ft (0.91 m) from the cylinder axis. If the vessel was previously filled with water to the top edge,

1. Will the water in the piezometer tube rise: (1a) 2.14 (0.65), (1b) 2.68 (0.82), (1c) 3.75 (1.14), or (1d) 4.82 ft (1.47 m) above the brim of the vessel?

2. Will the loss of water be: (2a) 6.72 (0.19), (2b) 10.09 (0.29), (2c) 11.77 (0.33), or (2d) 13.45 ft³ (0.38 m³)?

1.324 A flat circular sign board is 20 ft (6.10 m) in diameter. Weight of air is 0.0807 lb/ft³ (1.29 kg/m³). There is a drag coefficient of 1.12 for a wind force normal to the sign surface. For a wind velocity of 40 mph (64.36 km/h), will the normal force on the sign be: (a) 1480 (6583), (b) 1520 (6761), (c) 1600 (7117), or (d) 1720 lb (7651 N)?

1.325 Water leaves the toe of a spillway with a horizontal velocity of 30 fps (9.14 m/s) and a depth of 0.80 ft (0.24 m) flowing directly onto a level concrete apron.

1. If a jump is to occur on the apron, must the depth attained on the apron be: (1a) 6.3 (1.92), (1b) 6.8 (2.07), (1c) 7.2 (2.19), or (1d) 7.5 ft (2.29 m)?

2. Will the horizontal velocity of the water after the jump be: (2a) 2.65 (0.81), (2b) 3.81 (1.16), (2c) 4.97 (1.51), or (2d) 6.13 fps (1.87 m/s)?

3. If the stream is 200 ft (60.96 m) wide, will the power absorbed by the jump be: (3a) 4300 (3163), (3b) 4380 (3221), (3c) 4460 (3280), or (3d) 4540 hp (3339 kW)?

4. If the length L in which the jump will occur is approximately 4.8 times the depth after the jump occurs, will this length be: (4a) 36.00 (10.97), (4b) 34.56 (10.53), (4c) 32.64 (9.95), or (4d) 30.24 ft (9.22 m)?

MATHEMATICS AND MEASUREMENTS

1.401 An observer wants to determine the height of a tower. The observer takes sights at the top of the tower from points A and B, which are 50 ft (15.24 m) apart, at the same elevation, and on a direct line with the tower. The vertical angle at point A is 30° and at point B 40°. The height of the tower is: (*a*) 89.51 (27.28), (*b*) 92.54 (28.21), (*c*) 95.38 (29.07), or (*d*) 97.33 ft (29.67 m)?

1.402 Express the equation of the straight line passing through points $x_1 = 2$, $y_1 = 2$, and $x_2 = 4$, $y_2 = 3$ in the form $y = mx + b$.

1.403 Solve

$$x^2 + y^2 = 5z$$
$$x^2 - y^2 = 3z$$

How many and what numerical values for x, y, and z will satisfy these simultaneous equations?

1.404 The area of a circle is 89.42 in^2 (576.9 cm^2).

1. Will the diameter of the circle be: (1*a*) 10.50 (26.67), (1*b*) 10.67 (27.10), (1*c*) 10.81 (27.46), or (1*d*) 10.90 in (27.69 cm)?

2. Will the circumference be: (2*a*) 33.34 (84.68), (2*b*) 33.42 (84.89), (2*c*) 33.48 (85.04), or (2*d*) 33.52 in (85.14 cm)?

3. Will the length of a side of a regular hexagon inscribed in this circle be: (3*a*) 5.335 (13.55), (3*b*) 5.43 (13.79), (3*c*) 5.49 (13.94), or (3*d*) 5.52 in (14.02 cm)?

1.405 A vein of ore has a strike of N45°E.; that is, it intersects level ground along a line having this bearing. The vein dips westerly at an angle of 15° with the horizontal. A drift (slightly sloping mine opening following the vein) has been opened in the vein on a bearing of N30°E. The grade of this drift will be: (*a*) -6.44, (*b*) -6.74, (*c*) -6.94, or (*d*) -7.04 percent?

1.406 The distance between two points measured with a steel tape was recorded as 916.58 ft (279.373 m). Later the tape was checked and found to be only 99.9 ft (30.450 m) long. Will the true distance between the points be: (*a*) 915.66 (279.093), (*b*) 915.68 (279.099), (*c*) 915.69 (279.100), or (*d*) 915.70 ft (279.105 m)?

1.407 Solve for x (show all work):

$$4 + \frac{x + 3}{x - 3} - \frac{4x^2}{x^2 - 9} = \frac{x + 9}{x + 3}$$

1.408 A circular piece of tin 24 in (60.96 cm) in diameter has a triangular hole 9 (22.86) × 12 (30.48) × 15 in (38.10 cm). The vertex of the triangle at the intersection of the 9- (22.86-) and 12-in (30.48-cm) sides is at the center of the circle. The 9-in side is vertical, and the triangle is above and to the right of the circle's center.

1. Will the center of gravity measured vertically from the bottom edge of the circle be: (1a) 11.59 (29.44), (1b) 11.62 (29.51), (1c) 11.68 (29.67), or (1d) 11.77 in (29.90 cm)?

2. Will the center of gravity measured horizontally from the left edge of the circle be: (2a) 11.40 (28.96), (2b) 11.46 (29.11), (2c) 11.50 (29.21), or (2d) 11.52 in (29.26 cm)?

1.409 1. Will the moment of inertia of the area in Fig. 1.409 with respect to its centroidal x axis be: (1a) 588.0 (24,474), (1b) 620.0 (25,806), (1c) 644.0 (26,805), or (1d) 660.0 in⁴ (27,471 cm⁴)?

2. Will the moment of inertia of the area with respect to its centroidal y axis be: (2a) 277.0 (11,530), (2b) 296.0 (12,320), (2c) 313.0 (13,028), or (2d) 326.0 in⁴ (13,569 cm⁴)?

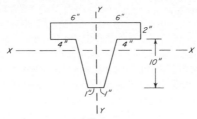

Fig. 1.409

1.410 Locate the centroids of Fig. 1.410 with respect to the axes.

1. Will the moments of inertia of the area with respect to the x axis be: (1a) 453.3 (18,868), (1b) 466.3 (19,409), (1c) 483.3 (20,116), or (1d) 506.3 in⁴ (21,074 cm⁴)?

2. Will the moments of inertia of the area with respect to the y axis be: (2a) 148.3 (6173), (2b) 167.3 (6964), (2c) 184.3 (7671), or (2d) 197.3 in⁴ (8212 cm⁴)?

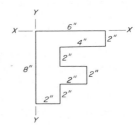

Fig. 1.410

1.411 Two planes leave Cleveland for Jacksonville, a distance of 900 mi (1448 km). The four-motored plane (A) travels at a ground speed of 90 mph (144.8 km/h) faster than the two-motored plane (B). Plane A arrives in Jacksonville 2 h 15 min ahead of plane B. Will the ground speed of plane B be: (a) 150 (241.4), (b) 200 (321.9), (c) 240 (386.2), or (d) 280 mph (450.6 km/h)?

1.412 Twice the sum of two numbers is 28. The sum of the squares of the two numbers is 100. The product of the two numbers is: (a) 42, (b) 48, (c) 54, or (d) 56?

1.413 An engineer was told that a survey has been made of a certain rectangular field but that the dimensions had been lost. An assistant remembered that if the field had been 100 ft (30.48 m) longer and 25 ft (7.62 m) narrower, the area would have been increased 2500 ft² (232.25 m²) and that if it had been 100 ft (30.48 m) shorter and 50 ft (15.24 m) wider, the area would have been decreased 5000 ft² (464.50 m²). Will the length of the field be: (a) 100 (30.48), (b) 150 (45.72), (c) 200 (60.96), or (d) 250 ft (76.2 m)?

1.414 If the outside diameter of a locomotive driving wheel minus its steel tire is 62.378 in (1.5844 m) and the inside diameter of the steel tire at 65°F (18.3°C) is 62.263 in (1.5814 m), the minimum temperature to which the tire must be heated to just fit the wheel is: (a) 323 (161.7), (b) 335 (168.3), (c) 349 (176.1), or (d) 362°F (183.3°C)?

1.415 A certain steel tape is known to be 100.000 ft (30.48 m) long at a temperature of 70°F (21.11°C). When the tape is at a temperature of 10°F (−12.22°C), will the tape reading corresponding to a distance of 90.000 ft (27.432 m) be: (a) 90.035 (27.443), (b) 90.040 (27.444), (c) 90.043 (27.445), or (d) 90.045 ft (27.446 m)?

1.416 Eight laborers can dig 150 ft (45.72 m) of trench in 7 h. Three laborers can backfill 100 ft (30.48 m) of the trench in 4 h. The time for 10 laborers to dig and fill 200 ft (60.96 m) of trench will be: (a) 9 h 52 min, (b) 10 h 1 min, (c) 10 h 24 min, or (d) 10 h 46 min?

1.417 A man owns two square lots of unequal size, together containing 15,025 ft² (1396 m²). If the lots were contiguous, it would require 530 ft (162.5 m) of fence to embrace them in a single six-sided enclosure. Will the side dimension of the smaller lot be: (a) 25 (7.62), (b) 36 (10.97), (c) 64 (19.51), or (d) 81 ft (24.69 m)?

1.418 Given the semicubical parabola $y^3 = x^3$,
  a. Find the area of the part enclosed by the line $x = 4$ and the x axis.
  b. Find the area under the curve $y = x^3 + 3x^2$ and the x axis between $x = 1$ and $x = 3$.
  c. Differentiate the following:

$$4x^2 + 17x \qquad \text{and} \qquad ax^2 + b^{1/2}$$

  d. Integrate the following:

$$(7x^3 + 4x^2)\,dx \qquad \text{and} \qquad x \cos(2x^2 + 7)\,dx$$

1.419 A line was measured with a steel tape when the temperature was 30°C. The measured length of the line was found to be 1256.271 ft (382.911 m). The tape was afterward tested when the temperature was 10°C and found to be 100.042 ft (30.493 m) long. The true length of the line if the coefficient of linear expansion of tape were 0.000011/°C is: (a) 1257.075 (383.156), (b) 1259.038 (383.755), (c) 1263.045, (384.976), or (d) 1275.042 ft (388.633 m)?

1.420 The sum of three numbers in arithmetical progression is 45. If 2 is added to the first number, 3 to the second, and 7 to the third, the new numbers are in geometrical progression. Find the numbers.

1.421 Solve algebraically:
a. $4x^2 + 7y^2 = 32$
  $11y^2 - 3x^2 = 41$
b. $x + 2y - z = 6$
  $2x - y + 3z = -13$
  $3x - 2y + 3z = -16$

1.422 a. Given $\cos 2A = 2\cos^2 A - 1$, find $\cos 75°$.
b. In a circle of unit radius, sketch a central angle B in the second quadrant and show lines that are equal in length to sin A, cos A, and tan A in (a).

1.423 A box is to be constructed from a piece of zinc 20 in² by cutting equal squares from each corner and turning up the zinc to form the sides. Will the volume of the largest box that can be so constructed be: (a) 591.11 (9681.52), (b) 592.59 (9710.77), (c) 593.33 (9722.90), or (d) 593.70 in³ (9728.96 cm³)? Solve by calculus.

1.424 Determine the volume of a right-truncated triangular prism with the following dimensions: The corners of the triangular base are defined by A, B, and C; length of AB = 10 ft (3.05 m), BC = 9 ft (2.74 m), and CA = 12 ft (3.66 m). The sides at A, B, and C are perpendicular to the triangular base and are 8.6 (2.62), 7.1 (2.16), and 5.5 ft (1.68 m) high, respectively. The volume will be: (a) 331 (9.37), (b) 319 (9.03), (c) 311 (8.81), or (d) 307 ft³ (8.69 m³)?

1.425 A 6 percent upgrade meets a 3 percent downgrade at elevation 100.00 at station 10 + 00. If the parabolic vertical curve is 600 ft long, will the station of the high point of the curve be: (a) 10 + 00, (b) 10 + 50, (c) 11 + 00, or (d) 11 + 50? Plot the profile of the curve, giving all pertinent information.

MECHANICS (KINETICS)

1.501 An elevator weighing 2000 lb (907.2 kg) attains an upward velocity of 16 fps (4.88 m/s) in 4 s with uniform acceleration. The tension in the supporting cable is: (a) 2165 (9630), (b) 2250 (10,008), (c) 2345 (10,431), or (d) 2478 lb (11,022 N)?

1.502 If the tension in the cable of the elevator described in question 1.501 is reduced so that the elevator comes to rest in a distance of 5 ft (1.524 m), will the tension be: (a) 390 (1735), (b) 400 (1779), (c) 410 (1824), or (d) 420 lb (1868 N)?

1.503 A body weighing 40 lb (18.14 kg) starts from rest and slides down a plane at an angle of 30° with the horizontal, for which the coefficient of friction $f = 0.3$.

1. To slide 60 ft (18.29 m), will it take: (1a) 3.75, (1b) 3.86, (1c) 3.94, or (1d) 4.00 s?

2. During the third second, will the body move: (2a) 19.33 (5.89), (2b) 19.66 (5.99), (2c) 20.00 (6.10), or (2d) 20.33 ft (6.20 m)?

1.504 A car and its load weigh 6000 lb (26,688 N), and the center of gravity is 2 ft (0.61 m) from the ground and midway between the front and rear wheels, which are 120 in (3.05 m) apart. The car is brought to rest from a speed of 30 mph (48.3 km/h) in 5 s by the brakes. Will the normal pressure on each of the front wheels and on each of the rear wheels be: (a) 1172 (5213), (b) 1336 (5943), (c) 1500 (6672), or (d) 1664 lb (7402 N)?

1.505 A body weighing 1000 lb (4448 N) falls 6 in (15.24 cm) and strikes a 2000 lb/in (3.502 N/cm) spring. The deformation of the spring is: (a) 3.0, (7.62), (b) 3.5 (8.89), (c) 3.8 (9.65), or (d) 4.2 in (10.67 cm)?

1.506 A freight car weighing 100,000 lb (444.8 kN) is moving with a velocity of 2 fps (0.61 m/s) when it strikes a bumping post. Assuming the drawbar spring on the car to take all the compression, so that the compression in the spring shall not exceed 2.5 in (6.35 cm), must the scale of the spring be: (a) 22,000 (38.53), (b) 22,925 (40.15), (c) 23,850 (41.77), or (d) 24,775 lb/in (43.39 kN/cm)?

1.507 If the coeffecent of friction under the two 100-lb (45.36-kg) blocks M and N in Fig. 1.507 (p. 42) is 0.20,

1. Will the acceleration of the blocks be: (1a) 1.505 (0.459), (1b) 1.610 (0.491), (1c) 1.715 (0.523), or (1d) 1.820 ft (0.555 m)/s²?

2. Will the time for the blocks to move 15 ft (4.57 m) be: (2a) 4.40, (2b) 4.47, (2c) 4.54, or (2d) 4.61 s?

3. Will the tension in the cord be: (3*a*) 63 (28.58), (3*b*) 67 (30.39), (3*c*) 70 (31.75), or (3*d*) 72 lb (32.66 kg)?

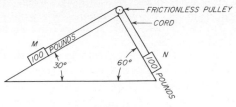

FRICTIONLESS PULLEY

CORD

M

100 POUNDS

30°

60°

N

100 POUNDS

Fig. 1.507

1.508 Bodies *X*, *Y*, and *Z* in Fig. 1.508 weigh 10 (4.54), 20 (9.07), and 30 lb (13.61 kg), respectively. If they are supported in the position shown and then simultaneously released,

Fig. 1.508

1. Will the velocities of the bodies (neglecting the mass of the cords and pulleys) 1 s later be: (1*a*) 1.9 (0.579), (1*b*) 5.7 (1.77), (1*c*) 9.51 (2.90), or (1*d*) 13.3 fps (4.05 m/s)?

2. Will the position of the bodies 1 s later be: (2*a*) 0.95 (0.289) down, (2*b*) 4.75 (1.45) down, (2*c*) 6.70 (2.04) up, or (2*d*) 9.65 ft (2.94 m) up?

1.509 A 60-ton (54,432-kg) freight car starts from rest and runs 100 ft (30.48 m) down a 1 percent grade, then strikes a bumping post.

1. If train resistance is 10 lb/ton, (5 kg/1000 kg), is the velocity of striking: (1*a*) 5.00 (1.52), (1*b*) 5.33 (1.62), (1*c*) 5.67 (1.73), or (1*d*) 6.00 fps (1.83 m/s)?

2. If the drawbar spring has a scale of 100,000 lb/in (17,855 kg/cm) and is assumed to take all the impact, will it be shortened: (2*a*) 3.450 (8.763), (2*b*) 3.565 (9.055), (2*c*) 3.680 (9.347), or (2*d*) 3.795 in (9.639 cm)?

1.510 A car weighing 40 tons (36,288 kg) is switched to a 2 percent upgrade with a velocity of 30 mph (48.28 km/min).

1. If train resistance is 10 lb/ton (5 kg/1000 kg), will it go up the grade: (1*a*) 556 (169.5), (1*b*) 880 (268.2), (1*c*) 1096 (334.1), or (1*d*) 1204 ft (367.0 m)?

2. If the car is then allowed to run back, will the velocity at the foot of the grade be: (2*a*) 34 (10.36), (2*b*) 36 (10.97), (2*c*) 39 (11.89), or (2*d*) 43 fps (13.11 m/s)?

3. If the track is then level, from the foot of the grade will the car run: (3*a*) 3400 (1036.3), (3*b*) 3506 (1068.6), (3*c*) 3612 (1100.9), or (3*d*) 3718 ft (1133.2 m)?

1.511 A 10° curve is one where a 100-ft (30.48-m)-long chord subtends an angle of 10° at its center.

1. Will the superelevation of the outer rail of a railroad track on a 10° curve to give equal rail pressures on a car when moving with a speed of 45 mph (72.42 km/h) be: (1a) 1.00 (0.305), (1b) 1.15 (0.351), (1c) 1.25 (0.381), or (1d) 1.30 ft (0.396 m)?

2. If a car weighing 100,000 lb (45,360 kg) with its center of gravity 5 ft (1.52 m) above the top of rail has a speed of 60 mph (96.56 km/h) on this curve and if the track now has a superelevation of only 8 in (20.32 cm), is the force of the outer rail on the car normal to the ties: (2a) 78,750 (35,721), (2b) 79,500 (36,001), (2c) 80,250 (36,401), or (2d) 81,000 lb (36,742 kg)?

1.512 A concrete highway curve with a 500-ft (152.4-m) radius is banked to give lateral pressure equivalent to $f = 0.15$. Will skidding impend for a speed of 60 mph (96.56 km/h) if the coefficient of friction is below: (a) 0.290, (b) 0.300, (c) 0.310, or (d) 0.320?

1.513 A girder weighing 20,000 lb (9072.0 kg) is suspended by a cable 100 ft (30.48 m) long.

1. Is the horizontal pull necessary to hold it 5 ft (1.52 m) from the vertical position: (1a) 1000 (453.6), (1b) 1001 (454.1), (1c) 1002 (454.5), or (1d) 1003 lb (455.0 kg)?

2. Is the tension in the cable as the girder is allowed to swing back through its vertical position: (2a) 20,000 (9072.0), (2b) 20,025 (9083.3), (2c) 20,050 (9094.7), or (2d) 20,075 lb (9106.0 kg)?

1.514 A shot is fired at an angle of 45° with the horizontal and a velocity of 300 fps (91.44 m/s).

1. Will the height of the projectile reach: (1a) 700 (213.4), (1b) 750 (228.6), (1c) 800 (243.8), or (1d) 850 ft (259.1 m)?

2. Will the range of the projectile reach: (2a) 2000 (609.6), (2b) 2500 (762.0), (2c) 2800 (853.4), or (2d) 3000 ft (914.4 m)?

1.515 1. Will the theoretical muzzle velocity required to give a projectile a maximum range of 40 mi (64.372 km) be: (1a) 2500 (762.00), (1b) 2600 (192.48), (1c) 2800 (853.44), or (1d) 3100 fps (944.88 m/s)?

2. Will the maximum height to which the projectile will rise be: (2a) 7.0 (11.265), (2b) 8.5 (13.679), (2c) 9.5 (15.288), or (2d) 10.0 mi (16.093 km)?

1.516 Cast iron weighs 450 lb/ft³ (7208 kg/m³). A cast-iron governor ball 3 in (7.62 cm) in diameter has its center 18 in (45.72 cm) from the point of support. Neglecting the weight of the arm itself and if the angle with the vertical axis is 60°,

1. Will the speed of rotation of the arm be: (1*a*) 62.6, (1*b*) 63.8, (1*c*) 65.3, or (1*d*) 67.0 rpm?

2. Will the tension in the arm be: (2*a*) 7.20 (32.0), (2*b*) 7.36 (32.7), (2*c*) 7.49 (33.3), or (2*d*) 7.58 lb (33.7 N)?

1.517 A car is moving on a horizontal track around a curve of 1000-ft (304.8-m) radius with a speed of 60 mph (96.56 km/h). A 25-lb (11.34-kg) weight is suspended from the ceiling by a 6-ft (1.83-m) cord.

1. Is the angle the cord makes with the vertical: (1*a*) 12.5, (1*b*) 13.8, (1*c*) 15.3, or (1*d*) 17.1°?

2. Is the horizontal displacement of the weight: (2*a*) 2.05 (0.625), (2*b*) 1.85 (0.564), (2*c*) 1.70 (0.518), or (2*d*) 1.58 ft (0.482 m)?

3. Is the tension in the cord: (3*a*) 25.9 (115.2), (3*b*) 26.4 (117.4), (3*c*) 27.3 (121.4), or (3*d*) 28.4 lb (126.3 N)?

1.518 The muzzle velocity of a projectile is 1500 fps (457.2 m/s) and the distance of the target is 10 mi (16.1 km). The angle of elevation of the gun must be: (*a*) 21°59′, (*b*) 22°41′, (*c*) 24°33′, or (*d*) 25°18′?

1.519 A cast-iron cylinder 24 in (60.96 cm) in outside diameter and 16 in (40.64 cm) high has a wall thickness of 1 in (2.54 cm). Cast iron weighs 450 lb/ft^3 (7208 kg/m^3). The moment of inertia of the cylinder with respect to its geometric axis is: (*a*) 10.2 (13.85), (*b*) 9.7 (13.15), (*c*) 9.2 (12.48), or (*d*) 8.7 lb·ft·s^2 (11.80 N·m·s^2)?

1.520 A steel disk 40 in (101.6 cm) in diameter and 4 in (10.16 cm) thick has a cylindrical hole 10 in (25.4 cm) in diameter at the center and another hole 12 in (30.48 cm) in diameter 12 in (30.48 cm) from the center. Steel weighs 490 lb/ft^3 (7848 kg/m^3). Is the moment of inertia with respect to the geometric axis of the 40-in (101.6-cm) disk: (*a*) 54.78 (74.28), (*b*) 55.88 (75.77), (*c*) 56.78 (76.99), or (*d*) 57.48 lb·ft·s^2 (77.94 N·m·s^2)?

1.521 A flywheel is brought from rest up to a speed of 1500 rpm in 1 min.

1. Was the average angular acceleration α (1*a*) 2.050, (1*b*) 2.300, (1*c*) 2.500, or (1*d*) 2.617 rad/s^2?

2. The number of revolutions made by the wheel in that 1 min were: (2*a*) 750, (2*b*) 1000, (2*c*) 1250, or (2*d*) 1500?

3. Was the angular velocity at the end of 40 s: (3*a*) 250, (3*b*) 500, (3*c*) 750, or (3*d*) 1000 rpm?

1.522 The rim of a 36-in (91.44-cm) wheel on a brakeshoe testing machine has a speed of 60 mph (96.56 km/h) when the brake is dropped. It comes to rest when the rim has traveled a tangential distance of 500 ft (152.4 m).

1. The number of radians of rotation to stop the wheel are: (1a) 267, (1b) 300, (1c) 333, or (1d) 367?

2. The angular deceleration is: (2a) 5.00, (2b) 5.17, (2c) 5.34, or (2d) 5.51 rad/s²?

1.523 A steel hemisphere 6 in (15.24 cm) in diameter is to rotate about an axis parallel to its axis of symmetry and 12 in (30.48 cm) distant from it. The axis of the rod that connects it to the axis of rotation is 0.5 in (1.27 cm) from the diametral plane.

1. For no bending moment in the rod, should it be placed above the diametral plane: (1a) 0.525 (1.334), (1b) 0.825 (2.096), (1c) 1.025 (2.604), or (1d) 1.125 in (2.858 cm)?

2. When the hemisphere is rotating at 300 rpm, will the induced bending moment in the rod be: (2a) 25.6 (34.7), (2b) 26.2 (35.5), (2c) 26.5 (35.9), or (2d) 26.6 ft·lb (36.1 N·m)?

1.524 A steel rod 24 in (60.96 cm) long rotates about a vertical axis through one end. If it stands at an angle of 60° with the axis, the speed at which it is rotating is: (a) 72, (b) 69, (c) 66, or (d) 64 rpm?

1.525 A cast-iron flywheel 12 ft (3.66 m) in outside diameter has a rim 4 in (10.16 cm) thick and 18 in (45.72 cm) wide. Cast iron weighs 450 lb/ft³ (7.208 kg/m³). If the flywheel is rotating at 240 rpm and the tension in the arms is neglected, will the unit centrifugal tensile stress in the rim be: (a) 2000 (13.79), (b) 2032 (14.01), (c) 2066 (14.25), or (d) 2100 psi (14.48 MPa)?

MECHANICS (STATICS)

1.601 A set of four vertical forces are in the same plane. The first force is 10 lb (44.5 N) upward; the next force, 5 ft (1.52 m) away, is 10 lb (44.5 N) downward; the third force, 4 ft (1.22 m) beyond the second, is 10 lb (44.5 N) downward; the fourth force, 3 ft (0.91 m) beyond the third, is 10 lb (44.5 N) upward.

1. Will the resultant be a couple: (1a) 20 (27.1) clockwise, (1b) 20 (27.1) counterclockwise, (1c) 30 (40.6) clockwise, or (1d) 30 ft·lb (40.6 N·m) counterclockwise?

2. If the direction of the fourth force is changed to downward, will the resultant force be located to the first force by: (2a) 10 (3.05), (2b) 11 (3.35), (2c) 13 (3.96), or (2d) 16 ft (4.88 m)?

1.602 A long wall stands on three rows of piles, with the rows spaced 3 ft (0.914 m) center to center. The spacings of piles parallel to the wall are 2 ft 6 in (0.762 m), 4 ft (1.219 m), and 6 ft (1.829 m) center to center in the first, second, and third rows, respectively. The resultant of the vertical loads is located midway between the first and second rows and is 20,000 lb (88.96 kN) per linear foot (0.305 m) of wall. In the front row, will the load per pile per linear foot of wall be: (a) 12,210 (54.31), (b) 8503 (37.82), (c) 5515 (24.53), or (d) 2253 lb/ft (10.04 kN/0.305 m)?

1.603 A beam 12 ft (3.66 m) long and simply supported at each end has a uniformly distributed load of 1000 lb/ft (14.59 kN/m) extending from the left end to a point 4 ft (1.22 m) away. There is also a clockwise couple of 10,000 ft·lb (13,560 N·m) applied at the center of the beam. Draw the shear and moment diagrams for the beam and give all the necessary values. Neglect the weight of the beam.

1.604 Draw shear and moment diagrams for a 20-ft (6.10-m) beam, simply supported at the ends. The beam carries a uniform load of 1000 lb/ft (14.59 kN/m) (including its own weight) and three concentrated loads of 2000 (8896), 4000 (17,792), and 6000 lb (26,688 N) acting respectively at the left, center, and right quarter points.

1.605 A beam 20 ft (6.10 m) long carries a uniform load of 2000 lb/lin ft (29.19 kN/m) including its own weight, on two supports 14 ft (4.27 m) apart. The right end of the beam cantilevers 2 ft (0.61 m), and the left end cantilevers 4 ft (1.22 m). Draw the shear and moment diagrams and compute the positions and amounts of maximum bending moments.

1.606 A beam 24 ft (7.32 m) long rests on two supports, one at its right end and the other 6 ft (1.83 m) from its left end. The beam carries a uniform load of 1000 lb/lin ft (14.59 kN/m) over its entire length and a concentrated load of 15,000 lb (66.72 kN) at the middle of the 18-ft (5.49-m) span. Calculate the reactions, draw the shear and moment diagrams, and calculate the position and amount of maximum bending moment.

1.607 The left-hand half of a beam on a 30-ft (9.14-m) span has a uniform load of 4000 lb/lin ft (58.37 kN/m). A concentrated load of 30,000 lb (133.44 kN) is located 10 ft (3.05 m) from the right end. Draw the shear and moment diagrams.

1.608 A beam carries a concentrated load P applied 6 ft (1.83 m) from the right end. The beam is partially restrained at the left end and simply supported at the right end. The bending moment at the partially restrained end is $PL/16$ in magnitude. The beam is 16 ft (4.88 m) long and has a 4 × 16 in (10.16 × 40.64 cm) cross section. The beam is made of wood, which has a modulus of elasticity $E = 16 \times 10^5$ psi (11,032 MPa). If the maximum induced bending stress is 1350 psi (9.31 MPa):

1. Will the magnitude of P be: (1a) 5060 (22.51), (1b) 5269 (23.44), (1c) 5478 (24.37), or (1d) 5687 lb (25.30 kN)?

2. Will the deflection at the point of load application be: (2a) 0.276 (0.701), (2b) 0.300 (0.762), (2c) 0.324 (0.823), or (2d) 0.348 in (0.884 cm)?

1.609 I is uniform throughout in Fig. 1.609.

1. The maximum positive moment is: (1a) 70.48 (95.57), (1b) 73.45 (99.60), (1c) 75.43 (102.28), or (1d) 78.41 ft·kips (106.32 kN·m)?

2. The maximum negative moment is: (2a) 53.85 (73.02), (2b) 58.65 (79.53), (2c) 60.75 (82.38), or (2d) 63.45 ft·kips (86.04 kN·m)?

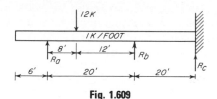

Fig. 1.609

1.610 Draw the shear and moment diagrams for the beam ABC in Fig. 1.610.

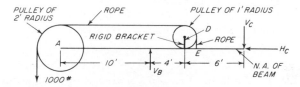

Fig. 1.610

1.611 A 1000-lb (454-kg) weight A is to be supported by two wires. The first wire BA runs to a ring at point B, 3 ft (0.91 m) above A and 4 ft (1.22 m) to the left. The second wire CA runs to a ring at C, 4 ft (1.22 m) above A and 3 ft (0.91 m) to the right. If points A, B, and C are in the same vertical plane, find the horizontal and vertical components of the stress in the wires at points B and C and the stress in each wire.

1.612 A painter's scaffold 20 ft (6.10 m) long weighing 150 lb (68.04 kg) is supported in a horizontal position by vertical ropes attached at equal distances from the ends of the scaffold. Will the greatest distance from the ends that the ropes may be attached to permit a 200-lb (90.72-kg) painter to stand safely at one end of the scaffold be: (*a*) 4.00 (1.22), (*b*) 4.29 (1.31), (*c*) 4.83 (1.47), or (*d*) 5.71 ft (1.74 m)?

1.613 A 20-ft (6.10-m) ladder stands on a rough horizontal floor and leans against a vertical smooth wall, the foot of the ladder being 8 ft (2.44 m) out from the wall. Halfway up the ladder is a 150-lb (68.04-kg) weight. Neglecting the weight of the ladder, the horizontal component of the reaction against the bottom end of the ladder is: (*a*) 31.5 (140.1), (*b*) 32.75 (145.7), (*c*) 35.83 (159.4), or (*d*) 39.85 lb (177.3 N)?

1.614 A rectangular masonry wall is 6 ft (1.83 m) thick and 15 ft (4.57 m) high. The masonry weighs 150 lb/ft³ (2403 kg/m³). Assuming there is no seepage causing uplift against the bottom of the wall and water weighs 62.5 lb/ft³ (1001 kg/m³), the height the water could rise behind the wall without causing an intensity of pressure at the toe greater than twice the average pressure on the base is: (*a*) 10.9 (3.32), (*b*) 11.4 (3.45), (*c*) 11.8 (3.60), or (*d*) 12.3 ft (3.75 m)?

1.615 In Fig. 1.615, is the wall safe against overturning about point *A* if the masonry weighs 150 lb/ft³ (2403 kg/m³) and the earth weighs 100 lb/ft³ (1602 kg/m³)? The horizontal pressure of the earth against the masonry wall is equal to *cwh*, where *cw* equals 30 psf (1.44 kPa) and *h* is the distance down from the top of the wall in feet (meters). Will the theoretical coefficient of friction between the wall and the base be: (*a*) 0.26, (*b*) 0.28, (*c*) 0.29, or (*d*) 0.30?

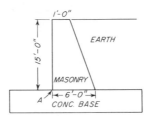

Fig. 1.615

1.616 A truck is placed on a 30-ft (9.14-m) span to give the greatest bending moment. The loads are 4 tons (3629 kg) on the front axle and 16 tons (14,515 kg) on the rear axle. The distance between axles is 14 ft (4.27 m). As the truck moves on the span,

1. The maximum bending moment induced by the loads is: (1*a*) 133.8 (362.9), (1*b*) 127.3 (345.2), (1*c*) 125.8 (341.2), or (1*d*) 123.3 ft·tons (334.4 kN·m)?

2. The maximum shear induced by the loads is: (2*a*) 18.13 (161.3), (2*b*) 19.83 (176.4), (2*c*) 21.73 (193.3), or (2*d*) 23.5 tons (209.1 kN)?

1.617 In Fig. 1.617, the tractor and semitrailer with axle loads is a live load on a 100-ft (30.48-m) simple span bridge girder.

1. When this load gives maximum moment, will the 8-kip (35.58-kN) load be distant to the right from the left end of the span by: (1*a*) 32.33 (9.85), (1*b*) 33.00 (10.06), (1*c*) 33.67 (10.26), or (1*d*) 34.33 ft (10.46 m)?

2. Will the maximum moment caused by this loading be: (2*a*) 1500 (2034), (2*b*) 1528 (2072), (2*c*) 1570 (2129), or (2*d*) 1626 ft·kips (2205 kN·m)?

3. Will the maximum shear caused by this load be: (3*a*) 65.3 (290.5), (3*b*) 66.8 (297.1), (3*c*) 68.8 (306.0), or (3*d*) 71.3 kips (317.1 kN)?

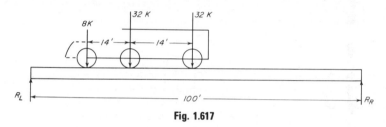

Fig. 1.617

1.618 Draw shear and moment diagrams for beams *AB*, *CD*, and *EF* in Fig. 1.618. Assume all beams as "simple beams."

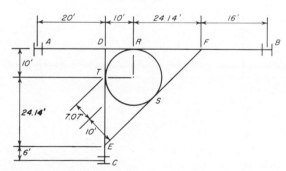

Fig. 1.618 Tank weighing 100 kips is supported at *R*, *S*, and *T* only.

1.619 Find the tension and compression in each member in Fig. 1.619.

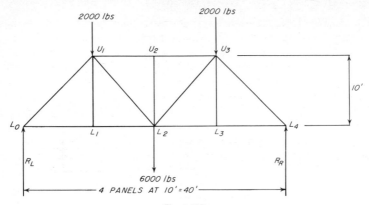

Fig. 1.619

1.620 In the roof truss in Fig. 1.620, determine the stresses in members L_0L_1, U_1U_2, and U_1L_2. State which members are in tension or compression.

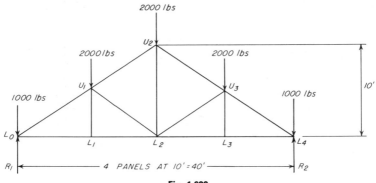

Fig. 1.620

1.621 Determine the stress in the members of the roof truss in Fig. 1.621.

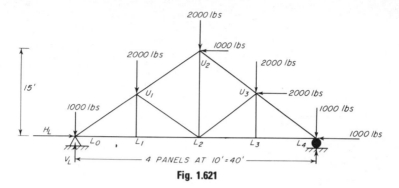

Fig. 1.621

1.622 For the deck truss in Fig. 1.622, compute the magnitude and direction (tension or compression) of the stress in the member U_2U_3 and the diagonals U_0L_1 and U_1L_2.

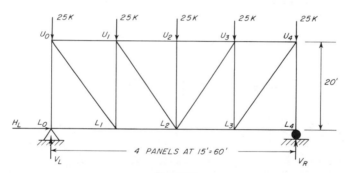

Fig. 1.622

1.623 Determine the amount and character of the stress in each member of the truss in Fig. 1.623. Place answers on your sketch of the truss.

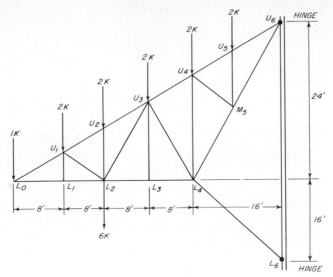

Fig. 1.623

1.624 Determine the kind and amount of stress in each member of the truss in Fig. 1.624.

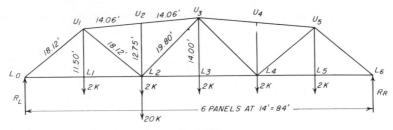

Fig. 1.624

1.625 The space structure in Fig. 1.625 supports a 2400-lb (10.68-kN) horizontal load. Determine the (a) vertical component of stresses in member AB, (b) X component of stress in member AD, (c) Y component of stress in member AC, (d) total stress in member AD.

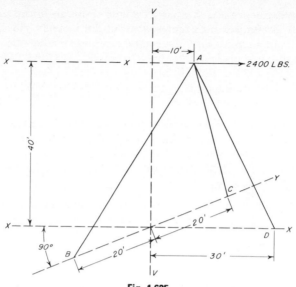

Fig. 1.625

THERMODYNAMICS

1.701   Define: (*a*) latent heat, (*b*) specific heat, (*c*) heat of fusion, (*d*) Btu, (*e*) absolute temperature, (*f*) calorie, (*g*) watt.

1.702   Thirty lb (13.6 kg) of ice at 32°F (0°C) is placed in 100 lb (45.4 kg) of water at 100°F (37.8°C). [The latent heat of ice may be taken as 144 Btu/lb (335.2 kJ/kg).] If no heat is lost or added to the mixture, the temperature when equilibrium is reached is: (*a*) 48 (8.9), (*b*) 49 (9.4), (*c*) 50 (10.0), or (*d*) 51°F (10.6°C)?

1.703   A small swimming pool 25 × 75 ft (7.62 × 22.86 m) is to be filled with water at a temperature of 70°F (21.1°C) to a depth of 5 ft (1.52 m). Hot water at 160°F (71.1°C) and cold water at 40°F (4.4°C) is available. Neglecting the thermal capacity of the tank itself, will the amount of hot or cold water be: (*a*) 2344 (66.38), (*b*) 4699 (133.08), (*c*) 7031 (199.12), or (*d*) 9340 ft³ (264.51 m³)?

1.704   A partly filled barrel contains 300 lb (136.08 kg) of water and 100 lb (45.36 kg) of ice at 32°F (0°C). Will the number of pounds of steam (212°F; 100°C) to be run into the barrel to bring its contents up to 80°F (26.7°C) be: (*a*) 30.0 (13.61), (*b*) 30.5 (13.83), (*c*) 31.5 (14.29), or (*d*) 33.0 lb (14.97 kg)?

1.705 Air is compressed in a diesel engine from an initial pressure of 13 psia (89.64 kPa abs) and a temperature of 120°F (48.9°C) to one-twelfth its initial volume. Assuming the compression to be adiabatic,

1. Will the final pressure be: (1*a*) 375 (2586), (1*b*) 400 (2758), (1*c*) 421 (2903), or (1*d*) 438 psi (3020 kPa)?

2. Will the final temperature be: (2*a*) 1100 (593), (2*b*) 1110 (599), (2*c*) 1130 (610), or (2*d*) 1160°F (627°C)?

1.706 An automobile tire is inflated to 32-psig (220.6-kPa gage) pressure at 50°F (15.6°C). After the car has been driven, the temperature rises to 75°F (23.9°C). Assuming that the volume remains constant, the final gage pressure is: (*a*) 33.0 (227.5), (*b*) 34.4 (237.2), (*c*) 37.3 (257.2), or (*d*) 39.2 psig (270.3 kPa gage)?

1.707 If 100 ft³ (2.83 m³) of atmospheric air (pressure 14.7 psi) (101.3 kPa) at 0°F (−17.8°C) is compressed to a volume of 1 ft³ (0.0283 m³) at temperature 200°F (93.3°C), will the pressure of the compressed air be: (*a*) 1900 (13.1), (*b*) 2000 (14.0), (*c*) 2070 (14.3), or (*d*) 2110 psia (14.6 MPa abs)?

1.708 *a.* Distinguish between higher heating and lower heating values of fuel.

b. Why does the lower heating value at constant volume differ from that at constant pressure?

1.709 Air at 90°F (32.2°C) dry bulb and 65°F (18.3°C) wet bulb travels across the surface of a pond of water that is at 70°F (21.1°C). Will any of the 70°F (21.1°C) water be evaporated into the air stream or will any of the water vapor in the air be condensed? Explain fully the physical laws upon which your answer is based.

1.710 *a.* If you want the temperature distribution within a room to be as even as possible, would you blow hot air into the room near the floor or the ceilings? State your reasons.

b. To heat the room most effectively, should the steam radiators be painted with black or aluminum paint? State your reasons.

1.711 A steam turbine carrying a full load of 50,000 kW uses 569,000 lb (258,098 kg) steam per hour. The engine efficiency is 75 percent and its exhaust steam is at 1 inHg abs and has an enthalpy of 950 Btu/lb (2209.7 kJ/kg).

1. At the throttle, will the temperature of the steam be: (1*a*) 400 (204.4), (1*b*) 450 (232.2), (1*c*) 476 (246.7), or (1*d*) 498°F (258.9°C)?

2. Will the pressure of the steam be: (2*a*) 243 (1675.5), (2*b*) 273 (1882.3), (2*c*) 293 (2020.2), or (2*d*) 303 psia (2089.2 kPa abs)?

1.712 The use of electricity for melting snow in a driveway 10 ft (3.048 m) wide × 50 ft (15.74 m) long is being considered. [Assume the following data: weight of snow = 10 lb/ft³ (160.17 kg/m³), temperature = 32°(0°C), efficiency of operation = 50 percent.] At 8 cents/kWh, will the cost of melting 6 in (15.24 cm) of snow be: (*a*) $16.88, (*b*) $17.08, (*c*) $17.23, or (*d*) $17.33?

1.713 A steam boiler on test generates 885,000 lb (401,436 kg) of steam in a 4-h period. The average steam pressure is 400 psia (2758 kPa abs), the average steam temperature is 700°F (371.1°C), and the average temperature of the feedwater supplied to the boiler is 280°F (137.8°C). If the boiler efficiency for the period is 82.5 percent and the coal has a heating value of 13,850 Btu/lb (32,151 kJ/kg) as fired, will the average amount of coal burned per hour be: (*a*) 19,000 (8614), (*b*) 20,200 (9163), (*c*) 21,000 (9571), or (*d*) 21,600 lb (9798 kg)?

1.714 A hot-water heater consists of a 20-ft (6.10-m) length of copper pipe ½ in (1.27 cm) average diameter and 1/16 in (0.1588 cm) thick. The outer surface of the pipe is maintained at 212°F (100°C). [Assume the conductivity of copper as 2100 Btu/(ft²)(°F)(h)(in) thickness (16,900 kJ/(m²)(°C)(h)(cm)]. If water is fed into the coil at 40°F (4.4°C) and expected to emerge heated to 150°F (65.6°C), will the capacity of the coil in gallons (liters) of water per minute be: (*a*) 1.75 (6.62), (*b*) 1.91 (7.23), (*c*) 2.10 (7.95), or (*d*) 2.35 (8.89)?

1.715 Coal having a heat of combustion of 14,000 Btu/lb (32,564 kJ/kg) is used in a heating plant of 50 percent efficiency. Steam of 50 percent quality and 212°F (100°C) temperature is made from this coal and from water whose initial temperature is 70°F (21.1°C). Will the pounds (kilograms) of steam per pound (kilogram) of coal be: (*a*) 11.2, (*b*) 11.7, (*c*) 12.1, or (*d*) 12.4?

1.716 A single-cylinder, double-acting, reciprocating steam engine has a 6-in (15.24-cm) bore, an 8-in (20.32-cm) stroke, and a 1.25-in (3.18-cm) piston-rod diameter. The average mean effective pressure found from the indicator cards is 62 psi (427.5 kPa) for each end of the cylinder. The engine operates at 300 rpm with a mechanical efficiency of 83 percent. If the engine is directly coupled to a generator having an efficiency of 92 percent, will the generator output be: (*a*) 11.00, (*b*) 11.50, (*c*) 11.85, or (*d*) 12.10 kW?

1.717 Steam is admitted to the cylinder of an engine in such a manner that the average pressure is 120 psi (827.4 kPa). The diameter of the piston is 10 in (25.4 cm) and the length of stroke is 12 in (30.5 cm). First find the work done during 1 revolution, assuming that steam is admitted to each side of the piston in succession. Will the power of the engine when it is making 300 rpm be: (a) 141.0 (105), (b) 150.0 (112), (c) 160.0 (119), or (d) 171.5 hp (128 kW)?

1.718 A volume of 400 cm³ of air is measured at a pressure of 740 mmHg abs and a temperature of 18°C. The volume at 760 mmHg abs and 0°C is: (a) 352, (b) 358, (c) 366, or (d) 369 cm³?

1.719 Will the temperature of 2 liters of water at 30°C after 500 cal of heat have been added to it be: (a) 29.75, (b) 30.00, (c) 30.25, or (d) 30.50°C?

1.720 A heat engine (Carnot cycle) has intake and exhaust temperatures of 157°C and 100°C, respectively. The engine's efficiency is: (a) 12.75, (b) 13.25, (c) 14.38, or (d) 16.33 percent?

1.721 A 300-W water heater is attached to a water faucet. The water runs at a rate that permits it to be heated from 60 to 120°F (15.6 to 48.9°C). Assume that 75 percent of the electrical energy will be utilized in heating the water. At 8 cents/kWh, to obtain 2 gal (7.57 liters) of heated water, will it cost: (a) 1, (b) 2, (c) 3, or (d) 4 cents?

1.722 By means of insulation, the loss in heat through a roof per square foot is reduced from 0.40 to 0.18 Btu (422 to 190J)/h for each °F [8.17 to 3.68 kJ/(m²)(°C)] of difference between inside and outside temperatures. The area of the roof is 10,000 ft² (929 m²), and the average difference between inside and outside temperatures is 35°F (19.4°C) during the heating seasons of 5000 h. If the heating value of coal is 13,000 Btu/lb (30238 kJ/kg) and the efficiency of the heating plant is 60 percent, will the value of the coal saved per season at $40/ton ($0.0441/kg) be: (a) $933, (b) $957, (c) $975, or (d) $987?

1.723 The quality of steam that gives up 475 Btu/lb (1105 kJ/kg) while condensing to water at a constant pressure of 20 psig (137.9 kPa gage) is: (a) 50.7, (b) 52.3, (c) 53.6, or (d) 56.2 percent?

1.724 A single-cylinder, double-acting, reciprocating steam engine has a 12-in (30.48-cm) diameter piston with an 18-in (45.72-cm) stroke. The piston-rod diameter is 2 in (5.08 cm). Indicator cards show a mean effective pressure of 70 psi (482.7 kPa) for both the heat and crank ends. The engine operates at 350 rpm with an efficiency of 92 percent. The power output is: (a) 228.5 (170.5), (b) 232.0 (173.1), (c) 237.5 (177.2), or (d) 245.0 hp (182.8 kW)?

1.725 A 300-hp (223.8-kW) engine is given a brake test. The brakes are water-cooled. Is the rate water at 80°F (26.7°C) must flow through the brakes if the water must not rise above 180°F (82.2°C): (*a*) 12.1 (45.8), (*b*) 15.3 (57.9), (*c*) 17.7 (67.0), or (*d*) 19.3 gal (73.1 liters)/min?

GENERAL

1.801 The phase-shifting network in Fig. 1.801 is used as a portion of a control circuit.

1. Is the value of R that will yield a voltage E_{AD} which leads the voltage E_{CA} by 45°: (1*a*) 2630, (1*b*) 2640, (1*c*) 2650, or (1*d*) 2660 ohms?

2. Is the magnitude of $|E_{AD}|$ as compared with $|E|$: (2*a*) 1.414, (2*b*) 1.500, (2*c*) 1.732, or (2*d*) 2.00?

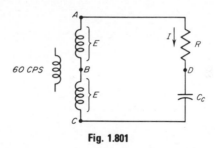

Fig. 1.801

1.802 A series RLC circuit resonates at a frequency of 10^6 cps. Its bandwidth between half-power points is 0.1 times the resonant frequency. If it draws 10 W from a 100-V rms source at resonant frequency:

1. Is the value of R: (1*a*) 1000, (1*b*) 1100, (1*c*) 1200, or (1*d*) 1250 ohms?

2. Is the value of L: (2*a*) 1.29, (2*b*) 1.59, (2*c*) 1.95, or (2*d*) 2.59 $\times$ 10^{-3} H?

3. Is the value of C: 15.9 $\times$ (3*a*) 10^{-3}, (3*b*) 10^{-6}, (3*c*) 10^{-9}, or (3*d*) 10^{-12} F?

1.803 Two transformers receive power from a three-phase three-wire 2200-V line. The load on the first transformer consists of induction motors that draw 80 kW at a power factor (pf) of 0.8. The second transformer supplies a balanced lighting load of 50 kW at 120 V and unity pf. Neglecting exciting currents of the transformers and their losses:

1. Is the total load: (1*a*) 80$\underline{/36.8°}$, (1*b*) 100$\underline{/36.8°}$, (1*c*) 130$\underline{/27.5°}$, or (1*d*) 143$\underline{/24.7°}$ kVA?

2. Is the pf: (2*a*) 0.866, (2*b*) 0.886, (2*c*) 0.900, or (2*d*) 0.910?

3. Is the primary current per phase: (3*a*) 35.0, (3*b*) 36.7, (3*c*) 37.6, or (3*d*) 38.5 A?

1.804 Figure 1.804 is the equivalent circuit of a series-connected coil with resistance and capacitor with leakage conductance. The values are: $L = 0.001$ H, $Q_{coil} = 2$. The product $LC = 10^{-12}$ s^2 and the time constant of the capacitor alone is 2×10^{-6} s.

1. Are the values of the circuit parameters R (ohms), C (F), G (mho), respectively: (1a) 500, 1×10^{-9}, 5×10^{-4}; (1b) 600, 1×10^{-9}, 5×10^{-4}; (1c) 700, 1×10^{-10}, 5×10^{-3}; or (1d) 500, 1×10^{-10}, 5×10^{-3}?

2. Is the series-resonant frequency: (2a) 0.856, (2b) 0.860, (2c) 0.865, or (2d) 0.885×10^6 rad/s?

3. Is the effective resistance of the circuit at resonance: (3a) 433, (3b) 500, (3c) 865, or (3d) 933 ohms?

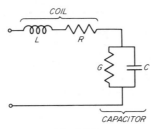

Fig. 1.804

1.805 A three-wire 60-cps three-phase power-transmission line has a balanced Y-connected load taking 1000 kW at 56.4 A, 0.85 pf lagging. The impedance of each conductor of the line is $1.0 + j10$ ohms.

1. Is the voltage required at the sending end of the line: (1a) 8800, (1b) 11,000, (1c) 12,700, or (1d) 13,200 V line to line?

2. Is the regulation of the line: (2a) 5.0, (2b) 5.2, (2c) 5.5, or (2d) 5.8 percent?

1.806 The two intersecting streets in Fig. 1.806 are to be connected by a simple curve. The centerline of the curve is to be so located that the distance from the curve to the corner x of the building is 150 ft (45.72 m). Is the radius: (a) 1040 (317.0), (b) 1125 (342.9), (c) 1210 (368.8), or (d) 1297 ft (395.3 m)?

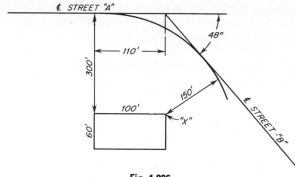

Fig. 1.806

1.807 Is the deflection angle alpha to the intersection of the centerline of a road (N85°E tangent) and of a property line bearing N9°E through coordinates 280N and 990E (Fig. 1.807): (*a*) 20°00′, (*b*) 20°29′, (*c*) 20°47′, or (*d*) 21°00′? (Coordinates are in feet, N and E.)

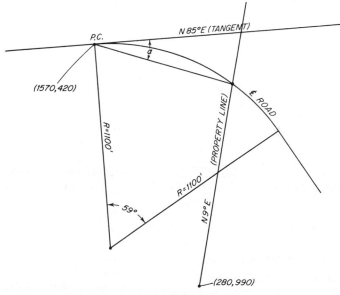

Fig. 1.807

1.808 If a string weighs 0.005 lb/ft and is under a tension of 25 lb, the velocity of a transverse pulse will be: (a) 157 (47.85), (b) 282 (85.95), (c) 365 (111.25), or (d) 400 fps (121.92 m/s)?

1.809 If air is of mean molecular weight, γ is taken as 1.40, R as 8.3 × 10^7 erg/mol °C, and T at 27°C, is the velocity of longitudinal waves in air (a) 1100 (335.3), (b) 1137 (346.7), (c) 1150 (350.5), or (d) 1160 fps (353.5 m/s)?

1.810 A railroad train is traveling at 60 mph (26.8 m/s).

1, 2. If the frequency of the note emitted by the locomotive whistle is 500 cps, will the wavelength of the sound waves in front of the locomotive be: (1a) 2.00 (0.610), (1b) 2.05 (0.625), (1c) 2.09 (0.637), or (1d) 2.12 ft (0.646 m)? Will the wavelength in back of the locomotive be: (2a) 2.05 (0.625), (2b) 2.12 (0.646), (2c) 2.26 (0.689), or (2d) 2.44 ft (0.744 m)?

3, 4. If a listener is standing at a crossing, will the frequency of the sound from the approaching train be: (3a) 5.38, (3b) 5.42, (3c) 5.46, or (3d) 5.50 cps? Will the frequency of the sound from the receding train be: (4a) 4.60, (4b) 4.62, (4c) 4.64, or (4d) 4.66 cps?

5, 6. If a listener is traveling on a train with a speed of 30 mph (13.41 m/s) toward the approaching first train, will the frequency of the sound from the approaching train be: (5a) 5.20, (5b) 5.28, (5c) 5.42, or (5d) 5.62 cps? Will the frequency of the sound from the receding train be: (6a) 4.46, (6b) 4.64, (6c) 4.80, or (6d) 4.82 cps?

1.811 1. The ratio of the intensity of two sound waves whose intensity levels differ by 10 dB is: (1a) 10, (1b) 20, (1c) 30, or (1d) 40?

2. If the difference of the intensity levels is doubled, is the ratio of the intensity of the two sound waves: (2a) 20, (2b) 40, (2c) 70, or (2d) 100?

3. Is the difference between the intensity levels of two sound waves whose ratio of intensity is 4 : 1 (3a) 3, (3b) 6, (3c) 9, or (3d) 12 dB?

4. Is the difference between the pressure levels of two sound waves whose ratio of pressure amplitudes is 6 : 1 (4a) 11.80, (4b) 14.81, (4c) 15.56, or (4d) 16.26 dB?

1.812 A 5-ft (1.524-m) diameter welded steel penstock pipe 1000 ft (304.8 m) long carries 300 ft³ (8.5 m³) of water per second from a reservoir to a turbine. The difference of elevation between the reservoir surface and the tailrace level of the turbine draft tube is 125 ft (38.1 m).

1. If the turbine develops 3500 bhp (2611 kW), will the overall efficiency for the plant be: (1a) 79.0, (1b) 80.6, (1c) 82.2, or (1d) 83.3 percent?

2. Will the efficiency of the turbine be: (2*a*) 88.0, (2*b*) 90.0, (2*c*) 91.0, or (2*d*) 91.5 percent?

1.813   A 12-in (30.48-cm) diameter steel pipeline traversing approximately level country carries oil having a viscosity of 400 Saybolt Seconds and an API gravity of 40 deg. Pumps develop a head of 1000 ft (304.8 m), and their inlet pressure must not fall below 6.5 psi (44.82 kPa). If the pumping rate is 2000 gal (7570 liters)/min, will the distance between pumping stations be: (*a*) 43,700 (13,320), (*b*) 50,500 (15,392), (*c*) 57,900 (17,648), or (*d*) 61,200 ft (18,654 m)?

1.814   Air flows through a smooth steel pipe 4 in (10.16 cm) in diameter at a temperature of 100°F (37.8°C). At two points 1000 ft (304.8 m) apart, the gage pressures are 100 and 60 psi (689.5 and 413.7 kPa). Is the weight of air flowing per second: (*a*) 1.82 (0.826), (*b*) 2.94 (1.334), (*c*) 4.06 (1.842), or (*d*) 5.36 lb (2.431 kg)?

1.815   Capstans are extensively used on ships to create a pull on a line, and they are essentially power-driven cylinders about which *n* turns of a line are wrapped, with the last wrapping held taut by a person pulling on it. If the coefficient of friction between the capstan and the rope can be taken as 0.12, the pull to be created is 4000 lb (17,793 N), and the pull exerted by a person is 50 lb (222.4 N), will the required number of turns of the rope be: (*a*) 6, (*b*) 8, (*c*) 9, or (*d*) 9.5?

1.816   An axle for a given machine is to transmit 80 hp (59.6 kW) at 300 rpm and at the same time support a 1000-lb (453.6-kg) load cantilevered 30 in (0.762 m) beyond the support. If axial forces are neglected, the maximum shear stress due to torque and moment loading is not to exceed 10,000 psi (68.95 MPa), and the maximum tension or compression due to torque and moment loading is not to exceed 20,000 psi (137.9 MPa), will the diameter of the shaft be: (*a*) 2.250 (5.71), (*b*) 2.375 (6.03), (*c*) 2.500 (6.35), or (*d*) 2.625 in (6.67 cm)?

1.817   A 36WF280 steel beam is cut as shown in Fig. 1.817*a* and welded to form the beam shown in Fig. 1.817*b*. Based on only the section modulus, is the new welded beam: (*a*) 1.41, (*b*) 1.52, (*c*) 1.63, or (*d*) 1.74 times as strong as the original beam? The properties of the original beam are area = 82.32 in²; depth out to out of flanges = 36.50 in; flange width

= 16.595 in; flange thickness = 1.570 in; web thickness = 0.885 in; moment of inertia = 18,819 in⁴; and section modulus = 1031 in³.

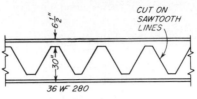

Fig. 1.817*a*

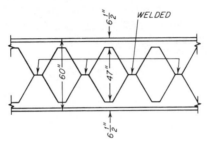

Fig. 1.817*b*

1.818 The hoisting cable for a vertical mine shaft is 600 ft (182.9 m) long from the drum in the head frame to the load at the bottom of the shaft.

1. If the cable has a cross section of 1 in² (6.45 cm²) and weighs 3.4 lb/ft (5.06 kg/m), will the cable have a modulus of elasticity of: (1*a*) 27 (186.2), (1*b*) 28 (193.1), (1*c*) 30.2 (208.2), or (1*d*) 33.5 × 10⁶ psi (231.0 GPa) if it stretches 0.222 ft (6.77 cm) in picking up a 5-ton (4536-kg) load?

2. Will the stretch due to its own weight be: (2*a*) 0.0113 (0.344), (2*b*) 0.0226 (0.689), (2*c*) 0.0452 (1.378 cm), or (2*d*) 0.0904 ft (2.755 cm)?

1.819 An endless steel ring 4 in (10.16 cm) wide and 0.25 in (0.64 cm) thick is 3 ft (0.914 m) in mean diameter. Its inside diameter is 1/32 in (0.079 cm) smaller, when unheated, than the diameter of the iron casting it is to encircle.

1. If the ring is to be heated above the shop temperature so that it will clear the casting by 1/32 in (0.079 cm), will the rise in temperature

be: (1a) 250 (121.1), (1b) 300 (148.9), (1c) 350 (176.6), or (1d) 400°F (204.4°C)?

2. Will the tension in the ring, when it cools, be: (2a) 12,000 (53.4), (2b) 20,000 (89.0), (2c) 26,000 (115.6), or (2d) 30,000 lb (133.4 kN)?

3. Will the pressure of the band against the casting be: (3a) 361 (2.49), (3b) 722 (4.98), (3c) 903 (6.23), or (3d) 994 psi (6.86 MPa)?

1.820 A cable consists of an equal number of steel and copper wire strands, with the cross sections of the steel and copper each equal to 1 in^2 (6.45 cm^2). The cable suspends a total load of 30,000 lb (133.4 kN) and all wires are taut, even, and unstressed (except for their own weight at the time of loading).

1. Do the steel strands carry: (1a) 17,720 (78.8), (1b) 18,270 (81.3), (1c) 19,170 (85.3), or (1d) 20,970 lb (93.3 kN) of load?

2. If the temperature rises 60°F (15.6°C) after loading, do the copper strands carry: (2a) 8130 (36.2), (2b) 9030 (40.2), (2c) 9930 (44.2), or (2d) 10,830 lb (48.2 kN) of load?

1.821 Four lb (1.81 kg) water at 200°F (93.3°C) is mixed with 2 lb (0.91 kg) water at 60°F (15.6°C) at 14.7 psia (101.4 kPa abs). Is the increase of entropy of the total mass of water due to the mixing process: (a) 0.041, (b) 0.062, (c) 0.093, or (d) 0.104 units?

1.822 An air turbine operates between a pressure of 60 (413.7) and 15 psia (103.4 kPa abs) and receives 1 lb (0.454 kg)/s of air at a temperature of 1200°F (648.9°C).

1. For an ideal turbine, is the developed power: (1a) 138.8 (103.5), (1b) 188.5 (140.6), (1c) 318.3 (237.5), or (1d) 381.5 hp (284.6 kW)?

2. When operating under the conditions above, the turbine develops 150 hp (111.9 kW) and has a discharge temperature of 300°F (148.9°C). The turbine blades are water-cooled with water entering at 50°F (10.0°C) and leaving at 100°F (37.8°C). With the constants for air assumed as c_p = 0.240 Btu/(°F)(lb)[1.006 kJ/(kg)(°C)], R = 53.3 ft/°R (29.24 m/K), and K = 1.4, is the rate of water flow under these conditions: (2a) 14.85 (56.21), (2b) 15.40 (58.29), (2c) 15.85 (59.99), or (2d) 16.20 gal (61.32 liters)/min?

1.823 Seven lb (3.18 kg) steam at atmospheric pressure, superheated to 242°F (117.8°C), is introduced simultaneously with 8 lb (3.63 kg) ice at 25°F (−3.9°C) into a copper calorimeter that weighs 5 lb (2.27 kg) and contains 50 lb (22.68 kg) of water at 60°F (15.6°C). The heats of fusion and vaporization for water are 144 and 970 Btu/lb (335 and 2256 kJ/kg), respectively. The thermal capacities in Btu/(lb)(°F)[4187J/(kg)(°C)] may be taken as: steam 0.48 [2.010 kJ/(kg)(°C)], ice 0.50 [2.093 kJ/(kg)(°C)], and copper 0.093 [389 J/(kg)(°C)]. Neglecting heat losses to all bodies other than the calorimeter itself, is the resulting temperature of the mixture: (*a*) 135 (57.2), (*b*) 148 (64.4), (*c*) 157 (69.4), or (*d*) 160°F (71.1°C)?

1.824 By means of insulation, the loss in heat through a roof per square foot (0.0929 m) is reduced from 0.45 to 0.25 Btu (474 to 263 J)/h for each degree difference between inside and outside temperatures. The area of the roof is 12,000 ft² (1114.8 m²), and the average difference between inside and outside temperature is 4.4°C during the heating season of 6000 h. If the heating value of coal is 14,000 Btu/lb (32564 kJ/kg) and the efficiency of the heating plant is 65 percent, is the value of the coal saved per season at $35/ton ($0.0386/kg): (*a*) $1114, (*b*) $1146, (*c*) $1194, or (*d*) $1258?

1.825 Ammonia enters the cooler of an ammonia refrigerating machine at 0°F (−17.8°C) and leaves it at 15°F (−9.4°C). When operating at 10-tons (9072-kg) ice-melting capacity and 15-hp (15 × 2545 Btu/h) input at the compressor:

1. For a temperature rise from 50 (10) at entrance to 70°F (21.1°C) at exit in the condenser, will the condensing water supplied be: (1*a*) 15.8 (59.8), (1*b*) 16.1 (60.9), (1*c*) 16.5 (62.5), or (1*d*) 16.9 gal (64.0 liters)/min?

2. Assuming the liquid ammonia leaves the condenser at 70°F (21.1°C), will the ideal coefficient of performance be: (2*a*) 6.57, (2*b*) 9.20, (2*c*) 25.5, or (2*d*) 30.6?

CHEMICAL ENGINEERING

2.01 A boiler is fired with a coal containing 75 percent carbon and 8 percent ash, burned under such conditions that the elimination of combustible matter from the refuse is complete. The air enters the furnace at 90°F (32.2°C) with a relative humidity of 80 percent. The vapor pressure of water at 90°F (32.2°C) is 36 mmHg (4.80 kPa). The flue gases go to the stack at 680°F (360°C). The average flue-gas analysis shows 12.6 percent CO_2, 6.2 percent O_2, and 1 percent CO.

1. The percent of excess air is: (1a) 41.5, (1b) 42.5, (1c) 43.5, or (1d) 44.5?

2. The complete analysis of the fuel indicates that C is: (2a) 65, (2b) 75, (2c) 80, or (2d) 83 percent?

3. The cubic feet (cubic meters) (1 ft³ = 0.0283 m³) of stack gas per pound (0.454 kg) of coal is: (3a) 400 (24.9), (3b) 412 (25.7), (3c) 418 (26.1), or (3d) 422 (26.3)?

4. The cubic feet (cubic meters) (1 ft³ = 0.0283 m³) of air per pound (0.454 kg) of coal is: (4a) 195 (12.2), (4b) 200 (12.5), (4c) 205 (12.8), or (4d) 210 (13.1)?

2.02 A mixture of ammonia and air at pressure 745 mmHg (99.3 kPa) and temperature 40°C contains 4.9 percent NH_3 by volume. The gas is passed at a rate of 100 ft³ (2.832 m³)/min through an absorption tower in which only ammonia is removed. The gases leave the tower at pressure

740 mmHg (98.7 kPa), temperature 20°C, and contain 0.13 percent NH_3 by volume.

Using the simple gas law:

1. The rate of flow of gas leaving the tower is: (1*a*) 88.7 (2.51), (1*b*) 89.5 (2.53), (1*c*) 91.8 (2.60), or (1*d*) 93.5 ft³ (2.65 m³)/min? (1 ft³ = 0.0283 m³.)

2. The weight of ammonia absorbed in the tower is: (2*a*) 0.193 (0.088), (2*b*) 0.225 (0.102), (2*c*) 0.251 (0.114), or (2*d*) 0.271 lb NH_3 (0.123 kg NH_3)/min?

2.03 A fractionating column is operating at 1 atm (101.3 kPa) to produce a product of ethanol and water that leaves the top plate at 78.41°C. The bottom product is to contain 1 mol percent ethanol. The feed contains 17 mol percent ethanol and is introduced at its boiling point. The feed rate is 44 mol/h. Reflux is returned to the top plate at a rate of 31 mol/h and at the top-plate temperature.

1. If plate efficiency is 60 percent, will the number of actual plates be: (1*a*) 13, (1*b*) 15, (1*c*) 17, or (1*d*) 19?

2. Will the rate of heat transfer in the condenser be: (2*a*) 648,000 (684), (2*b*) 672,000 (709), (2*c*) 690,000 (728), or (2*d*) 702,000 Btu (741 MJ)/h?

3. Will the rate of heat transfer in the reboiler be: (3*a*) 623,000 (657), (3*b*) 671,000 (708), (3*c*) 695,000 (733), or (3*d*) 707,000 Btu (746 MJ)/h?

2.04 An absorption tower packed with 0.50-in (12.7-mm) Berl saddles is to be used for scrubbing a very small amount of ammonia out of air at 70°F (21.1°C) and about 1 atm pressure (101.3 kPa), with water at 70°F (21.1°C) as the scrubbing liquid. The tower is 12 in (30.48 cm) in internal diameter and has a packed height of 7 ft (2.13 m). The water rate is 5 gal (18.92 liters)/min. Experiment has shown that, below the loading point, the resistance offered by the wetted packing at this water rate is 1.5 times that of dry packing of similar size and shape and at the same rate of flow of gas. The tower is to be operated with a gas velocity of 1.5 fps (0.457 m/s) based upon the empty tower; this is below the loading point.

Would the expected drop in pressure of water through the tower at 70°F (21.1°C) be: (*a*) 0.90 (22.9), (*b*) 1.20 (30.5), (*c*) 1.40 (35.6), or (*d*) 1.50 in (38.1 mm)?

2.05 It is desired to strip ammonia from a water solution that is 20 percent by weight of ammonia; 8000 ft³ (226.6 m³)/h of air-ammonia mixture 10 percent ammonia by volume (dry basis) is to be produced. The entering air contains no ammonia and the exhausted solution contains 2 percent ammonia by weight. The gas and liquor rates are such that $K_g a$ is 7 lb·mol/(h)(ft³)(atm) [1.11 kg·mol/(h)(m²)(kPa)].

1. If the operation is isothermal at 20°C, will the volume of packing required be: (1a) 2.28 (0.065), (1b) 3.24 (0.092), (1c) 3.72 (0.105), or (1d) 3.96 ft³ (0.112 m³)?

2. If $K_g a$ varies as $G^{0.8}$ and the gas and liquor rates are decreased by 25 percent, will the volume of packing be: (2a) 3.50 (0.099), (2b) 4.50 (0.127), (2c) 5.00 (0.142), or (2d) 5.25 ft³ (0.149 m³)?

2.06 A multipass cooler for a continuous ethyl alcohol still consists of a copper tube of 1/3/16-in (30.2-mm) inside diameter with walls 1/16 in (1.6 mm) thick and 24 ft (7.32 m) long, surrounded by a standard 2-in (50.8-mm) steel pipe. The hot alcohol flows through the inner tube, and the cooling water flows through the outer pipe countercurrent to the alcohol. The cooler consists of a number of these double pipes in series. If the alcohol is to be cooled from 172 to 70°F (77.8 to 21.1°C), using cooling water entering at 50°F (10°C) and leaving at 80°F (26.7°C), the number of pipes in series needed to handle 200 gal (0.757 m³)/h of alcohol will be: (a) 6, (b) 8, (c) 10, or (d) 12?

2.07 A 50 mol percent solution of methanol in water is to be rectified to give a distillate containing 95 mol percent methanol and a residue containing 95 mol percent water. The feed enters the column at 20°C. A reflux ratio 40 percent greater than the minimum is to be used.

1. Will the minimum reflux ratio be: (1a) 0.615, (1b) 0.723, (1c) 0.789, or (1d) 0.812?

2. Will the number of perfect plates needed for the desired separation under the above conditions be: (2a) 5, (2b) 7, (2c) 9, or (2d) 11?

3. If the feed is 2000 lb (907 kg)/h, will the product be: (3a) 1252 (568), (3b) 1275 (578), (3c) 1283 (582), or (3d) 1293 lb (587 kg)/h?

4. Will the residue be: (4a) 763 (346), (4b) 758 (344), (4c) 748 (339), or (4d) 746 lb (338 kg)/h?

5. The gage pressure of steam at 5 psi (34.5 kPa) required per hour will be: (5a) 800 (363), (5b) 825 (374), (5c) 850 (386), or (5d) 875 lb (397 kg)/h?

2.08 The heats of combustion $(-\Delta H)$ of crystalline succinic acid and succinic anhydride are 356.2 and 369.4 k·cal/g·mol, respectively, and the entropy of the acid is 42.0 entropy units at 25°C. The entropy of the anhydride is estimated as 35.0 units at 25°C. Calculate the partial pressure of water in the equilibrium system

$$\underset{\text{Solid}}{C_4H_6O_4} \rightarrow \underset{\text{Solid}}{C_4H_4O_3} + \underset{\text{Gas}}{H_2O}$$

at 500 and 700 K. Based upon these calculations, does the straight thermal dehydration of the acid to the anhydride appear to be: (*a*) commercially feasible *without* the use of any powerful desiccating agents, (*b*) commercially feasible *with* the use of powerful desiccating agents, or (*c*) commercially unfeasible *with or without* any powerful desiccating agents? Assume the molal heat capacity of the acid equal to that of the anhydride.

2.09 An experimental run made in a pilot-plant flow reactor having an internal volume of 5 ft³ (0.14 m³) converted 50 percent of material *A* according to the homogeneous gas reaction $2A \rightarrow R$. The conditions were atmospheric pressure 500°C and a feed rate of 1 lb·mol (0.454 kg·mol)/h, gas at standard conditions.

For design purposes, the volume of reactor required to treat 10,000 ft³ (283.2 m³)/h of feed gases at 5 atm (507 kPa) and 300°C with 25 percent conversion will be: (*a*) 4960 (140.5), (*b*) 5020 (142.2), (*c*) 5085 (144.0), or (*d*) 5105 ft³ (144.6 m³)?

Assume the gases behave as perfect gases, isothermal conditions for the reactor, and an activation energy of 30,000 cal (125.6 kJ)/g·mol for the homogeneous reaction.

2.10 A storage tank at 100°F (37.8°C) is partly full of a liquid mixture that contains 50 mol percent *n*-butane and 50 mol percent *n*-pentane. The vapor space contains only butane and pentane vapors. At 100°F (37.8°C), pure *n*-butane has a vapor pressure of 3.0 atm (304 kPa); at 100°F (37.8°C), pure *n*-pentane has a vapor pressure of 0.9 atm (91.3 kPa).

1. The pressure in the vapor spaces will be: (1*a*) 1.55 (157), (1*b*) 1.75 (177), (1*c*) 1.95 (198), or (1*d*) 2.05 atm (208 kPa)?

2. The composition of the vapors will be: (2*a*) Y*n*-butane 0.769, Y*n*-pentane 0.231; (2*b*) Y*n*-butane 0.783, Y*n*-pentane 0.217; (2*c*) Y*n*-butane 0.815, Y*n*-pentane 0.185; or (2*d*) Y*n*-butane 0.833, Y*n*-pentane 0.167?

2.11 The autoclave process for hydrolyzing monochlorbenzene to phenol at absolute pressure of 1500 psi (10.34 MPa) and 350°C in the presence of NaOH and CuCl is apparently a unimolecular reaction with a velocity constant of 0.0098 min⁻¹. Will the reaction time required to obtain a conversion of 99 percent using a batch of 1000-lb (454-kg) monochlorbenzene be: (*a*) 440, (*b*) 460, (*c*) 470, or (*d*) 475 min?

2.12 Outline the safety rules and practices for protecting life, health, and property that should be observed in a plant producing toxic and highly flammable chemicals.

2.13   *a.* Describe how glycerin is usually obtained from fats.
     b. Account for the presence of sodium chloride in the raw glycerin.

2.14   Will the weight of steel containing 0.020 percent manganese, so that the final solution will contain 10 ppm manganese, be: (*a*) 2, (*b*) 5, (*c*) 7, or (*d*) 8 percent?

2.15   What are the products of the following reactions?
    a. Oxidation of formaldehyde   *d.* Saponification of a fat
    b. Hydrolysis of an ether   *e.* Inversion of sucrose
    c. Hydrolysis of the CN group

2.16   A plate-and-frame filter press with frames 1 in (25.4 mm) thick is filtering a noncompressible slurry under constant pressure. The cake is washed with three successive portions of wash water of about the same viscosity as the filtrate; the pressure drop through the press during washing is the same as that during filtration. The time required for filtration only, to form a cake that fills the frame completely, is 2 h. The time required for washing is 30 min. The time required for dumping, cleaning, and reassembling the press is 30 min. The resistance offered by the press and the cloth can be considered negligible in comparison with the resistance offered by the cake. The percentage and the direction in which the substitution of frames 1.50 in (38.1 mm) thick change the average output from the press, in gallons (cubic meters) of filtrate per hour of total operating time, will be: (*a*) reduced by 26.5 percent, (*b*) increased by 29.8 percent, (*c*) reduced by 31.8 percent, or (*d*) increased by 32.7 percent? Assume that the composition of the sludge, the filtration pressure, and the cleaning time are the same in both cases and that the volume of wash water used is proportional to the volume of filtrate.

2.17   A mixture of the following composition is to be distilled in a continuous still:

Benzene	60 percent
Toluene	25 percent
Xylene	15 percent

The overhead product is to be substantially pure benzene; the bottoms product is to be substantially pure xylene. From some plate in the column a side stream is to be taken off to a smaller stripper still that delivers substantially pure toluene as a bottoms product.

    a. Draw a neat diagram showing, at least qualitatively, the concentrations of the several components on the plate of the main still as ordinates and as abscissas the plate numbers in the main still. Indicate the position of the feed plate and the plate from which the side stream is taken.

b. Draw a neat diagram showing the main still with feed line, side-stream drawoff, and overhead and waste lines; and the stripper column with feed line, overhead lines, and drawoff line. Where would the overhead from the stripper still be sent?

2.18 Consider the following startup problem: A vessel 10 ft (3.048 m) high and 4 ft (1.22 m) in diameter is to be filled with liquid to a level of 6 ft (1.83 m). It is initially empty. The filling is to be accomplished automatically, by using a control loop as shown in Fig. 2.18.

Vessel transfer function: $G(s) = 1/4\pi s$ ft/(ft^3)(min) [m/(m^3)(min)]

Linear valve: $G_v(s) = 3$ (ft^3)(min)/psi [4.1 (m^3)(min)/MPa]

Proportional controller: $G_c(s) = 6$ psi/psi (6 Pa/Pa)

Measuring element: $H(s) = 0.5$ psi/ft (11.31 kPa/m)

a. Write the equation that expresses the liquid level Y as a function of the controller set point (the reference input) R.

b. It is suggested that a simple way to fill the vessel to the desired level is to turn up the controller set point in a stepwise manner and allow the system to respond. If the size of this step is γ_0 psi, find the time domain response $y(t)$. What value of γ_0 would you suggest?

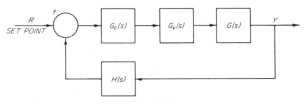

Fig. 2.18

2.19 Two different strengths of *n*-butanol and water are accumulated continuously in intermediate tankage and then processed for 10 days in a butanol recovery plant generating *n*-butanol having a purity of 99.3 wt percent. The intermediate tankage facilities contain two tanks, one containing a rich butanol solution (77.9 wt percent *n*-butanol) and the other a water-rich solution (92.3 wt percent water).

The proposed process configuration to recover the butanol is shown in Fig. 2.19 giving the flow arrangement. Note that since the butanol solutions to be processed are of radically different strengths, there are two feed streams to the columns of the recovery system, rather than a single feed to the decanter. The water purge to the dirty water sewer has been set at 500 ppm by weight (exclusive of further dilution by stripping steam).

For the proposed process configuration operating at essentially a constant pressure of 760 mmHg (101.32 kPa), what are the:

   a. Recycle steam rates and compositions from the decanter?

   b. Condenser and reboiler heat loads?

   c. Water column stripping steam rate predicted on a butanol-rich-solution feed rate of 4640 lb (2104 kg)/h and a water-rich-solution feed rate of 1530 lb (694 kg)/h exclusive of the recycle requirements?

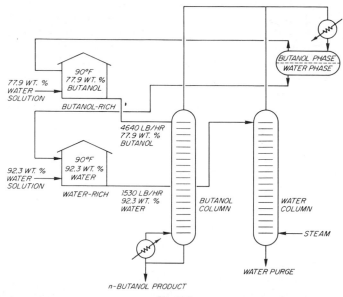

Fig. 2.19

2.20 What are the raw materials used in the Fischer-Tropsch process for motor-fuel manufacture? What main reaction is involved? What by-products result?

2.21 The sulfur-burning process in a sulfuric acid plant comprises the following operations:

   a. Sulfur is conveyed from storage pile to melting tank.

   b. Sulfur is melted by steam coils.

   c. Sulfur is burned in a spray-type burner with dry air.

Indicate the equipment needed for adequate automatic control of the process.

2.22 Fully discuss the following problems relating to phosphate utilization:

   a. The use of CO_2 as an oxidizing agent to oxidize $P_{4(gas)}$ to the oxide for preparation of phosphoric acid

   b. The use of steam and silica for defluorinating phosphate for animal nutritional purposes

   c. The use of $Na_3PO_4 \cdot 12H_2O$ in boiler waters

2.23 A paper plant requires $CaCO_3$ essentially free of all soluble impurities. Powdered $CaCO_3$ is available containing 1 percent NaOH. A two-step, continuous-countercurrent washing system with 24,000 gal (90.84 m³)/day of pure water is to be used for removing the NaOH when feeding 10 tons (9072 kg)/day of powdered $CaCO_3$, with the slurry discharge containing 0.091 tons water/ton (0.091 kg water/kg) of $CaCO_3$ removed in underflow. What is the caustic content of the washed and dried $CaCO_3$?

2.24 A shell and tube heat exchanger has four tube passes and one shell pass. Each tube pass has 13 copper tubes, 1 in (25.4 mm) OD, 16 BWG (Birmingham wire gage), each tube 6 ft (1.83 m) long. The exchanger is to be used to heat 60,000 lb (27.22 Mg)/h of an aqueous solution entering at 60°F (15.56°C). Saturated steam at gage pressure in pounds per square inch (kilopascals) will be used as the heating medium; the steam condenses outside the tubes.

Assume negligible heat losses and no scale deposits. Properties of the solution are the same as for water. Steam side coefficient of heat transfer is 2000 Btu/(h)(ft²)(°F) [40.92 MJ/(h)(m²)(°C)]. Will the approximate temperature of the solution leaving the exchanger be: (a) 112 (44.4), (b) 142 (61.1), (c) 162 (72.2), or (d) 172 (77.8)°F (°C)?

2.25 Water at 68°F (20°C) is supplied from a mountain lake to a pipeline consisting of 4-in (101.6-mm) Schedule 80 steel pipe. The pipeline consists of 1000 equivalent feet (304.8 m) of pipe from the lake surface to a point B, which is 400 ft (121.9 m) lower in elevation. At this point the pipe branches into two lines both 3000 ft (914.4 m) long (equivalent length). One of these branch lines discharges into an open water tower at a point 500 ft (152.4 m) below the lake surface. The other line feeds into the nozzles of a turbine in a powerhouse at a point 700 ft (213.4 m) below the lake surface. The pressure at the inlet of the turbine nozzles must be maintained at gage pressure of 80.6 psi (555.74 kPa). All expansion and contraction losses and all changes in kinetic energy may be neglected. In a single line of 1000 ft (304.8 m) feeding the two branches, will the total flow of water be: (a) 625 (2366), (b) 635 (2403), (c) 645 (2441), or (d) 655 (2479) gal/min (liters/min)?

2.26   It is required to heat 64,000 lb (29,030 kg)/h of air from 60 to 160°F (15.6 to 71.9°C) at atmospheric pressure. It is decided to use a heater made of 3/4-in (19.1-mm) ASA Schedule 40 steel pipes with saturated steam at 235°F (112.8°C), condensing inside the vertical pipes, which are 10 ft (3.05 m) long. The air will be blown across these tubes arranged on an equilateral triangular pitch of 3 in (76.2 mm). The side walls of the casing will be spaced so that there will be a clearance of 1½ in (38.1 mm) from the nearest tube (no clearance at the top and bottom).

A safety factor of 1.25 will be used. The specific heat and viscosity of the air may be considered constant at 0.25 and 0.0455 lb/(h)(ft) [0.372 and 0.068 kg/(h)(m)], respectively. The average between the mean air density and the density of the blower outlet can be taken as 0.072 lb/ft³ (1.15 kg/m³).

1. Would the optimum number of tubes per row be: (1*a*) 12, (1*b*) 10, (1*c*) 8, or (1*d*) 6?

2. Would the optimum number of rows of tubes be: (2*a*) 48, (2*b*) 54, (2*c*) 60, or (2*d*) 66?

2.27   A dilute mixture of ethyl acetate in air is the product of a certain drying operation. Outline possible methods of recovering the ethyl acetate for reuse as a solvent.

2.28   In a hydroforming process, toluene, benzene, and other aromatic materials are produced from naphtha feed. After the toluene has been separated from other components, it is condensed and cooled as shown in the flow sheet in Fig. 2.28. Into the system is fed 48,000 lb (21,773 kg)/day of naphtha feed. For every 100 lb (45.4 kg) of feed, 27.5 lb (12.5 kg) of a toluene and water mixture (9.1 percent by weight water) are

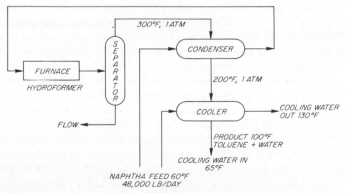

Fig. 2.28

produced as overhead vapor and condensed in the condenser, using the feed stream as a cooling medium.

1. Will the temperature of the feed stream after it leaves the condenser be: (1*a*) 170 (76.7), (1*b*) 180 (82.2), (1*c*) 190 (87.8), or (1*d*) 200°F (93.3°C)?

2. Will the amount of cooling water required per hour in the cooler be: (2*a*) 340 (154.2), (2*b*) 360 (163.3), (2*c*) 375 (170.1), or (2*d*) 385 (174.6) lb (kg)/h?

Additional data are:

Component	C_p, Btu/(lb)(°F)	NBP, °F	Latent heat of vaporization, Btu/lb
Water (liquid)	1.0	212	970
Steam	0.5		
Toluene (liquid)	0.4	260	100
Toluene (vapor)	0.3		
Naphtha feed liquid	0.5		

2.29 A tower 4 ft (1.22 m) in diameter is packed to a depth of 30 ft (9.14 m) with 1-in (25.4-mm) Berl saddles. The bed porosity (volumn fraction of voids) is 0.69. For this packing, the particle surface-to-particle volume ratio is 76 ft^{-1} (23.16 m^{-1}). Air at absolute pressure of 14.7 psi (101.32 kPa) and 70°F (21.1°C) is to be forced through the bed at a rate of 1000 ft^3 (28.32 m^3)/min. Will the pressure drop through the bed be: (*a*) 5.21 (13.23), (*b*) 5.31 (13.49), (*c*) 5.46 (13.87), or (*d*) 5.66 in (14.38 cm) H$_2$O?

2.30 An absorption tower containing wooden grids is to be used for absorbing SO$_2$ in a sodium sulfite solution. A mixture of air and SO$_2$ will enter the tower at a rate of 70,000 ft^3 (1982 m^3)/min at temperature 250°F (121.1°C) and pressure 1.1 atm (111.5 kPa). The concentration of SO$_2$ in the entering gas is specified, and a given fraction of the entering SO$_2$ must be removed in the absorption tower. The molecular weight of the entering gas mixture is assumed to be 29.1. Under the specified design conditions, the number of transfer units necessary varies with the superficial gas velocity as follows:

Number of transfer units = $0.32 G_s^{0.18}$, where G_s is entering gas velocity, lb/(h)(ft^2) [4.882 kg/(h)(m^2)], based upon the cross-sectional area of the empty tower. The height of a transfer unit is constant at 15 ft (4.57 m). The cost for the installed tower is $1/ft^3 ($35.31/m^3) of inside volume, and annual fixed charges amount to 20 percent of the initial cost. Variable operating charges for the absorbent, blower, and pumping powers are represented by the following equation:

$$\text{Total variable operating costs (\$/h)} = 1.8G_s^2 \times 10^{-8} + \frac{81}{G_s} + \frac{4.8}{G_s^{0.18}}$$

The unit is to operate 8000 h/year. Under conditions of minimum annual cost,

1. Would the diameter of the absorption tower be: (1*a*) 12.6 (3.84), (1*b*) 13.1 (3.99), (1*c*) 13.5 (4.11), or (1*d*) 13.8 ft (4.21 m)?
2. Would the height of the absorption tower be: (2*a*) 18.5 (5.64), (2*b*) 19.0 (5.79), (2*c*) 20.0 (6.10), or (2*d*) 21.5 ft (6.55 m)?

2.31 A gas system has two parallel lines 50 mi (80.5 km) long, one 20 in (50.8 cm) and the other 22 in (55.9 cm) in diameter. Gas at gage pressure of 555 psi (3.83 MPa) and 40°F (4.4°C) is fed to the system, amounting to 6 million ft³ (169.9 km³)/h measured at 30 in (76.2 cm) and 60°F (15.6°C). The gas' specific gravity is 0.65. The viscosity at 500 psi (3.45 MPa) and 40°F (14.4°C) is 2.5×10^{-7} lb·s/ft² (12.21×10^{-7} kg·s/m²). The deviation from Boyle's law is $+0.00022P$ for the weight of 1 ft³ (0.02832 m³) at P psi (6.895 kPa). The 20-in (50.8-cm) line has an efficiency of 92 percent and the other line 88 percent. How would the outlet pressure be determined?

2.32 Two similar vertical-tube evaporators have been arranged as a double effect in the evaporation of a caustic soda liquor from 10 to 16 percent, the first effect being operated as the high pressure one and the second at 28 inHg (94.6 kPa) vacuum. The steam to the first effect is at 5 psi (34.5 kPa) and the feed liquor is at 70°F (21.1°C). Operation is continuous. Given an adequate supply of steam (5 psi) (34.5 kPa) and cooling water, and having available some old heat exchangers, an old water-tube condenser of small capacity, pumps, pipes, fittings, etc., how would you adapt the equipment for a 10 percent increase in production? If you had to adapt the equipment to handle a temporary 50 percent overload, how would you do so? Substantiate your proposed setup.

2.33 An evaporator (single-effect) is removing 500 lb (227 kg)/h of water from a colloidal suspension that deposits scale on the steam chest. During one 8-h shift each week it is cleaned, which increases the overall heat-transfer coefficient from 50 to 225. Assuming that the scale is deposited uniformly throughout the week and none flakes off, calculate:

a. The mean heat-transfer coefficient for the week
b. How often the evaporator should be cleaned for maximum capacity

2.34 Specify the type of operation and type of slurry for which each filter would be adapted: (a) plate and frame, (b) rotary continuous, (c) Sweetland, (d) Moore.

2.35 Briefly discuss the processes of age hardening and precipitation hardening.

2.36 It is desired to recover ethanol from an air-ethanol mixture 5 percent by volume of ethanol and 95 percent by volume of air at 90°F (32.2°C), by compressing the mixture to gage pressure of 100 psi (689.5 kPa) and then cooling the gas at constant pressure to 78.8°F (26°C). The recovery of the ethanol will be: (a) 72.6, (b) 78.3, (c) 80.6, or (d) 83.7 percent?

2.37 Six-thousand lb (2722 kg) of a material goes through a crusher and grinder per hour in succession (on the same power drive). Screen analysis from the crusher shows a surface area of product of 500 ft²/lb (102.4 m²/kg). Screen analysis of the grinder product indicates a surface area of 4200 ft²/lb (860.2 m²/kg). The Rittinger number of the material processed is 163 in²/ft·lb (7604.3 cm²/m·kg). The efficiency of the crusher is estimated to be 25 percent, that of the grinder 30 percent. Will the total power delivered to the equipment be: (a) 38.4 (28.6), (b) 40.4 (30.1), (c) 42.8 (31.9), or (d) 44.4 hp (33.1 kW)?

2.38 Tank storage is to be provided for the following chemicals and solutions: (a) concentrated nitric acid, (b) 15 percent sodium chloride solution, (c) benzol, (d) 50 percent caustic soda solution, and (e) 90 percent mixed acid. Specify the material construction of each tank.

2.39 A leaf filter press is operating at constant pressure, building up a 1-in (25.4-mm) noncompressible cake in 5 h, and is delivering 3.8 ft³ of filtrate/ft² (1.158 m³/m²) of filtering surface. Cleaning requires 30 min. If it is necessary to wash the cake with a volume of water equal to one-half the volume of the filtrate, what will be the maximum capacity of the press obtainable by varying the cake thickness? The nature of the sludge and the pressure remain unchanged.

2.40 A solid granular product is to be dried in recirculating air. The moist air bled off from the recirculating stream is at 120°F (48.9°C) and has a dew point of 110°F (43.3°C). The air admitted to the dryer has a dry-bulb temperature of 100°F (37.8°C) and a wet-bulb temperature of 75°F (23.9°C).

Assuming no loss of heat from the dryer, will the heat that must be supplied for each pound (kilogram) of water removed from the cake be: (a) 1125 (2617), (b) 1154 (2684), (c) 1172 (2726), or (d) 1184 Btu/lb (2754 kJ/kg)? (Neglect heat of wetting of the solid.)

2.41   The thermal system in Fig. 2.41 is controlled by a proportional controller; these data apply:

W = flow rate of liquid through tanks, 400 lb (181.44 kg)/min

Density of liquid = 50 lb/ft³ (800.95 kg/m³)

V = holdup volume of each tank, 4 ft³ (0.113 m³)

Transducer: change of 1°F (0.56°C) causes controller pen to move 0.25 in (6.35 mm)

Final control element: change of 1 psi (6.895 kPa) from controller changes heat input q by 400 Btu (422.4 kJ)/min

Heat capacity of liquid = 1 Btu/(lb)(°F) [4.19 kJ/(kg)(°C)]; temperature of the inlet stream may vary

    a. Draw a block diagram of the control system, with appropriate transfer function in each block.

    b. From the block diagram, determine the overall transfer function relating the temperature T in tank 2 to a change in the set point.

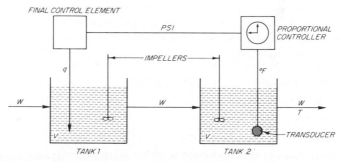

Fig. 2.41

2.42   For absorption-tower design, give or explain:

    a. The data required to determine the packed volume

    b. The procedure to follow to determine the packed volume

    c. The way the design of a stripping tower differs from that of an absorption tower

    d. The advantages and disadvantages of using the height-of-transfer-unit method instead of the method of operating lines

    e. For a stripping tower followed by a rectifying tower, the economic factors that must be considered and the optimum value that must be determined

2.43   As an engineer, you are asked to design the evaporating system to concentrate 50,000 lb (22,680 kg)/day of caustic solutions from 8 to

50 percent. Briefly outline the procedure to be followed, including the data required, method of determining the best setup, and materials of construction.

2.44 In a mercury-steam binary cycle, the saturated mercury vapor leaves the boiler at absolute pressure of 60 psi (413.7 kPa) [t = 836.1°F (446.7°C)], and it is sent through the mercury turbine, where it leaves at an absolute pressure of 2 inHg (50.8 mm) (6.77 kPa). The mercury turbine is adiabatic and reversible. From the mercury turbine the mercury goes through the mercury condenser-steam boiler, where it leaves as saturated liquid at an absolute pressure of 2 inHg (50.8 mmHg) (6.77 kPa).

In the condenser-boiler, saturated steam is generated at absolute pressure of 360 psi (2482 kPa), and the steam turbine exhausts at absolute pressure of 1.513 inHg (38.4 mmHg) (5.12 kPa). The steam turbine is adiabatic and reversible. Saturated liquid mercury at absolute pressure of 2 inHg (50.8 mmHg) (6.77 kPa) is pumped into the mercury boiler, and saturated liquid water at absolute pressure of 1.513 inHg (38.4 mmHg) (5.12 kPa) is pumped into the condenser-boiler. Pump work can be considered negligible.

The properties of mercury are:
At absolute pressure of 60 psi (413.7 kPa):

h_f = 30 Btu/lb (69.8 kJ/kg) h_{fg} = 118.6 Btu/lb (275.8 kJ/kg)
S_g = 0.1277

At absolute pressure of 2 inHg (50.8 mmHg) (6.77 kPa):

h_f = 15.85 Btu/lb (36.9 kJ/ kg) h_{fg} = 126.95 Btu/lb (295.3 kJ/kg)
S_f = 0.02323 S_{fg} = 0.1385

All entropy values are in Btu/(lb)(°F) [4190 kJ/(kg)(°C)].

1. Would the work of the mercury cycle be: (1a) 36.90 (85.83), (1b) 37.00 (86.06), (1c) 37.15 (86.41), or (1d) 37.35 Btu/lb (86.88 kJ/kg) mercury?

2. Would the work of the steam cycle be: (2a) 382.0 (888.53), (2b) 383.5 (892.02), (2c) 385.5 (896.67), or (2d) 388.0 Btu/lb (702.49 kJ/kg) steam?

3. Would the pounds (kilograms) of mercury per pound (kilogram) of steam be: (3a) 11.98, (3b) 12.48, (3c) 12.88, or (3d) 13.18?

4. Would the overall thermal efficiency of the binary cycle be: (4a) 46.7, (4b) 48.4, (4c) 50.2, or (4d) 52.1 percent?

2.45 Product S is formed according to the liquid phase reaction $A \rightarrow S$, with $-r_A = kC_A$, k = 2 min⁻¹. A stream containing 5 mol/liter of A is available at cost of $10/liter. The product S is sold at $2.50/mol. Operating cost is $10/day and no recycling of unreacted A can be used. Given a backmix reactor, the conversion of A to be used to yield the maximum profit per day is: (a) 0.69, (b) 0.75, (c) 0.89, or (d) 0.98 percent?

2.46 a. Describe the use of exchange resins in demineralizing water.

  b. Compare the operation of two- and mixed-bed units.

2.47 A mixture of hydrocarbons contains 20 mol percent methane, 20 mol percent ethane, 30 mol percent propane, 10 mol percent n-butane, 10 mol percent isobutane, and 10 mol percent n-pentane. If the material is flashed at 100°F (37.8°C) at absolute pressure of 115 psi (792.9 kPa), what fraction leaves the separator as a liquid? What is the composition of the liquid and vapor?

2.48 A solution containing 35 wt percent ethanol (and 65 percent water) is supplied to a fractioning column. The feed is a saturated liquid. The top product contains 80 wt percent ethanol; the bottom product contains 5 wt percent ethanol. The condenser operates as a total condenser. The reflux enters the top plate at 120°F (48.9°C).

  a. If 2 mol liquid at 120°F (48.9°C) is returned for each mole of product removed, determine the slope of the operating line to be used in a McCabe-Thiele diagram.

  b. Compare the amount of heat that must be supplied in the reboiler per pound of feed when the reflux is returned to 120°F (48.9°C) with that required when the reflux is returned as saturated liquid.

2.49 A reactor 10 ft (3.05 m) high with an internal diameter of 1 ft (0.305 m) is to be used for contacting a spherical catalyst with a gas mixture of density 0.500 lb/ft³ (8.01 kg/m³) and viscosity 0.03 cP (0.00003 N·s/m²). The catalyst density is 152 lb/ft³ (2435 kg/m³), its diameter is 0.174 in (4.4 mm), and the static-bed porosity is 0.40.

  1. The gas velocity necessary to give a porosity of 0.6 is: (1a) 2 (0.61), (1b) 3 (0.91), (1c) 4 (1.22), or (1d) 5 fps (1.52 m/s)?

  2. If a factor of 100 percent "free space" is allowed for disengagement of solids, the amount of static bed that can be accommodated is: (2a) 3 (0.91), (2b) 4 (1.22), (2c) 5 (1.52), or (2d) 6 ft (1.83 m)?

2.50 A single-stage single-acting compressor has an 8-in (203.2-mm) bore and a 10-in (254-mm) stroke and turns at 200 rpm. The compressor takes in dry saturated ammonia vapor at 0°F (−17.8°C) and compresses the ammonia adiabatically but irreversibly to an absolute pressure of 140 psi (965.3 kPa). The actual shaft work of the compressor is 20 percent more than if the compression were reversible. The volumetric efficiency of the compressor is 88 percent. The ammonia leaving the compressor enters a condenser, where it is cooled and condensed, leaving the condenser as a liquid at 70°F (21.1°C). The potential and kinetic energy changes are negligible. Using ammonia tables or charts:

1. The amount of ammonia per minute handled by the compressor is (1*a*) 4.52 (2.05), (1*b*) 5.61 (2.54), (1*c*) 6.14 (2.79), or (1*d*) 7.15 lb (3.24 kg)/min?

2. The power input to the compressor if the mechanical efficiency is 85 percent is: (2*a*) 17.5 (13.05), (2*b*) 18.75 (13.98), (2*c*) 19.37 (14.44), or (2*d*) 20.25 hp (15.10 kW)?

3. The amount of heat per minute removed in the condenser is: (3*a*) 2980 (3144), (3*b*) 3380 (3566), (3*c*) 3495 (3687), or (3*d*) 3600 Btu (3798 kJ)/min?

4. The quality of the resulting vapor-liquid mixture if the ammonia leaving the condenser is throttled to an absolute pressure of 30.42 psi (209.7 kPa) is: (4*a*) 13.64, (4*b*) 14.34, (4*c*) 14.98, or (4*d*) 15.62 percent?

2.51 An atmospheric rotary dryer handles 10 tons (9072 kg) of wet crystalline salt per day, reducing the moisture content from 10 to 1 percent by a countercurrent flow of hot air entering at 225°F (107.2°C) dry bulb and 110°F (43.3°C) wet bulb and leaving at 150°F (65.6°C) dry bulb.

1. The amount of dry product (salt) per 24 h is: (1*a*) 9.09 (8246), (1*b*) 10.5 (9526), (1*c*) 11.3 (10,251), or (1*d*) 11.9 tons (10,796 kg)?

2. The amount of water removed from the salt per hour is: (2*a*) 62.4 (28.30), (2*b*) 68.3 (30.98), (2*c*) 75.8 (34.38), or (2*d*) 78.1 lb (35.43 kg)?

3. The humidity of the air entering the dryer is: (3*a*) 0.0285, (3*b*) 0.0315, (3*c*) 0.0355, or (3*d*) 0.0405 lb (kg) water/lb (kg) dry air?

4. The humidity of the air leaving the dryer is: (4*a*) 0.096, (4*b*) 0.068, (4*c*) 0.047, or (4*d*) 0.033 lb (kg) water/lb (kg) dry air?

2.52 What is the power cost for the production of electrolytic copper sheet weighing 8 oz/ft^2 (74 Pa) if a sheet is 36 in (0.91 m) wide and 100 ft (30.48 m) long, is produced at 2.5 V for the deposition, the current efficiency of the operation is 85 percent, the power is 5 cents/kWh on the transformer, transformer losses are 3 percent, and there is 90 percent conversion efficiency at the germanium rectifier? (*a*) \$8.65, (*b*) \$9.25, (*c*) \$9.65, or (*d*) \$9.85.

2.53 The anodes are 3 × 3 ft × 1.5 in (0.91 × 0.91 m × 38.1 mm) and are fabricated from graphite. The current is furnished by a germanium rectifier supplied with 1320-V three-phase alternating current, through aluminum bus bars. The chlorine is produced at 700 mmHg (93.3 kPa), contains 98 percent chlorine, 1 percent oxygen, and is saturated with water vapor. The current efficiency is 92 percent and the energy efficiency is 46 percent. The feed to the cell is saturated with sodium chloride at 60°F (15.6°C). The amount of chlorine that would be produced per 24-h day from a 20,000-A Hooker cell operating at 6 V and 190°F (87.8°C) is: (a) 1633 (741), (b) 1438 (652), (c) 1286 (583), or (d) 1175 lb (533 kg)/day?

2.54   In a batch process, 100 lb (145.36 kg) of carbon monoxide gas are compressed adiabatically from 80°F (26.7°C) and an absolute pressure of 15 psi (103.4 kPa) to a final temperature of 600°F (315.6°C). The heat capacity in Btu/(lb·mol)(°F) [J/(kg·mol)(°C)] is given as $c_p = 9.46 - (3.29 \times 10^3)/T + (1.07 \times 10^6)/T^2$ where T is in °R. The gas is considered to be ideal.

1. The work of compression is: (1a) 8900 (9389), (1b) 9300 (9811), (1c) 9700 (10,233), or (1d) 9900 Btu (10,444 kJ)?

2. The final pressure (absolute) if the pressure is reversible is: (2a) 162.3 (1119), (2b) 165.4 (1140), (2c) 170.6 (1176), or (2d) 175.8 psi (1212 kPa)?

2.55   One-thousand lb (453.6 kg) of a solution, containing 50 percent by weight of component A and 50 percent by weight of component C, is to be extracted with a solvent B to remove at least 80 percent of the component C initially present in the 1000 lb (453.6 kg) of solution. The extraction is to be carried out isothermally by using 200 lb (90.72 kg) of solvent B in each stage of a multistage cocurrent extraction. Determine the number of stages required based upon the following data:

Equation

Point	Weight, percent A	Weight, percent B	Weight, percent C
1	85	2	13
2	65	5	30
3	46	8	46
4	35	10	55
5	20	20	60
6	8	35	57
7	3	47	50
8	2	55	43
9	1	79	20

Tie Line

	Weight percent	Point ①	Point ②	Point ③
Raffinate	*A*	59	74.1	88.4
	B	6	3.9	1.6
	C	35	22.0	10.0
Extract	*A*	5	1.8	1.3
	B	40	58.0	73.2
	C	55	40.2	25.5

Civil Engineering

3.01 Using the information given in the following table,

1. Will the cost of completing the job in the least possible time be: (1*a*) $14,000, (1*b*) $15,000, (1*c*) $16,000, or (1*d*) $17,000?

2. The duration of the job in days will be: (2*a*) 11, (2*b*) 13, (2*c*) 15, or (2*d*) 17 days?

Activity	Normal cost	Crash cost	Immediately follows activity	Normal duration, days	Crash duration, days
A	$1000	$1100	None	3	2
B	2000	2000	A	4	4
C	1600	1600	B	4	4
D	250	300	C	3	2
E	500	600	A	2	1
F	600	800	B, E	3	1
G	800	1100	F, I	3	1
H	3000	3800	A	6	2
I	400	700	H	2	1
J	1000	1000	I	4	4
K	500	500	C, G, J	2	2
L	500	1000	D	2	1
M	600	600	K, L	2	2

3.02 For each lettered item, select the numbered description that most nearly identifies or describes it. Each numbered description may be used only once.

a. Freeway
b. Outer connection
c. Induced traffic
d. Practical capacity
e. Critical density

f. Stream friction
g. Overall speed
h. Internal study
i. Live parking
j. Desire line

1. Traffic increase due to population growth
2. Retarding effect caused by intersecting streams of traffic
3. Right-turn ramp from one through roadway to a second at a grade separation
4. Retarding effect by units traveling in the same direction
5. Maximum hourly vehicle volume without unreasonable delay or hazard or feeling of constriction
6. Total distance traversed divided by travel time
7. Traffic density just before complete stagnation
8. Traffic increase due to presence of a new facility
9. Total distance traversed divided by running time
10. O and D study by home interview
11. Expressway with fully controlled access
12. O and D study inside cordon line
13. Traffic density at maximum flow
14. Parking other than all day
15. A connecting highway which skirts a congested area
16. Plot of vehicles per capita versus average vehicle cost
17. Highway on which no toll is charged
18. Maximum hourly vehicle volume under prevailing conditions
19. Straight line between origin and destination
20. Parking with operator in attendance

3.03 An aircraft-runway pavement is to be designed for a maximum single-wheel load of 60,000 lb (266.9 kN) with a tire pressure of 200 psi (1379 kPa). The pavement is to be a hot-mix bituminous concrete. The following materials are available to be used in the base course: (a) well-graded sandy gravel, compacted CBR 80; (b) silty sandy gravel, compacted CBR 55; (c) silty sand, compacted CBR 35; (d) sandy clay, compacted CBR 15. The subgrade is a clayey sand with a compacted CBR of 10. The relative cost of these materials is given in the accompanying table. The cost is for a compacted layer 4 in (10.16 cm) thick or thinner. Thicker layers must be constructed in two or more lifts. The relative pavement cost is for a compacted layer 3 in (7.62 cm) thick or thinner.

Thicker pavements must be constructed in two or more lifts.

Material	*Relative cost in place*
Compacted subgrade	1.0
Sandy clay	2.0
Silty sand	4.5
Silty sandy gravel	5.5
Well-graded sandy gravel	7.0
Hot-mix bituminous concrete	20.0

1. Design a suitable cross section for the pavement and base courses. Maximum economy is desirable within the limits of good design practice.

2. Describe the pavement course or courses.

3.04 A road is to be built that will run above an existing cable conduit from station 97 + 00 (2956.56 m) to about station 103 + 50 (3154.68 m). The pipe is horizontal in grade, and its top is at elevation 94.42 ft (28.78 m). The PVI of a symmetrical parabolic vertical curve will be at station 100 + 00 (3048 m) and at elevation 95.00 (28.96 m). The highway grade descends at a rate of -2 percent to the PVI and then rises at a rate of $+1.5$ percent leaving the PVI. Will the length of a symmetrical parabolic vertical curve, such that the minimum cover above the conduit will be 4.00 ft (1.22 m), be: (a) 700 (213.4), (b) 750 (228.6), (c) 800 (243.8), or (d) 850 ft (259.1 m)?

3.05 Answer (a) to (j) true or false.

a. Correct determination of superelevation designed to take care of centrifugal force on curves varies directly as the radius and inversely as the square of the speed.

b. All state plane coordinate systems in the United States use Lambert or transverse Mercator projections.

c. If a transit is set up at the midpoint of a highway curve, and a backsight is taken to the quarter point, the deflection angle to the point of tangency (PT) is one-half of the angle formed at the center of the curve by radii drawn to the quarter point and to the PT.

d. A manhole at the quarter point of a 60-ft (18.28-m)-wide street with an 8-in (20.32-cm) parabolic crown is 2 in (5.08 cm) below the centerline of the street.

e. In a polyconic projection map, all meridians are shown curved.

f. In a survey using the New Jersey state coordinate system, all the bearings shown would refer to the same meridian.

g. A rising curve on a highway mass diagram represents fill and a descending part represents cut.

h. Building lines are usually staked out to show the center of walls.

i. Zero percent grades on streets are undesirable in most road layouts.

j. The average velocity of the water in a vertical section taken in a stream is usually found between the center and the surface.

3.06 A car skidded going into an intersection, struck a pedestrian, and continued until it hit a tree. Based upon the damage to the front of the car, it is estimated that the car was doing 5 mph (8.05 km/h) at impact with the tree. The length of the skid marks was measured at 130 ft (39.62 m). The road is on a downhill grade of −5 percent. A test car skidded 45 ft on the same section of road when braked from a speed of 25 mph (40.23 km/h) to a halt. Was the probable speed of the car involved in the accident when the brakes were applied: (*a*) 42.2 (67.9), (*b*) 42.7 (68.7), (*c*) 43.1 (69.6), or (*d*) 43.4 mph (69.8 km/h)?

3.07 On a certain section of a busy four-lane highway, it has been observed that 1 out of every 12 motorists who make illegal left turns are caught and ticketed by the police.

1. Will the probability of being apprehended once a work week (5 days), if you commit the violation twice each day, be: (*1a*) 36.9, (*1b*) 37.4, (*1c*) 37.8, or (*1d*) 38.1 percent?

2. Will the probability of never being caught in any given week be: (*2a*) 41.9, (*2b*) 42.4, (*2c*) 42.8, or (*2d*) 43.1 percent?

3.08 Select from group B the item described in group A. Select carefully since group B items can be used only once.

Group A

1. A method of trip distribution involving the use of existing volumes of interzonal traffic to measure friction prior to expanding the volume between two points in proportion to an interactance factor

2. The allocation of traffic flows among routes available

3. A diamond highway interchange in which the intersecting conflicts are changed to weaving conflicts

4. A street that serves internal traffic movements within an area and connects this area with major arterials

5. A method of signal timing whose purpose is to first clear the vehicles desiring to turn left at an intersection

6. A signal system in which the signal faces that control a given street will be green according to a time schedule which will permit continuous operation of vehicles as is possible along the street

7. The maximum number of passenger cars that can pass a given point per hour under the most nearly ideal roadway and traffic conditions

8. The difference between the observed speed and the standard speed for that particular type of street

9. A name applied to the value obtained by dividing the hourly volume of traffic by the average speed

10. A value that may be used as a good guide for establishing upper speed limits

Group B

a.	Collector street	*l.*	Bridge rotary
b.	Simple progression	*m.*	Possible capacity
c.	Local street	*n.*	Desire line method
d.	Growth factor method	*o.*	Travel speed
e.	Two-quadrant cloverleaf	*p.*	Volume
f.	Traffic assignment	*q.*	Gravity method
g.	Parkway speed	*r.*	15th percentile value
h.	85th percentile value	*s.*	Delay rate
i.	Simultaneous progression	*t.*	Density
j.	Basic capacity	*u.*	Advance green
k.	Scramble system		

3.09 A town having a present population of 10,000 is planning to construct a rapid-sand-filtration plant. From the following data calculate the (1) size of the filter units, (2) pounds of alum required per day, and (3) size of the distribution reservoir required for fire protection.

a. Estimated future population is 14,000.

b. Consumption per capita is 100 gal (379 liters).

c. Maximum rate of consumption is 225 percent of average rate.

d. Filters to operate at a rate of 125 mgd/acre [1169 kl/(day)(m²)].

e. Washwater equals 4 percent of total water filtered.

f. Average time of filter operation between washings equals 7½ h.

g. Time required to wash filter and restore it to satisfactory operating condition equals ½ h.

h. Average alum dosage is 4½ gr/gal (0.077 g/liter).

i. Fire protection to meet requirements of National Board of Fire Underwriters.

3.10 A grit chamber installation is to be provided for a city with an average dry-weather flow of 1 mgd (3785 kl/day) and a storm flow of 3 mgd (11,355 kl/day). What values would you choose for: (*a*) minimum detention period, (*b*) maximum velocity of flow, (*c*) length of units, (*d*) top width of units, (*e*) maximum depth of units, (*f*) number of units, (*g*) grit storage capacity, (*h*) method of cleaning, (*i*) average interval between cleanings. Give reasons for your choices.

3.11 A sewage has a suspended-solids content of 250 ppm. A sedimentation tank with a retention period of 1.5 h is planned.

1. The reduction in suspended solids will be: (1a) ±138, (1b) ±148, (1c) ±158, or (1d) ±168 ppm in the effluent?

2. If the average sewage flow is 500,000 gal (1893 kl)/day, the capacity of the tank must be: (2a) 29,375 (111.2), (2b) 31,250 (118.3), (2c) 32,850 (124.3), or (2d) 33,782 gal (127.9 kl)?

3. The dimensions you would use if the tank is to be rectangular in plan are: (3a) $w = 12$ (3.61), $L = 48$ (14.63), $D = 8$ (2.44); (3b) $w = 10$ (3.05), $L = 43$ (13.11), $D = 7$ (2.13); (3c) $w = 13$ (3.96), $L = 37$ (11.27), $D = 9$ (2.74); or (3d) $w = 11$ ft (3.35 m), $L = 39$ ft (11.88 m), $D = 10$ ft (3.05 m)?

4. If a hopper-bottom tank is to be used and the sludge is to be removed once daily, the capacity of the sludge hoppers, assuming the water content of the sludge as 95 percent, will be: (4a) 140 (3.96), (4b) 145 (4.11), (4c) 150 (4.25), or (4d) 155 ft^3 (4.39 m^3)?

3.12 *a.* Describe the separate sludge-digestion process used for sewage treatment.

b. Name two methods of dewatering sludge. Give advantages of each method.

3.13 In parts 1 to 10, select only one correct answer.

1. In the active-sludge process of sewage treatment, it is essential to have: (1a) a dosing tank, (1b) an adequate supply of air, (1c) an Imhoff tank, (1d) flocculation, (1e) an acid sludge?

2. For protection of aquatic life in a fresh-water stream, sewage effluent should never lower the dissolved-oxygen content lower than: (2a) 1, (2b) 5, (2c) 10, (2d) 15, (2e) 20 ppm?

3. The biochemical treatment of sewage effluents is essentially a process of: (3a) reduction, (3b) dehydration, (3c) polymerization, (3d) oxidation, (3e) alkalinization?

4. The process of lagooning is primarily a means of: (4a) increasing the capacity of storage reservoirs, (4b) increasing flow of sewage through Imhoff tanks, (4c) reducing the excessive flow in sewers, (4d) disposing of sludge, (4e) rendering sludge suitable for fertilizing purposes?

5. In treating turbid waters, a popular coagulant is: (5a) calcium sulfate, (5b) chlorine, (5c) iodine, (5d) ferric sulfate, (5e) pulverized coke?

6. In most well-designed sewer systems: (6a) manholes are generally equipped with regulators; (6b) sewers never flow full; (6c) sewage ejectors are used to accelerate flow and prevent deposition at low-flow periods; (6d) catch basins are not usually essential; (6e) manholes are always placed over the centerlines of the sewer below?

7. Chlorine is used in the treatment of sewage to: (7a) cause bulking of activated sludge, (7b) help grease separation, (7c) aid flocculation, (7d) increase the biochemical oxygen demand, (7e) reduce the production of ecologies?

8. The quantity of grit in sewage depends largely upon the: (8a) extent to which the sewered area is built up, (8b) number of industrial plants in the area, (8c) chemicals used in the precipitation process, (8d) strength of the sewage, (8e) period of detention?

9. Septic tanks are primarily used for: (9a) the aerobic decomposition of deposited sewage solids, (9b) separation of deposited solids, (9c) separation of oil and grease scums, (9d) anaerobic decomposition of deposited solids, (9e) the nitrification of raw sewage?

10. The gas from Imhoff tanks is composed mainly of: (10a) carbon dioxide, (10b) methane, (10c) nitrogen, (10d) ethane, (10e) hydrogen sulfide?

3.14 Show by sketch the plan and sections for a septic tank and subsurface irrigation system (nitrification field) for a country home housing 10 people—soil is fairly porous. Give dimensions and data for design.

3.15 Describe how malaria is transmitted. If you were appointed sanitary engineer to a county health unit, describe just what steps you would take to reduce the number of cases of malaria in your county.

3.16 A state has a population of 2 million people. For the year 1968, the physicians of the state reported 2891 cases of tuberculosis, with 1304 deaths from this disease.

1. Will the case rate be: (1a) 1.446, (1b) 1.542, (1c) 1.583, or (1d) 1.603 per thousand?

2. Will the death rate be: (2a) 0.344, (2b) 0.589, (2c) 0.652, or (2d) 0.704 per thousand?

3. Will the fatality rate be: (3a) 41, (3b) 43, (3c) 45, or (3d) 48 percent?

4. Define morbidity as used in vital statistics.

3.17 a. What diseases are prevented by protecting the dairy herd, and what measures are applied in such protection?

b. What are the protective measures applied to the "milk line?"

c. Name a disease thus prevented.

d. What is the principal danger of dirty milk, and what specific measures can be taken to eliminate this danger?

e. Name two methods of milk pasteurization that require the use of recording thermometers. Why is one method used more commonly than the other?

3.18 Give the method of transmission of:

a. Bubonic plague
b. Murine typhus fever
c. Malaria
d. Dengue fever
e. Smallpox

f. Undulant fever
g. Tularemia
h. Yellow fever
i. Rocky Mountain spotted fever

3.19 When investigating a typhoid fever outbreak in a county of 10,000 population, what steps would you take and what data would you collect to determine the sources of the infection?

3.20 Name two methods of municipal sanitary garbage disposal applicable to a city or town. Compare the two methods, giving the application of each with advantages and disadvantages. *Note:* Open refuse dumps and feeding garbage to hogs are not to be considered.

3.21 Explain how to calculate the (a) yield of a watershed for power or water-supply purposes and (b) necessary storage capacity to provide a given rate of delivery.

3.22 Discuss the (a) importance of leakage and waste in a water-supply distribution system, (b) most satisfactory methods to reduce such water loss, and (c) effectiveness of these methods or the percentage of waste reduction that may be realized from them.

3.23 Describe the general features of a rapid-sand-water-filtration plant.

3.24 a. Explain the chemical reactions that occur in a zeolite softener.
   b. What is the effect of hydrogen sulfide in water that is to be chlorinated?
   c. Explain the lime-soda method of softening water, and write the chemical equations to show just what chemical reactions take place.

3.25 For what purpose is each of the following chemicals used in the process of water treatment?

a. Chloramine
b. Sodium hexametaphosphate
c. Sodium bisulfite
d. Sulfur dioxide
e. Activated carbon

f. Aluminum sulfate
g. Hydrated lime
h. Sodium chloride
i. Sodium carbonate

3.26 a. What do the following terms mean when applied to pumps: (1) single-stage, (2) centrifugal, (3) multistage, (4) positive displacement, (5) reciprocating, (6) double acting, (7) rotary?

b. For what service conditions would you recommend the use of a single-stage centrifugal pump?

c. State the service conditions for which you would recommend a two-stage instead of a single-stage pump.

3.27 *a.* A report on a proposed municipal water supply shows the presence of *B. coli, coli-aerogenes,* or gives a colon index. What is the significance of these terms?

b. Under what conditions do anaerobic bacteria thrive? Under what conditions do aerobic bacteria thrive?

c. The report on a prospective water supply shows carbonate hardness of 100 ppm. What is the significance of this information? Will softening be necessary for municipal use?

3.28 When carrying out a triangulation network for the horizontal control of an extended survey, several factors influence the accuracy with which the distances between stations can be computed in addition to the precision with which the angles in the scheme are measured. State what these factors are and the manner in which each factor affects the determination of the lengths.

3.29 A water-purification plant is situated on the bank of a large river. It is a run-of-the-river plant without reservoir storage. The raw water has considerable color and is quite turbid but reasonably soft. Upstream, the river passes through several shallow lakes that have a heavy algal growth during warm weather and that are much used for swimming in season. Below the lakes, the effluent from two sewage-treatment plants, using the activated-sludge process and chlorination, flow into the river. The water to be treated is diverted through a short intake canal to low-lift pumps that lift it sufficiently to provide gravity flow through the plant. Following treatment, high-lift pumps force the water to a large distribution reservoir. Outline your suggestions for a plant to render this water potable, listing all necessary units, processes, and facilities.

3.30 A dam for a water storage is to be constructed across a river. The available length of spillway is controlled by geologic considerations. When designing the spillway section, it is necessary to determine the maximum head of water over it during the design flood. Studies of past floods have provided data for a distribution graph at the site. From these data and from analysis of the expected runoff from the peak storm, determined from long-term precipitation records, the inflow hydrograph for design has been plotted. Assuming the reservoir to be full at the beginning of the flood period, show, using the mass-curve method, each step in determining the crest height on the spillway during the passage of this flood.

3.31 Indicate whether the following statements, which refer to water purification and treatment, are true or false.

a. Color is fully removed by slow sand filters.

b. Activated carbon may be used for taste and odor control without subsequent filtration.

c. Turbidity is normally removed by adding a coagulant prior to sedimentation.

d. Copper sulfate is used in the shallow portions of reservoirs for algae control.

e. Activated carbon is applied to water supplies to reduce hardness.

f. The usual rate of flow through rapid-sand filters is 2 gal/ft^2 (81.5 liters/m^2) of surface area per minute.

g. Bacteria counts of coliform organisms are usually greater in well supplies than in surface supplies.

h. The zeolite process reduces the hardness of a water supply by no more than 50 percent.

i. Aeration of water is effective in CO_2 removal.

j. Recent installations have tended to utilize slow- rather than rapid-sand filters.

k. Sodium carbonate or sodium sulfate does not cause hardness in water.

l. Freezing does not necessarily kill spore-forming pathogens.

m. Boiling for at least 1 h kills all pathogenic bacteria.

n. The presence of *E. coli* in a water supply proves the presence of typhoid bacilli.

o. The confirmed test for *E. coli,* when positive, indicates intestinal pollution.

p. Hard surface waters that require filtration for removal of pollutants are normally treated by zeolite process for the removal of hardness.

q. Newly laid water mains should be disinfected before being placed in service.

r. A raw water treated with alum requires the addition of soda ash when the natural alkalinity is high.

s. Fluoridation of water in the amount of 1 ppm effectively reduces the incidence of dental caries in children.

t. A water with a pH below 7 is alkaline.

u. The lime-soda ash process is well-adapted to the softening of hard surface supplies.

v. The most commonly used coagulant in water treatment is ferric chloride.

w. Highly turbid water is usually highly colored and normally requires treatment for both conditions.

x. The major waterborne diseases are typhoid, malaria, encephalitis, and scarlet fever.

y. Typhus is not a waterborne disease.

3.32 *a.* What are the major causes of precipitation in northern New Jersey, and why is there no marked seasonal variation in precipitation in this area?

b. When determining the precipitation on a watershed from the records of Weather Bureau stations in or near the watershed, what methods may be used, and what conditions control the method selected?

c. What causes the marked seasonal variation in precipitation in the San Francisco area?

d. Why does the Upper Mississippi Valley receive most of its precipitation during the crop-growing season?

3.33 *a.* Discuss in detail the effects of:
1. Discharging high-temperature wastes into a sewer
2. Discharging sawdust from a mill into a stream
3. Discharging precipitates from lime-soda water-softening plants into a stream
4. Discharging extremely fine suspended solids from an industrial plant into a river
5. Discharging oily wastes into a sewerage system

b. Discuss possible measures to prevent undesirable results in situations 1 to 5.

3.34 *a.* When a relatively small volume of raw sewage is disposed of by dilution into a large body of water, what are the physical, chemical, and biological processes that result in the eventual clarification and purification of the polluted water?

b. What considerations limit the disposal of sewage by dilution?

3.35 Classify sewage-treatment processes under the following headings:

a. Methods or devices that separate floating and settleable solids from the liquid

b. Processes for the aerobic oxidation of the finer suspended, colloidal, and dissolved organic compounds

c. Disinfection of sewage effluents

d. Digestion of sewage sludges

e. Methods for dewatering, drying, and disposing of either raw or digested sludge

f. Final disposal of sewage effluents

3.36 With reference to sewerage and sewage treatment:

1. You find that at a given disposal plant the Department of Health requires complete removal of all floating solids, at least 60 percent removal of suspended solids, and an *E. coli* count of less than 100/cm³. Would you consider that the river below the point of discharge was used for: (1*a*) a potable water supply, (1*b*) bathing, (1*c*) irrigation, (1*d*) industrial purposes, or (1*e*) none of the above?

2. Upon entering the storage area at a plant, you find the floor partially covered with an oily brown liquid that has leaked out of a crack in the lining of a rubber-lined tank. What is the liquid, and for what use was it intended?

3. What would you conclude from observing a series of light hollow metal balls, varying in size from small to large, being dropped, one after the other, into a sewer at a manhole?

4. Why would copper sulfate be dumped into a sewer?

5. What would you infer if you saw a lighted candle being lowered into a sewer?

6. For about what minimum velocity, sewer flowing half full, would you design a sanitary sewer?

7. Would you lay sewer pipe upgrade or downgrade?

8. When designing sewers, what governs the location and spacing of manholes?

9. Why does a circular gravity sewer not flow full under peak flows?

10. Why are septic tanks not commonly used in treating a municipal sewage? Where are they used?

3.37 A community has decided to use an activated-sludge-treatment plant to treat the sewage. The BOD is 300 mg/liter (mg/liter = ppm or part/million) and the design flow is 5 mgd (million gallons per day) (18.9 Ml/day). [Also, 1 part/million gal is 1 gal or 8.33 lb/million gal (1 kg/1.002 Ml)]. The aeration tanks are required to have a minimum detention period of 6 h at 125 percent of the design flow. The maximum BOD loading of each aeration tank is not to exceed 38 lb/(day)(1000 ft³)[609 g/m³] of aeration-tank capacity. Use four aeration tanks, each having a design flow of 1.25 mgd (4.73 Ml/day). The primary settling tank is designed to remove 30 percent of the raw sewage BOD. Aeration capacity at standard pressure and temperature shall be at least 1.5 ft³/gal (0.0112 m³/liter) of incoming raw sewage. The air-diffuser system shall be able to deliver 150 percent of normal requirements. Consider such requirements to be 1000 ft^3/lb (62.4 m^3/kg) BOD to be removed from the sewage entering the aeration tanks.

1. Will the volume of each aeration tank be: (1a) 51,250 (1451), (1b) 52,500 (1487), (1c) 54,000 (1529), or (1d) 55,750 ft^3 (1579 m^3)?

2. Will the aeration capacity be: (2a) 4700 (133.1), (2b) 5000 (141.6), (2c) 5200 (147.3), or (2d) 5300 ft^3 (150.1 m^3)/min?

3. Will the air diffuser capacity be: (3a) 2280 (64.6), (3b) 2320 (65.7), (3c) 2390 (67.7), or (3d) 2480 ft^3 (70.2 m^3)/min?

3.38 Four-hundred mgd (17.6 m^3/s) of sewage is discharged into a stream that has a minimum rate of flow of 2000 ft^3 (56.6 m^3)/s. The temperature of the sewage and the diluting water is 15°C. The 5-day BOD at 20°C of the sewage is 150 ppm and 0.5 ppm for the river diluting water. The sewage contains no dissolved oxygen, and the stream diluting water is 100 percent saturated with oxygen. Determine the dissolved oxygen deficit 80 mi (128.7 km) downstream from the point of loading if the velocity of the stream is 20 mi/day. $K_1 = 0.08$ and $K_2 = 0.28$ at 15°C; $K_1 = 0.10$ at 20°C. Will the dissolved oxygen deficit be: (a) 7.50, (b) 7.80, (c) 8.20, (d) 8.70 ppm?

3.39 Select the one choice that best completes statements 1 through 10.

1. Liquid waste is considered: (1a) A Newtonian fluid; (1b) a non-Newtonian fluid; (1c) an ideal plastic; (1d) a real fluid?

2. In a situation where the topography is flat and wastes must be pumped to the treatment plant, the type of system to use is: (2a) separate, (2b) combined, (2c) storm, (2d) lateral?

3. The estimate of demand for the design basis of a treatment plant should be about: (3a) 5, (3b) 30, (3c) 50, (3d) 70 years?

4. Storm and sanitary sewers:

    4a. May be placed in the same trench

    4b. Should not be in the same trench

    4c. Should never be needed in separate systems

    4d. Should be located in easements through backyards

5. Good practice in the design of storm sewers for residential areas would indicate the use of: (5a) 10-year, (5b) 50-year, (5c) 100-year, or (5d) all storm figures?

6. When sewers are placed in rock cuts,

    6a. Steel pipe must be used.

    6b. Concrete foundations are used.

    6c. Trenches may remain open for weeks.

    6d. Neither steel pipe, concrete foundations, nor leaving the trench open for weeks is practical.

7. Water-pollution control on navigable waters is a responsibility of
   7a. A civilian agency of the federal government
   7b. The U.S. Army
   7c. The U.S. Navy
   7d. The individual states

8. The strength of a sewage is related to
   8a. Its grit content
   8b. The nuisance-producing items possessed
   8c. The BOD
   8d. The turbidity

9. The organic portion of sewage may best be estimated by measuring the (9a) total solids, (9b) volatile solids, (9c) fixed solids, or (9d) pH?

10. Extra width in a sewer trench over and above that necessary for placing the pipe will:
    10a. Benefit the alignment
    10b. Increase the loads on the pipe
    10c. Help support the trench
    10d. Reduce the loads on the pipe

3.40 Calculate the carbonate and noncarbonate hardness (in mg/liter of $CaCo_3$) of a water that has been analyzed as follows:

$$Na^{2+} = 20 \text{ mg/liter} \qquad Cl^- = 40 \text{ mg/liter}$$
$$Ca^{2+} = 20 \text{ mg/liter} \qquad NO_3^- = 1 \text{ mg/liter}$$
$$Mg^{2+} = 15 \text{ mg/liter} \qquad HCO_3^- = 5 \text{ mg/liter}$$
$$Sr^{2+} = 5 \text{ mg/liter} \qquad SO_4^{2-} = 16 \text{ mg/liter}$$
$$CO_3^{2-} = 10 \text{ mg/liter}$$

3.41 Name and describe four soil tests and tell what each measures.

3.42 Answer briefly:

a. How would you take care of a minor leak in the sides of a single-wall interlocking steel-sheet cofferdam?

b. What precautions would you take prior to unwatering the cofferdam in (a)?

c. If the lunch whistle blew at a time when a pile-driving crew had only partly driven a pile, would you permit the crew to stop for lunch? Why?

d. If you were ordering stone to riprap a stream bank near a bridge, what size stone would you try to get?

e. What is the advantage of prefabricating timber that is to be pressure creosoted? How would you treat the exposed timber in holes to be bored in the creosoted timbers?

3.43 Answer the following briefly:

a. When digging a trench for a sewer in hardpan and in the dry, would it be necessary to use sheeting and bracing? Would this apply to any depth trench?

b. How would you protect steel from rusting if the steel is to be welded at a later date?

c. Is "oiling the forms" harmful or beneficial to the concrete?

d. What is meant by "gunite" and how is it applied?

e. In structural welding, what is the advantage of using an ac machine over a dc machine?

3.44 A consolidation test has been run in a consolidation apparatus whose internal dimensions are 1.258 in (31.9 mm) in height and 4.23 in (107.4 mm) in diameter. At the end of the test, a determination of moisture content gives the following data:

Weight of containers plus wet soil	663.2 g
Weight of container plus dry soil	582.0 g
Weight of container	194.4 g

A specific-gravity determination on part of the test specimen gives the following:

Weight of bottle plus water plus soil	723.22 g
Weight of bottle plus water	687.39 g
Weight of dish plus dry soil	175.50 g
Weight of dish	118.77 g

The test specimen has a wet weight of 468.8 g. The consolidation-test results are summarized as follows:

Applied pressure, kg/cm²	Dial reading, in	Specimen height, in
0	0.4200	
¼	0.4131	
½	0.4087	
1	0.4010	1.235
2	0.3828	
4	0.3427	1.178
8	0.2914	

Draw the voids ratio versus pressure diagram for this test.

3.45 A concentrated wheel load of 7000 lb (3.75 kg) on the surface of an embankment has cartesian coordinates 0.0 ft (m). The horizontal top of a culvert has corner coordinates (all in feet) $(+2, +3)$ $(+0.61, +2.13$ m), $(+2, +7)$ $(+0.61, +0.91$ m), $(+5, +3)$ $(+1.52 + 0.91$ m), and $(+5, +7)$ $(+1.52, +2.13$ m). If the top of the culvert is 9 ft (2.74 m) beneath the surface of the embankment, will the total load imposed on the culvert by the wheel, assuming the soil horizontally layered, be: (a) 117 (53.1), (b) 118 (53.5), (c) 119 (54.0), or (d) 120 lb (54.4 kg)?

3.46 It is necessary to make foundation designs for two reinforced-concrete columns that support a 12-story apartment house. The columns are spaced 22 ft (6.71 m) on centers. One column carries a load of 350 kips (1557 kN) and one a load of 450 kips (2002 kN). Depth to firm bedrock beneath the 350-kip (1557-kN) column is 9 ft (2.74 m), beneath the 450-kip (2002-kN) column 13 ft (3.96 m). Discuss the possible alternate foundation designs, explaining how you would select trial dimensions, what each foundation consists of, the stresses that must be investigated, permissible values for these stresses, where critical sections occur, how these stresses may be computed, where reinforcing steel is necessary and how it should be placed, how the necessary quantity may be computed, etc.

3.47 A sample of soil weighing 587.4 lb (266.44 kg) is removed from a test pit; 298.1 lb (135.22 kg) water will just fill the pit. A sample of the soil weighing 112.4 g is dried in an oven, and its weight after drying is 103.6 g. Assume the specific gravity of solids is 2.67. Maximum attainable unit weight of the soil is 117.4 lb/ft^3 (1880 kg/m^3), and minimum attainable is 103.9 lb/ft^3 (1664 kg/m^3). Find (a) wet unit weight, (b) dry unit weight, (c) moisture content, (d) voids ratio, (e) porosity, (f) degree of saturation, (g) relative density.

3.48 The following excerpts are from a geologist's report on the route of a proposed interstate highway. Discuss the engineering significance and the effect, if any, on the costs of construction, possible slowdowns, and the availability of construction materials.

   a. Section 4 will cross an ancient meander floodplain from station 19 + 60 (597.4 m) to about station 26 + 00 (792.5 m)

   b. ...the section in the area near the town of Carlton is being built on a glacial lake bed. Station 97 + 00 (2956.6 m) is in the center of a beach ridge extending to station 102 + 50 (3124.2 m)

   c. Most of the section northeast of Carlton is glacial till...obviously young and bedrock-controlled. There is every indication that limestone is beneath the till.

d. This section passes through the barchanes (sand dunes), which are quite extensive.

e. The mile-long section of proposed highway outlined in red is traversing along an esker about 60 ft (18.29 m) deep.

3.49 1. Will the required depth of penetration in the cantilever sheet piling in Fig. 3.49 be: (1*a*) 10.7 (3.26), (1*b*) 11.4 (3.47), (1*c*) 12.01 (3.66), or (1*d*) 12.85 ft (3.92 m)?

2. Will the maximum bending moment in the sheet piling be: (2*a*) 30,218 (40,976), (2*b*) 32,750 (44,409), (2*c*) 34,340 (46,565), or (2*d*) 36,478 ft·lb (49,464 N·m)?

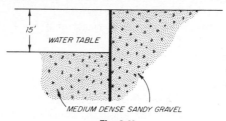

Fig. 3.49

3.50 In reinforced concrete:

a. What is meant by "balanced design?"

b. What is meant by "the transformed section?"

c. What locates the neutral axis

    1. In a new design?

    2. In an investigation of an existing beam?

d. What assumption is made as to the variation of intensity of fiber stress in a beam?

e. In column design, how much additional concrete coverage of outermost steel is usually specified for fireproofing?

f. In a T beam, what part of the cross-sectional area carries the shear?

g. What function do the stirrups play?

h. Why are stirrups not needed in slabs?

i. Distinguish between beam shear and punching shear.

j. How far above the ground should you place the bottom reinforcement in a footing?

3.51 Lists 1 and 2 are various kinds of tests and short statements about the value of each test. For each test in list 1, select a statement from list 2 that is most appropriate.

List 1

1. Tension tests
2. Compression tests
3. Transverse tests
4. Impact tests
5. Shear tests
6. Torsion tests

List 2

a. With brittle materials a criterion of strength, called the modulus of rupture, together with the flexibility and toughness may be determined.

b. With ductile materials the strength, ductility, toughness, and modulus of elasticity may be determined. These tests are the most valuable of all mechanical tests for ductile materials.

c. A valuable measure of shock resistance for brittle and ductile materials.

d. These tests yield an imperfect measure due to the existence of bending stresses.

e. This test is of great value in determining the strength of such materials as wood, concrete, cast iron, and brick.

f. The only test wherein the shearing modulus of elasticity may be determined for ductile materials.

3.52 Answer or define briefly:

In plate girders:

a. What is meant by a sole plate and what should its minimum thickness be?

b. State three limiting conditions for spacing stiffener angles.

c. Where must stiffener angles have milled ends?

d. How much of the web area may be used for flange area?

e. Anchor bolts are usually of what minimum diameter and embedded to what minimum length?

In truss design:

f. What is meant by "combined stresses?"

g. What is meant by "alternate stresses?"

h. What is a gusset plate?

i. What is a filler plate?

j. What are stay plates?

3.53 At both approaches to a bridge it is necessary to construct fills of a maximum height of 20 ft (6.10 m) and a base width of 49 ft (14.9 m) with side slopes of 2 : 1.

a. Describe at least three methods of compacting fills, and discuss the advantages and disadvantages of each method.

b. What is meant by the term "shrinkage" when used in connection with earthwork? What determines the amount of shrinkage? Why is it generally necessary to use a shrinkage factor when balancing cuts and fills? How is this factor used?

  c. A temporary pavement is to be laid on the approaches. Describe in detail the construction of the type of pavement you would use. Upon what do the strength and durability of your pavement depend? Why would you recommend constructing this type of pavement in preference to some others that may be used?

3.54 Show by suitable ruled sketches and full explanation the procedure you would use to lay out a control survey system for a bridge of four spans across a wide river. Both banks are unobstructed. Piers are to be constructed within rectangular cofferdams. Suitable range flags must be located from which the piles for the cofferdams can be driven on proper lines in both directions. The control must be adequate to permit the location of finished work within the cofferdams with the necessary precision for steel arch construction.

3.55 Describe the method you would use to carry an important traverse along a concrete highway. The traverse is to connect a survey of valuable property with a pair of Geodetic Control Survey stations about ½ mi away so that the coordinates of the property corners can be determined. It is desired to have the linear measurements correct to 1 in 20,000. List the equipment and personnel you would use to carry out this work with the required accuracy and safety. List the corrections it would be necessary to apply to the tape measurements as actually made.

Electrical Engineering

4.01 Design a lighting layout for a parking lot in a downtown area. This lot measures 200 by 300 ft (60.96 by 91.44 m). The projectors may be mounted on 25-ft (7.62-m) poles located around the outer edge of the lot.

4.02 Two 500-W, bare incandescent lamps, each having symmetrical downward intensity of 800 candlepower, are mounted outdoors 10 ft (3.05 m) above a rectangular table. The table measures 6 by 16 ft (1.83 by 4.88 m), and the lamps are located directly above the long centerline and 4 ft (1.22 m) in from each end. Will the horizontal illumination at:

1. The center of the table be: (1a) 12.8 (137.8), (1b) 15.0 (161.5), (1c) 17.2 (185.1), or (1d) 19.5 fc (209.9 lm/m²)?

2. The center of each end of the table be: (2a) 6.0 (64.6), (2b) 8.5 (91.5), (2c) 11.0 (118.4), or (2d) 13.5 fc (145.3 lm/m²)?

3. Each corner of the table be: (3a) 5.4 (58.1), (3b) 6.5 (70.0), (3c) 7.6 (81.8), or (3d) 8.7 fc (93.6 lm/m²)?

4.03 A retail store maintains a 20 fc (215.3 lm/m²) illumination at counter height. The luminaires use 750-W PS Mazda lamps. (Take fluorescent lumens per watt as 42.) If the existing luminaires were replaced with fluorescent units and the level raised to 30 fc (322.9 lm/m²), would the percent saving of power cost be: (a) 24.0, (b) 25.5, (c) 27.0, or (d) 28.5 percent?

4.04 Would you permit two type T #6 wires and one bare #6 (4.11-mm diam.) wire to be run in a 3/4-in (19.1-mm) conduit as service conductors? Length of run is 90 ft (27.43 m), and the equivalent of three right-angle bends are included.

4.05 A three-phase transmission line 50 mi (80.5 km) long has an impedance of 5 + j40 ohms per phase. Its "nominal" voltage rating is 345 kV (line-to-line voltage). The terminal voltages at both ends of the line are held at the following levels:

$$|\tilde{V}_1| = 345 \text{ kV} \qquad |\tilde{V}_2| = 360 \text{ kV}$$

Assume that $\tilde{V}_1$ leads $\tilde{V}_2$ by 10°.

    a. Compute the real and reactive line powers in each end of the line.

    b. Compute the total line losses, $P_{loss} + jQ_{loss}$. Express all powers in total three-phase values.

4.06 A 250-V 10-kW dc generator is separately excited. It has an effective armature circuit resistance of 0.5 ohm and inductance of 0.1 H when it is supplying rated current. Suddenly the terminals beyond its protective circuit breaker are short-circuited with a short-circuit resistance of 0.2 ohm. The breaker operates 0.02 s after the fault occurs. Neglecting saturation of the magnetic circuit, will the maximum current to which the generator is subjected be: (*a*) 80, (*b*) 85, (*c*) 90, or (*d*) 95 A?

4.07 The field of a 120-V dc generator draws 8.4 A. Resistance is required of a field-discharge resistor such that the induced voltage, upon disconnecting the field from the 120-V line, will be limited to one-third of the standard high-potential test voltage. Will the resistance be: (*a*) 29.4, (*b*) 32.7, (*c*) 34.9, or (*d*) 36.0 ohms?

4.08 A symmetrial three-conductor cable is enclosed in a grounded metal sheath. The capacitance between the three conductors connected together and the sheath is 0.5 μF. The capitance between one conductor and the other two connected together and also connected to the sheath is 0.6 μF. Will the charging current per conductor when a 60-cycle three-phase 25-kV voltage between conductors is applied to the cable be: (*a*) 4.00, (*b*) 4.10, (*c*) 4.25, or (*d*) 4.45 A?

4.09 The U.S. Navy possesses a self-cooled transformer, intended for use in arctic surroundings and having the hypothetical ratings 500 kVA, 60 Hz, 1000/500 V. If this transformer is used in a tropical climate with the same voltage as before but at 50 Hz, will the new hypothetical rating be: (a) 370, (b) 380, (c) 395, or (d) 415 kVA? Work with the following assumptions:

1. In the new installation, only 80 percent of the total losses accepted in the arctic installation are tolerated.

2. In the original installations the total losses amounted to 0.5 percent of rated power and were distributed as follows:

Copper loss	1.25 kW measured at 100 percent current
Eddy-current loss	0.625 kW measured at 100 percent voltage
Hysteresis loss	0.625 kW measured at 100 percent voltage

4.10 A 60-cycle 200-kVA three-winding transformer is rated at 2400 V primary voltage, and there are two secondary windings, one rated at 600 V and the other at 240 V. There are 200 primary turns. The rating of each secondary winding is 100 kVA, one-half that of the transformer. Determine (a) turns in each secondary winding; (b) rated primary current at unity power factor; (c) rated primary current at 0.8 pf, lagging current; (d) rated current of the 600- and 240-V secondary windings; (e) primary current when the rated current, pf = 1, flows in the 240-V winding and the rated current, pf = 0.7 lagging, flows in the 600-V winding.

4.11 A 15-kVA 2300/230-V transformer is given a short-circuit test by impressing 65 V on the high-tension side with the 230-V side short-circuited. The power input for rated current is 350 W. If the core loss is 245 W:

1. Will the percent voltage regulations at full load and unity power factor be: (1a) 2.10, (1b) 2.22, (1c) 2.34, or (1d) 2.46 percent?

2. Will the maximum efficiency that could be expected from this transformer at full load be: (2a) 94.0, (2b) 96.2, (2c) 97.3, or (2d) 97.3 percent?

4.12 Two transformers, each rated at 100 kVA, 2200/220 V, and 60 cycles, operate in parallel to supply a 200-kVA load at a lagging power factor of 0.80. The short-circuit tests on these transformers show that, with power supplied to the 2200-V windings, there is rated current in each transformer under the following conditions:

Transformer A	1000 W at 72 V
Transformer B	1100 W at 75 V

Neglect the exciting currents.

1. Will the current in the 220-V coil of transformer B in percentage of the total be: (1a) 98, (1b) 100, (1c) 102, or (1d) 104 percent?

2. Will the current in the 220-V coil of transformer A in percentage of the total be: (2a) 98, (2b) 100, (2c) 102, or (2d) 104 percent?

4.13 Two transformers are in parallel sharing a common load. Unit 1 is rated at 45 MVA, 120 to 24 kV, $\Delta - Y$ with an equivalent impedance of $j0.08$ ohm per unit on its own base. Unit 2 is rated at 30 MVA, 120 to 24 kV, $\Delta - Y$, with $Z_e = j0.07$ per unit on its own base. To prevent either unit from being overloaded, the total MVA load limit to the transformers should be: (a) 30.0, (b) 39.4, (c) 52.4, or (d) 69.4 MVA? Give the unit split in each unit.

4.14 A turbine-driven ac generator is to deliver power to an electric furnace located 10 mi (16.093 km) from the turbine. The turbine generator is 60-cycle three-phase star-connected, with 11,000 V between terminals. The electric furnace is three-phase, operates at 200 V between electrodes, and takes 3000 kVA at 80 percent lagging power factor. Specify the number, size, type, and location of the transformers required; and the voltage, size, and spacing of conductors of the high-tension line, the loss of which is not to exceed 5 percent of the delivered power. (Draw diagram and show all essential calculations.)

4.15 A load of 25,000 kW at 85 percent pf is located 25 mi (40.23 km) from a large substation at which 60-cps power is available at 12.5, 66, and 115 kV. Choose voltage of transmission line and size of conductor you would recommend and give reasons for your choice. Voltage regulation and power loss are not to exceed 8 and 6 percent, respectively.

4.16 An industrial plant is supplied from a three-phase transmission line having a capacity of 10,000 kVA. The present plant load is 5000 kW, balanced, three-phase, and at a lagging power factor such that the transmission line is loaded to capacity. The plant must increase its load and this will be done with 30 induction motors which run at 80 percent pf. To make use of the line to its full capacity in active power a synchronous condenser having losses of 300 kW (at the load at which it will operate) will be installed in the plant.

a. Determine the maximum kVA in new induction motors that can be added without overloading the line.

b. What is the kVA rating of the synchronous condenser required for this service?

c. Draw a power vector diagram.

4.17 An industrial load consists of:
One 50-hp induction motor; load, 30 hp; efficiency, 0.86; and pf, 0.70.
One 100-hp induction motor; load, 75 hp; efficiency, 0.89; and pf, 0.80.
Two 15-hp induction motors; load, 15 hp; efficiency, 0.92; and pf, 0.85.
One 300-hp induction motor; load, 310 hp; efficiency, 0.92; pf, 0.85.
Lighting load, 33 kW.
Synchronous motor (to be added): 500 hp, 0.80 pf, leading current; load, 300 hp; and efficiency, 0.925, exclusive of field loss.

Determine for load, overall (1 hp = 0.746 kW), (a) kilowatts; (b) kilo-voltamperes; (c) kilovar; (d) power factor.

4.18 A 15-MVA 8.5 kV three-phase generator has a subtransient reactance of 20 percent. It is connected through a $\Delta - Y$ transformer to a high-voltage transmission line having a total series reactance of 70 ohms. At the load end of the line is a $Y - Y$ step-down transformer. Both transformer banks are composed of single-phase transformers connected for three-phase operation. Each of the three transformers composing each bank is rated 6667 kVA, 10 to 100 kV, with a reactance of 10 percent. The load, represented as impedance, is drawing 10,000 kVA at 12.5 kV and 80 percent pf lagging. Choose a base of 10 MVA, 12.5 kV in the load circuit and draw the impedance diagram showing all impedances in the unit. Will the voltage at the terminals of the generator be: (a) 7.37, (b) 7.43, (c) 7.55, or (d) 7.79 kV?

4.19 A telephone line consisting of two #12 (2.05 mm-diam.) standard copper wires spaced 1 ft (30.48 cm) apart has the following parameters: $r = 10.44$ ohms/loop mile (1.6093 km), $l = 0.00366$ H/loop mile, $c = 0.00838 \times 10^{-6}$ F/loop mile, and $g = 0.300 \times 10^{-6}$ mho/loop mile.

Determine: (a) the wavelength in loop miles; (b) the phase velocity as a percentage of the speed of light; (c) the loss in decibels per loop mile; and (d) the inductive loading per loop mile required for producing distortionless transmission.

4.20 An ideal transmission line 15.25 wavelengths long connects a generator, with an internal impedance of $100 + j0$ ohms, to a load of $200 + j200$ ohms. The load is matched to the line by a stub line bridged across the load. The frequency of operation is 600 Mc. The line and the stub are to have the same characteristic impedance. Calculate the characteristic impedance of the line and the location and the minimum length of the stub for a proper termination of the line.

4.21 It is desired to transmit over a single-phase line 10 mi (16.093 km) long, a load of 1200 kW 0.85 pf lagging current, the loss not exceeding 7.5 percent of the power delivered. The conductors are spaced 18 in (45.72 cm) on centers. The voltage at the load is 11,000 V 50 cycles. (a) Draw a vector diagram. Determine; (b) the smallest size AWG solid-copper conductor, (c) the resistance per wire, (d) the reactance per wire, (e) the sending-end voltage, (f) the line regulation, and (g) the efficiency.

4.22 A 60-cycle three-phase 100-mi (160.93-km) transmission line with a resistance and reactance, respectively, of 0.35 and 0.8 ohm/mi (1.6093 km) of wire, draws a charging current of 0.5 A/mi when the line is tested at 100,000 V to neutral. Calculate the efficiency when this line delivers 50,000 kW at 132,000 V and when the power factor is 0.80 lagging.

4.23 A Y-connected generator rated at 220 V has 0.2-ohm resistance and 2.0-ohms reactance per phase. The generator is connected by lines each having an impedance of $2.06\underline{/29.05°}$ ohms to a Y-Y transformer bank. Each transformer has a total equivalent impedance referred to the high side of $100\underline{/60°}$ ohms, and the transformer bank is connected through lines each of which has a resistance of 50 ohms and an inductive reactance of 100 ohms. If the ratio of transformation is 6 and the low-voltage side is connected to the generator lines, will the actual fault current for a three-phase symmetrical short circuit at the load be: (a) 2.63, (b) 2.75, (c) 2.90, or (d) 3.08 A?

4.24 The data for a 335-hp 2000-V three-phase six-pole 50-cycle Y-connected induction motor are as follows: ohmic resistance per phase of stator, 0.165 ohm; rotor, 0.0127 ohm; and ratio of transformation, 4:1. No-load test; line voltage, 2000 V; line current, 15.3 A; and power, 10,100 W. The friction and windage of the motor are 2000 W. Blocked test: line voltage. 440 V; line current, 170 A; power, 40,500 W. When the slip is 0.015, determine: (a) the current in stator and rotor, (b) motor output, (c) speed, (d) torque developed by rotor and torque at pulley, (e) power factor, and (f) efficiency.

4.25 A shunt-wound generator rated at 35 kW 125 V is to be converted into a series generator to produce the same terminal voltage at rated load. The machine has six poles, and each field spool has 480 turns and carries 11 A. Will the number of turns on each spool of the series winding, neglecting drop in series field, be: (a) 19.0, (b) 19.5, (c) 20.0, or (d) 20.5 turns per pole?

4.26 A 2500-kVA three-phase 60-cycle 6600-V alternator has a field resistance of 0.43 ohm and an armature resistance of 0.072 between each terminal and the neutral. The windings are Y-connected. The field current at full-load unity power factor is 200 A, and at full-load 0.80 pf lagging it is 240 A. The friction loss is 35 kW and the core loss, 47.5 kW. Assume friction and core loss constant at either unity power factor or 0.80 pf lagging.

1. Will the full-load efficiency at unit power factor be: (1a) 94.8, (1b) 95.4, (1c) 95.8, or (1d) 96.0 percent?

2. Will the full-load efficiency at 0.80 pf be: (2a) 92.8, (2b) 93.6, (2c) 94.1, or (2d) 94.4 percent?

4.27 For the periodic wave given in Fig. 4.27, what are the indications on the following types of voltmeters? Neglect meter losses.

a. Dynamometer

b. Iron vane

c. Peak above average, calibrated in rms of sine wave

d. Peak above average, calibrated in rms of a sine wave with leads reversed from that of part c.

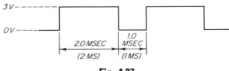

Fig. 4.27

4.28 A 10-hp 550-V 60-cps three-phase induction motor has a starting torque of 160 percent of full-load torque and a starting current of 425 percent full-load current.

1. To limit the starting current to full-load value, will the voltage be: (1a) 130, (1b) 135, (1c) 145, or (1d) 160 V?

2. If the motor is used on a 440-V 60-cps system, will the starting torque expressed in percent of full-load values be: (2a) 100, (2b) 102, (2c) 105, or (2d) 110 percent of rated torque?

3. Will the starting current expressed in percent of full-load values be: (3a) 300, (3b) 325, (3c) 340, or (3d) 350 percent of rated current?

4.29 A 120-V dc motor rated at 5 hp has a field resistance of 50 ohms and an armature circuit resistance of 0.8 ohm. If the armature current at starting is to be limited to 200 percent of rated armature, will the required starter resistance be: (a) 0.895, (b) 0.920, (c) 0.940, or (d) 0.955 ohm?

4.30 A 208-V 60-cycle four-wire three-phase source of power supplies a 208-V balanced motor load of 28.8 kW with a lagging power factor of 80 percent. In addition, it supplies three 120-V 60-A resistive loads to neutral. If the circuit breaker should disconnect the phase load from one line to neutral, what would be the total currents in the four lines?

4.31 An amplifier with a voltage gain of 100 has a maximum variation in gain of 10 percent. To limit the gain variation, two of these amplifiers were cascaded, and the overall gain brought back down to 100 by placing a negative feedback loop around the two similar amplifiers. Will the maximum variation in gain for the cascaded amplifier be: (*a*) 0.1, (*b*) 0.2, (*c*) 0.4, or (*d*) 0.7 percent?

4.32   The input to a 600-V 100-hp synchronous motor is measured by the two-wattmeter method. One wattmeter reads plus 62.5 kW, and the other reads plus 30.5 kW. The motor is known to be taking a leading current. If the efficiency of the motor at this load is 0.90 exclusive of the dc field loss:

    1. Will the line current be: (1*a*) 104.3, (1*b*) 104.8, (1*c*) 105.5, or (1*d*) 106.6 A in each line?

    2. Will the power output be: (2*a*) 111.0 (82.77), (2*b*) 112.2 (83.67), (2*c*) 114.6 (85.46), or (2*d*) 118.2 hp (88.14 kW)?

4.33   Two three-phase alternators are operating in parallel to supply a 4000-kW unity power-factor load at 6600 V. The current delivered by alternator no. 1 is 200 A at 0.85 pf leading. Find the power, power factor, and current for alternator no. 2.

4.34   Two identical three-phase 13.8-kV 100-MVA 60-Hz turbine generators operate in parallel to supply a load of 150 MVA, 0.8 pf, current lagging, at rated voltage, and rated frequency. The synchronous reactance of each machine is 1.1 per unit. Both machines share the real and reactive power equally.

    1. Will the generated phase voltage be: (1*a*) 13.0, (1*b*) 13.5, (1*c*) 13.8, or (1*d*) 13.9 kV per phase?

    2. Will the torque angle be: (2*a*) 23.00, (2*b*) 23.50, (2*c*) 23.85, or (2*d*) 24.10°?

4.35   A permanent-magnet loudspeaker is equipped with a 10-ohm 10-turn voice coil wound on a tube 1 in (2.54 cm) in diameter. The flux density in the air gap is 10,000 lines/in^2 (6.45 cm^2). If the impedance of the voice coil is a pure resistance, what will be the maximum thrust delivered to the cone if the impressed signal is 20 dB above 0.005 W?

4.36   Given: $\beta = 40$, $V_{BE} = 0.7$ V, $I_{CO} = 0$, $R_a = 90$ kΩ, $R_b = 10$ kΩ, $R_c = 5$ kΩ, and $R_e = 1$ kΩ.

    1. Will the quiescent value of I_c be: (1*a*) 0.86, (1*b*) 0.94, (1*c*) 1.00, or (1*d*) 1.04 mA?

    2. Will the quiescent value of V_{CE} be: (2*a*) 13.76, (2*b*) 14.00, (2*c*) 14.12, or (2*d*) 14.18 V? (See Fig. 4.36, p. 110.)

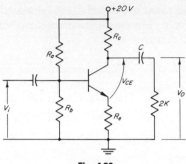

Fig. 4.36

4.37 Given: $V_{BE} = 0.7$ V, $I_{CO} = 0$, and $\beta = 10$. To have an operating point of $V_{CE} = 10$ V and $I_C = 2$ mA:

1. Will R_c be: (1a) 2.3, (1b) 2.6, (1c) 2.8, or (1d) 2.9 kΩ?
2. Will R_a be: (2a) 20.0, (2b) 21.0, (2c) 21.5, or (2d) 21.75 kΩ?

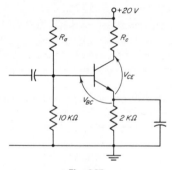

Fig. 4.37

4.38 A stack of selenium rectifier disks is rated at 2 A average and 20 V inverse peak per disk. A rectifier to operate from a 208-V three-phase four-wire without transformers is desired.

Specify the number of disks per stack, and the number of single stacks required, together with the voltage and current capacity assuming 15 percent voltage regulation from no load to full load, for the following cases:

a. The rectifier that will give the largest output without using stacks in parallel.

b. The rectifier that gives the smallest voltage ripples.

c. The rectifier that requires the smallest number of disks and still draws a balanced load. How does the ripple in this case compare with that of part *b*?

4.39 A single-phase condenser motor, operating at rated load, takes 2.50 A from a 220-V line; the current in the 6.75-μF condenser is 1.30 A, and the current in the main winding of the motor is 1.45 A. The total power input is 550 W.

a. What is the apparent impedance of the two windings of the motor?
b. What is the power delivered to each winding of the motor?

4.40 Given: β = 40 and r_i = 1 kΩ (transistor input impedance). r_0 can be neglected. Capacitors have negligible impedance.

1. Draw the small-signal model.
2. Will the current gain i_0/i_s, at midfrequency be: (2*a*) −4, (2*b*) −12, (2*c*) −16, or (2*d*) −18?
3. Will the voltage gain v_0/v_{be} at midfrequency be: (3*a*) −40, (3*b*) −70, (3*c*) −90, or (3*d*) −100?

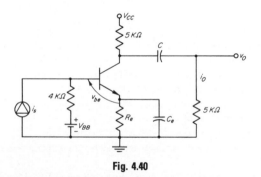

Fig. 4.40

4.41 The capacitor in Fig. 4.41 is initially charged in the direction shown. The capacitor and resistor are then connected to a dc voltage of 200 V by closing the switch at *t* = 0. Derive the equation for the current *i(t)*, and sketch its graph. Will the current 2.25 ms after switching occurs be: (*a*) 0.86, (*b*) 0.88, (*c*) 0.92, or (*d*) 0.98 A?

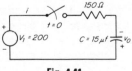

Fig. 4.41

4.42 A wave trap to block carrier current frequencies as furnished by the manufacturer consists of an air core coil and a variable capacitor pack connected in parallel. The inductance of the coil is 260 μH.

Calculate the capacitance in microfarads required to tune the trap for 60 kc.

4.43 Given: $g_m = 10^{-2}$ mho, $R_d = 20$ kΩ, $r_d = 30$ kΩ, and $C = 0.01$ μF. For the FET in Fig. 4.43:

 a. Draw the small-signal model neglecting the transistor capacitances.
 b. Find the voltage gain v_0/v_{gs} in terms of angular frequency w.
 c. Sketch A_{dB} versus w with values shown.

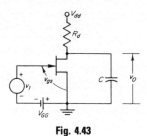

Fig. 4.43

4.44 The coil in the circuit of Fig. 4.44 has a series resistance of 12.6 ohms and the initial current of 9.2 A is present when it has been connected to the dc source for a long time. A make-before-break switch is thrown to the lower position at $t = 0$.

 a. At 2 s after the switch is thrown, the voltage at terminals a-b is 20 V. Find the inductance of the coil.
 b. What is the induced voltage at this time?
 c. What is the greatest voltage across the terminals a-b resulting from switching?
 d. How long before the transient has decreased to 1 percent?

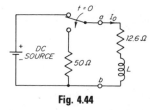

Fig. 4.44

4.45 A two-element 115-V 5-A polyphase watthour meter, having a basic watthour constant of K_H = ⅔, is connected to a three-phase three-wire circuit through 100/5-A current transformers and 2300/115-V potential transformers. A stopwatch check shows that the meter disk is making 15 revolutions in 50 s. Will the load on the circuit be: (a) 276, (b) 288, (c) 312, or (d) 348 kW?

4.46 The four corners of a bridge circuit are marked, respectively, A, B, C, and D. Between A and B there is a resistor of 1000 ohms in parallel with a capacitor of 0.053 μF; between B and C there is a resistor of 1500 ohms in series with a capacitor of 0.53 μF; and between D and A there is a capacitor of 0.265 μF. An impedance, for insertion between C and D, is to be designed so that the bridge will balance when it is supplied from a voltage source which has a frequency of 796 cps.

  a. Determine the coefficients of the impedance which must be inserted between C and D for balance.

  b. After the bridge is balanced by inserting the proper impedance between C and D, the value of the capacitance between D and A is changed to 0.270 μF. An infinite impedance detector is connected across B and D. Determine the voltage across the detector as a percent of the voltage supplied to the bridge across A and C.

4.47 The following data are for the two machines of a properly matched motor-generator set:

  Motor: 10 hp, 1480 rpm full load, 220 V, 42-A line full load, shunt field resistance = 110 ohms, armature resistance = 0.50 ohm, brush voltage = 1.0 V total, stray power = 500 W.

  Generator: 6 kW, 48 V, 1480 rpm, shunt field resistance = 8.0 ohms, armature resistance = 0.01528 ohm, brush voltage = 2.0 V total, stray power = 648 W.

  a. Determine the overall efficiency of the complete unit when the set is operating at full load.

  b. Compute the motor power, line current, and speed when the set is operating at no load.

  c. If the feeder which supplies the motor is 150 ft (45.72 m) from the main, what size wire should be used?

4.48 A multistage amplifier is shown in Fig. 4.48. All capacitors have negligible impedence. The effect of R_a and R_b can be neglected. Draw the small-signal model and find the voltage gain v_2/v_1 at midfrequencies. Transistors are identical with a current gain of β and an input resistance of r_i. (See p. 114 for Fig. 4.48.)

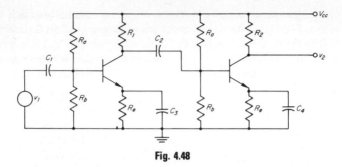

Fig. 4.48

4.49 Figure 4.49 shows a single-line plan of a three-phase electric-power system consisting of two generating stations and three transmission lines. The problem is to determine the short-circuit currents which will

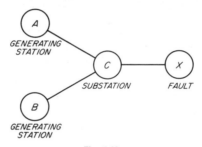

Fig. 4.49

flow in the system if a three-phase fault occurs at the location indicated. The ratings, voltages, and reactances of the generators and transmission lines are as follows:

Item	kVA rating	kV line to line	Ohms reactance
Generator A	30,000	13.2	1.10
Generator B	35,000	13.2	0.95
Line AC	30,000	13.2	0.50
Line BC	35,000	13.2	0.30
Line C to fault	65,000	13.2	0.35

4.50 On the local 208-V three-phase 60-cycle network distribution system with neutral, a manufacturer connects a three-phase 208-V motor and a single-phase 120-V motor. The three-phase motor is rated as follows: 15 hp, 208 V, 1740 rpm, 87 percent efficiency, and 86.6 percent pf. The single-phase motor is rated as follows: 3.5 hp, 115 V, 1750 rpm, 85 percent efficiency, and 80 percent pf. When the machines are operating with full loads, how much current is in each line and in the neutral?

4.51 A three-phase Y-connected power system, with neutral grounded, supplies the load shown in Fig. 4.51. The line voltages are as follows: $E_{AB} = 1000\underline{/0°}$, $E_{BC} = 1000\underline{/120°}$, $E_{CA} = 1000\underline{/240°}$. The generator and line impedances are negligible. If the neutral connection at the load were accidentally open-circuited, what would be the voltage E_{OG}?

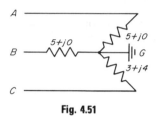

Fig. 4.51

4.52 The class A power amplifier shown in Fig. 4.52 has $I_C = 40$ mA and is designed to give greatest power to the load. Operation is under maximum signal conditions.

1. Sketch the ac and dc load lines showing the Q point and the values at each end of the ac load line.

2. Will the resistance of R_L be: (2a) 20, (2b) 22, (2c) 25, or (2d) 30 ohms?

3. Will the power to the load be: (3a) 0.3, (3b) 0.4, (3c) 0.6, or (3d) 0.9 W?

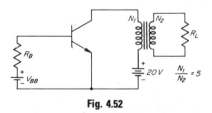

Fig. 4.52

4.53 When 2200 V, 60 cps, is applied to a transformer, the core loss is 240 W. When the frequency is changed to 25 cps and the flux density is maintained constant, the iron loss falls to 75 W.

1. Will the magnitude of the applied voltage at 25 cps be: (1*a*) 875, (1*b*) 900, (1*c*) 917, or (1*d*) 925 V?

2. Will the eddy-current loss at 60 cps be: (2*a*) 98, (2*b*) 103, (2*c*) 113, or (2*d*) 128 W?

3. Will the hysteresis loss at 60 cps be: (3*a*) 137, (3*b*) 135, (3*c*) 130, or (3*d*) 120 W?

4.54 For the tuned amplifier: $R_L = 100$ kΩ, $L = 1$ mH, r_{DS} is large, $g_m = 5 \times 10^{-3}$ mho, $w_0 = 10^6$ rad/s center frequency, $r = 10$ ohms (coil resistance). Neglect transistor capacitances. Coupling and bypass capacitors are large.

a. What is the value of C?

b. Find the bandwidth.

c. Find the voltage gain at the resonant frequency.

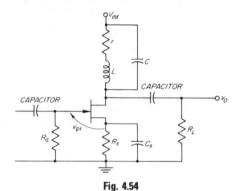

Fig. 4.54

4.55 Design a low-pass audio-frequency filter of "constant-K" type with T configuration which will match a 600-ohm resistance load and have a cutoff frequency of 1000 cps.

Engineering Economics and Business Relations

5.01 A small manufacturer with a maximum factory output of 1200 machines per year prepared the following estimate of manufacturing expenses at various outputs:

Output	Direct labor and material	Indirect factory costs	Selling and administrative costs	Total cost	Unit cost
0	0	$28,000	$26,000	$ 54,000	
200	$16,000	30,000	27,000	73,000	$365.00
400	32,000	32,000	28,000	92,000	230.00
600	48,000	34,000	29,000	111,000	185.00
800	64,000	36,000	30,000	130,000	162.50
1000	80,000	38,000	31,000	149,000	149.00
1200	96,000	40,000	32,000	168,000	140.00

At the start of the year, after obtaining contracts with reliable buyers for 600 machines at $160 each, it becomes evident that due to business conditions it will not be possible to sell any more machines in the domestic market for the remainder of the year regardless of price.

A reliable foreign purchaser offers to buy 600 machines at $100 each.

Disregarding all questions of long-run policy with respect to foreign sales, should the offer be accepted? Why?

5.02 A shoe manufacturer produces a pair of shoes at a labor cost of $6.00 a pair and a material cost of $5.50 a pair. The fixed charges on the business are $420,000 a month and the variable costs are $2.50 a pair. If the shoes sell for $20.00 a pair, will the number of pairs that must be produced each month for the manufacturer to break even be: (a) 60,000, (b) 65,000, (c) 70,000, or (d) 75,000?

5.03 Define depreciation rate and give the general formula for determining it by (a) the straight-line method, and (b) the sinking-fund method. The Interstate Commerce Commission prescribes the former in maintaining the accounts of public utilities, while engineers customarily use the latter in determining the economic propriety of major undertakings. Explain why, although there is a radical difference in the rates developed by the two formulas, both these choices are appropriate and adequate for their respective purposes.

5.04 An old, light-capacity highway bridge may be strengthened at a cost of $60,000 or it may be replaced by a modern bridge of sufficient capacity at a cost of $250,000. The present net salvage value of the old bridge is $25,000. It is estimated that the reinforced bridge will last for 5 years after which replacement will be necessary. At the end of the 5 years, the net salvage value of the reinforced bridge will be $10,000, and the net salvage value of the new bridge will be $100,000 after 30 years. The additional cost of maintenance and inspection of the old bridge will be $1500/year. Assuming interest at 8 percent and depreciation on a straight-line basis, state whether it is more economical to reinforce the old bridge or to replace it. Show the calculations on which you base your reply.

5.05 Which is the more economical: (a) a building costing $200,000 that will have to have an addition built 10 years from now at an additional cost of $200,000 or (b) a complete building constructed now at a cost of $300,000? Interest at 10 percent. Assume life of the building is so long that depreciation may be neglected.

5.06 Two proposed types of furnishings are to be compared as to capitalized cost at 8 percent interest in perpetuity. Type A has a life of 10 years; initial cost and cost of renewal is $10,000; annual maintenance is $100; repairs every 5 years are $500; no salvage value. Type B has a life of 15 years; initial cost and cost of renewal is $15,000; no annual maintenance cost; repairs every 5 years are $200; salvage value is $3000. Determine the capitalized cost of each.

5.07 A debt of $1000 is to be paid off in five equal yearly payments, each payment combining an amortization installment and interest at 8 percent on the previously unpaid balance of the debt. What should be the amount of each payment?

5.08 A syndicate wishes to purchase an oil well which, estimates indicate, will produce a net income of $500,000/year for 20 years. What should the syndicate pay for the well if, out of this net income, a return of 10 percent on the investment is desired and a sinking fund will be established at 6 percent interest to recover the investment?

5.09 Two methods, A and B, of conveying water are being studied. Method A requires a tunnel, first cost $1,000,000; life perpetual, annual operation, and upkeep are $25,000. Method B requires a ditch plus a flume. First cost of ditch is $350,000; life perpetual, annual operation, and upkeep are $20,000. First cost of the flume is $200,000; life, 10 years; salvage value is $30,000; annual operation and upkeep are $40,000. Compare the two methods for perpetual service, assuming an interest rate of 8 percent.

5.10 A $1,000,000 issue of 6 percent 15-year bonds was sold at 90. If miscellaneous initial expenses of the financing were $20,000 and a yearly expense of $2000 is incurred, what is the true cost to the nearest 0.1 percent that the company is paying for the money it borrowed?

5.11 A corporation has total outstanding stock consisting of 10,000 shares of common and 1000 shares of simply participating, $100 par, 5 percent cumulative preferred. The corporation distributed $2500 in dividends in 1965 but suffered a slight loss in 1966 and just broke even in 1967. During 1966 and 1967, it paid no dividends. In 1968, it had a net income of $78,500 applicable to dividends and in 1969, $77,000.

1. Will the payment on each share of common stock for the year 1968 be: (1a) $5.00, (1b) $6.00, (1c) $6.50, or (1d) $6.75?

2. Will the payment on each share of preferred stock for the year 1969 be: (2a) $5.00, (2b) $6.00, (2c) $7.00, or (2d) $8.00?

If the incorporation and stock certificates carry no specifications beyond that given in the above statement:

3. Will the number of voting shares be: (3a) 1000, (3b) 3000, (3c) 7000, or (3d) 11,000?

4. Will the number of redeemable shares be: (4a) none, (4b) 1000, (4c) 7000, or (4d) 11,000?

5.12 A manufacturer analyzing its inventory control finds that 15,000 units of an item are purchased every year. The average daily requirement is 50 units, and the normal time for delivery after placement of a standard order of 2000 units is 30 days. No more than 1800 units or less than 200 units are used in any 30-day period.

1. Should the normal ordering point in units be: (1*a*) 1800, (1*b*) 1850, (1*c*) 1900, or (1*d*) 2000?

2. Should the maximum number of units of storage space be: (2*a*) 1800, (2*b*) 2400, (2*c*) 3600, or (2*d*) 4200?

3. Should the normal maximum inventory limit in units be: (3*a*) 2000, (3*b*) 2300, (3*c*) 2600, or (3*d*) 3000?

4. Is the average inventory in number of units: (4*a*) 2250, (4*b*) 1800, (4*c*) 1550, or (4*d*) 1300?

5. Is the annual inventory rate of turnover: (5*a*) 11.28, (5*b*) 11.40, (5*c*) 11.54, or (5*d*) 11.70?

5.13 In addition to the data in question 5.12, assume the cost of handling a purchase is $7.50, the storage cost for average inventory per unit per annum is $0.30, and the carrying charges (percentage of average annual inventory valuation) are 30 percent. If the item in question may be purchased in lots of 1500, 2000, 2500, or 3000 units at unit prices of $2.00, $1.85, $1.75, or $1.70, respectively:

1. Would the value of the average inventory if a lot of 2000 units is purchased be: (1*a*) $2405, (1*b*) $2475, (1*c*) $2545, or (1*d*) $2615?

2. Would the unit buying expense if a lot of 3000 units is purchased be: (2*a*) $0.00500, (2*b*) $0.00375, (2*c*) $0.00300, or (2*d*) $0.00250?

3. Would the cost to store 2000 units for 1 year at the average inventory per unit be: (3*a*) $400, (3*b*) $500, (3*c*) $600, or (3*d*) $700?

4. Would the inventory carrying charge for an order of 2500 units be: (4*a*) $813.75, (4*b*) $836.50, (4*c*) $859.25, or (4*d*) $882.00?

5.14 In keeping with the data in questions 5.12 and 5.13, would the economic order size be: (*a*) 2500, (*b*) 3000, (*c*) 3500, or (*d*) 4000 units?

5.15 A manufacturer has the following commodity inventory account:

June	1	Balance on hand	500 units at $100/unit
June	12	Purchased	500 units at $125/unit
June	25	Delivered to shop	600 units
July	6	Purchased	500 units at $110/unit
July	15	Delivered to shop	600 units
Aug.	3	Purchased	500 units at $105/unit
Aug.	19	Delivered to shop	600 units

1. If the last-in first-out method of inventory valuation is used, would the value of the inventory on August 20 be: (1a) $17,000, (1b) $18,000, (1c) $19,000, or (1d) $20,000?

2. If the first-in first-out method of inventory valuation is used, would the value of the inventory on August 20 be: (2a) $19,000, (2b) $21,000, (2c) $24,000, or (2d) $28,000?

3. If the weighted-average method of inventory valuation is used, would the value of the inventory on July 31 be: (3a) $33,333.33, (3b) $37,500, (3c) $41,000, or (3d) $45,000?

5.16 A company is planning to buy a special fixture for $1000. Their data indicated that the estimated saving in direct labor cost would be 5 cents, the interest rate would be 6 percent, the rate for fixed charges would be 6 percent, the rate for upkeep would be 13 percent, the life of the equipment could be assumed to be 4 years, and the overhead saving due to direct labor saved would be 40 percent. The estimated cost of each setup should average $50.

1. To return the cost out of earnings in 4 years, must the number of pieces to be put through in one lot per year be: (1a) 7675, (1b) 7857, (1c) 8040, or (1d) 8224?

2. If the company plans to produce 9200 pieces per year, will the number of years to amortize the cost of the equipment be: (2a) 1, (2b) 2, (2c) 3, or (2d) 4?

3. If the fixture produces 10,000 pieces per year, will the resulting gross profit per year be: (3a) $110, (3b) $120, (3c) $135, or (3d) $150?

4. If 12,000 pieces are to be manufactured in four lots per year, could the money invested in this fixture be: (4a) $1155, (4b) $1215, (4c) $1280, or (4d) $1350?

5.17 The balance sheet of a company is as follows:

Assets		Liabilities	
Current:		Current:	
Cash	$ 2,000	Accounts payable	$ 8,000
Accounts receivable	3,000	Notes payable	1,000
Inventory	8,000	Accrued taxes	1,000
Total current assets	$13,000	Total current liabilities	$10,000
Fixed:		Capital:	
Land	$ 1,000	Common stock, $10 par,	
Buildings less depreciation	15,000	2500 shares outstanding	$25,000
Machinery less		Surplus	4,000
depreciation	9,000	Tangible net worth	$29,000
Total fixed assets	$25,000		
Deferred charges	1,000		
Total assets	$39,000	Total liabilities	$39,000

1. Would the value of the ratio of the total current liabilities to the tangible net worth be: (1*a*) 0.345, (1*b*) 0.350, (1*c*) 0.355, or (1*d*) 0.360?

2. Would the fixed assets to tangible net worth be: (2*a*) 0.856, (2*b*) 0.861, (2*c*) 0.866, or (2*d*) 0.871?

3. If the net sales for the balance period were $30,000, would the ratio of net sales to net working capital be: (3*a*) 8.5, (3*b*) 9.0, (3*c*) 10.0, or (3*d*) 11.5?

5.18 A chemical is purchased for use as a raw material in a manufacturing plant. The costs involved in making a purchase are $21 per purchase order regardless of the size of the order. During the year 3000 gal of this chemical is consumed at a fairly uniform rate. The chemical is purchased and stored in 50-gal drums, and the purchase price per gallon, including freight, is $3.30. Annual storage costs are estimated as 50 cents per drum of maximum inventory and the annual carrying charges are estimated as 12 percent on average inventory. To assure continuous operations, at least 200 gal should be maintained on hand at all times as an emergency stock. Is it most economical to purchase drums in lots of (*a*) 8, (*b*) 10, (*c*) 12, or (*d*) 15?

5.19 A general contractor is required to install and operate a temporary well-point system during a 6-month phase of construction of a riverside powerhouse, from April 1 through September 30, 1973. The necessary equipment will cost $1000/month to rent. A pump operator will have to be in attendance continuously and must be paid an hourly wage of $9.00 for each 8-h weekday shift, $13.50 for each 8-h Saturday shift, and $18.00 for each 8-h Sunday and legal holiday shift. Payroll taxes and insurance are 13 percent of wages. Fuel is estimated at $40/day. Overhead and maintenance charges are 15 percent of wages, fuel, and rental charges. Payment for successful completion of the well-pointing operation will be one lump sum at the end of the 6-month period. If financing costs 8 percent/annum and the contractor desires a profit and contingency of 10 percent of costs, would the lump-sum bid for the well-pointing operation be: (*a*) $85,432, (*b*) $87,345, (*c*) $89,123.45, or (*d*) $90,238.93?

5.20 A parent wishes to develop a fund for a newborn child's college education. The fund is to pay $5000 on the eighteenth, nineteenth, twentieth, and twenty-first birthdays of the child. The fund will be built up by the deposit of a fixed sum on the child's first to seventeenth birthdays. If the fund earns 4 percent:

1. Will the yearly deposit into the fund be: (1*a*) $796.48, (1*b*) $821.93, (1*c*) $877.17, or (1*d*) $902.45?

2. Will there be: (2*a*) 15, (2*b*) 16, (2*c*) 17, or (2*d*) 18 equal payments?

5.21 A lot was purchased in January 1960 for $5000. Taxes and assessments were charged at the end of each year as follows: $100 for 1960, $100 for 1961, $200 for 1962, $100 for 1963, $100 for 1964, and $100 for 1965. The owner paid the charges for 1960 and 1961 but not for subsequent years. At the end of 1965 the lot was sold for $10,000, the seller paying back charges at 7 percent interest compounded annually and also paying a commission of 5 percent to an agent for handling the sale. Will the rate of return realized on the investment be: (a) 8.5, (b) 8.8, (c) 9.1, or (d) 9.4 percent?

5.22 A man owns a building on which there is a $100,000 mortgage which earns 6 percent/annum. The mortgage is being paid in 20 equal year-end payments. After making 8 payments, the man desires to reduce his payments by refinancing the balance of the debt with a 30-year mortgage at 8 percent, to be retired by equal annual payments. Would the reduction in the yearly payment be: (a) $2001.20, (b) $2112.40, (c) $2224.80, or (d) $2339.60?

5.23 A sprinkler system for a building is designed to be renewed every 20 years. It will have a salvage value equal to 10 percent of its cost and will save $2000 a year in insurance premiums. If money is worth 10 percent, the maximum cost of the sprinkler system should not exceed: (a) $17,284, (b) $18,186, (c) $18,640, or (d) $18,866?

5.24 A new boiler has just been installed. It is expected that there will be no maintenance charges until the end of the eleventh year, when $500 will be spent on the boiler and $500 will be spent at the end of each successive year until the boiler is scrapped at the end of its nineteenth year of service. Will the sum of money to be set aside at the time of installation of the boiler at 4 percent to take care of all maintenance expenses for the boiler be: (a) $2296.93, (b) $2304.89, (c) $2403.48, or (d) $2511.56?

5.25 An automobile costs $3000. It is run approximately the same distance each year, with transportation costs at a minimum. If annual expenses for maintenance are $100 at the end of the first year and increase $100/year each year thereafter, and if the trade-in value is $1800 at the end of the first year and this decrease uniformly $200 each year thereafter, should the car be traded in at the end of the: (a) third, (b) fourth, (c) fifth, or (d) sixth year?

Mechanical Engineering

6.01 At the beginning of compression in a cold-air Otto cycle, the temperature is 100°F (37.78°C), the pressure is 13.75 psia (94.81 kPa abs), and the volume is 1 ft³ (0.02832 m³). At the end of compression, the pressure is 155 psia (1068.7 kPa abs). The heat supplied to the cycle is 50 Btu (52.8 kJ).

1. Will the compression ratio be: (1a) 5.642, (1b) 5.842, (1c) 5.942, or (1d) 5.992?

2. Will the percent of clearance be: (2a) 19.5, (2b) 21.5, (2c) 22.5, or (2d) 23.0?

3. Will the temperature just after combustion be: (3a) 5218 (2899), (3b) 5418 (3010), (3c) 5518 (3066), or (3d) 5568°R (3093 K)?

4. Will the pressure just after combustion be: (4a) 412.8 (2846), (4b) 613.6 (4231), (4c) 714.0 (4923), or (4d) 764.2 psia (5269 kPa abs)?

5. Will the network per cycle be: (5a) 18,412 (24,963), (5b) 19,432 (26,346), (5c) 19,942 (27,037), or (5d) 20,197 ft·lb (27,308 N·m)?

6. Will the heat rejected from the cycle be: (6a) 25.02 (26.40), (6b) 31.08 (32.79), (6c) 34.11 (35.99), or (6d) 35.52 Btu (37.47 kJ)?

7. Will the m.e.p. be: (7a) 146 (1007), (7b) 158 (1089), (7c) 164 (1131), or (7d) 167 psi (1151 kPa)?

Show sketches of *pv* and *Ts* diagrams, noting pertinent points in cycle.

6.02 A 6-in (15.24-cm) steel main has an inside diameter of 4.897 in (12.44 cm) and an outside diameter of 6.625 in (16.83 cm). It is insulated at the outside with asbestos. The steam temperature is 300°F (148.9°C), and the air temperature outside is 70°F (21.1°C).

$$h_s(\text{steam}) = 20 \text{ Btu/(ft}^2)(\text{h})(°\text{F}) \text{ [113.5 W/(m}^2)(\text{h})(°\text{C})]$$
$$h_a(\text{air}) = 6 \text{ Btu/(ft}^3)(\text{h})(°\text{F}) \text{ [34 W/(m}^2)(\text{h})(°\text{C})]$$
$$k(\text{asbestos}) = 0.06 \text{ Btu/(ft)(h)(°F) [0.10 W/(m)(h)(°C)]}$$
$$k(\text{steel}) = 30 \text{ Btu/(ft)(h)(°F) [51.89 W/(m)(h)(°C)]}$$

To limit the heat loss to 10 Btu/(ft^2)(h)(°F) [56.75 W/(m^2)(h)(°C)], should the thickness of the asbestos be: (a) 8.25 (20.96), (b) 8.50 (21.58), (c) 8.75 (22.23), or (d) 9.00 in (22.86 cm)?

6.03 A turbine generator operates on the reheat-regenerative cycle with one reheat and two stages of feedwater heating, with pressures and temperatures as indicated below. Heater no. 1 (nearest the condenser) is the "open" or direct-contact type. Heater no. 2 is the "closed" or surface type and drains from this heater are "cascaded," i.e., flow back to heater no. 1. Operation may be assumed to be ideal except for departures indicated below.

Steam is generated at 700 psia (4826.5 kPa abs) and some superheat, and reaches the turbine without heat loss at p = 655 psia (4516.2 kPa abs) and t = 740°F(393.3°C). Resuperheating takes place at p = 109 psia (751.6 kPa abs) to a temperature of 700°F (371.1°C). Extractions for feedwater heating occur $at\ p$ = 109 psia (751.6 kPa abs) and 15 psia (103.4 kPa abs). Condenser pressure is 1 psia (6.895 kPa abs). In the closed (no. 2) heater the temperature difference between the condensed steam leaving and the feedwater leaving is 10°F (5.55°C).

To reduce the number of values to be looked up in the steam tables, the following partial list of values may be assumed to be correct.

h(heat)	= Btu/lb
At entrance to boiler	296.8
At first extraction point	1188.7
After resuperheating	1378.4
At second extraction point	1177.6
At entrance to condenser	973.0
Condensate leaving condenser	69.7
Feedwater leaving no. 1 heater	181.1
Drains leaving no. 2 heater	305.0

To be done: Assuming a flow of 1 lb (0.4536 kg) of steam to the turbine:

a. Sketch a line diagram of apparatus, naming each piece. Use a consistent system of numbers or letters to indicate state entering or leaving each piece of apparatus.

b. Sketch an *hs* diagram for the cycle, identifying pertinent points in terms of the symbols used in (*a*).

c. Determine the initial temperature of the steam leaving the boiler.

d. Determine the temperature of the feedwater leaving the last heater.

e. Determine the amount of steam that must be extracted from the turbine for heater no. 1, assuming that extraction for heater no. 2 = 0.1262 lb/lb (kg/kg) of steam supplied to the turbine.

f. If the internal efficiency of the turbine before and after resuperheating is 80 percent, estimate the quality of the steam at exhaust.

g. For the Rankine cycle of operation with no resuperheating and no extracting, what would be the *h* of the exhaust steam?

6.04 Will the heat loss through a wall construction of 8-in (20.32-cm) cinder block with 2 in (5.08 cm) of cork and ½ in (1.27 cm) of plaster on one side and ½ in (1.27 cm) of plaster on the other side be: (*a*) 4.16 (23.61), (*b*) 4.30 (24.40), (*c*) 4.42 (25.08), or (*d*) 4.50 Btu/(ft²)(h)(°F) [25.54 W/(m²)(h)(°C)]? Assume still air on both sides having a temperature difference of 40°F (22.22°C).

6.05 The cross section of a wall of a cold storage room consists of 16 in (40.64 cm) of brick ($K = 3.6$), and air space equivalent to the resistance of about 8 in (20.32 cm) of brick, 8 in more of brick, 4 in (10.16 cm) of corkboard ($K = 0.3$), and 1.5 in (3.81 cm) of cement plaster ($K = 2.3$).

When the temperatures of the inner and outer wall surfaces are 30 and 70°F (-1.1 and $+21.1$°C), respectively, will the quantity of heat transmitted through the wall by conduction be: (*a*) 35.0 (198.6), (*b*) 39.8 (225.9), (*c*) 43.0 (244.0), or (*d*) 44.6 Btu/(ft²)(°F) [253.1 W/(m²)(°C) per day?

6.06 Air passes through a duct at the rate of 1800 ft³ (50.98 m³)/min, the dry-bulb temperature being 70°F (21.1°C) and the relative humidity 40 percent. It is proposed to raise the humidity of the air to 75 percent by blowing into it (*a*) water spray at 50°F (10°C) or (*b*) saturated steam at atmospheric pressure. How many pounds of water or steam will be required and what will be the final dry- and wet-bulb temperatures in each case? Assume adiabatic conditions and thorough mixing so as to produce uniform conditions.

6.07 Water enters a cooling tower at 126°F (52.22°C) and leaves at 80°F (26.67°C). Air enters at 85°F (29.44°C) and relative humidity 47 percent and leaves in a saturated condition at 115°F (46.11°C).

1. Will the volume of air needed per pound of water cooled be: (1a) 8.68 (0.542), (1b) 9.13 (0.570), (1c) 9.43 (0.588), or (1d) 9.58 ft³/lb (0.598 m³/kg)?

2. Will the amount of water that can be cooled with 2000 ft³/min (56.64 m³/min) of free air be: (2a) 140 (63.50), (2b) 176 (79.76), (2c) 200 (90.72), or (2d) 212 lb (96.16 kg)/min?

6.08 An auditorium seating 1800 people is to be maintained at 78°F (25.56°C) dry-bulb and 67°F (19.44°C) wet-bulb temperature when outdoor air is at 90°F (32.22°C) dry-bulb and 75°F (23.89°C) wet-bulb. Solar load is 120,000 Btu/h.

1. Will the amount of outdoor air required for ventilation be: (1a) 13,500 (382.3), (1b) 14,000 (396.5), (1c) 14,500 (410.6), or (1d) 15,000 ft³ (424.8 m³)/min?

2. Will the volume of conditioned air at 65°F (18.33°C) dry-bulb that must be circulated to carry the total sensible heat load be: (2a) 37,000 (1047.8), (2b) 37,400 (1059.2), (2c) 37,800 (1070.5), or (2d) 38,200 ft³ (1081.8 m³)/ min?

3. Will the wet-bulb temperature of the conditioned air to absorb the moisture load be: (3a) 59.2 (15.11), (3b) 59.7 (15.39), (3c) 60.2 (15.67), or (3d) 60.7°F (15.94°C)?

6.09 In an auditorium maintained at a temperature not to exceed 75°F (23.89°C) and a relative humidity not to exceed 60 percent, a sensible heat load of 450,000 Btu (131.9 kW)/h, and 1,200,000 gr (77.76 kg) moisture/h must be removed. Air is supplied to the auditorium at 65°F (18.33°C).

1. Will the amount of air per hour to be supplied be: (1a) 175,000 (79.38), (1b) 187,500 (85.05), (1c) 200,000 (90.72), or (1d) 212,500 lb (96.39 Mg)/h?

2. Of the entering air, will the dew-point temperature be: (2a) 55.4 (13.0), (2b) 56.6 (13.67), (2c) 57.8 (14.33), or (2d) 59.0°F (15.0°C)? Will the relative humidity be: (2e) 70, (2f) 73, (2g) 75, or (2h) 76 percent?

3. Will the latent heat load picked up in the auditorium be: (3a) 171.2 (77.66), (3b) 174.8 (79.29), (3c) 177.2 (80.38), or (3d) 178.4 lb (80.92 kg) moisture/h?

6.10 A two-pass surface condenser is to be designed using an overall heat-transfer coefficient of 480 Btu/(h)(°F)(ft²)[2723.8 W/(m²)(°C)] of outside tube surface. The tubes are to be 1 in (2.54 cm) outside diameter with 1/16-in (1.59-mm) walls. Entering circulating water velocity is to be 6 fps (1.829 m/s). Steam enters the condenser at a rate of 1,000,000 lb (453.6 Mg)/h at a pressure of 1 psia (6.895 kPa abs) and an enthalpy of

1090 Btu/lb (2537.5 kJ/kg). Condensate leaves as saturated liquid at 1 psia (6.895 kPa abs). Circulating water enters the condenser at 85°F (29.44°C) and leaves at 95°F (35.0°C).

1. Will the required number of tubes be: (1a) 17,000, (1b) 18,200, (1c) 19,000, or (1d) 19,400?

2. Will the length of the tubes be: (2a) 19.5 (5.944), (2b) 20.0 (6.096), (2c) 20.3 (6.187), or (2d) 20.5 ft (6.248 m)?

6.11 Exhaust steam at 1 psig (108.6 kPa gage) pressure and containing 12 percent moisture is fed into an open-type feedwater heater. The barometric pressure is 30 inHg (76.2 cmHg) (101.35 kPa). Will the amount of this steam required to raise the temperature of 1000 lb (453.6 kg) of water from 60 to 200°F (15.56 to 93.33°C) be: (a) 161 (73.03), (b) 165 (74.84), (c) 174 (78.93), or (d) 199 lb (90.27 kg)?

6.12 A tubular-type air heater has 31,154 ft² (2894.2 m²) of heating surface and heats 357,000 lb (161,935 kg)/h of air from 80 to 479°F (26.67 to 248.33°C). Flue gas enters at 725°F (385°C) and leaves at 405°F (207.22°C).

1. Will the mean temperature difference between the flue gas and the air be: (1a) 270 (132.22), (1b) 284 (140.00), (1c) 312 (155.56), or (1d) 354°F (178.89°C)?

2. Will the heat-transfer rate between the flue gas and the air be: (2a) 32.4 × 10⁶ (94.9 × 10⁵), (2b) 33.0 × 10⁶ (96.7 × 10⁵), (2c) 34.2 × 10⁶ (100.2 × 10⁵), or (2d) 36.0 × 10⁶ Btu/h (105.5 × 10⁵ W)?

3. Will the weight of flue gas passing through the heater be: (3a) 345,000 (156,492), (3b) 396,000 (179,626), (3c) 430,000 (195,048), or (3d) 447,000 lb (202,759 kg)?

6.13 A single-stage impulse turbine receives steam having an available energy of 100 Btu/lb (232.6 kJ/kg). Five percent of the maximum theoretical nozzle exit velocity is lost through nozzle friction; 10 percent of the entering relative velocity is lost by the steam in passing through the blading. The nozzle makes an angle of 20° with the plane of the wheel. The blading is symmetrical, i.e., equal entrance and exit angles. Blade speed is 900 ft (274.3 m)/s. Assume nozzle inlet velocity negligible. Calculate:

   a. Steam velocity leaving nozzle.
   b. Relative velocity entering blade.
   c. Relative velocity leaving blade.
   d. Absolute exit velocity.
   e. Blade entrance angle.

f. Force exerted on blade by each pound (kilogram) of steam flowing per second.

g. Blade horsepower if 1800 lb (816.5 kg) steam flow/h.

h. Shaft horsepower. Friction between the steam and the wheels absorbs 6 hp (4.4474 kW). Bearing and governors take 2 hp (1.491 kW). Assume this 2 hp does not go into the steam.

i. Steam entering the turbine has an enthalpy of 1200 Btu/lb (2793.6 kJ/kg). Neglecting radiation losses from the turbine, what should be the enthalpy of the steam leaving the turbine if the energy due to residual velocity goes to increasing the enthalpy of the steam?

6.14 A steam-electric generating station has four 10,000-kW turbogenerators. Steam is supplied at 250 psia (1723.75 kPa abs) and 600°F (315.56°C). Exhaust is at 2 inHg (5.08 cmHg) abs. Daily load factor is 75 percent. Assume average steam rates and efficiencies.

1. Will the coal required per day be: (1*a*) 461 (418.2), (1*b*) 468 (424.6), (1*c*) 479 (434.5), or (1*d*) 496 tons (450.0 Mg)?

2. Will the condenser cooling water per minute be: (2*a*) 55,000 (208.2), (2*b*) 56,200 (212.7), (2*c*) 58,600 (221.8), or (2*d*) 62,200 gal (235.4 kl)?

6.15 An ideal refrigeration cycle using ammonia as the refrigerant compresses isentropically from saturated vapor at 20 to 190 psi (137.9 to 1310.1 kPa). The temperature at the high-pressure side of the expansion valve is 80°F (26.67°C). Draw the *TS* diagram. Considering 1 lb (0.4536 kg):

1. Will (*a*) the network, (*b*) the heat rejected to the condenser water, and (*c*) the refrigeration be: (1*a*) 146.8 (341.6), (1*b*) 316.0 (735.0), (1*c*) 474.2 (1103.0), or (1*d*) 621 Btu/lb (1444.4 kJ/kg)?

2. Will the coefficient of performance (cop) be: (2*a*) 3.13, (2*b*) 3.23, (2*c*) 3.38, or (2*d*) 3.58?

3. Will the amount of NH_3 circulating per quantity of refrigeration be: (3*a*) 0.372 (0.186), (3*b*) 0.396 (0.198), (3*c*) 0.412 (0.206), or (3*d*) 0.422 lb/ton (0.211 kg/Mg) per min of refrigeration?

6.16 A test of an oil-fired boiler indicated 12.77 lb (5.79 kg) of water evaporated/lb of oil. The boiler pressure was 200 psia (1379 kPa abs), superheat was 87°F (30.56°C), feedwater temperature 93°F (33.89°C), and boiler and furnace efficiency 82.8 percent. Will the calorific value of the oil be: (*a*) 18,400 (42,835.2), (*b*) 18,700 (43,533.6), (*c*) 19,300 (44,930.4), or (*d*) 20,200 Btu/lb (47,025.6 kJ/kg)?

6.17 A steam boiler on a test generates 885,000 lb (401,436 kg) of steam in a 4-h period. The average steam pressure is 400 psia (2758 kPa abs), the average steam temperature is 700°F (371.11°C), and the average temperature of the feedwater supplied to the boiler is 280°F (137.78°C). If the boiler efficiency for the period is 82.5 percent, and if the coal has a heating value of 13,850 Btu/lb (32,242.8 kJ/kg) as fired, will the average amount of coal burned be: (*a*) 10.00 (9072.0), (*b*) 10.75 (9752.4), (*c*) 11.25 (10,206.0), or (*d*) 11.50 short tons (10,432.8 kg)/h?

6.18 1. Is a boiler horsepower equal to: (1*a*) 33,000 (9666), (1*b*) 33,300 (9754), (1*c*) 33,475 (9805), or (1*d*) 33,550 Btu/h (9827 W/h)?

2. Is the ratio of a boiler horsepower to a mechanical horsepower: (2*a*) 12.40, (2*b*) 12.75, (2*c*) 13.00, or (2*d*) 13.15?

3. Calculate the horsepower of a boiler generating 4000 lb (1814.4 kg) dry steam/h, with a factor of evaporation of 1.06. The latent heat of steam at 212°F (100°C) is 970 Btu/lb (2258 kJ/kg). Will the boiler horsepower be: (3*a*) 123 (1207), (3*b*) 134 (1314), (3*c*) 145 (1422), or (3*d*) 156 (1530 kW)?

6.19 A fan whose static efficiency is 40 percent has a capacity of 60,000 ft^3 (1699.2 m^3)/h at 60°F (15.56°C) and a barometer of 30 inHg (76.2 cmHg), and gives a static pressure of 2 in (5.08 cm) of water column on full delivery. Should the size of electric motor to drive this fan be: (*a*) 0.75 (0.560), (*b*) 1.00 (0.746), (*c*) 1.25 (0.933), or (*d*) 1.5 hp (1.119 kW)?

6.20 A fan discharged 9300 ft^3 (263.4 m^3)/min of air through a duct 3 ft (0.9144 m) in diameter against a static pressure of 0.85 in (2.16 cm) of water. The gage fluid density was 62.1 lb/ft^3 (994.7 kg/m^3), the air temperature 85°F (29.44°C), and the barometric pressure 28.75 inHg (73.03 cmHg). If the power input to the fan is measured as 3.55 hp (2.65 kW):

1. Will the overall mechanical efficiency of the fan be: (1*a*) 38.0, (1*b*) 38.7, (1*c*) 39.2, or (1*d*) 39.5 percent?

2. Will the static efficiency of the fan be: (2*a*) 26, (2*b*) 30, (2*c*) 33, or (2*d*) 35 percent?

6.21 Air is moving at 20 ft (6.096 m)/min through a 2 × 3 ft (0.61 × 0.914 m) duct under a pressure of 3 in (7.62 cm) water gage. Will the horsepower required to move this air be: (*a*) 0.0566 (0.0422), (*b*) 0.0590 (0.0440), (*c*) 0.0626 (0.0467), or (*d*) 0.0674 (0.0503 kW)?

6.22 A fan whose efficiency is 40 percent has a capacity of 300,000 ft^3 (8496 m^3)/h at 60° (15.56°C), a barometer of 30 in (76.2 cm), and gives a pressure of 2 in (5.08 cm) of water column of full delivery. Should the size electric motor to drive this fan be: (*a*) 2 (1.49), (*b*) 2.5 (1.86), (*c*) 3 (2.24), or (*d*) 4 hp (2.98 kW)?

6.23   A refrigeration plant is rated at 20-ton (253,440-kJ) capacity.

1. Will the air per hour needed for cooling from 90 to 70°F (32.22 to 21.11°C) at constant pressure be: (1a) 40,000 (18,144), (1b) 50,000 (22,680), (1c) 55,000 (24,948), or (1d) 58,000 lb (26,309 kg)?

2. Will the approximate engine horsepower required to operate the plant be: (2a) 21.0 (15.66), (2b) 22.5 (16.78), (2c) 23.5 (17.52), or (2d) 24.0 hp (17.90 kW)?

6.24   A single-acting twin-cylinder 12 × 12 in (30.48 × 30.48 cm) compressor receives saturated ammonia vapor at 31.16 psia (214.85 kPa abs) and discharges it at 180 psia (1241.1 kPa abs). Saturated liquid enters the expansion valve. The water to be frozen is at 80°F (26.67°C) and the manufactured ice is at 16°F (−8.89°C). The speed of the compressor is 150 rpm and volumetric efficiency is 82 percent. The ratio of the ideal to brake work of the compressor is 80 percent. Assume specific heat of ice is 0.5 Btu/lb (1.164 kJ/kg).

1. Will the capacity be: (1a) 35.5 (32.21), (1b) 43.0 (39.01), (1c) 48.0 (43.55), or (1d) 50.65 short tons (45.95 Mg)?

2. Will the amount of ice per 24 h, neglecting radiation losses, be: (2a) 36.4 (33.02), (2b) 38.0 (34.47), (2c) 41.2 (37.38), or (2d) 46.0 tons (41.73 Mg)?

3. Will the brake horsepower of the compresser be: (3a) 70.0 (52.20), (3b) 71.2 (53.09), (3c) 73.6 (54.88), or (3d) 77.2 hp (57.57 kW)?

6.25   In the two-speed planetary reduction gear shown schematically in Fig. 6.25, either gear A or internal gear C can be clutched-in at will directly to a shaft rotating at constant speed, the other gear meanwhile being held stationary by a brake. (The clutch and brake mechanisms are not shown.) The gears are of 12 diametral pitch. Gear A has a pitch diameter of 3 in (7.62 cm), and planet gears BB each have 18 teeth. Will the ratio between the high and low speeds of the follower shaft D be (a) 1, (b) 1.5, (c) 2.0, or (d) 3?

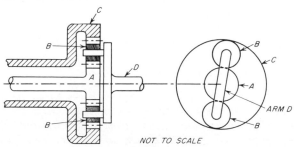

Fig. 6.25

6.26 An airplane engine having 14 cylinders and a 5⅜-in (13.65-cm) bore by 6-in (15.24-cm) stroke develops a torque of 2100 ft·lb (2847.2 N·m) at 2500 rpm, peak power. The clearance volume of each cylinder is 23.4 in³ (383.5 cm³). Total engine weight is 1450 lb (657.7 kg).

1. Is the compression ratio: (1a) 6.10:1, (1b) 6.46:1, (1c) 6.70:1, or (1d) 6.82:1?

2. Is the b_{mep} at peak power: (2a) 166 (1.145), (2b) 182 (1.255), (2c) 190 (1.310), or (2d) 194 psig (1.338 MPa gage)?

3. Is the bhp/lb (3a) 0.677 (1.113), (3b) 0.689 (1.133), (3c) 0.695 (1.143), or (3d) 0.698 hp/lb (1.148 kW/kg)?

6.27 *a.* A shaft made of SAE 1045 steel [ultimate = 97,000 psi (668.8 MPa) and yield point = 58,000 psi (399.9 MPa)] is subjected to a torque varying from 20,000 in·lb (2260 N·m) clockwise to 6000 in·lb (678 N·m) counterclockwise. What would be the diameter of the shaft if the factor of safety is 2 based on the yield point and the endurance strength in shear?

b. Find the diameter of the shaft when it transmits a constant torque of 20,000 in·lb (2260 N·m) and the factor of safety is 2, as in part *a*.

c. Explain the reason for the difference in diameters obtained in parts *a* and *b*.

6.28 A Corliss engine 28 in (71.12 cm) in diameter by 54-in (137.16-cm) stroke, driving a three-roller cane mill through double-reduction machine-cut spur-gearing runs at 60 rpm, receives live steam at the throttle at 150 psi (1034.25 kPa) saturated steam and exhausts against 15 psi (103.43 kPa) pressure. Valve gear is set for ⅜ cutoff. Piston rod is 5 in (12.7 cm) in diameter. Engine does not have tail rod. Assuming a transmission loss of 10 percent in each gear reduction and an additional loss of 15 percent for misalignments in the mill coupling and crownwheels:

1. Will the mean effective pressures at head and crank ends be: (1a) 96.8 (667.4), (1b) 88.8 (612.3), (1c) 82.3 (567.5), or (1d) 76.8 psia (529.5 kPa abs)?

2. Will the horsepower delivered by engine at crankshaft be: (2a) 583 (434.7), (2b) 683 (509.3), (2c) 733 (546.6), or (2d) 758 hp (565.2 kW)?

3. Will the horsepower delivered to mill be: (3a) 330 (246.1), (3b) 430 (320.7), (3c) 480 (357.9), or (3d) 505 hp (376.6 kW)?

6.29 A car engine rated at 12 hp (8.95 kW) gives a maximum torque of 6516 in·lb (736.3 N·m). The clutch is of a single-plate type, and both sides of the plate are effective. The coefficient of friction is 0.3, the axial pressure is limited to 12 psi (82.74 kPa), and the external radius of friction surface is 1.25 times the internal radius. Find the dimensions of the plate and the total axial pressure which must be exerted by the springs.

6.30 A four-cylinder engine is equipped with a flywheel which may be considered to have an effective weight of 50 lb (22.68 kg) concentrated in the rim at a mean radius of 7 in (17.78 cm). The excess energy delivered to the flywheel, which increases its speed from minimum to maximum, is estimated to be 20 percent of the indicated work per revolution. At an output of 20 hp (14.91 kW) the mean speed is 1800 rpm and the mechanical efficiency of the engine is 80 percent. Will the total speed variation be: (a) 6.4, (b) 8.8, (c) 10.0, or (d) 10.6 rpm?

6.31 A test on a 1-cylinder Otto 4-cycle (typical gasoline engine) internal-combustion engine yielded the following data: torque, 700 ft·lb (949.2 N·m); mean effective pressure, 110 psi (758.45 kPa); bore, 11 in (27.94 cm); stroke, 12 in (30.48 cm); speed, 300 rpm; fuel consumption, 24 lb (10.89 kg)/h; and lower heating value of fuel, 18,000 Btu/lb (41,904 kJ/kg).

1. Will the thermal efficiency of the engine be: (1a) 25, (1b) 26, (1c) 28, or (1d) 31 percent?

2. Will the mechanical efficiency be: (2a) 75.8, (2b) 80.6, (2c) 83.0, or (2d) 84.2 percent?

3. Will the fuel cost of horsepower per hour with fuel at $1.30/gal be: (3a) $0.125 ($0.167), (3b) $0.131 ($0.176), (3c) $0.135 ($0.181), or (3d) $0.137/hp ($0.184/kW)?

6.32 A built-up cylinder consists of an outer steel ring of thickness ⅛ in (3.175 mm) press-fitted on a copper ring of thickness ¹⁄₁₆ in (1.588 mm) as shown in Fig. 6.32. The copper ring has an outside diameter of 6 in (15.24 cm) and after assembly is under a compressive stress of 2000 psi

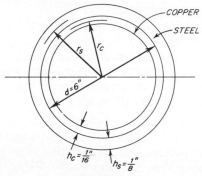

Fig. 6.32

(13,790 kPa). Steel density = 490 lb/ft³ (7848.3 kg/m³) and E = 30,000,000 psi (206.85 GPa) Copper density = 558 lb/ft³ (8937.5 kg/m³) and E = 16,000,000 psi (110.32 GPa). Will the stress in the copper ring become zero at a speed of: (a) 6450, (b) 6600, (c) 6675, or (d) 6740 rpm?

6.33 A screw press has the dimensions shown in Fig. 6.33. Two rods A are of steel 1.00 in (2.54 cm) in diameter. A brass tube B is 2.50 in (6.36 cm) in outside diameter and 1.50 in (3.81 cm) inside diameter. The screw is turned until the unit stress in the brass tube is 5000 psi (34,475 kPa) at 70°F (21.11°C). Assume no deformation of the heads of the press, or of the screw and bolts. The coefficients of thermal expansion for brass and steel are, respectively, 0.0000105 and 0.0000065/(unit)(°F)[0.0000189 and 0.0000117/(unit)(°C)]. The stress in the brass will be 10,000 psi (68,950 kPa) when the temperature reaches (a) 430 (221.1), (b) 454 (234.4), (c) 466 (241.1), or (d) 472°F (244.4°C)? E_{steel} = 30,000,000 psi (206.85 GPa), and E_{brass} = 14,500,000 psi (99.98 GPa).

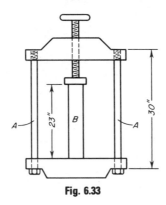

Fig. 6.33

6.34 A steam engine develops 80 hp (59.66 kW) at 150 rpm against a steady load. The flywheel weighs 3 tons (2721.6 kg), and its radius of gyration is 5 ft (1.524 m). If the load suddenly changes to 0.1 of the initial value and there is no change in steam supply for 5 revolutions after the reduction of the load, will the change in speed from beginning to end of this period be: (a) 3, (b) 7, (c) 10, or (d) 12 rpm?

6.35 1. An automobile motor has a torque of 230 ft·lb (311.9 N·m) driving the rear axles through a differential having a ratio of 4.11, and an efficiency of 97 percent. Will the tractive effort at the rear wheels, their diameters being 30 in (76.2 cm), be: (1a) 669 (2976), (1b) 705 (3136), (1c) 723 (3216), or (1d) 732 lb (3256 N)?

2. When the speed is 40 mph (64.37 km/h), will the rear wheel develop: (2*a*) 78.1 (58.24), (2*b*) 93.7 (69.87), (2*c*) 104.1 (77.63), or (2*d*) 109.3 hp (81.51 kW)?

6.36 A 6-cylinder automobile motor delivers 30 bhp (22.37 kW) at 1800 rpm, the ratio of stroke to bore being 1.4. Assume the mean effective pressure in the cylinder to be 90 psi (620.6 kPa) and the mechanical efficiency 85 percent.

1. Will the length of stroke be: (1*a*) 4.00 (10.16), (1*b*) 4.15 (10.54), (1*c*) 4.25 (10.80), or (1*d*) 4.30 in (10.92 cm)?

2. Will the bore be: (2*a*) 2.67 (6.78), (2*b*) 2.85 (7.24), (2*c*) 2.97 (7.54), or (2*d*) 3.03 in (7.70 cm)?

6.37 A composite shaft is made of solid-steel rod 1.50 in (3.81 cm) of outside diameter on which is shrunk an aluminum tube of 2.25 in (5.72 cm) outside diameter. E_{shear} for steel = 12,000,000 psi (82.74 GPa), and E_{shear} for aluminum = 3,500,000 psi (24.13 GPa). The shaft is 5 ft long and carries a torque of 15,000 in·lb (1690 N·m).

1. Will the total angle of twist of the shaft be: (1*a*) 0.0693, (1*b*) 0.0700, (1*c*) 0.0713, or (1*d*) 0.0736 rad?

2. Will the maximum shearing stress in the steel and in the aluminum be: (2*a*) 2260 (15.58), (2*b*) 4520 (31.30), (2*c*) 7460 (51.44), or (2*d*) 10,400 psi (71.71 MPa)?

6.38 A screw press, having two screws of ½-in (1.27-cm) pitch and exerting a pressure of 50 tons (444.8 kN), is to be driven by an electric motor running at 1700 rpm and connected to the screws by a suitable train of gears, the successive ratios of which are not to exceed 6 : 1. The pressure head is moved at the rate of 1 in (2.54 cm)/min. The overall efficiency is assumed to be 40 percent. Sketch a suitable gear train and calculate the horsepower delivered by the motor when the press is operating at 50-tons (444.8-kN) pressure.

6.39 Find suitable pitch and width of face for a pair of 14½° involute spur gears to transmit 8 hp (5.97 kW). The velocity ratio is to be 2 : 3, and the distance between centers is 10 in (25.4 cm). The pinion is to be of mild steel and is to run at 150 rpm. The gear is to be bronze.

6.40 A slugging or striking wrench is one designed for tightening bolted connections in restricted quarters with the aid of a sledge or heavy hammer. The free end of this wrench is hit with a hammer in a tangential direction while the socket engages the nut. This alloy-steel wrench is designed to accommodate the impact caused by a 5-lb (2.268-kg) hammer striking the free end of the wrench at a velocity of 12 fps (3.66 m/s). The wrench handle is 18 in (45.72 cm) long and of uniform diameter. Will the diameter of the circular section at the root of the wrench be: (a) 1.750 (4.45), (b) 1.875 (4.76), (c) 2.000 (5.08), or (d) 2.125 in (6.40 cm)?

6.41 A single-stage double-acting reciprocating air compressor is guaranteed to deliver 500 ft³ (14.16 m³)/min of free air with clearance of 3 percent and suction conditions of 14.7 psia (101.4 kPa abs) and temperature of 70°F (21.11°C) and discharge pressure of 105 psia (724.0 kPa abs). When tested under these conditions, it satisfied the manufacturer's guarantee, and test results show that compression and reexpansion curves follow pv^n = constant, with n = 1.34.

1. Will the piston displacement of the unit be: (1a) 500 (14.16), (1b) 555 (15.72), (1c) 590 (16.71), or (1d) 605 ft³ (17.13 m³)/min?

2. If the compression ratio is held constant and the unit is operated at an altitude of 6000 ft (1828.8 m) where barometric pressure and temperature are 23.8 inHg (60.45 cmHg) and 70°F, respectively, will the discharge pressure be: (2a) 83.5 (575.7), (2b) 87.0 (599.9), (2c) 94.0 (648.1), or (2d) 108.0 psia (744.7 kPa abs)?

3. Assuming that the bore, stroke, and clearance remain unchanged, what would you suggest doing to the compressor under part 2 to make it deliver the same weight of air per minute at the same discharge pressure as under guaranteed sea-level test conditions?

6.42 A dc motor-driven pump running at 100 rpm delivers 500 gal (1892.5 liters)/min of water against a total pumping head of 90 ft (27.43 m) with a pump efficiency of 60 percent.

1. Will motor horsepower required be: (1a) 15.18 (11.32), (1b) 17.43 (13.00), (1c) 18.93 (14.12), or (1d) 19.93 hp (14.86 kW)?

2. If the pump rpm is increased to produce a pumping head of 120 ft (36.58 m) with no change in efficiency.

    a. Will the rpm be: (2a) 105, (2b) 110, (2c) 114, or (2d) 116?

    b. Will the capacity be: (2e) 550 (2082), (2f) 580 (2195), (2g) 598 (2270), or (2h) 609 gal (2305 liters)/min?

3. Can a 25-hp motor be used under conditions indicated by part 2?

6.43 Will the weight which can be lifted by a screw jack that has an efficiency of 80 percent, if it is operated by a 50-lb (222.4-N) force at the end of a 30-in (762.2-cm) lever and the pitch of the screw is ½ in (1.27 cm), be: (a) 10,300 (4672), (b) 12,400 (5625), (c) 14,000 (6350), or (d) 15,100 lb (6849 kg)?

6.44 A hoist consists of a drum 24 in (60.96 cm) in diameter fastened to a gear wheel with 80 teeth. The gear wheel is turned by a double-threaded worm. The hoist can raise a 4000-lb (1814.4-kg) weight on a

⅞-in (22.23-mm) rope at the rate of 100 fpm (30.48 m/s). The mechanical efficiency is 60 percent.

1. Will the horsepower be: (1a) 20.2 (15.06), (1b) 21.4 (15.96), (1c) 23.8 (17.75), or (1d) 27.4 hp (20.43 kW)?

2. Will the torque be: (2a) 170 (230.5), (2b) 173 (234.6), (2c) 178 (241.4), or (2d) 187 ft·lb (253.6 N·m)?

6.45 A safety-valve spring having 9½ coils has the ends squared and ground. The outside diameter of the coil is 4½ in (11.43 cm), and the wire diameter is ½ in (1.27 cm). It has a free length of 8 in (20.32 cm).

1. Will the length to which this spring must be initially compressed to hold a boiler pressure of 205 psig (1,413.5 kPa gage) on a 1½-in (3.175-cm)-diameter valve seat, with $E_s = 11.5 \times 10^6$ psi (79.29 GPa), be: (1a) 6.56 (16.66), (1b) 6.61 (16.79), (1c) 6.66 (16.92), or (1d) 6.68 in (16.97 cm)?

2. If the spring is compressed to its solid length, will the stress in the spring be: (2a) 48,000 (331), (2b) 53,400 (368), (2c) 57,000 (393), or (2d) 58,800 psi (405 MPa)?

6.46 A reverted gear train is required to have an output-to-input speed ratio of exactly 1131 : 2000. The input and output gear shafts are colinear. Make a sketch of the gear train and specify the number of teeth on each gear. Use gears of not less than 20 nor more than 50 teeth.

1. Will the number of teeth of the input gear be: (1a) 39, (1b) 40, (1c) 41, or (1d) 42?

2. Will the number of teeth of the output gear be: (2a) 40, (2b) 45, (2c) 48, or (2d) 50?

6.47 Two helical gears of the same hand are used to connect two shafts that are 90° apart. The smaller gear has 24 teeth and a helix angle of 35°. The nominal circular pitch is 0.7854 in (19.95 mm). If the speed ratio is ½, will the center distance between the shafts be: (a) 14.000 (35.560), (b) 14.125 (35.876), (c) 14.375 (36.513), or (d) 14.750 in (37.465 cm)?

6.48 Body 1, a 2-in-diameter disk, is driven by body 2. Body 2 is a 2-in square rotating clockwise at a constant speed of 10 rad/s as shown in Fig. 6.48. When the center of the circle is vertically upward:

1. Will the angular velocity of body 1 be: (1a) 8.5, (1b) 9.0, (1c) 9.5, or (1d) 10.0 rad/s?

2. Will the rotational direction be: (2a) clockwise or (2b) counterclockwise?

3. Will the angular acceleration of body 1 be: (3a) 471.7, (3b) 472.2, (3c) 473.2, or (3d) 475.0 rad/s²?

4. Will the rotational direction be: (4a) clockwise or (4b) counterclockwise?

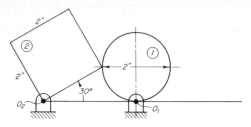

Fig. 6.48

6.49 A rotor has two unbalances as shown in Fig. 6.49. Determine the necessary ounce-inch corrections and their angular positions in the end planes if the rotor is to be dynamically balanced.

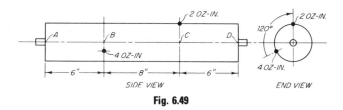

Fig. 6.49

6.50 Figure 6.50 is a schematic diagram of a viscous-damped phonograph arm. What is the value of the damping constant C if the arm is to be critically damped?

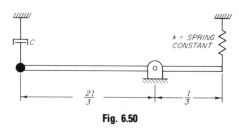

Fig. 6.50

6.51 A rectangular steel beam is subjected to a bending moment of 240,000 lb·ft (325,440 N·m). The yield point of the steel is equal to 35,000 psi (241.3 MPa). If the beam is 6 in (15.24 cm) wide and 8 in (20.32 cm) deep, and the theoretical stress-strain diagram as shown in Fig. 6.51 is used, will the distance from the outer surface that a plastic state exists in the beam be: (*a*) 1.381 (3.51), (*b*) 1.431 (3.63), (*c*) 1.461 (3.71), or (*d*) 1.471 in (3.74 cm)?

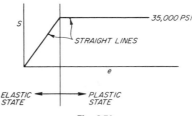

Fig. 6.51

6.52 4,000 gal/min (0.253 m³/s) of oil having a specific gravity of 0.85 flows through a clear wrought-iron pipe for a distance of 10,000 ft (3048 m) with an energy loss of 75 ft·lb/lb (22.86 m·kg/g). Upstream and downstream elevations are 200 and 50 ft (60.96 and 15.24 m), respectively. The pressure upstream is 40 psi (275.8 kPa). The oil has a kinematic viscosity of 0.0001 ft² (0.00000929 m²)/s.

1. Will the theoretical inside diameter of the pipe be: (1*a*) 16.0 (40.64), (1*b*) 16.6 (42.16), (1*c*) 17.0 (43.18), or (1*d*) 17.2 in (43.69 cm)?

2. Will the downstream pressure be: (2*a*) 60.0 (413.7), (2*b*) 62.5 (430.9), (2*c*) 65.0 (448.2), or (2*d*) 67.5 psi (465.4 kPa)?

6.53 Tests of airfoils in a wind tunnel are to be used to determine the lift and drag of hydrofoils with a 6-in (15.24-cm) chord length for a boat designed to travel at 60 fps (18.29 m/s) in water at 70°F (21.1°C). To avoid compressibility effects in the wind tunnel, assume standard air is to be used at a maximum velocity of 200 fps (60.96 m/s). The hydrofoils are to run deep, so that only viscous effects need be considered. For the model airfoils, will the chord length L_m be: (*a*) 2.00 (60.96), (*b*) 2.24 (68.28), (*c*) 2.36 (71.93), or (*d*) 2.42 ft (73.76 cm)?

6.54 The curved beam shown in Fig. 6.54 (p. 140) has a uniform thickness of 1 in (2.54 cm). If the resultant stresses normal to the vertical cross section *A*-*B* are numerically equal at points *A* and *B*, will the vertical force *F* be: (*a*) 1400 (6227), (*b*) 1424 (6334), (*c*) 1440 (6405), or (*d*) 1448 lb (6441 N)?

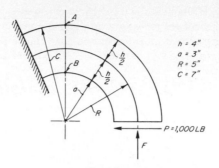

Fig. 6.54

6.55 An E6010 electrode is used and the factor of safety by the maximum shear theory of failure is to be equal to 2. For the welded connection in Fig. 6.55, will the permissible load P be: (a) 25,000 (111.2), (b) 27,200 (121.0), (c) 28,100 (125.0), or (d) 28,650 lb (127.4 kN)?

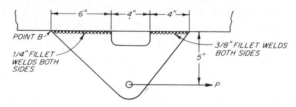

Fig. 6.55

STRUCTURAL ENGINEERING

7.01 A steel hanger consists of a plate $5 \times \frac{3}{4}$ in $(127 \times 6.35$ mm$)$ held at the top by four $\frac{3}{4}$-in $(19.05$-mm$)$ rivets in single shear. The rivets are arranged in diamond form so that the first net section has one hole and the second net section has two holes. Use unit stresses of 16,000 psi (110.32 MPa) for tension, 12,000 psi (82.74 MPa) for shear, and 24,000 psi (165.48 MPa) for bearing.

1. Will the joint fail in: (1a) shear, (1b) bearing, (1c) tension through section 1-1, or (1d) tension through section 2-2?

2. Will the efficiency of the joint be: (2a) 80.0, (2b) 83.3, (2c) 85.0, or (2d) 85.8 percent?

7.02 A tension member is made up of $\frac{3}{8}$-in (9.53-mm) thick steel plates 9 in (228.6 mm) wide and spliced by a lap joint made with three rows of $\frac{3}{4}$-in (19.05-mm) bolts snug fit in drilled holes and arranged in the pattern shown in Fig. 7.02. The permissible stresses in the bolts and the plates are 20,000 psi (137.90 MPa) for tension in the net section, 16,000 psi (110.32 MPa) for shear, and 32,000 psi (220.64 MPa) for bearing.

1. Will the maximum permissible tension in the plates be: (1a) 56,200 (249, 978), (1b) 57,000 (253,540), (1c) 60,000 (266,880), or (1d) 67,500 lb (300,240 N)?

2. Will the efficiency of the joint be: (2a) 80.0, (2b) 83.3, (2c) 85.0, or (2d) 85.8 percent?

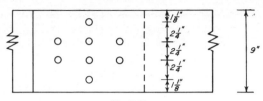

Fig. 7.02

7.03 A double-covered riveted butt seam has main plates ½ in (12.7 mm) thick and ⁵⁄₁₆-in (7.94-mm) cover plates. The rivets are ¾ in (19.05 mm) in diameter [the holes ¹³⁄₁₆ in (20.65 mm)] and are arranged in four rows, two on each side of the seam. The pitch in the rows next to the seam is three rivet diameters, and in the outer rows six rivet diameters. The permissible working stresses in the rivets and plates are 16,000 psi (110.32 MPa) for tension, 10,000 psi (68.95 MPa) for shear, and 20,000 psi (137.90 MPa) for bearing.

1. Will the allowable load per inch (25.4 mm) be: (1a) 4000 (17,792), (1b) 5000 (22,240), (1c) 5500 (24,464), or (1d) 5750 lb (25,576 N)?

2. Will the working efficiency of the seam be: (2a) 71.2, (2b) 66.2, (2c) 62.5, or (2d) 60.0 percent?

7.04 A 5-ft (15.24-m) diameter tank is to be designed to withstand an internal working pressure of 500 psi (3448 kPa). All rivets are to be the same size for both the transverse and longitudinal double-cover-plated butt joints. Rivet holes are drilled ¹⁄₁₆-in (1.59-mm) larger diameter than the rivets. Working stresses for plates and rivets are 20,000 psi (137.90 MPa) for tension; 15,000 psi (103.425 MPa) for shear; and 24,000 psi (165.48 MPa) s.s., 30,000 psi (206.85 MPa) d.s. for bearing. If the longitudinal rivet pattern is as shown in Fig. 7.04,

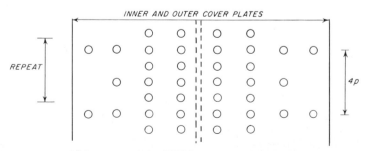

Fig. 7.04

    1. Will the pitch for the inside row of rivets be: (1*a*) 3.0 (76.2), (1*b*) 3.5 (88.9), (1*c*) 4.0 (101.6), or (1*d*) 4.5 in (114.3 mm)?

    2. Will the thickness of the main plate be: (2*a*) 0.75 (19.05), (2*b*) 0.8125 (20.64), (2*c*) 0.875 (22.23), or (2*d*) 1.00 in (25.4 mm)?

    3. Will the thickness of the splice plates be: (3*a*) 0.4375 (11.11), (3*b*) 0.500 (12.70), (3*c*) 0.5625 (14.29), or (3*d*) 0.625 in (15.88 m)?

    4. Will the efficiency of the joint be: (4*a*) 88.4, (4*b*) 90.0, (4*c*) 90.8, or (4*d*) 91.2 percent?

7.05   For the transverse plate splice in question 7.04 and for the rivet pattern in Fig. 7.05,

    1. Will the rivet pitch be: (1*a*) 4.75 (120.65), (1*b*) 5.00 (127.00), (1*c*) 5.25 (133.35), or (1*d*) 5.50 in (139.70 mm)?

    2. Will the splice plate thickness be: (2*a*) 0.250 (6.35), (2*b*) 0.3125 (7.94), (2*c*) 0.375 (9.53), or (2*d*) 0.4375 in (11.11 mm)?

    3. Will the efficiency of the splice be: (3*a*) 47.4, (3*b*) 55.6, (3*c*) 59.8, or (3*d*) 61.9 percent?

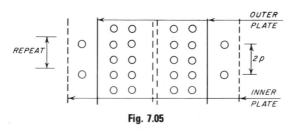

Fig. 7.05

7.06   The longitudinal quadruple-riveted butt joint in Fig. 7.06 has ¾-in (19.05-mm)-thick main plates and ½-in (12.70-mm) splice plates. Rivets are 1⅛ in (28.58 mm) in diameter and holes are 1³⁄₁₆ in (30.16

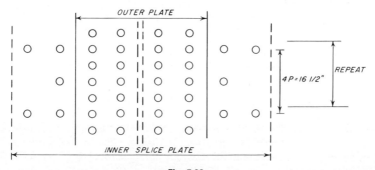

Fig. 7.06

mm). If this splice were designed by using boiler code unit stresses of 11,000 psi (75.845 MPa) in tension, 8800 psi (60.676 MPa) in shear, and 19,000 psi (131.005 MPa) in bearing,

1. Will the splice fail in: (1a) shear, (1b) bearing, (1c) tension in the main plate, or (1d) tension in the inner cover plate?

2. Will the efficiency of the splice be: (2a) 76.0, (2b) 80.0, (2c) 82.0, or (2d) 83.0 percent?

7.07 Design a longitudinal lap joint for a tank 72 in (1.829 m) in diameter carrying 125 psi (862 kPa) internal pressure. The allowable working stresses are 16,000 psi (110.32 MPa) in tension, 12,000 psi (82.74 MPa) in shear, and 24,000 psi (165.48 MPa) in bearing.

1. Will the thickness of the tank plate be: (1a) ⅜ (9.53), (1b) ⁷⁄₁₆ (11.11), (1c) ½ (12.70), or (1d) ⁹⁄₁₆ in (14.29 mm)?

2. Will the splice fail in: (2a) shear, (2b) bearing, or (2c) tension?

3. Will the efficiency of the splice be: (3a) 58.0, (3b) 64.0, (3c) 67.0, or (3d) 68.5 percent?

7.08 Design a welded joint as in Fig. 7.08 to develop the full strength of the angle in tension. The allowable stress in the throat of the fillet welds is 15,000 psi (103.43 MPa), and the tensile stress in the angle is 20,000 psi (137.90 MPa).

1. If all welds are the same size, will the total theoretical length of welds be: (1a) 15.0 (381.0), (1b) 16.0 (406.4), (1c) 17.0 (431.8), or (1d) 18.0 in (457.2 mm)?

2. Will the theoretical length of each of the two welds along the back of the 3-in (76.2-mm) leg of the angle be: (2a) 5.10 (129.5), (2b) 5.46 (138.7), (2c) 5.70 (144.8), or (2d) 5.86 in (148.9 mm)?

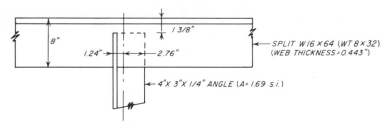

Fig. 7.08

7.09 Design a welded joint to develop the full strength of the T in tension (Fig. 7.09). The allowable stresses in fillet welds are 2000 lb/in (350 N/mm) for ¼-in (6.35-mm) fillet weld; 2500 lb/in (438 N/mm) for ⁵⁄₁₆-in (7.94-mm); 3000 lb/in (525 N/mm) for ⅜-in (9.53-mm); and 3500 lb/in

(613 N/mm) for ⁷⁄₁₆-in (11.11-mm). Tensile stress in the T is 20,000 psi (137.90 MPa).

1. If one ¼-in (6.35-mm) weld is used along the stem of the T, will its theoretical length be: (1a) 5.50 (139.7), (1b) 5.80 (147.3), (1c) 6.10 (154.9), or (1d) 6.40 in (162.6 mm)?

2. If three ⅜-in (9.53-mm) welds are used along the flange of the T, will the theoretical length of each of the three welds be: (2a) 5.31 (134.9), (2b) 5.47 (138.9), (2c) 5.55 (141.0), or (2d) 5.59 in (142.0 mm)?

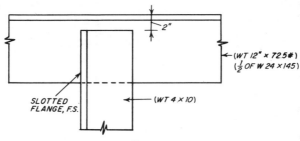

Fig. 7.09

7.10 A single plate 9 × 26 in (228.5 × 660.4 mm) is fastened to the flange of a column with six bolts snug fit in drilled holes and arranged as shown in Fig. 7.10. The working stresses in the bolts and the plate may be 16,000 psi (110.32 MPa) in shear, 32,000 psi (220.62 MPa) in bearing, and 20,000 psi (137.90 MPa) in tension. If the maximum load on a corner rivet shall not exceed 9500 lb (42,256 N):

1. Will the load P be: (1a) 8200 (36,474), (1b) 8850 (39,365), (1c) 9250 (41,144), or (1d) 9500 lb (42,256 N)?

2. Will the size of the bolts be: (2a) ⅝ (15.88), (2b) ¾ (19.05), (2c) ⅞ (22.23), or (2d) 1 in (25.40 mm)?

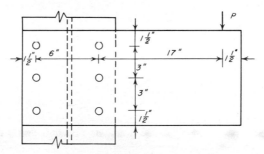

Fig. 7.10

3. Will the thickness of the plate be: (3*a*) ⅝ (15.88), (3*b*) ¹¹⁄₁₆ (17.46), (3*c*) ¾ (19.05), or (3*d*) ¹³⁄₁₆ in (20.64 mm)?

7.11 For the bracket shown in Fig. 7.11:

1. Will the force in the outermost rivet under combined shear and torsion be: (1*a*) 8200 (36,474), (1*b*) 8700 (38.698), (1*c*) 9100 (40,477), or (1*d*) 9400 lb (41,811 N)?

2. Will the maximum stress in the plates be: (2*a*) 16,850 (116.18), (2*b*) 17,300 (119.28), (2*c*) 18,200 (125.49), or (2*d*) 20,000 psi (137.90 MPa)?

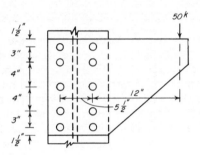

Fig. 7.11 Two ½-in plates, one on each flange of column. All rivets are 1 in diameter.

7.12 Find the force per inch (25.4 mm) in the outermost inch of fillet weld (Fig. 7.12) under combined shear and torsion of the following bracket of one plate and 32 in (812.8 mm) of ¼-in (6.35-mm) fillet weld.

Is this a good arrangement of welds?

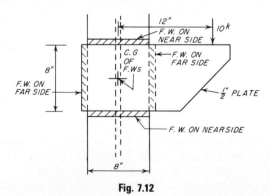

Fig. 7.12

7.13   An A4 standard beam connection is made up of two angles $4 \times 3\frac{1}{2} \times \frac{3}{8}$ in (101.6 × 88.9 × 9.53 mm), with 4-in (101.6-mm) legs outstanding. Rivets are $\frac{7}{8}$ in (22.23 mm) in diameter. Will the tension in the top rivet under a maximum shear loading be: (a) 4650 (20,683), (b) 4950 (22,018), (c) 5150 (22,907), or (d) 5250 lb (23,352 N)?

7.14   An A10 standard beam connection is made up of two angles $4 \times 3\frac{1}{2} \times \frac{7}{16}$ in (101.6 × 88.9 × 11.11 mm), with 4-in (101.6-mm) legs outstanding. Rivets are $\frac{7}{8}$ in (22.23 mm) in diameter. Will the tension in the top rivet under a maximum shear loading be: (a) 2110 (9385), (b) 2550 (11,342), (c) 2770 (12,321), or (d) 2880 lb (12,810 N)?

7.15   A bracket in the form of a plate $12 \times \frac{5}{8} \times 18$ in (304.8 × 15.88 × 457.2 mm) is to be welded to the 12 × 1 in (304.8 × 25.4 mm) flange of a WF column section. The plate is to be placed with the 18-in (457.2-mm) side horizontal and symmetrically located with respect to the centerline of the flange. The bracket is to support a pair of 100,000-lb (444,800-N) vertical forces, each spaced 16 in (406.4 mm) apart and symmetrically located with respect to the flange. If the allowable shearing stress through the throat of the weld is 13,000 psi (89.64 MPa) and $\frac{1}{2}$-in (12.7-mm) welds are used, will the shearing stress in the weld be: (a) 4200 (736), (b) 4600 (806), (c) 4800 (841), or (d) 4900 lb/in (858 N/mm)?

7.16   The main material of a plate girder consists of a web plate $42 \times \frac{3}{8}$ in (1066.8 × 9.5 mm), four angles $6 \times 4 \times \frac{5}{8}$ in (152.4 × 101.6 × 15.9 mm) placed $42\frac{1}{2}$ in (1079.5 mm) back to back of 6-in (152.4-mm) legs, and two cover plates $14 \times \frac{1}{2}$ in (355.6 × 12.7 mm). If the span is 30 ft, (9.144 m),

1. Will the maximum permissible spacing of $\frac{3}{4}$-in (19.05-mm) rivets in the vertical legs of the flange angles at a point where the total shear is 75,000 lb (333,600 N) be: (1a) 3.0 (76.2), (1b) 4.5 (114.3), (1c) 6.0 (152.4), or (1d) 7.15 in (181.6 mm)?

2. Will the total uniform load that can be placed on this girder be: (2a) 9200 (134,256), (2b) 10,400 (151.767), (2c) 11,000 (160,523), or (2d) 11,292 lb/ft (164,784 N/m)?

7.17   A simply supported 50-ft (15.24-m)-long plate girder $36\frac{1}{2}$ in (927.1 mm) deep back to back of angles and without cover plates is to be designed for a single moving concentrated load of 70 kips (311,360 N) (including impact). The top flange is considered to be laterally supported. The web plate has already been selected at $36 \times \frac{3}{8}$ in (914.4 × 9.5 mm). Select proper flange angles and compute the maximum end of girder flange to web rivet pitch for $\frac{3}{4}$-in (19.05-mm) rivets. The bending stress is 18,000 psi (124.11 MPa), shear is 15,000 psi (103.43 MPa), and bearing is 40,000 psi (275.8 MPa).

7.18 A plate girder having a simple span of 60 ft (18.29 m) carries a total uniform load of 5000 lb/lin ft (72,965 N/m). Design a section to take the maximum moment.

7.19 In Fig. 7.19, a channel and a W beam form a T beam of 30-ft (9.144-m) simple span. What concentrated load P may be supported at the center of the T beam if the shear in the rivets is not to exceed 15,000 psi (103,43 MPa), the fiber stress in the W is not to exceed 20,000 psi (137.90 MPa) in tension, compression in the channel is not to exceed 17,050 psi (117.56 MPa), and the deflection for the concentrated load is not to exceed $\frac{1}{360}$ of the span? *Channel*: weight = 33.9 lb, d = 15 in, a = 9.90 in², I = 312.6 and 8.2 in⁴, x = 0.79 in, web t = 0.400 in; *W section*: d = 18.16 in, a = 28.22 in², I = 1674.7 and 206.8 in⁴, weight = 96 lb; *rivets*: diameter = ⅝ in, spacing = 6-in centers longitudinally.

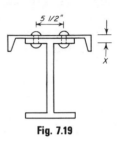

Fig. 7.19

7.20 A crane-runway girder of 21-ft (6.40-m) span is simply supported at the ends. The girder carries a pair of moving wheel loads with axles 11 ft 4 in (3.453 m) on centers. Each wheel load carries 38,500 lb (171,248 N). The girder section is made up of a W 24 × 80, on top of which is fastened a 12-in × 20.7-lb (304.8-mm × 9.39-kg) channel, with the flanges turned down. Two rows of ¾-in (19.05-mm) diameter rivets, one on each side of the top beam flange, hold the two sections together.

Allowable rivet shearing stress = 15,000 psi (103.43 MPa), bearing stress = 32,000 psi (220.64 MPa). *W section*: depth = 24 in, area = 23.54 in², I = 2230 and 82.4 in⁴, flange thickness = 0.727 in, web thickness = 0.455 in; *channel*: d = 12 in, a = 6.03 in², I = 128.1 and 3.9 in⁴, x = 0.70 in, flange t = 0.501 in, and web t = 0.280 in.

1. Due to vertical loading only, will the tension in the W section and/or the compression in the channel be: (1*a*) 9130 (62.95), (1*b*) 13,170 (90.81), (1*c*) 16,390 (113.01), or (1*d*) 18,000 psi (124.11 MPa)?

2. Will the maximum permissible rivet pitch for the two rows of flange rivets be: (2*a*) 3.0 (76.2), (2*b*) 4.5 (114.3), (2*c*) 6.0 (152.4), or (2*d*) 12.0 in (304.8 mm)?

7.21 A steel beam 24 ft (7.32 m) long carrying a total uniformly distributed load of 10,000 lb/ft (145.9 kN/m) is supported at the two ends and the center. The center reaction settles 0.50 in (12.7 mm). I of beam = 1373 in⁴ (57,148 cm⁴), and E of beam = 30,000,000 psia (206.9 GPa). Will the change in bending moment over the center support be: (*a*) 207 (280.7), (*b*) 231 (313.2), (*c*) 243 (329.8), or (*d*) 249 ft·kips (337.6 kN·m)?

7.22   A simply supported beam has a rectangular cross section 1.0 in (25.4 mm) wide, 6.0 in (152.4 mm) deep, and 6 ft long (1.83 m). It carries loads of 4800 (21,350) and 2400 lb (10,675 N) located 2 ft (0.61 m) from the left and right ends, respectively. At a section 1.5 ft (0.46 m) from the left end and 1.5 in (38.1 mm) from the bottom face,

    1. Will the maximum normal stress be: (1*a*) 5910 (40.75), (1*b*) 5970 (41.16), (1*c*) 6030 (41.58), or (1*d*) 6090 psi (41.99 MPa)?

    2. Will the maximum normal stress be: (2*a*) tension or (2*b*) compression?

    3. Will the minimum normal stress be: (3*a*) 90 (621), (3*b*) 100 (690), (3*c*) 110 (758), or (3*d*) 120 psi (827 kPa)?

    4. Will the minimum normal stress be: (4*a*) tension or (4*b*) compression?

    5. Will the maximum shearing stress be: (5*a*) 2960 (20.41), (5*b*) 3030 (20.89), (5*c*) 3090 (21.31), or (5*d*) 3140 psi (21.65 MPa)?

7.23   What would be the sizes required for the three W steel beams of the horizontal frame in question 1.618 if only the tank loads are considered? What AISC standard end connections are needed? Neglect deflection considerations.

7.24   Find the amount and direction of the deflection (in terms of EI) at the 4P load (Fig. 7.24).

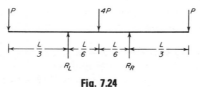

Fig. 7.24

7.25   Find the location and amount of the maximum deflection (in terms of EI) for the beam in Fig. 7.25.

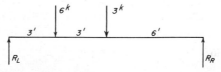

Fig. 7.25

7.26 A 20-ft (6.096-m) beam of I moment of inertia has top and bottom cover plates over the center 10 ft (3.048 m) of the beam, giving a moment of inertia of $2I$ for the builtup section. The beam carries a uniform load of 1 kip/ft (14,593 N/m) over the full length of the beam and 10 kips (44.48 kN) of load concentrated at the center. Find the deflection at the center in terms of EI.

7.27 An elevated water tank has legs consisting of two channels, one cover plate, and one open side adequately latticed. These legs are continuous for a height of 100 ft (30.48 m), but they are fixed at their ends and braced both ways at each 20-ft (6.096-m) interval. The allowable unit stress is given by $f_s = 17,000 - 0.485\,(L/r)^2$. One plate is 13 × ⅜ in (330.2 × 9.53 mm). There are two 8-in (203.2-mm) channels (Fig. 7.27), 11.5 lb (5.22 kg) each.

Fig. 7.27

Properties of one channel are $A = 3.36$ in², $I_{xx} = 32.3$ in⁴, $I_{yy} = 1.3$ in⁴, $x = 0.58$ in. Will the maximum allowable load on one leg be: (a) 149,600 (665.4), (b) 158,000 (702.8), (c) 162,200 (721.5), or (d) 164,300 lb (730.8 kN)?

7.28 A top chord member of a bridge truss is made up of two 15-in (381-mm) 50-lb (22.7-kg) channels [placed 12 in (305 mm) clear distance with flanges turned out], with one 22 × ¾ in (559 × 19 mm) cover plate and one open side adequately latticed. The member is supported in either a horizontal or vertical plane by members or bracing at panel points 25 ft (7.62 mm) on centers, and the chord may be considered fixed at these points. Is the safe axial load, applied at the center of gravity of the cross section, that the chord may carry (neglecting deflection), when the unit stress is governed by the formula $15,000 - L^2/4r^2$, be: (a) 565,000 (2513), (b) 625,000 (2780), (c) 655,000 (2913), or (d) 680,000 lb (3025 kN)?

7.29 A 15-ft (4.57-m) H column is made up of four angles 4 × 3 × ⅜ in (101.6 × 76.2 × 9.53 mm) and a web plate 8 × ⅜ in. The angles are 8½ in (215.9 mm) back to back, with long legs outstanding. Use AISC column formula $F_y = 36,000$ psi. Will the permissible concentric load be: (a) 137,600 (612.0), (b) 157,600 (701.0), (c) 167,600 (745.5), or (d) 172,600 lb (767.7 kN)?

7.30 A steel column unbraced for a length of 20 ft (6.096 m) is made up of a W 14 × 287 section and two cover plates 22 × 2 in (558.8 × 50.8 mm). The column is braced against sidesway and its ends are unrestrained. Will this column carry a concentrated load of 1250 kips (5560 kN) placed upon the axis of the web of the W section but acting upon the inner face of one of the cover plates? Use the AISC specifications.

For the W 14×287, $A = 84.37$ in^2, $d = 16.81$ in, $I_{1-1} = 3912$ in^4, $I_{2-2} = 1466.5$ in^4.

7.31 A W 14×167 column is subjected to a total axial load of 500 kips (2224 kN), a moment of 220 ft·kips (298.3 kN·m) at the top and a moment of 200 ft·kips (271.2 kN·m) at the bottom. The column is 15 feet 0 in (4.57 m) long and subject to sidesway. The effective length in the plane of loading has been found to be $K_x = 1.80$. In the perpendicular plane there is bracing and $K_y = 0.75$. Determine if the column is adequate. Use A36 steel and the AISC code.

7.32 The bottom chord of a roof truss is to be made of two steel angles with vertical legs ½ in (12.7 mm) clear back to back. The angles are 10 ft (3.048 m) long center to center of truss panel points and carry a 65,000-lb (289.1-kN) load in tension. The angles are connected to a ½-in (12.7-mm) thick gusset plate at each end by a single row of ⅞-in (22.22-mm) rivets passing through the vertical legs. Assuming two angles $8 \times 4 \times 1$ in ($203.2 \times 101.6 \times 25.4$ mm) are to be used and considering both the eccentricity of the connection and bending due to the weight of the angles, will the maximum tensile stress be: (*a*) 20,140 (138.9), (*b*) 21,070 (145.3), (*c*) 21,690 (149.6), or (*d*) 22,000 psi (151.7 MPa)?

7.33 The bottom chord of a bridge truss is to be made of two channels with flanges outstanding and webs vertical and 12 in (304.8 mm) clear distance apart. Assume two 15-in (381-mm) channels at 33.9 lb/ft (50.4 kg/m), adequately laced together and held with ⅞-in (22.22-mm) rivets. The length of the chord is 25 ft (7.62 m) center to center of panel points. If the chord is subjected to a 350,000-lb (1557-kN) axial load and to bending due to its own weight, will the maximum stress be: (*a*) 22,000 (151.7), (*b*) 21,940 (151.3), (*c*) 21,820 (150.3), or (*d*) 21,580 psi (148.8 MPa)?

7.34 The top chord of a roof truss, having a very slight slope, is to be made of a structural T (WT) section and is continuous over panel points 8 ft (2.44 m) on centers. Purlins bring 4000-lb (17.8-kN) concentrations of load to the roof truss at the panel points and halfway between these points. This is to be a welded truss with center of gravity of members matching the joint-line diagram of the truss. Design the chord member, by AISC standards, to carry an axial compressive load of 60,000 lb (266.9 kN).

7.35 If the top chord in question 7.34 had carried a compressive load of 75,000 lb (333.6 kN) and been made up of an angle and plate set symmetrically on each side of ½-in (12.7-mm) guesset plates and fastened with ⅞-in (22.22-mm) rivets, what would be a satisfactory cross section for the chord?

7.36 The top chord of a roof truss has a slope of 1 vertical to 2 horizontal. What size standard I-beam purlin is required to satisfy the following conditions if the combined bending stress is not to exceed 20,000 psi (137.9 MPa)?

Span of purlin is 20 ft (6.10 m). Purlin is simply supported at its ends, with the bottom flange resting on top chord of roof truss.

The purlin is supported in the plane of the top chord of the roof truss by sag rods at the third points of the purlins running to the ridge purlins. There is a load normal to the top flange of the purlin of 90 lb/ft (1313 N/m) of purlin and a vertical load of 160 lb/ft (2335 N/m) including the weight of the purlin.

7.37 Design a beam 24 ft 0 in (7.32 m) long to support a uniformly distributed load of 5 kips/ft (72.96 kN/m). Lateral support is provided only at the ends. Use A36 steel and the AISC code. Will the beam be: (a) W 14 × 136, (b) W 14 × 142, (c) W 14 × 150, or (d) W 14 × 158?

7.38 If a 1½-in (38.1-mm) diameter bolt is subjected to a tension of 20,000 lb (88.96 kN) and a torque (for tightening) of 6000 in·lb (678 N·m), will the normal stress be: (a) 23,750 (163.8), (b) 24,250 (167.2), (c) 24,500 (168.9), or (d) 24,625 psi (169.8 MPa)?

7.39 A 4 × 6 in (101.6 × 152.4 mm) timber beam has a 4 × ½-in (101.6 × 12.7 mm) steel plate connected to the lower surface. E_s = 30,000,000 psi (206.8 GPa) and E_w = 1,500,000 psi (10.34 GPa). Using the nominal sizes for the timber when the wood is stressed to 600 psi (4.14 MPa), will the steel be stressed to: (a) 3502 (24.15), (b) 4000 (27.58), (c) 4249 (29.30), or (d) 4374 psi (30.16 MPa)?

7.40 W24 × 160 stringers are spaced 5 ft (1.52 m) on centers and have a 7-in (177.8-mm) concrete slab bonded to the top flange of the steel stringers by Z-bar shear adapters. The top of the concrete is 5.50 in (139.7 mm) above the top of the W. The stringers make 40-ft (12.19-m) simple spans and carry the full dead weight. If the stress in the steel is not to exceed 20,000 psi (138.0 MPa) and the stress in the concrete is 1350 psi (9.31 MPa), will the composite section carry a static concentrated load at the center of each stringer equal to: (a) 63,705 (283.4), (b) 68,445 (304.4), (c) 70,815 (315.0), or (d) 72,000 lb (320.3 kN)?

7.41 In a composite beam in question 7.40, the end shear could be 33,000 lb (146.8 kN). If the Z-bar shear adapters are fastened to the top flange of the W stringer with two ⅞-in (22.22-mm) rivets, will the maximum spacing at the end of the beam be: (a) 10 (254.0), (b) 14 (355.6), (c) 16 (406.4), or (d) 17 in (431.8 mm)?

7.42 A vertical retaining wall made of reinforced concrete is 18 ft (5.49 m) high and rigidly supported at the base. The wall is loaded by earth pressure that varies from 0 at the top to 540 psf (25.86 kPa) at the base. Use $f_c' = 3500$ psi (24.13 MPa), $f_y = 50,000$ psi (344.8 MPa), and ACI specifications. Use ultimate-strength method and a 1-ft (0.305-m) wall length.

    1. Will the reinforcement per foot (0.305 m) of wall at its base be: (1a) 1 (645), (1b) 1.2 (774), (1c) 1.4 (903), or (1d) 1.6 in² (1032 mm²)?

    2. Must the thickness of the wall at the base be: (2a) 10.0 (254.0), (2b) 10.5 (266.7), (2c) 11.0 (279.4), or (2d) 12.0 in (304.8 mm)?

7.43 A reinforced concrete slab is simply supported on two parallel supports 10 ft (3.05 m) center to center and carries a uniform live load of 400 psf (19.15 kPa). Design the slab by ACI specifications if $f_c' = 3000$ psi (20.69 MPa) and $f_y = 50,000$ psi (344.8 MPa). Use ultimate-strength method.

7.44 A concrete beam as in Fig. 7.44 has a simple span of 15 ft (4.57 m). It was designed according to ACI specifications, with $f_c' = 3000$ psi (20.69 MPa) and $f_y = 40,000$ psi (275.8 MPa). Assuming that the web reinforcement is satisfactory, what concentrated live load may be carried at the center of the span if the deflection due to this load is not to exceed ⅟₈₀₀ of the span? Use ultimate-strength method and the ACI specifications.

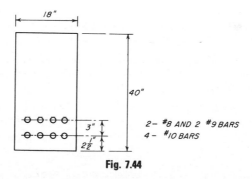

Fig. 7.44

7.45 Design the section of a single reinforced concrete beam that is simply supported on a 20-ft 0-in (6.10-m) span and carries a dead load of 0.60 kips/ft (8.76 kN/m) (including an estimated value for its own weight) and a live load of 1.2 kips/ft (17.51 kN/m). Use the ACI code. Use f'_c = 4000 psi (27.59 MPa) and f_y = 50,000 psi (344.8 MPa). Use ultimate-strength method.

7.46 Design a single reinforced, three-span continuous beam, each span 20 ft (6.10 m) long. The beam carries a dead load of 0.60 kips/ft (8.76 kN/m) (including an estimated value for its own weight) and a live load of 1.2 kips/ft (17.51 kN/m). Use the ACI code. Use f'_c = 4000 psi (27.59 MPa) and f_y = 50,000 psi (344.8 MPa). Use ultimate-strength method.

7.47 The double-reinforced concrete beam in (Fig. 7.47) is now to carry a maximum bending moment of 50,000 ft·lb (67.78 kN·m). The concrete tested better than 3000 psi (20.69 MPa). Steel may be stressed to 20,000 psi (137.9 MPa). Assume adequate web reinforcement. Is this bending moment permissible? Use the alternate method of ACI.

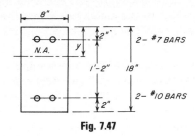

Fig. 7.47

7.48 A simply supported concrete beam 12 in (304.8 mm) wide by 15.5 in (393.7 mm) deep to the centroid of two rows of tension steel is to carry a dead load moment of 75 ft·kips (101.7 kN·m), including an estimate of its own weight, and a live load moment of 120 ft·kips (162 kN·m). If f'_c = 5000 psi (34.48 MPa) and f_y = 50,000 psi (344.8 MPa), design the required flexural steel reinforcement using ultimate-strength methods and the ACI specifications.

7.49 The dimensions for the concrete beam designed in question 7.45 must be retained, but the beam must now carry an additional live load of 600 lb/ft (8.76 kN/m). What will be the new schedule of reinforcing steel?

7.50 A 4-in (101.6-mm) slab is supported by T beams 45 in (1143 mm) on centers. The T beams have a 20-ft 0-in (6.10-m) span. The web is 12 in (304.8 mm) and there is an overall depth of 24 in (609.6 mm). What tensile reinforcement is required so that the beam can resist an ultimate moment of 985 ft·kips (1336 kN·m)? Use $f_c' = 4$ ksi (27.59 MPa) and $f_y = 60$ ksi (413.7 MPa). Use the ACI code.

7.51 A T beam has a 40-ft (12.2-m) span and is simply supported. The slab is 6 in (152.4 mm) thick and the T-beam stem is 24 in (609.6 mm) wide. The T-beam stems are 6 ft 0 in (1.83 m) on centers. The live loading is a truck that runs with one front wheel and one rear wheel on the longitudinal axis of the T beam and produces at the midpoint of the T beam a moment 172 ft·kips (233.2 kN·m). Impact adds moment of 52 ft·kips (70.5 kN·m). If d is 24 in (609.6 mm) from top of slab to bottom of stem, design the necessary steel, in tension and/or compression, to carry the live + impact + dead moment and in the web to carry the shear [live + impact = 22 kips (97.86 kN)]. Check design by using $f = Mc/I$. Use $f_c' = 3000$ psi (20.69 MPa) and $f_s = 20,000$ psi (137.9 MPa).

7.52 A floor is composed of T beams, 8 ft (2.44 m) on centers, of 16-ft (4.88-m) span center-to-center bearings, with a partial restraint at the ends equal to 20 percent of the simple T-beam moment. The live loading is 1000 psf (47.88 kPa). If the slab is 4.5 in (114.3 mm) of total thickness, design the T beam, showing sizes of all steel required and work out a suitable stirrup spacing. Use $f_c' = 4000$ psi (27.59 MPa) and $f_s = 20,000$ psi (137.9 MPa).

7.53 A T beam has a flange 25 in (635 mm) wide and 4 in (101.6 mm) thick, an effective depth of 24 in (609.6 mm), and a web width of 10 in (254 mm). Tensile reinforcement consists of six #9 (28.65 mm diameter) bars placed in two rows. Using $f_c' = 4$ ksi (27.59 MPa) and $f_y = 60$ ksi (413.7 MPa) and the ACI specifications for ultimate-strength design, will the moment capacity of the beam be: (a) 569.5 (772.2), (b) 578.0 (783.8), (c) 585.0 (793.3), or (d) 590.5 ft·kips (800.8 kN·m)?

7.54 Design a 14-ft (4.27-m)-long round spirally reinforced concrete column to carry an axial 266,000-lb (1.18-MN) dead load and 530,000-lb (2.36-MN) live load. Use the ACI specifications, with $f_c' = 4000$ psi (27.59 MPa) and $f_y = 60,000$ psi (413.7 MPa). Use ultimate-strength design.

7.55 Redesign the column in question 7.54, using a square section but with longitudinal steel circularly arranged within a spiral. Use same loads, stresses, and ultimate-strength design.

7.56 A rectangular tied column is to support a dead load of 210 kips (934.1 kN) and a live load of 140 kips (622.7 kN). Dead and live load moments of 63 and 42 ft·kips (85.4 and 57.0 kN·m), respectively, are applied normal to the major axis. The column is braced against sidesway.

The unsupported length of the column is 8 ft 0 in (2.44 m). The total moment at the top of the column is 2.5 times that at the bottom. The column bends in double curvature along the major axis. Architectural considerations limit the size of the column to 14 × 20 in (355.6 × 508 mm). Using $f_c' = 4000$ psi (27.59 MPa), $f_y = 60,000$ psi (413.7 MPa), and the ACI specifications, design the required reinforcement. Use ultimate-strength method.

7.57 Given a 14 × 22 in (355.6 × 558.8 mm) short rectangular tied column with a longitudinal reinforcement of four #10 (32.3 mm diameter) bars placed 2.5 in (63.5 mm) from the faces of the column to the center of the steel. The eccentricity of the compressive force is 16 in (406.4 mm) from the centerline of the column along the long axis. Use $f_c' = 4000$ psi (27.59 MPa) and $f_y = 60,000$ psi (413.7 MPa) and the ACI specifications for ultimate-strength design. Will the usable load be: (*a*) 207 (921), (*b*) 219 (974), (*c*) 234 (1040), or (*d*) 250 kips (1112 kN)?

7.58 A 24 × 24 in (609.5 × 609.5 mm) column reinforced with eight #10 (32.3 mm) bars, four in two opposite faces, has an unsupported length of 17 ft 0 in (5.18 m). The column is subjected to the following loads:

Dead load	350 kips (1557 kN)
Live load	153 kips (681 kN)
Dead load moment at top	50 ft·kips (67.8 kN·m)
Live load moment at top	50 ft·kips (67.8 kN·m)
Dead load moment at bottom	20 ft·kips (27.1 kN·m)
Live load moment at bottom	15 ft·kips (20.3 kN·m)

The column bends in double curvature and is not braced against sidesway. Determine whether the column is adequate. Use $f_c' = 5000$ psi (34.48 MPa), $f_y = 60,000$ psi (413.7 MPa), the ACI specifications, and the ultimate-strength method.

7.59 A reinforced concrete tied column 15 × 24 in (381 × 609.5 mm) has a normal load N of 150,000 lb (667.2 kN) applied 5 in (127 mm) eccentrically along the short axis. If $f_c' = 3000$ psi (20.69 MPa) and $f_y = 50,000$ psi (344.8 MPa), check the adequacy of the column on the basis of a cracked section using elastic relationships. There are three #9 (28.7 mm) bars centered 2 in (50.8 mm) from each of the 24-in (609.5-mm)-long faces.

1. Will the compression in the concrete be: (1*a*) 750 (5.17), (1*b*) 975 (6.72), (1*c*) 1050 (7.24), or (1*d*) 1125 psi (7.76 MPa)?

2. Will the tension in the bars be: (2*a*) 2.57 (17.72), (2*b*) 3.74 (25.79), (2*c*) 4.91 (33.85), or (2*d*) 6.08 ksi (41.92 MPa)?

3. Will the compression in the bars be: (3*a*) 11.9 (82.05), (3*b*) 13.0 (89.64), (3*c*) 14.1 (97.22), or (3*d*) 15.2 ksi (104.8 MPa)?

7.60 Determine the section of a composite column to carry an axial load of 1000 kips (4448 kN). Use alternate method, the ACI specifications, with f'_c = 3000 psi (20.69 MPa), f_y = 40,000 (275.8 MPa), and A36 (248.2 MPa) structural steel.

7.61 A steel W 12 × 65 column is encased in an 18 × 18 in (457.2 × 457.2 mm) concrete section and unsupported for its 20-ft (6.10-m) height. Determine the axial load this column may carry. Use alternate method, the ACI specifications, f'_c = 3000 psi (20.69 MPa), and A36 (248.2 MPa) structural steel (248.2 MPa). May this column carry an axial load of: (*a*) 474 (2.11), (*b*) 538 (2.39), (*c*) 570 (2.54), or (*d*) 586 kips (2.61 MN)?

7.62 A pipe is filled with concrete and unsupported for its 20-ft (6.10-m) height. Determine the pipe required to support an axial load of 350 kips (1.56 MN). Use alternate method, the ACI specifications, f'_c = 3000 psi (20.69 MPa), and A36 (248.2 MPa) structural steel.

7.63 Design a spread footing for a 20 × 20 in (508 × 508 mm) square column that carries a dead load of 350 kips (1.56 MN) and a live load of 125 kips (0.56 MN). The bearing capacity of the soil is 2.75 tons/ft² (263.3 kPa). The bottom of the footing is 4 ft 6 in (1.37 m) below grade. The weight of the soil is 120 pcf (1922 kg/m³). Use f'_c = 4000 psi (27.59 MPa), f_y = 50,000 psi (344.8 MPa), and the ACI specifications. Use ultimate-strength method.

7.64 Design a rectangular concrete spread footing for a centrally placed column load of 700,000 lb (3.11 MN), where the short dimension of the footing is 9 ft (2.74 m). The column is 28 × 28 in (711 × 711 mm) in cross section. f'_c = 3000 psi (20.69 MPa) and f_y = 50,000 psi (344.8 MPa). Soil bearing = 3 tons/ft² (287.3 kPa). Use working-stress method.

7.65 A 12 × 12 in (305 × 305 mm) concrete column carrying a load of 75,000 lb (333.4 kN) and a 15 × 15 in (381 × 381 mm) concrete column carrying a load of 120,000 lb (533.8 kN) are 12 ft (3.66 m) center to center. If f'_c = 3000 psi (20.69 MPa) and f_y = 50,000 psi (344.8 MPa), design a combined footing to support these loads if the soil bearing = 2 tons/ft² (191.5 kPa). Use working-stress method.

7.66 If the timbers in Fig. 7.66 are short-leaf southern pine with working stresses of P = 1200 psi (8.27 MPa) compression parallel to the grain,

$Q = 380$ psi (2.62 MPa) compression perpendicular to the grain, and N psi compression at an angle θ with the grain, is this joint satisfactory when $N = \dfrac{PQ}{P \sin^2 \theta + Q \cos^2 \theta}$?

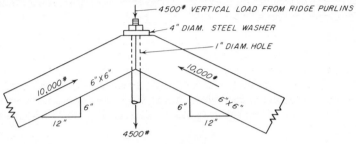

Fig. 7.66

7.67 A 4×6 in (101.6×152.4 mm) timber diagonal of a timber roof truss is cut square-ended and notched into a 6×6 in (152.4×152.4 mm) bottom chord timber. If the unit stresses are the same as in question 7.66, will the bearing stress in the 6×6 in timber be: (*a*) 260 (1.79), (*b*) 440 (3.03), (*c*) 838 (5.78), or (*d*) 1200 psi (8.27 MPa)?

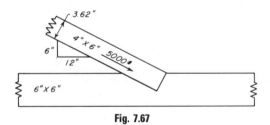

Fig. 7.67

7.68 Timber joists of 3×12 in (76.2×304.8 mm) nominal dimension are on 16-in (406.4-mm) centers and their ends rest upon steel beams of 6-in (152.4-mm) flange width and upon 14-ft (4.27-m) centers. Dressed timbers are used and the following stresses are not to be exceeded: shear at 120 psi (827 kPa), bending at 1600 psi (11.03 MPa), bearing at 380 psi (2.62 MPa), and deflection not to exceed ¹⁄₃₆₀ span with $E = 1,600,000$ psi (11.03 GPa). Including the weight of the joist and floor, will the allowable working load per square foot be: (*a*) 641 (30.69), (*b*) 259 (12.40), (*c*) 236 (11.30), or (*d*) 216 psf (10.34 kPa)?

7.69   A wood floor with a 14-ft (4.27-m) span is to carry 250 psf (12.0 kPa), including its weight. An opening 5 ft (1.52 m) square is to be left in the center of the floor. Sketch a plan showing how the floor joists should be placed, and specify the size of the floor joists if the stress conditions in question 7.68 are not to be exceeded.

7.70   Four 2 × 8 in (50.8 × 203.2 mm) dressed planks are to be framed to give the maximum stiffness as a beam. Will the moment of inertia be: (*a*) 228 (9490), (*b*) 495 (20,603), (*c*) 628 (26,139), or (*d*) 761 in⁴ (31,675 cm⁴)?

7.71   Four 2 × 8 in (50.8 × 203.2 mm) dressed planks are to be framed to carry the maximum bending shear. Will the usable area for shear be: (*a*) 14.3 (92.2), (*b*) 32.4 (209.0), (*c*) 36.8 (237.4), or (*d*) 48.6 in² (313.5 cm²)?

7.72   The top chord of a wooden roof truss is continuous, supported, and braced at 6-ft (1.83-m) centers. The chord carries a maximum axial load of 60,000 lb (267 kN), and purlins bring loads to the truss only at panel points. Will the nominal depth of a 6-in (152.4-mm)-wide dressed section of dense southern yellow pine be: (*a*) 6 (152.4), (*b*) 8 (203.2), (*c*) 10 (254.0), or (*d*) 12 in (304.8 mm)?

7.73   A timber column 12 ft long (3.66 m) carries a compressive axial load of 27,500 lb (122.7 kN) and is to be of one piece. Will the dressed size of dense southern yellow pine as required be: (*a*) 4 × 4 (102 × 102), (*b*) 4 × 6 (102 × 152), (*c*) 6 × 6 (152 × 152), or (*d*) 6 × 8 in (152 × 203 mm)?

7.74   A timber column 12 ft (3.66 m) long carries a compressive axial load of 75,000 lb (333.6 kN). The column is to be a hollow box made of four 2-in (51-mm) planks. Will the nominal size of dressed planks of dense southern yellow pine be: (*a*) 2 × 6 (51 × 152), (*b*) 2 × 8 (51 × 203), (*c*) 2 × 10 (51 × 254), or (*d*) 2 × 12 in (51 × 305 mm)?

7.75   Two 2 × 4 in (51 × 102 mm) and one 4 × 8 in (102 × 203 mm) dressed dense southern yellow pine timbers are nailed together to form an approximate 8 × 8 in (203 × 203 mm) cross (in cross section). If all three timbers are 10 ft (3.05 m) long, will this composite column carry a safe axial load of (*a*) 38,500 (171.2), (*b*) 44,500 (197.9), (*c*) 47,500 (211.3), or (*d*) 49,000 lb (218.0 kN)?

PART THREE

ANSWERS

to Questions from Professional Engineers' Examinations

Basic Fundamentals

CHEMISTRY, COMPUTER AND MATERIALS SCIENCE, AND NUCLEONICS

1.101 A liter of hydrochloric acid (density 1.201) contains $1000 \times 1.201 \times 0.4009 =$ **481.5** g HCl.

Molecular weights:

$$HCl = 1.008 + 35.46 = 36.47$$
$$NaCl = 23 + 35.46 = 58.46$$
$$H_2SO_4 = 2 \times 1.008 + 32 + 4 \times 16 = 98.02$$
$$Na_2SO_4 = 2 \times 23 + 32 + 4 \times 16 = 142.00$$

Check equation:

$$2 \times 58.46 + 98.02 = 214.92 = 142.00 + 2 \times 36.47$$

Weight of NaCl required $= 481.5 \times 58.46/36.47 =$ **771.7** g

Weight of (100 percent) H_2SO_4 required $= \dfrac{481.5}{36.47} \dfrac{98.02}{2}$

$$= \textbf{646.8 g} \qquad \textbf{b}$$

Basis: 1 mol NaCl $\rightarrow$ 1 mol HCl
½ mol $H_2SO_4 \rightarrow$ 1 mol HCl

1.102 *a.* Oxidation and reduction occur simultaneously in a chemical reaction. In the process of oxidation, one or more electrons are lost by an atom or an ion; in reduction, one or more electrons are gained by an atom or an ion. The oxidizing agent is reduced by the gain of electrons and the reducing agent is oxidized by the loss of electrons.

b. The observations of Newlands, Lotharmeyer, and Mendeleev indicated to Mendeleev that when the elements were arranged in the order of increasing atomic weights, they could be divided on the basis of chemical behavior into series or periods, not all of which, however, contained the same number of elements. The properties of the elements were thought to be a periodic function of their atomic weights. Later investigations have shown that most of the physical and chemical properties of the elements are periodic functions of their atomic numbers.

c. Radioactivity has been defined as the spontaneous disintegration of atoms accompanied by the emission of "rays." In the case of radium, the rays have been identified as alpha (helium with a double positive charge), beta (high-velocity electrons), and gamma (x-rays with wavelengths from 10^{-9} to 10^{-8} cm).

d. Acidity is customarily considered a function of the concentration of the hydrogen ion (H^+) or oxonium ion (H_3O^+). The more general definition of an acid, however, is that the ion or molecule may act as a proton donor.

The hydrogen-ion concentration in any solution is a function of the nature of the acid under consideration and the nature of the solvent used.

In aqueous solutions, the greater the concentration of hydrogen ion (H^+)—more recent concept, oxonium ion (H_3O^+)—the greater the acidity of the solution.

The concentration of the hydrogen ion is usually expressed as pH, where $pH = 1/H^+$ per liter. That is, $pH = 1$ when $H^+ = 10^{-1}$ mol/liter.

1.103 One ton ore that shows 60 percent copper contains 1200 lb copper per ton ore. One pound CuS contains 0.666 lb copper. Therefore, 1 ton ore contains $1200/0.666 - 1200$, or 600 lb sulfur.

Sulfur available for H_2SO_4 production equals $600 - 60$, or 540 lb. Atomic weight of $H_2SO_4 = 2 \times 1 + 32 + 4 \times 16 = 98$. One lb 93 percent H_2SO_4 contains $0.93 \times {}^{32}/_{98}$ or 0.304 lb sulfur. Therefore, 1 ton ore will produce $540/0.304 = 1775$ lb 93 percent H_2SO_4.   **c**

1.104 The production of sodium hydroxide (caustic soda) from soda ash (Na_2CO_3) is known as the lime-soda process. Caustic soda prepared from soda ash as manufactured by the Solvay process is characterized by its low chlorine content.

The reaction depends on the low solubility product of calcium carbonate.

$$Na_2CO_3 + Ca(OH)_2 \rightleftharpoons 2NaOH + CaCO_3 \downarrow$$

The degree of conversion is a function of the initial concentration of soda ash, a 12 percent solution giving approximately 95 percent conversion and a 15 percent solution giving only 90 percent conversion.

The reaction equilibrium is insensitive to temperature changes, but the reaction rate and rate of settling of $CaCO_3$ are favored by increased temperatures. The commercial process may be carried out as a batch or as a continuous operation. The process, equipment, and engineering problems are shown in the accompanying table.

Step	Equipment	Engineering problems
1. Preparing soda-ash solution	Tanks with agitation	Economic consideration of cost of processing dilute solutions with high degree of conversion as compared to cost of processing more concentrated solution with a lower degree of conversion
2. Blending soda solution with reburned and makeup lime	Classifier shaker	Feed rates, screen size, etc.
3. Soda solution causticized with slight excess of lime	Tanks with agitation	Corrosion, agitation, retention time, temperature of reaction
4. Separating liquids and solids in first of two thickeners	Dorr thickener sludge pump	Settling rates, retention time, etc.
5. Overflow from 4 containing 10 to 11% NaOH is fed to evaporator	Evaporator and auxiliary equipment	Materials of construction, heat transfer, economic operating conditions
6. Underflow from 4 pumped to second thickener. Filtrate from 7 and water is added	Same as 4	Same as 4
7. Underflow from 6 filtered on continuous filter	Continuous filter	Optimum operating conditions; that is, vacuum, filter media, etc.
8. Overflow from 6 used as weak liquor in 1		
9. Filter cake from 7 returned to lime kiln		

1.105 Equal numbers of molecules of different (nonvolatile and none-lectrolyte) solutes dissolved in equal weights of the same solvent normally depress the vapor pressure, raise the boiling point, and lower the freezing point by constant amounts.

When 1 g molecular weight nonvolatile nonelectrolyte is dissolved in 1000 g benzene, the freezing point of benzene is lowered 5.12°C.

Twenty g unknown substance in 1000 g benzene lowers the freezing point (5.49 − 4.82) or 0.67°C.

Therefore, (5.12/0.67) × 20, or 152.8, g would be required to lower the freezing point by 5.12°C. The experimental value of the molecular weight is 152.8.   **d**

$$M = \frac{1000 \times K_f \times W}{W_0 \times \Delta T}$$

where M = molecular weight
K_f = molal lowering of freezing point
W = weight of unknown substance
W_0 = weight of solvent used
ΔT = experimental freezing point lowering

1.106 The freezing-point–solubility curve of binary systems (H_2O–Na_2SO_4 in this instance) should be studied carefully in problems of this type. The transition temperature between Na_2SO_4 and H_2O lies at 32.38°C. At temperatures above this point, the decahydrate becomes unstable with respect to the anhydrous salt. In this problem, the solution was saturated at 32°C so that only the decahydrate need be considered. Analysis of solubility data indicates that Na_2SO_4 is soluble to the extent of 46.25 g/100 g water when equilibrium is attained with the decahydrate as the solid phase.

Basis of solution: 46.25 g Na_2SO_4 and 100 g H_2O at 32°C.

Molecular wt: $Na_2SO_4 = 142$
$H_2O = 18$
$Na_2SO_4 + 10H_2O = 142 + 180 = 322$

Total wt solution $= 146.25$ g
Wt to be removed $= 146.25 \times 0.6 = 87.75$ g
Wt solution crystallized/g $Na_2SO_4 = {}^{322}\!/_{142} = 2.27$
Wt Na_2SO_4 crystallized $= 87.75/2.27 = 38.6$
Wt H_2O removed from solution $= 38.6 \times {}^{180}\!/_{142} = 49.0$

$$\text{Conc. liquor} = \frac{\text{g Na}_2\text{SO}_4}{100 \text{ g H}_2\text{O}} (46.25 - 38.6) \frac{100}{100 - 49}$$

$$= \frac{15 \text{ g Na}_2\text{SO}_4}{100 \text{ g H}_2\text{O}}$$

Inspection of solubility curve indicates temperature of 16.6°C.   **b**

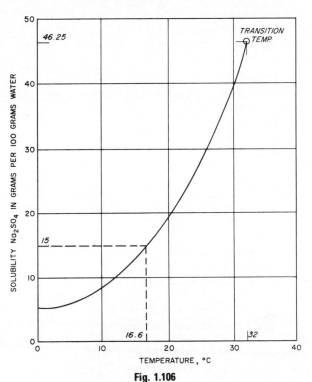

Fig. 1.106

1.107 $\text{Fe}_3\text{O}_4 + \text{CO} \rightarrow 3\text{FeO} + \text{CO}_2$

$$3\text{FeO} + 3\text{CO} \rightarrow 3\text{Fe} + 3\text{CO}_2$$

Each lb·mol Fe_3O_4 requires 4 lb·mol CO for the reduction of Fe_3O_4 to Fe.

$$\text{Atomic wt } Fe_3O_4 = 3 \times 55.84 + 4 \times 16 = 231.52$$
$$3FeO = 215.5 \quad \text{and} \quad CO = 28$$
$$2000/231.52 = 8.63 \text{ lb} \cdot \text{mol in 1 ton } Fe_3O_4$$
$$8.63 \times 0.84 = 7.25 \text{ lb} \cdot \text{mol } Fe_3O_4 \text{ in 1 ton ore}$$
$$n = 7.25 \times 4$$
$$= 29.00 \text{ lb} \cdot \text{mol CO required to reduce } Fe_3O_4 \text{ in 1 ton ore}$$

Using the perfect-gas law, which states that 1 lb·mol perfect gas occupies 359 ft^3 at atmospheric pressure (14.7 psi) and 492°R (Rankin or absolute temperature), which are standard conditions, and correcting for temperature and pressure by

$$VP = nRT$$

$$V(\text{vol}) = \frac{29.00 \times 10.71 \times (460 + 85)}{100 + 14.7} = \frac{1472 \text{ ft}^3 \text{ CO}}{1 \text{ ton ore}}$$

where $R = 10.71 =$ gas constant or unit conversion factor $(14.7 \times 359)/(1 \times 492)$
[To change from (psi × ft^3)/°R to (psf × ft^3)/°R, multiply R(10.71) by 144, and new $R = 1544$.]
Since the gas used in the reduction is only 28 percent CO, the total volume of gas required is $1472 \div 0.28 = 5230$ ft^3 gas at 85°F and 100 psi.   **d**

1.108 $Al = 27, 2Al$ or $Al_2 = 54$
$Fe = 56, 2Fe$ or $Fe_2 = 112$
$SO_4 = 32 + 4 \times 16 = 96, (SO_4)_3 = 288$

$$?Al + ?Fe_2(SO_4)_3 \rightarrow Al_2(SO_4)_3 + ?FeSO_4$$
$$2Al + 3Fe_2(SO_4)_3 \rightarrow Al_2(SO_4)_3 + 6FeSO_4$$
$$2 \times 27 + 3 \times (112 + 288) \rightarrow (54 + 288) + 6 \times (56 + 96)$$
$$54 + 1200 \rightarrow 342 + 912$$

Valences: $Al = 0$, $Fe_2(SO_4)_3 = (+3) \times 2, (-2) \times 3$
$Al_2(SO_4)_3 = (+3) \times 2, (-2) \times 3$
$FeSO_4 = (+2)(-2)$
Valence change: $Al + O \rightarrow +3 =$ loss of 3 electrons
$Fe + 3 \rightarrow +2 =$ gain of 1 electron
Therefore: $2Al$ loses 6 electrons when $1Al_2(SO_4)_3$ is formed.
$3Fe_2(SO_4)_3$ gains 6 electrons when $6FeSO_4$ is formed.

1.109   **c** Lime

1.110   **e** Methane (approximately 60 to 75 percent)

1.111   **a** 0–25

1.112   **c** Relative stability

1.113   **b** Saprophytic

1.114   **e** Maximum velocity of flow is 1 ft/s.

1.115   **d** Sludge and raw sewage are not mixed.

1.116   **1a** Lines 2 and 6
          2a Lines 2 and 4

1.117   **1b** Line 2
          2a Line 4
          3b Line 30

1.118   **1d** Line 25
          2c Line 25
          3d There are two digits after the decimal point.

1.119   **1b** Line 1
          2c Line 9
          3c 4 decimal points

1.120   **1b** Line 10
          2a Line 4

1.121   **d** Composition

1.122   **b** Coarse-grained

1.123   **d** Inhibits further deterioration

1.124   **b** $2m_0C^2$

1.125   **c** Gamma radiation

ELECTRICITY

1.201 $5 \text{ mph} = \dfrac{5 \times 5280}{3600} = 7.33 \text{ fps}$

$$\text{Horsepower} = \frac{200 \times 7.33}{550} = 2.67 \text{ hp}$$

$$\text{Current} = \frac{2.67 \times 746}{0.70 \times 110} = \textbf{25.8 A} \qquad \textbf{a}$$

1.202 12 and 20 ohms in parallel.

$$\frac{1}{R_1} = \frac{1}{12} + \frac{1}{20} = \frac{32}{240} = \frac{1}{7.5}$$

$$R_1 = 7.5$$

Resistance A to B = 16 ohms = R_{AB}:

$$\frac{1}{R_{AB}} = \frac{1}{16} = \frac{1}{R_2 + 7.5} + \frac{1}{30} = \frac{30 + R_2 + 7.5}{30(R_2 + 7.5)}$$

$$1.87R_2 + 14.08 = R_2 + 37.5$$
$$0.87R_2 = 23.42, R_2 = \textbf{26.8} \text{ ohms} \qquad \textbf{b}$$

1.203 Three resistors in parallel:

$$\frac{1}{R_1} = \frac{1}{5} + \frac{1}{2} + \frac{1}{10} = \frac{8}{10} \qquad R_1 = 1.25$$

Two resistors in parallel:

$$\frac{1}{R_2} = \frac{1}{4} + \frac{1}{4} = \frac{2}{4} \qquad R_2 = 2.00$$

Two lines of resistors: Line 1. $6.75 + 1.25 = 8$ ohms
Line 2. $2.00 + 6.00 = 8$ ohms

$$\frac{1}{R_3} = \frac{1}{8} + \frac{1}{8} = \frac{2}{8}$$

$$R_3 = 4 \text{ ohms}$$

$$\text{Current} = \frac{V}{R} = \frac{12}{4 + 1} = \textbf{2.4} \text{ A from battery} \qquad \textbf{c}$$

1.204 Two resistors in parallel:

$$\frac{1}{R} = \frac{1}{5} + \frac{1}{17} = \frac{22}{85}$$

$$R = 3.86$$

$$\text{Current} = \frac{V}{R} = \frac{6}{3.86} = 1.553 \text{ A}$$

Each coil has 6 V.

Coil *1* takes $\%_5$ = **1.2** A   **1b**
Coil *2* takes $\%_{17}$ = **0.353** A   **2d**
                       Σ = 1.553 A (*check*)

1.205   Increase in cathode weight $= 300 - 15 = 285$ lb (129.28 kg). At 382.52 Ah/lb (843.30 Ah/kg), this requires $285 \times 382.52 = $ **109,000** Ah.   **1a**

$$\text{Current} = \frac{109,000}{15 \times 24} = \textbf{303 A} \qquad \textbf{2c}$$

$$\text{Plate area} = \frac{40 \times 36}{144} = 10 \text{ ft}^2 \text{ (0.929 m}^2)$$

Current density $= 303/10 = $ **30.3** A/ft^2 **(326.2** A/m^2)   **3b**

Reference: "Standard Handbook for Electrical Engineers."

1.206   Fifty kVA at 90 percent pf $= 0.9 \times 50 = 45$ kW

$$\text{Input} = 45 + \frac{400 + 600}{1000} = 46 \text{ kW}$$

$$\text{Efficiency} = {}^{45}\!/_{46} = \textbf{97.83} \text{ percent} \qquad \textbf{1d}$$

Maximum efficiency occurs when copper loss is equal to core loss. To reduce the full-load copper loss to equal the core loss requires that the load be reduced to

$$\sqrt{{}^{400}\!/_{600}} = \sqrt{0.667} = 0.816 \text{ or } \textbf{81.6} \text{ percent of full load} \qquad \textbf{2c}$$

1.207   The equivalent resistance of similar lamps in parallel varies inversely as the number of lamps in the group. The voltage across groups in series varies directly as the equivalent resistance, by Ohm's law. Therefore, the voltage across each group will vary inversely as the number of lamps in the group.

$$\text{Voltage across 25-lamp group} = 220 \, \frac{15}{15 + 25} = \textbf{82.5 V} \qquad \textbf{1a}$$

$$\text{Voltage across 15-lamp group} = 220 \, \frac{25}{15 + 25} = \textbf{137.5 V} \qquad \textbf{2c}$$

Assuming constant resistance, the current through each lamp varies directly as the voltage.

$$\text{Current through each lamp in 25 group} = \frac{82.5}{110} \, 0.9$$

$$= \textbf{0.675 A} \qquad \textbf{3a}$$

$$\text{Current through each lamp in 15 group} = \frac{137.5}{110} \, 0.9$$

$$= \textbf{1.12 A} \qquad \textbf{4d}$$

1.208 Wire tables in handbook show that copper wire of 0.204-in diameter has a resistance of 0.253 ohms/1000 ft at 25°C. The two wires from the generator to the load 500 ft away will then have a resistance of 0.253 ohms. The load requires a current of $10,000/120 = 83.3$ A. Generator voltage must be greater than the load voltage by the IR drop in the line:

$$V_g = 120 + 83.3 \times 0.253 = 120 + 21.1 = \mathbf{141.1} \text{ V} \qquad \mathbf{1a}$$

Power loss in the line will be I^2R, or

$$P_L = (83.3)^2 \times 0.253 = \mathbf{1755} \text{ W} \qquad \mathbf{2c}$$

1.209 Charge is the product of capacitance and voltage, or $Q = CE$. The voltage across capacitors in series varies inversely as the capacitance:

$$\text{Voltage across } 8\mu\text{F} = 120 \, \frac{12}{8 + 12} = \mathbf{72} \text{ V} \qquad \mathbf{1d}$$

$$\text{Voltage across 2 and 10 } \mu\text{F} = 120 \, \frac{8}{8 + 20} = \mathbf{48} \text{ V} \qquad \mathbf{2b}$$

Charge of 8 μF = $8 \times 10^{-6} \times 72 = \mathbf{576 \times 10^{-6}}$ C
Charge of 2 μF = $2 \times 10^{-6} \times 48 = \mathbf{96 \times 10^{-6}}$ C
Charge of 10 μF = $10 \times 10^{-6} \times 48 = \mathbf{480 \times 10^{-6}}$ C
Energy stored = ½ charge × voltage
Total energy = ½ × $576 \times 10^{-6} \times 72 + ½(96 + 480)10^{-6} \times 48$
$$= 0.0207 + 0.0138 = \mathbf{0.0345} \text{ J} \qquad \mathbf{3a}$$

1.210 a. By Ohm's and Kirchhoff's laws:

$$E_1 - I_1R_1 = E_2 - I_2R_2 \qquad (1)$$
$$I_1 + I_2 = 20 \qquad (2)$$

Substituting Eq. (2) in Eq. (1),

$$E_1 - I_1R_1 = E_2 - (20 - I_1)R_2$$

$$I_1 = \frac{E_1 - E_2 + 20R_2}{R_1 + R_2} = \frac{2.1 - 2.0 + 20 \times 0.05}{0.05 + 0.05} = \frac{1.10}{0.10}$$
$$= \mathbf{11} \text{ A} \qquad \mathbf{1b}$$

$$I_2 = 20 - I_1 = 20 - 11 = \mathbf{9} \text{ A} \qquad \mathbf{2d}$$

b. As a lead storage cell discharges, the following reaction takes place:

$$\underset{\substack{\text{Positive} \\ \text{plate}}}{\text{PbO}_2} + \underset{\substack{\text{Negative} \\ \text{plate}}}{2\text{H}_2\text{SO}_4} + \underset{\substack{\text{Positive} \\ \text{plate}}}{\text{Pb}} \rightarrow \underset{}{\text{PbSO}_4} + \underset{\substack{\text{Negative} \\ \text{plate}}}{\text{PbSO}_4} + 2\text{H}_2\text{O}$$

The sulfuric acid combines with the lead peroxide of the positive plate and the lead of the negative plate to form lead sulfate and water. The specific gravity of the sulfuric acid electrolyte therefore decreases as this reaction takes place.

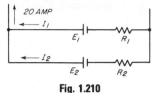

Fig. 1.210

1.211 The equation for emf of a dc generator can be found in any handbook:

$$E = \frac{ZPNBA}{60a10^8} \text{ V}$$

where Z = no. of armature conductors
P = no. of poles
N = speed, rpm
B = average flux density under poles, lines/in^2
A = area of each pole face, in^2
a = no. of parallel paths through armature

Substituting the values given:

$$E = \frac{360 \times 4 \times 1200 \times 40,000 \times (6 \times 6)}{60 \times 4 \times 10^8} = \mathbf{103.7} \text{ V} \quad \mathbf{d}$$

1.212 By Ohm's and Kirchhoff's laws:

$$IR = 15 - 2I_2 = 10 - 1I_1 \quad (1)$$
$$I_1 + I_2 = I \quad (2)$$

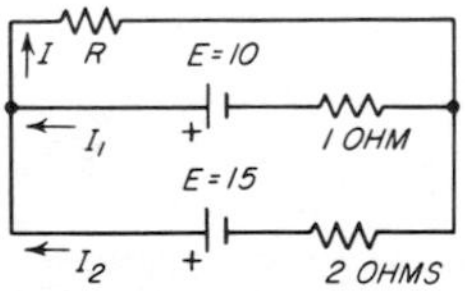

If $I_1 = I_2$, then $I_1 = I/2$ and $I_2 = I/2$ [from Eq. (2)]. Substituting in Eq. (1),

Fig. 1.212

$$IR = 15 - 2\frac{I}{2} \quad \text{or} \quad I(1 + R) = 15 \quad (3)$$

and

$$IR = 10 - \frac{I}{2} \quad \text{or} \quad I(0.5 + R) = 10 \quad (4)$$

Dividing Eq. (3) by Eq. (4),

$$\frac{1 + R}{0.5 + R} = \frac{15}{10} \quad \text{and} \quad 10 + 10R = 7.5 + 15R$$

$$5R = 2.5, R = \mathbf{0.5} \text{ ohm} \quad \mathbf{c}$$

1.213 The 60-W lamp could be connected in series with the parallel combination of the two 30-W lamps across 240 V for normal operation as shown in Fig. 1.213. Since resistance = V^2/W, the resistance of two 30-W 120-V lamps in parallel would be equal to the resistance of the 60-W 120-V lamp and the voltage across each lamp would be normal.

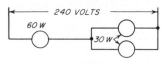

Fig. 1.213

1.214 Total equivalent resistance:

$$R = 15 + \frac{7 \times 11}{7 + 11} = 15 + 4.28 = 19.28 \text{ ohms}$$

The total current, by Ohm's law, of $I = V/R$:

$$I = 110/19.28 = \textbf{5.7 A} \qquad \textbf{1a}$$

This is the current through the 15-ohm resistance. The total current divides inversely as the resistances through each branch of the parallel combination:

$$\text{Current through 7-ohm branch} = 5.7 \frac{11}{7 + 11} = \textbf{3.48 A} \qquad \textbf{2b}$$

$$\text{Current through 11-ohm branch} = 5.7 \frac{7}{7 + 11} = \textbf{2.22 A} \qquad \textbf{3d}$$

1.215 $I = \dfrac{\text{hp} \times 746}{V \times \text{efficiency}} = \dfrac{5 \times 746}{220 \times 0.85} = 19.9 \text{ A}$

IR permitted = $0.05 \times 220 = 11$ V
R for 300 ft = $11/19.9 = 0.553$ ohm
$R/1000$ ft = $1000/300 \times 0.553 = 1.84$ ohms
Size wire = **#12** AWG or **20.5** mm in diam   **c**
Current capacity check = 25 A

1.216 $P = V \times I = 115 \times 15 = 1725$ W. Therefore, **17** 100-W lamps can be used.   **b**

1.217 $V_2 = \dfrac{20,000}{16,000} V_1 = 1.25 \times 60 = 75$ V

$V = V_1 + V_2 = 60 + 75 = \textbf{135 V} \qquad \textbf{c}$

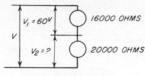

Fig. 1.217

1.218 $R_{tot} = 1 + 7 + \dfrac{3 \times 6}{3 + 6} = 10$ ohms

$I_{tot} = {}^{30}\!/_{10} = 3.0$ A

1. Voltage across 7-ohm device $= 3 \times 7 = $ **21** V   **1a**
2. Voltage across 3- and 6-ohm devices $= 3 \times 2 = 6$ V

Power expended on 3-ohm device $= I^2R = 6 \times {}^{6}\!/_{3} = $ **12** W   **2b**

3. Current in 6-ohm device $= \dfrac{V}{R} = \dfrac{6}{6} = $ **1.0** A   **3c**

1.219 $I_{line} = \dfrac{\text{hp} \times 746}{V \times \text{efficiency}} = \dfrac{10 \times 746}{220 \times 0.83} = 40.8$ A

$I_{starting} = 2 \times 40.8 = 81.6$ A
$I_a = 81.6 - {}^{220}\!/_{100} = 79.4$ A

$\qquad = \dfrac{V}{R_1 + Ra} = 79.4 = \dfrac{220}{R_1 + 0.25}$

$\qquad R_1 + 0.25 = 2.77 \qquad R_1 = $ **2.5** ohms   **b**

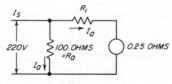

Fig. 1.219

1.220 $I = \dfrac{V}{Z}$ and $Z = R + jX_L - jX_c$

$\qquad Z = 63 + j377 \times 0.15 - j\,\dfrac{10^6}{377 \times 15}$

$\qquad = 63 + j56.6 - j177 = 63 - j120 = 135\underline{/-62.2°}$

$I = {}^{110}\!/_{135}\underline{/62.2} = $ **0.815**$\underline{/62.2°}$ A   **d**

(See p. 176 for Fig. 1.220.)

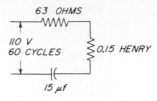

Fig. 1.220

1.221   Power input $= \dfrac{\text{hp} \times 746}{\text{efficiency}} = \dfrac{8 \times 746}{0.82} = 7280$ W. Apparent power = power divided by power factor and $\text{VA}_{\text{input}} = 7280/0.70 = 10,400$ VA. Because the motor is a three-phase balanced load, one-third or 3460 VA is applied to each phase. The three phases are Y-connected with a line-to-line voltage of 220 V. Hence each phase voltage is $220/\sqrt{3} =$ **127** V.   **1a**

Since the apparent power per phase is 3460 VA and the phase voltage is 127 V, $I\varphi = 3460/127 =$ **27.3** A current per phase and also per line in a Y connection.   **2d**

1.222   The plate-characteristic curves, commonly found in tube manuals, are a family of graphs, each drawn for a constant value of control grid bias voltage, E_c, in which the vertical axis is plate current, I_p, and the horizontal axis is plate voltage, E_p. Suppose we want to find r_p, g_m, and μ for a 6J5 triode at the point $E_c = -9$ V, $I_p = 5.5$ mA, and $E_p = 240$ V. To find r_p, find the change in E_p for a small change in I_p, with E_c held constant at -9 V. Find that at $E_c = -9$ and $I_p = 6$, $E_p = 246$, and that at $E_c = -9$ and $I_p = 5$, $E_p = 236$. Hence

$$r_p = \frac{\Delta E_p}{\Delta I_p} = \frac{246 - 236}{(6 - 5)10^{-3}} = \textbf{10,000} \text{ ohms}$$

To find g_m, find the change in I_p, occasioned by a small change in E_c with E_p held constant at 240 V. Find that at $E_p = 240$ and $E_c = -10$, $I_p = 3.5$, and that at $E_p = 240$ and $E_c = -8$, $I_p = 7.5$. Hence

$$g_m = \frac{\Delta I_p}{\Delta E_c} = \frac{(7.5 - 3.5)10^{-3}}{-8 - (-10)} = 2 \times 10^{-3} \quad \text{or} \quad \textbf{2000} \text{ μmho}$$

To compute μ, find the change in E_p caused by a small change in E_c with I_p held constant at 5.5 mA. Find that at $I_p = 5.5$ and $E_c = -8$, $E_p =$

222, and that at $I_p = 5.5$ and $E_c = -10$, $E_p = 262$. Hence

$$\mu = -\frac{\Delta E_p}{\Delta E_c} = -\frac{262 - 222}{-10 - (-8)} = \mathbf{20}$$

Check: $\qquad \mu = g_m \times r_p = 2 \times 10^{-3} \times 10^4 = 20$

1.223 The internal resistance of the cell on short circuit is found from Ohm's law

$$R_i = \frac{E_i}{I_{\text{short circuit}}} = \frac{1.5}{25} = \mathbf{0.06} \text{ ohms} \qquad \mathbf{1a}$$

When connected to an external resistance R_e, and using Ohm's law again,

$$I = \frac{E_i}{R_i + R_e} = \frac{1.5}{0.06 + 1.00} = \mathbf{1.414} \text{ A} \qquad \mathbf{2c}$$

The voltage across the battery terminals when connected to the 1-ohm load will be the internal voltage minus the IR drop,

$$E_t = E_i - IR_i = 1.5 - 1.414 \times 0.06 = \mathbf{1.414} \text{ V} \qquad \mathbf{3d}$$

This voltage is also that across the 1-ohm load,

$$E_t = IR_e = 1.414 \times 1 = \mathbf{1.414} \text{ V } (check)$$

1.224 1. Period of any oscillation is the time required per cycle. Hence the period

$$T = \frac{1}{f} = \frac{1}{5} = \mathbf{0.2} \text{ s} \qquad \mathbf{1a}$$

2. In the absence of resistance, the natural undamped frequency of transient oscillations is the same as the ac resonant frequency; that is, the angular frequency is

$$\omega_0 = \frac{1}{\sqrt{LC}} \qquad \text{whence } C = \frac{1}{\omega_0^2 L} = \frac{1}{(5 \times 2\pi)^2 \times 100}$$

$$C = 10.1 \times 10^{-6} \text{ F} \qquad \text{or} \qquad \mathbf{10.1} \ \mu\text{F} \qquad \mathbf{2c}$$

1.225 *a.* Differential equation for a series LCR circuit during discharge is, by Kirchhoff's voltage law,

$$Ri + L\frac{di}{dt} + \frac{1}{C}\int_{-\infty}^{t} i \, dt = 0$$

In this problem the initial charge Q is the value of the last term at $t = 0$.

The equation then becomes

$$Ri + L\frac{di}{dt} + \frac{1}{C}\int_0^t i\,dt - \frac{Q}{C} = 0$$

b. In the equation as first written, each term represents a voltage drop. Thus $R \times i$ is the instantaneous resistance drop and $L\,di/dt$ is the instantaneous inductive voltage drop. The voltage drop across the capacitance, because of the voltage due to the initial charge, is

$$e_c = \frac{1}{C}\int_0^t i\,dt - \frac{Q}{C}$$

FLUID MECHANICS

1.301

Depth cylinder $= 60$ ft (18.29 m)
Diam. cylinder $= 40$ ft (12.19 m)
Weight water $= 62.5$ lb/ft^3 (1001 kg/m^3)
Intensity water pressure at base $= 60 \times 62.5 = 3750$ psf (179.6 kPa)
Total pressure on base $= \pi(20)^2 \times 3750 = 4,710,000$ lb
$= $ **2355** tons (**20.95** MN)   **1a**
Total pressure on side $= \pi \times 40 \times 60 \times 3750/2$
$= 14,130,000$ lb $= $ **7065** tons (**62.85** MN)   **2d**
Total normal pressure on base and sides $= 18,840,000$ lb
$= $ **9420** tons (**83.80** MN)

1.302

Sum of moments of forces about hinge $= 0$

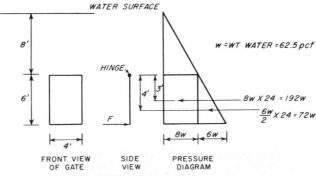

Fig. 1.302

$$+ 3 \times 192w + 4 \times 72w - 6F = 0$$
$$F = 96w + 48w = 144 \times 62.5 = \mathbf{9000} \text{ lb } (\mathbf{40.0} \text{ kN}) \qquad \mathbf{b}$$

1.303 $\quad x_c - x_0 = \dfrac{l_1^2}{12x_0} \quad$ or $\quad \dfrac{1}{12} = \dfrac{9}{12x_0}$

$$x_0 = \mathbf{9} \text{ ft } (\mathbf{2.74} \text{ m})$$

$x = 9 \text{ ft} - 1.5 \text{ ft} = \mathbf{7.5} \text{ ft } (\mathbf{2.29} \text{ m}) = $ distance top edge of plate is below water surface.   **c**

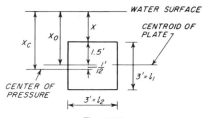

Fig. 1.303

1.304 $\quad v_{10} \times d_{10}^2 = v_{12} \times d_{12}^2$
$$v_{10} = 6 \times (^{12}\!/_{10})^2 = \mathbf{8.64} \text{ fps } (\mathbf{2.63} \text{ m/s}) \qquad \mathbf{1d}$$

$$w = 62.22 \text{ lb/ft}^3 \text{ (at } 80°\text{F)}$$

$$\mu = \frac{0.00003716}{0.4712 + 0.01435T + 0.0000682T^2}$$

For $T = 80°$F, $\mu = 0.000018 \text{ lb}\cdot\text{s/ft}^2$ $(0.000862 \text{ Pa}\cdot\text{s})$.

$$\rho = \frac{w}{g} = \frac{62.22}{32.17} = 1.934 \ \frac{\text{lb/ft}^3}{\text{ft/s}^2} \left(\text{or } \frac{\text{lb}\cdot\text{s}^2}{\text{ft}^4}\right)\left(\frac{101.7 \text{ kg}\cdot\text{s}^2}{\text{m}^4}\right)$$

Re (Reynolds number) $= \dfrac{6 \times 1 \times 1.934}{0.000018} = \mathbf{645{,}000} \qquad \mathbf{2c}$

1.305   For the oil

$$\rho = \frac{0.80 \times 62.4}{32.17} = 1.55$$

$$\frac{v \times 0.833 \times 1.55}{0.000042} = 645{,}000v = \mathbf{21.0} \text{ fps } (\mathbf{6.40} \text{ m/s}) \qquad \mathbf{a}$$

1.306 $\qquad$ Width barge $= 20 \text{ ft } (6.10 \text{ m})$
Wt water displaced $=$ wt barge $= 250T \ (226.8 \text{ Mg})$
$$d = \text{draft, ft}$$

$$\frac{250 \times 2000}{62.4} = \left(65 + \frac{7.5d}{12}\right) d \times 20$$

$$d^2 + 104d = 641 \qquad d = \mathbf{5.84} \text{ ft } (\mathbf{1.78} \text{ m}) \qquad \mathbf{c}$$

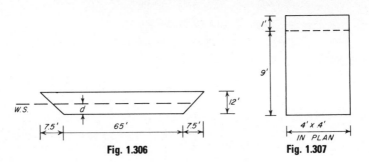

Fig. 1.306   Fig. 1.307

1.307   Wood cube $2 \times 2 \times 2$ ft will displace $2 \times 2 \times 1$ ft or 4 ft³ of water.

Water will rise in tank h ft.

$$h \times 4 \times 4 = 4 \qquad h = 0.25 \text{ ft } (0.076 \text{ m})$$

Orig. water pressure on one side tank $= 4 \times 9.00 \times \frac{1}{2}(0 + 62.4 \times 9)$
$$= 10,109 \text{ lb}$$

New water pressure $= 4 \times 9.25 \times \frac{1}{2}(0 + 62.4 \times 9.25)$
$$= 10,678 \text{ lb}$$

Increase in water pressure per side $= \mathbf{569}$ lb $(\mathbf{2.53}$ kN) $\qquad \mathbf{a}$

1.308   Pressure head + velocity head + elevation + lost head (if any) = constant = energy equation. All heads and elevations are in feet.

$$Q = v \times a \qquad 49.5 \text{ ft}^3/\text{s} = v \text{ fps} \times 7.07 \text{ ft}^2$$

$$v = 7 \text{ fps} \quad \text{and} \quad \frac{v^2}{2g} = \frac{49}{64.4} = 0.76 \text{ ft } (0.232 \text{ m})$$

P_R = pressure head at summit $= 2.31 \times 30 = 69.3$ ft $(21.12$ m)
Total head of pump $= 69.3 + 0.7 + 300 + 10 = 380$ ft $(115.84$ m)

$$\text{Energy of pump} = \frac{380 \text{ ft} \times 49.5 \text{ ft}^3/\text{s} \times 62.4 \text{ lb/ft}^3}{550 \text{ fps}}$$

$$= \mathbf{2130} \text{ hp } (\mathbf{1588} \text{ kW}) \qquad \mathbf{b}$$

1.309 $V_A = \dfrac{\text{discharge}}{A \text{ of 12-in pipe}} = \dfrac{2}{0.785} = 2.55 \text{ fps } (0.78 \text{ m/s})$

$V_D = \dfrac{2}{0.196} = 10.2 \text{ fps } (3.11 \text{ m/s})$

Equate energy of points A and B.

$-5 \times 2.31 + \dfrac{2.55 \times 2.55}{2 \times 32.2} + 0 + \text{pump head}$

$= +24 \times 2.31 + \dfrac{10.2 \times 10.2}{64.4} + 5$

Pump head $= +55.4 + 11.6 + 1.60 - 0.10 + 5 = 73.5 \text{ ft } (22.4 \text{ m})$

$\text{hp output of pump} = \dfrac{73.5 \times 2 \times 62.4}{550}$

$= \mathbf{16.7} \text{ hp } (\mathbf{12.45} \text{ kW})$ **c**

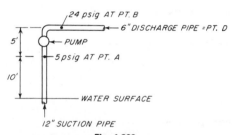

Fig. 1.309

1.310 This is the case of water passing through an orifice under constant head. The barge sinks as fast as the water rises within the barge.

$v = \sqrt{2gh} = \sqrt{64.4 \times 4} = 16.06 \text{ fps } (4.90 \text{ m/s})$

Rate water enters barge $= 0.60 \times 0.196 \times 16.06$
$= 1.88 \text{ ft}^3 (0.0532 \text{ m}^3)/\text{s}$

Time needed to sink barge $= \dfrac{20 \times 10 \times 2}{1.88} = \mathbf{213} \text{ s}$ **d**

1.311 Francis' formula for a contracted weir with no velocity of approach is given by

$Q = 3.33 \left(b - \dfrac{2H}{10}\right) H^{3/2}$ $b = 10 \text{ ft}, H = 0.875 \text{ ft}$

$$Q(\text{discharge of ft}^3/\text{s}) = 3.33 \left(10 - \frac{2 \times 0.875}{10}\right) \sqrt{0.875^3}$$
$$= 3.33(10 - 0.175) \sqrt{0.67}$$
$$= 3.33 \times 9.825 \times 0.819$$
$$= \textbf{26.8} \text{ ft}^3 \ (\textbf{0.76} \text{ m}^3)/\text{s} \quad \textbf{b}$$

[Coefficient 3.33 is approx. (1.87) for metric use.]

1.312 Francis' formula for a suppressed weir with no velocity of approach is given by

$$Q = 3.33(bH^{3/2}) = (3.33)(7)(0.819) = \textbf{19.1} \text{ ft}^3 \ (\textbf{0.54} \text{ m}^3)/\text{s} \quad \textbf{a}$$

1.313 Use energy equation (all items in feet).

Elev. = entrance loss + discharge loss + friction loss

$$100 = 0.5\frac{v^2}{2g} + \frac{v^2}{2g} + f\frac{l}{d}\frac{v^2}{2g}$$

f for clear smooth pipe and assumed velocity of 5 to 6 fps from tables = 0.0182.

$$l = 12{,}000 \text{ ft} \qquad d(\text{diam. pipe, ft}) = 1 \text{ ft}$$
$$100 = \left(0.5 + 1.0 + \frac{0.0182 \times 12{,}000}{1}\right)\frac{v^2}{64.4}$$

$$v^2 = \frac{6440}{220} = 29.2 \qquad v = 5.41 \text{ fps } (1.65 \text{ m/s})$$

$$Q = \text{average} = 0.785 \times 5.41 = \textbf{4.25} \text{ ft}^3 \ (\textbf{0.12} \text{ m}^3)/\text{s} \quad \textbf{c}$$

1.314 7500 gpm $= \dfrac{7500}{7.5 \times 60} = 16.7 \text{ ft}^3 \ (0.473 \text{ m}^3)/\text{s}$

$$Q = \text{average} \qquad v = \frac{16.7}{1.77} = 9.45 \text{ fps } (2.88 \text{ m/s})$$

Elev. = entrance loss + velocity loss + pipe loss + pressure head

$$180 = \left(0.5 + 1.0 + 0.0160\frac{4000}{1.5}\right)\frac{v^2}{2g} + h(\text{pressure})$$

$$h(\text{pressure}) = 180 - 44.2 \times \frac{9.45 \times 9.45}{64.4} = 180 - 61.3$$

$$= 118.7 \text{ ft } (36.18 \text{ m})$$

$$\text{Pressure in pipe} = \frac{118.7}{2.31} = \textbf{51.4} \text{ psi } (\textbf{354.4} \text{ kPa}) \quad \textbf{a}$$

1.315   Pressure head (pump) = total elev. pumped + total pipe losses.

$$Q = \text{average} \qquad v = \frac{2}{0.196} = 10.2 \text{ fps } (3.11 \text{ m/s})$$

$$h(\text{pump}) = 1250 - 1000 + \frac{0.0225 \times 8000}{0.5} \frac{v^2}{2g} + \frac{1.5v^2}{2g}$$

$$= 250 + 361.5 \frac{10.2 \times 10.2}{64.4} = 834 \text{ ft } (254.2 \text{ m})$$

$$834 = 1100 - 1000 + \frac{0.0225 \times 3200}{0.5} \frac{10.2 \times 10.2}{64.4}$$

$$+ \text{ pressure head}_{3200}$$

Pressure head at 3200 = 834 − 100 − 232.6 = 501.4 ft (152.8 m)
Pressure at 3200 = 501.4/2.31 = **217** psi (**1496** kPa)   **b**

1.316   Equate energy between A and B and between A and M.
A and B:

$$h(\text{pump}) = 900 - 800 + 0.02 \times \frac{4000}{1} \frac{v^2}{2g} + \frac{1.5v^2}{2g}$$

$$= 100 + 81.5 \frac{v^2}{2g}$$

A and M:

$$\left(100 + 81.5 \frac{v^2}{2g} \right) + 100 = 100 \times 2.31 + 0.02 \frac{2000}{1} \frac{v^2}{2g} + \frac{1.5v^2}{2g}$$

$$40 \frac{v^2}{2g} = 31 \qquad v^2 = \frac{31 \times 64.4}{40} = 50 \qquad v = 7.07 \text{ fps } (2.15 \text{ m/s})$$

$$Q = \text{average} = 0.785 \times 7.07 = \textbf{5.55} \text{ ft}^3 \textbf{ (0.16 m}^3\textbf{)}/\text{s} \qquad \textbf{1d}$$

$$\text{Pump hp output} = \frac{5.55 \times 62.4}{550} \left(100 + 81.5 \frac{31}{41.5} \right)$$

$$= 0.629(100 + 61)$$

$$= \textbf{101} \text{ hp } \textbf{(74.3} \text{ kW)} \qquad \textbf{2c}$$

1.317   Write energy relationship between points A, B, C, and Y. Hydraulic gradient at point Y will be called Y.

1. A and Y: $1000 - Y = \dfrac{0.02 \times 3000}{0.67} \dfrac{V_a^2}{2g} = 90 \dfrac{V_a^2}{2g}$

2. Y and B: $Y - 850 = \dfrac{0.02 \times 2000}{0.5} \dfrac{V_b^2}{2g} = 80 \dfrac{V_b^2}{2g}$

3. Y and C: $Y - 875 = \dfrac{0.02 \times 1000}{0.5} \dfrac{V_c^2}{2g} = 40 \dfrac{V_c^2}{2g}$

Combining steps 1 and 2:

$$150 = 90 \frac{V_a^2}{2g} + 80 \frac{V_b^2}{2g} \qquad (a)$$

Combining steps 1 and 3:

$$125 = 90 \frac{V_a^2}{2g} + 40 \frac{V_c^2}{2g} \qquad (b)$$

$$Q(\text{pipe } A) = Q(\text{pipe } B) + Q(\text{pipe } C)$$
$$0.349V_a = 0.196V_b + 0.196V_c \qquad (c)$$

Equations (a) to (c) can best be solved by assuming values of V_a, and then V_b and V_c can be found from Eqs. (a) and (b). If these values satisfy Eq. (c), the assumed value of V_a was good. Thus, if V_a is taken as 8.03 fps (2.45 m/s), then $V_b = 6.9$ fps (2.13 m/s) and $V_c = 7.4$ fps (2.26 m/s).

$$Q_a = 2.80 \text{ ft}^3 \ (0.79 \text{ m}^3)/\text{s} = Q_b + Q_c$$
$$= 1.35 \text{ ft}^3 \ (0.38 \text{ m}^3)/\text{s} + \textbf{1.45} \text{ ft}^3 \ (\textbf{0.041} \text{ m}^3)/\text{s} \qquad \textbf{b}$$

1.318 $Q_R = aV_R$, $7 = 1.767V_R$, $V_R = 3.96$ fps, $\dfrac{V_R^2}{2g} = 0.244$

Let elevation of hydraulic gradient at Y be Y. Equate energy between R and Y, Y and A, Y and B.

1. $400 - Y = 0.02(2000/1.5) \times 0.244 = 6.5$ ft (1.98 m)
   $Y = 393.5$ ft (119.94 m)

2. $393.5 - 250 = 143.5 = 0.02 \dfrac{13{,}000}{1} \dfrac{V_a^2}{2g} = 4.04V_a^2$

   $V_a^2 = 35.6 \qquad V_a = 5.96$ fps (1.83 m/s)

   $Q_R - Q_A = Q_B = 7 - 0.785 \times 5.96 = 2.32 = \dfrac{\pi d^2}{4} \times V_b$

   $d^2 = \dfrac{2.95}{V_b}$

3. $393.5 - 50 = 343.5 = 0.02 \dfrac{4000}{2g} \dfrac{V_b^2}{d}$

   $\dfrac{V_b^2}{d} = 276 \qquad \dfrac{V_b^4}{d^2} = 76{,}200$

   $V_b^5 = 2.95 \times 76{,}200 = 225{,}000$

by trial: $11^5 = 161,000$ $12^5 = 248,500$

$11.9^5 = 239,000$ $11.8^5 = 228,500$ $11.75^5 = 224,000$

Now
$$d^2 = \frac{2.95}{11.75} = 0.251$$

$$d = 0.5 \text{ ft} = \mathbf{6} \text{ in } (\mathbf{15.24} \text{ cm}) \qquad \mathbf{a}$$

1.319 The loss in head between points A and B in either the 36- or the 24-in pipe must be the same.

$$0.022 \frac{2000}{3} \frac{V_3^2}{2g} = 0.023 \frac{6000}{2} \frac{V_2^2}{2g}$$

$$\frac{44}{3} V_3^2 = 69V_2^2 \qquad V_3 = V_2 \sqrt{\frac{3 \times 69}{44}} = 2.17V_2$$

$$Q_4 = Q_2 + Q_3$$
$$75.4 = 3.14V_2 + 7.07V_3 = 3.14V_2 + 7.07 \times 2.17V_2$$
$$V_2 = \frac{75.4}{3.14 + 15.33} = 4.08 \text{ fps } (1.24 \text{ m/s})$$

Now
$V_3 = 8.85 \text{ fps } (2.70 \text{ m/s})$
$Q_2 = 3.14 \times 4.08 = \mathbf{12.9} \text{ ft}^3 \ (\mathbf{0.365} \text{ m}^3)/\text{s} \qquad \mathbf{c}$
$Q_3 = 7.07 \times 8.85 = \mathbf{62.5} \text{ ft}^3 \ (\mathbf{1.770} \text{ m}^3)/\text{s}$
$\Sigma = Q_4 \qquad\quad = \mathbf{75.4} \text{ ft}^3 \ (\mathbf{2.135} \text{ m}^3)/\text{s}$

1.320 Let p = pipe and n = nozzle.

$$A_p \times V_p = 8 \times A_n \times V_n$$
$$19.63 \times V_p = 8 \times 0.049 \times V_n = 0.392V_n$$
$$V_n = 50V_p$$

Equate energy of reservoir and nozzles.

$$600 = 0.5 \frac{V_p^2}{2g} + 0.017 \frac{25,000 \times V_p^2}{5 \times 2g} + \left\{ 1 + \left[\frac{1}{(0.95)^2} - 1 \right] \right\} \frac{2500 \; V_p^2}{2g}$$
$$38,600 = 85.5V_p^2 + 2770V_p^2 = 2855.5V_p^2$$
$$V_p^2 = 13.5 \qquad V_p = 3.68 \text{ fps } (1.12 \text{ m/s})$$
$$V_n = 184.0 \text{ fps } (56.08 \text{ m/s})$$
$$Q_p = 3.68 \times 19.63 = 72.3 \text{ ft}^3 \ (2.05 \text{ m}^3)/\text{s}$$
$$\text{hp eight jets} = \frac{72.3 \times 62.4}{550} \frac{184.0 \times 184.0}{64.4}$$
$$= \mathbf{4312} \text{ hp } (\mathbf{3171} \text{ kW}) \qquad \mathbf{b}$$

1.321 Let t = throat and p = pipe.

$$\frac{A_t}{A_p} = \frac{1}{9} \qquad \frac{d_t}{d_p} = \frac{1}{3} = \frac{8 \text{ in}}{24 \text{ in}}$$

h_t and h_p = recorded pressure heads, ft

$$Q = \frac{C \times A_t}{\sqrt{1 - (d_t/d_p)^4}} \sqrt{2g(h_t - h_p)}$$

$$= \frac{0.99 \times 0.349 \sqrt{64.4(36 - 9)}}{\sqrt{1 - \frac{1}{81}}} = \frac{0.346 \times 8.03 \times \sqrt{27}}{\sqrt{80}/9}$$

$$= \frac{9 \times 2.78 \times 5.2}{8.94} = \mathbf{14.53} \text{ ft}^3 \ (\mathbf{0.411}/\text{m}^3)/\text{s} \qquad \mathbf{d}$$

1.322 Area $= \dfrac{2 \times 4 \times 12}{2} + 4 \times 30 = 168 \text{ ft}^2 \ (15.62 \text{ m}^2)$

Wetted perimeter $= 30 + 2\sqrt{16 + 144}$
$$= 30 + 2 \times 12.65 = 55.3 \text{ ft } (16.86 \text{ m})$$

Hydraulic radius $= R = \dfrac{A}{P} = \dfrac{168}{55.3} = 3.04 \qquad R^{1/2} = 1.743$

$$S = \text{slope of 1 in } 1000 = 0.001$$

$$n = 0.02 \qquad \frac{n}{R^{1/2}} = 0.01145$$

Kutter's $C = \dfrac{41.65 + 0.00281/S + 1.811/n}{1 + (41.65 + 0.00281/S)\,(n/R^{1/2})}$

$$= \frac{41.65 + 2.81 + 90.55}{1 + 44.46 \times 0.01145} = \frac{135.01}{1.51}$$

$$= 89.4$$

$v = C \times R^{1/2} \times S^{1/2} = 89.4 \times 1.743 \times 0.0316 = 4.92 \text{ fps } (1.50 \text{ m/s})$

$$Q = 168 \times 4.92 = \mathbf{826} \text{ ft}^3 \ (\mathbf{23.4} \text{ m}^3)/\text{s} \qquad \mathbf{a}$$

1.323 $y = \dfrac{\omega^2 x^2}{2g}$ where $\omega = \dfrac{2\pi(56)}{60}$

$$= 5.88 \text{ rad/s}$$

$$y_1 = \frac{(5.88)^2(3)^2}{64.4} = 4.82 \text{ ft } (1.47 \text{ m})$$

$$y_2 = \frac{(5.88)^2(2)^2}{64.4} = 2.14 \text{ ft } (0.65 \text{ m})$$

$$y_3 = y_1 - y_2 = \mathbf{2.68} \text{ ft } (\mathbf{0.82} \text{ m}) \qquad \mathbf{1b}$$

[Check using parabola: $y_2 = \dfrac{4}{9}(4.82) = 2.14.$]

$$\text{Vol. water lost} = \frac{Ay}{2} = \frac{\pi(2)^2(2.14)}{2} = \textbf{13.45} \text{ ft}^3 \textbf{ (0.38 m}^3\textbf{)} \qquad \textbf{2d}$$

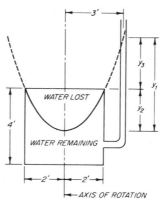

Fig. 1.323

1.324 40 mph = 58.67 fps.

$$\text{Force} = C_D \times w \times A \frac{v^2}{2g}$$

$$= (1.12)(0.0807)(100\pi)\frac{(58.67)^2}{64.4} = \textbf{1520} \text{ lb } \textbf{(6761 N)} \qquad \textbf{b}$$

1.325 Let point 1 be at the toe of the spillway and point 2 be at the point the jump has occurred.

$$d_1 d_2 \frac{d_1 + d_2}{2} = \frac{Q^2}{g}$$

$$d_1 = 0.8 \text{ ft} \qquad \text{width} = 1.0 \text{ ft} \qquad a_1 = 0.8 \text{ ft}^2$$
$$v_1 = 30 \text{ fps} \qquad Q = av = 0.8 \times 30 = 24 \text{ ft}^3/\text{s}$$

$$0.8 d_2 \frac{0.8 + d_2}{2} = \frac{(24)^2}{32.2} = 17.9$$

$$d_2(0.8 + d_2) = d_2^2 + 0.8 d_2 = 44.8$$
$$d_2^2 + 0.8 d_2 + 0.16 = 44.8 + 0.16$$
$$d_2 + 0.4 = \sqrt{44.96} = 6.7$$
$$d_2 = \textbf{6.3} \text{ ft } \textbf{(1.92 m)} \qquad \textbf{1a}$$

$$v_2 = \frac{24}{1 \times 6.3} = \textbf{3.81} \text{ fps } (\textbf{1.16} \text{ m/s}) \quad \textbf{2b}$$

$$E_{s1} = \frac{v_1^2}{2g} + d_1 = \frac{900}{64.4} + 0.8 = 14.75 \text{ ft·lb/lb } (4.50 \text{ N·m/N})$$

$$E_{s2} = \frac{3.81^2}{64.4} + 6.3 = 6.52 \text{ ft·lb/lb } (1.99 \text{ N·m/N})$$

$$E_{s1} - E_{s2} = 14.75 - 6.52 = 8.23 \text{ ft·lb/lb } (2.51 \text{ N·m/N})$$

$$\text{Energy} = \frac{24 \times 200 \times 62.4 \times 8.23}{550} = \textbf{4460} \text{ hp } (\textbf{3280} \text{ kW}) \quad \textbf{3c}$$

$$\frac{L}{d_2} = 4.8 \qquad L = 4.8 \times 6.3 = \textbf{30.24} \text{ ft } (\textbf{9.22} \text{ m}) \quad \textbf{4d}$$

MATHEMATICS AND MEASUREMENTS

1.401 By law of sines

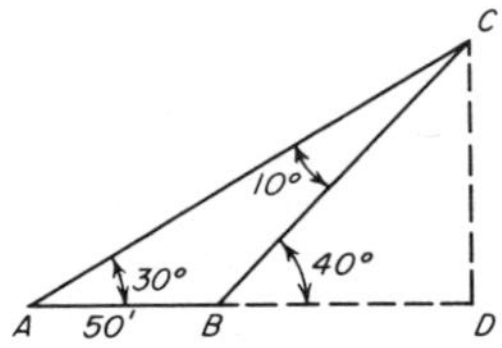

$$\frac{BC}{AB} = \frac{\sin 30°}{\sin 10°}$$

Fig. 1.401

$BC = 50 \times 0.5000/0.1737$
  $= 143.97 \text{ ft } (43.88 \text{ m})$

$CD = 143.97 \times \sin 40°$
  $= 143.97 \times 0.6428 = 92.54 \text{ ft } (28.21 \text{ m})$

$BD = 143.97 \times \cos 40° = 143.97 \times 0.7660 = 110.29 \text{ ft } (33.62 \text{ m})$

$$\tan 30° = \frac{CD}{AB + BD}$$

$CD = 160.29 \times 0.5774 = \textbf{92.54} \text{ ft } (\textbf{28.21} \text{ m}) \; (check) \quad \textbf{b}$

1.402 Equation of line is in the form $y = mx + b$ when $x = 0$, $mx = 0$, and $y = b = 1$.

$$m = \text{slope of line} = \frac{y_2 - y_1}{x_2 - x_1} = \frac{3 - 2}{4 - 2} = \frac{1}{2} = 0.5$$

$$y = \textbf{0.5}x + 1$$

Check: When $x = 2$, $y = 0.5 \times 2 + 1 = 2$
When $x = 4$, $y = 0.5 \times 4 + 1 = 3$

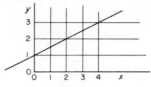

Fig. 1.402

1.403 $x^2 + y^2 = 5z$ (1)
                                $x^2 - y^2 = 3z$ (2)

Eqs. (1) + (2): $2x^2 = 8z$ and if $z = w^2$
                        $x^2 = 4z = 4w^2$
                        $x = 2w$ and $y = w$ or $x = 2y$

If $z = 1^2$, then $x = 2$ and $y = 1$
  $z = 2^2$, then $x = 4$ and $y = 2$
  $z = 3^2$, then $x = 6$ and $y = 3$
  $z = n^2$, then $x = 2n$ and $y = n$

There are an infinite number of values that will satisfy.

1.404   1. $A = 89.42 = 0.785d^2$
               $d =$ **10.67** in (**27.10** cm)   **1b**
     2. Circumference $= 3.14 \times 10.67$
                       $=$ **33.52** (**85.14** cm)   **2d**
     3. Length of side of regular inscribed hexagon $= d/2$
                                       $=$ **5.335** in (**13.55** cm)
                                                        3a

1.405

Grade (percent) $AB' = -6.94 \times 100/100 = $ **-6.94** percent   **c**

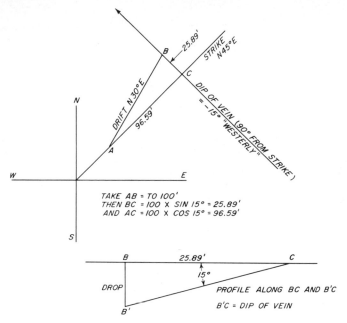

TAKE AB = TO 100'
THEN BC = 100 × SIN 15° = 25.89'
AND AC = 100 × COS 15° = 96.59'

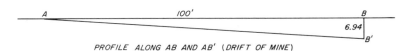

DROP B B' = 25.89 × TAN 15° = 25.89 × 0.2680 = 6.94'

PROFILE ALONG AB AND AB' (DRIFT OF MINE)

Fig. 1.405

1.406 Tape was used 9.17 times. If the tape is short, the distance measured is too long.

Total correction $= -0.10 \times 9.17 = -0.92$ ft (0.280 m)

Corrected or true distance taped $= 916.58 - 0.92$
$$= \textbf{915.66} \text{ ft } (\textbf{279.093} \text{ m}) \quad \textbf{a}$$

1.407 $4 + \dfrac{x + 3}{x + 3} - \dfrac{4x^2}{x^2 - 9} = \dfrac{x + 9}{x + 3}$

To clear fractions, multiply by $x^2 - 9$:

$$4x^2 - 36 + x^2 + 6x + 9 - 4x^2 = x^2 + 6x - 27$$

Combining terms,

$$-27 = -27 \quad \text{or} \quad 0 = 0$$

Since this is an identity, there are an infinite number of $+$ or $-$ values for x.

A special case is for $x = \pm 3$.

$x = +3$:

$$+4 + \frac{6}{0} - \frac{36}{0} = \frac{12}{6} \quad \text{or} \quad +4 + \infty_1 - \infty_2 = 2$$

$x = -3$:

$$+4 - \frac{0}{-6} - \frac{36}{0} = +\frac{6}{0} \quad \text{or} \quad +4 - 0 - \infty_2 = \infty_1$$

If we approach ± 3 as a limit from 2 or 4 and use sufficient significant figures, the two very large values, here called infinity, will differ by enough to show that the equation is an identity.

1.408 The center of gravity of a circle is at its center and that of a triangle is one-third of the height above the base (or side).

Find center of gravity of 24-in (60.96-cm) diam. circle with 9×12 in (22.86×30.48 cm) triangular hole.

Above OX, take moments of areas about OX.

Part	Area	$\times$	Arm	$=$	Moment	
Circle	452.4 in^2	$\times$ 12 in			$= +5428.8$ in^3	
Triangle	$- 54.0$	$\times$ 15			$= - 810.0$	
Net area	398.4 in^2	$\times$ (**11.59** in) (**29.44** cm)			$= +4618.8$ in^3	**1a**

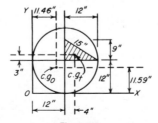

Fig. 1.408

To right of OY take moments of areas about OY.

Circle	$+452.4$	$\times\ 12$	$=\ +5428.8$
Triangle	$-\ 54.0$	$\times\ 16$	$=\ -\ 864.0$
Net area	$398.4\ \text{in}^2$	$\times\ (\mathbf{11.46}\ \text{in})\ (\mathbf{29.11}\ \text{cm})$	$=\ +4564.8\ \text{in}^3$ **2b**

1.409 The center of gravity of this area lies on the YY axis. To find X, take moment of areas (of right half only because of symmetry) about top of figure.

Area no.	Area	$\times$	Arm	$=$ Moment
1	2×6 $= 12\ \text{in}^2$	$\times\ 1\ \text{in}$		$=\ 12.0$
2	1×10 $= 10$	$\times\ 7$		$=\ 70.00$
3	$\frac{1}{2} \times 1 \times 10 =\ 5$	$\times\ 5.33$		$=\ 26.67$
Total	27	$\times\ (\mathbf{4.02}\ \text{in})\ (\mathbf{10.21}\ \text{cm})$		$=\ 108.67$

The moment of inertia of an area about an axis other than that through its center of gravity $= I$ about center of gravity plus area times distance between center-of-gravity axis and new axis squared.

$$I_{xx} = \mathbf{644.0}\ \text{in}^4\ (\mathbf{26.805}\ \text{cm}^4)$$

Double area	I_{cg}	$+$	Area $\times\ d^2$	$=$	New I
1	$\frac{1}{12} \times 12 \times 2 \times 2 \times 2$	$+\ 12 \times 2 \times 3.02 \times 3.02$		$=$	$8.0 + 218.9$
2	$\frac{1}{12} \times 2 \times 10 \times 10 \times 10$	$+\ 2 \times 10 \times 2.98 \times 2.98$		$=$	$166.7 + 177.6$
3	$\frac{1}{36} \times 2 \times 10 \times 10 \times 10$	$+\ \frac{1}{2} \times 2 \times 10 \times 1.31 \times 1.31$	$=$		$55.6 +\ 17.2$
Total			$\mathbf{644.0}\ \text{in}^4$	$=$	$230.3 + 413.7$ **1c**

$$I_{yy} = \mathbf{313.0}\ \text{in}^4\ (\mathbf{13,028}\ \text{cm}^4)\ \ \ \mathbf{2c}$$

Double area	I_{cg}	$+$	Area $\times\ d^2$	$=$	New I
1	$\frac{1}{12} \times 2 \times 12 \times 12 \times 12$	$+\ 0$		$=$	288.00
2	$\frac{1}{12} \times 10 \times 2 \times 2 \times 2$	$+\ 0$		$=$	6.67
3	$2 \times \frac{1}{36} \times 10 \times 1 \times 1 \times 1$	$+\ 2 \times \frac{1}{2} \times 1 \times 10 \times 1.33 \times 1.33$		$=$	0.55
					17.78
Total					$313.00\ \text{in}^4$

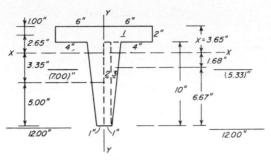

Fig. 1.409

1.410 X distance to center of gravity $= 2.14$ in to right. Moments about YY axis:

Area no.	Area	$\times$	Arm	$=$ Moment
1	16 sq in.	$\times$ 1 in		$= 16$ in³
2	8	$\times$ 4 in		$= 32$
3	$\underline{4}$	$\times$ 3 in		$= \underline{12}$
Total	28 in²	$\times$ **(2.143** in) **(5.44** cm)		$= 60$ in³

Fig. 1.410

Y distance to center of gravity $= 3.29$ in down from xx. Moments about xx:

Area no.	Area	$\times$	Arm	$=$ Moment
1	16 in²	$\times$ 4 in		$= 64$ in³
2	8	$\times$ 1		$= 8$
3	$\underline{4}$	$\times$ 5		$= \underline{20}$
Total	28 in²	$\times$ **(3.286** in) **(8.35** cm)		$= 92$ in³

$$I_{xx} = \textbf{453.3} \text{ in}^4 \ (\textbf{18,868} \text{ cm}^4) \qquad \textbf{1a}$$

Area	I_{cg}	$+$	$A \times d^2$	$=$	New I
1	$\frac{1}{12} \times 2 \times 8 \times 8 \times 8$	$+$	$16 \times 4 \times 4$	$=$	$85.3 + 256$
2	$\frac{1}{12} \times 4 \times 2 \times 2 \times 2$	$+$	$8 \times 1 \times 1$	$=$	$2.7 + 8$
3	$\frac{1}{12} \times 2 \times 2 \times 2 \times 2$	$+$	$4 \times 5 \times 5$	$=$	$\underline{1.3 + 100}$
				453.3 in⁴ $=$	$89.3 + 364$

$$I_{yy} = \mathbf{197.3} \text{ in}^4 \; (\mathbf{8212} \text{ cm}^4) \qquad \mathbf{2d}$$

Area	I_{cg}	$+ A \times d^2$	$= $ New I
1	$\frac{1}{12} \times 8 \times 2 \times 2 \times 2 +$	$16 \times 1 \times 1 =$	$5.3 + 16$
2	$\frac{1}{12} \times 2 \times 4 \times 4 \times 4 +$	$8 \times 4 \times 4 =$	$10.7 + 128$
3	$\frac{1}{12} \times 2 \times 2 \times 2 \times 2 +$	$4 \times 3 \times 3 =$	$\underline{1.3 + 36}$
		$197.3 \text{ in}^4 =$	$17.3 + 180$

1.411 Let x = rate (mph) of plane B and $x + 90$ = rate (mph) of plane A.

$$\frac{900}{x + 90} + 2.25 = \frac{900}{x} \qquad \text{hours} + \text{hours} = \text{hours}$$

Multiply through by $x(x + 90)$ to clear fractions:

$$900x + 2.25(x^2 + 90x) = 900x + 81{,}000$$
$$x^2 + 90x = 36{,}000$$

Complete square:

$$x^2 + 90x + 45 \times 45 = (x + 45)^2 = 36{,}000 + 2025$$
$$x + 45 = 195, \qquad x = 150 \text{ mph (241.4 km/h) for plane } A$$
$$x + 90 = \mathbf{240} \text{ mph } (\mathbf{386.2} \text{ km/h}) \text{ for plane } B \qquad \textbf{(c)}$$

Check: $900/240 + 2.25 = 6.00 \text{ h} = 900/150$

1.412 Let A be one number and B the other. Given

$$A^2 + B^2 = 100 \qquad 2(A + B) = 28$$

This suggests a 3-4-5 triangle, where

$$A = 6 \qquad B = 8 \qquad C = 10$$

Thus

$$36 + 64 = 100 \qquad 2(6 + 8) = 28 \qquad AB = 48$$

By direct solution

$$A + B = 14 \qquad A^2 + 2AB + B^2 = 196$$

But $A^2 + B^2 = 100$, so $2AB + 100 = 196$ and $AB = 48$. For a check, solve for A and B.

$$A = 14 - B \qquad A^2 = 196 - 28B + B^2$$
$$196 - 28B + B^2 + B^2 = 100 \qquad B^2 - 14B + 48 = 0$$

By quadratic equation

$$B = \frac{-(-14) \pm \sqrt{+196 - 192}}{2} = \frac{+14 \pm 2}{2}$$

$$= +8 \text{ or } +6 \qquad A = +6 \text{ or } +8$$

The product of $A \times B = $ **48.**   **b**

1.413 Area of field $= xy$

(1) $\qquad\qquad +xy + 2500 = (x + 100)(y - 25)$
(2) $\qquad\qquad +xy - 5000 = (x - 100)(y + 50)$
$\qquad\qquad\quad xy + 2500 = xy + 100y - 25x - 2500$

(a) $\qquad\qquad +100y - 25x = +5000$
$\qquad\qquad\quad\; xy - 5000 = xy - 100y + 50x - 5000$
(b) $\qquad\qquad -100y + 50x = 0$
(a) $\qquad\qquad\underline{+100y - 25x = +5000}$
(a) + (b) $\qquad\qquad\quad +25x = +5000$

$$\text{Length} = x = +\textbf{200} \text{ ft } (\textbf{60.96} \text{ m}) \qquad \textbf{c}$$
$$y = +100$$

Check: $\qquad\qquad\qquad xy = 20{,}000 \text{ ft}^2 \ (1828 \text{ m}^2)$
$$xy + 2500 = 22{,}500 = 300 \times 75$$
$$xy - 5000 = 15{,}000 = 100 \times 150$$

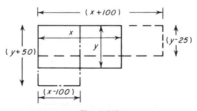

Fig. 1.413

1.414 Tire must be expanded

$$(62.378 - 62.263)\pi = 0.115\pi \text{ in}$$

Coefficient of expansion of steel $= 0.0000065$ units/(unit) (°F)

$t = $ degrees above 65°F that tire must be heated. Inside circumference of tire $= 62.263\pi$.

$$0.115\pi = 62.263\pi \times 0.0000065 \times t$$

$$t = \frac{0.115}{0.405 \times 1/1000} = 284°F$$

Temperature of tire must be raised to 284 + 65 = **349**°F (**176.1**°C) or above.   **c**

1.415   A -60°F change in temperature will shorten a 90.000-ft length of steel tape by $90 \times 60 \times 0.0000065 = 0.035$ ft (0.011 m).

90.035 ft (**27.443** m) must be read on the tape to lay out a 90.000-ft (27.432-m) distance on the ground.   **a**

1.416   10 laborers can dig $1\tfrac{1}{4} \times 150 = 187.5$ ft of trench in 7 h.

10 laborers can dig 200 ft of trench in $(7 \times 200)/187.5 = 7.467$ h.

3 laborers can backfill 200 ft of trench in 8 h.

10 laborers can backfill 200 ft of trench in $\tfrac{3}{10} \times 8 = 2.4$ h.

10 laborers can dig and backfill 200 ft of trench in 9.867 h = **9 h 52 min.**   **a**

1.417
$$x^2 + y^2 = +15{,}025 \tag{1}$$
$$2x + 4y = +530 \quad \text{or} \quad x = +265 - 2y \tag{2}$$

$$x^2 = +70{,}225 - 1060y + 4y^2$$
$$y^2 + 4y^2 - 1060y + 70{,}225 = +15{,}025 \tag{1}$$
$$y^2 - 212y + 11{,}040 = 0$$

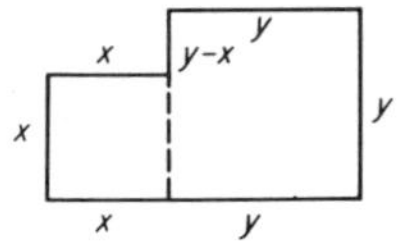

Fig. 1.417

By quadratic equation

$$y = \frac{-(-212) \pm \sqrt{212 \times 212 - 44{,}160}}{2}$$

$$= \frac{+212 \pm \sqrt{784}}{2} = \frac{+212 \pm 28}{2} = +106 \pm 14 = 120 \text{ or } 92$$

When $y = 120$, $x = $ **25** ft (**7.62** m) and $14{,}400 + 625 = 15{,}025$.   **a**

When $y = 92$, $x = $ **81** ft (**24.69** m) and $8464 + 6561 = 15{,}025$.   **d**

Both solutions satisfy.

1.418   *a.* $y^3 = x^3$ and $y = \sqrt[3]{x^3} = x$ or a straight line

Area $= \tfrac{1}{2} \times 4 \times 4 = $ **8**

b. Area $= \int_1^3 \left[(x^3 + 3x^2) \, dx = \frac{x^4}{4} + \frac{3x^3}{3} \right]_1^3$

$= 8\tfrac{1}{4} + 27 - \tfrac{1}{4} - 1 = 20 + 26 = $ **46**

c. $\dfrac{d}{dx}(4x^2 + 17x) = $ **8x + 17**

and $\dfrac{d}{dx}(ax^2 + b^{1/2}) = $ **2ax**

d. $\int (7x^3 + 4x^2)\, dx = \dfrac{7x^4}{4} + \dfrac{4x^3}{3} + \mathbf{c_1}$

and $\int x \cos (2x^2 + 7)\, dx = \dfrac{\sin}{4}(2x^2 + 7) + \mathbf{c_1}$

1.419 Increase in length of 100-ft tape from 10 to 30°C = $100 \times (30 - 10) \times 0.000011 = 0.022$ ft. Tape at 30°C = $100.042 + 0.022 = 100.064$ ft. Measurement of 1256.271 ft (382.911 m) is short because it was made with a long tape.

Correction at $+0.064$ ft/100 ft = $12.563 \times 0.064 = (0.245$ m$) + 0.804$ ft (0.245 m).

$$\text{True length of line} = 1256.271 + 0.804$$
$$= \mathbf{1257.075} \text{ ft } (\mathbf{383.156} \text{ m}) \qquad \mathbf{a}$$

1.420 Arithmetical progression:
$$x + (x + a) + (x + 2a) = 45$$
$$3x + 3a = 45 \qquad x + a = 15 \qquad a = 15 - x$$

If $x = 10$, then $a = 5$ and the three numbers are 10, 15, and 20. Geometrical progression:
$$10 + 2 = 12 = b \qquad r = 1.5$$
$$15 + 3 = 18 = br = 12 \times 1.5$$
$$20 + 7 = 27 = br^2 = 12 \times 1.5 \times 1.5$$

By direct solution:

$$x + a = +15 \qquad \text{or} \qquad a = +15 - b + 2 = 17 - b \qquad (1)$$
$$x + 2 = b \qquad x = b - 2 \qquad\qquad\qquad\qquad\qquad\qquad (2)$$

$$x + a + 3 = br = 18 \qquad r = \frac{18}{b} \qquad\qquad\qquad\qquad (3)$$

$$x + 2a + 7 = br^2 \qquad (b - 2) + 2(17 - b) + 7 = 18 \times \frac{18}{b} \qquad (4)$$

Equation (4) becomes

$$b^2 - 39b + 324 = 0 \qquad b = \frac{-(-39) \pm \sqrt{1521 - 1296}}{2}$$

$$b = \frac{+39 \pm 15}{2} = 12 \qquad x = 10 \qquad a = 5 \qquad r = 1.5$$

1.421 (a) $+4x^2 + 7y^2 = +32$, or $+12x^2 + 21y^2 = +96$ (1)

          $-3x^2 + 11y^2 = +41$, or $\underline{-12x^2 + 44y^2 = +164}$ (2)

Eqs. (1) + (2): $+65y^2 = +260$

       $y^2 = 4$ $y = +2$ $x = \pm 1$ $y = -2,$ $x = \pm 1$

(b)

(1) $+x + 2y - z = +6$

$2 \times$ (2) $+4x - 2y + 6z = -26$

(3) $\underline{+3x - 2y + 3z = -16}$

(a) = (1) + 2 × (2), $+5x + 5z = -20$, or $+x + z = -4$

(b) = (1) + (3) $+4x + 2z = -10$, or $\underline{+x + 0.5z = -2.5}$

(a') − (b') = $+0.5z = -1.5$

From (a'), $x = -1$; and from (1), $y = +2$ $z = -3$

Check using (2) and (3).

1.422 (a) $\cos 2A = 2\cos^2 A - 1$

          $2\cos^2 75° - 1 = \cos 150° = -\cos 30°$

$$\cos 30° = \frac{\sqrt{3}}{2} = \frac{1.732}{2} = 0.866$$

$$\cos^2 75° = \frac{1 - 0.866}{2} = \frac{+0.133}{2} = 0.0666$$

$$= \sqrt{0.0666} = 0.258$$

(b) Angle $A = 75° = AoX$ (first quadrant)

        $\sin A = Aa$

        $\cos A = oa$

        $\tan A = \dfrac{Aa}{oa}$

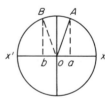

Fig. 1.422b

Angle $B = BoX = 105°$ (second quadrant)

    $\cos B = bo = -oa$

    $\sin B = Bb = Aa$

    $\tan B = \dfrac{Bb}{bo} = -\dfrac{Aa}{oa}$

1.423 Vol. $= x(20 - 2x)^2$

        $= 400x - 80x^2 + 4x^3$

(For V max) $\dfrac{dV}{dx} = 0 = +400 - 160x + 12x^2$

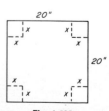

Fig. 1.423

or
$$+3x^2 - 40x + 100 = 0$$

Using quadratic equation,

$$x = \frac{-(-40) \pm \sqrt{1600 - 1200}}{6} = \frac{+40 \pm 20}{6} = +10 \text{ or } 3.33$$

V min (0) when $x = 10$, so use $x = 3.33$

$V = \mathbf{13.33} \times \mathbf{13.33} \times \mathbf{3.33} = \mathbf{592.59}$ in³ (**9710.77** cm³)   **b**

x may be found by trial (carry x to $\frac{1}{100}$ in).
1. Try values of $x = 1, 2, 3,$ and 4.
2. Try values of $x = 3.1$ to 3.5.
3. Try values of $x = 3.31$ to 3.35.

1.424 $s = \frac{1}{2}(a + b + c) = \frac{1}{2}(9 + 12 + 10) = 15.5$

$$r = \sqrt{\frac{(s - a)(s - b)(s - c)}{s}} = \sqrt{\frac{6.5 \times 3.5 \times 5.5}{15.5}}$$

$$= \sqrt{8.07} = 2.84$$

$\tan \dfrac{1}{2} A = \dfrac{r}{s - a} = \dfrac{2.84}{6.5} = 0.437,$ $\dfrac{1}{2} A = 23.6°,$ $A = 47.2°$

$\tan \dfrac{1}{2} B = \dfrac{r}{s - b} = \dfrac{2.84}{3.5} = 0.812,$ $\dfrac{1}{2} B = 39.05°,$ $B = 78.1°$

$\tan \dfrac{1}{2} C = \dfrac{r}{s - c} = \dfrac{2.84}{5.5} = 0.517,$ $\dfrac{1}{2} C = 27.35°,$ $C = 54.7°$

Check: $\overline{180.0°}$

Area $ABC = \frac{1}{2} \times 12 \times 10 \times \sin 47.2° = 60 \times 0.7337$

$= 44.0$ ft² (4.09 m²)

Vol. prism $= \dfrac{h_a + h_b + h_c}{3}$ 44.0

$= \dfrac{8.6 + 7.1 + 5.5}{3}$ 44.0

$= \dfrac{21.2}{3}$ 44.0

$= \mathbf{311}$ ft³ (**8.81** m³)   **c**

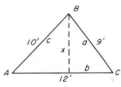

Fig. 1.424

1.425

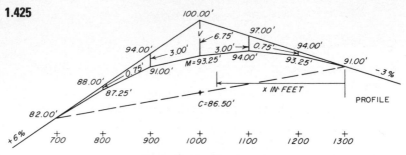

Fig. 1.425   Profile of curve.

Elevations Along Grade Tangents and Parabola

Station	Vertex elevation	−	Drop to tangent	=	Elevation of tangent	−	Drop to parabola	=	Elevation of parabola
7 + 00	100.00	−	3 × 6	=	82.00	−	0.00	=	82.00
8 + 00	100.00	−	2 × 6	=	88.00	−	0.75	=	87.25
9 + 00	100.00	−	1 × 6	=	94.00	−	3.00	=	91.00
10 + 00				=	100.00	−	6.75	=	93.25
11 + 00	100.00	−	1 × 3	=	97.00	−	3.00	=	94.00
12 + 00	100.00	−	2 × 3	=	94.00	−	0.75	=	93.25
13 + 00	100.00	−	3 × 3	=	91.00	−	0.00	=	91.00

Point C has an elevation the mean of that at 7 + 00 and 13 + 00.
Point M has an elevation the mean of 100.00 and C. $V = 6.75$ ft.
Distance down (feet) from tangent to prabola is

$$\left(\frac{x}{\frac{1}{2} \text{ length parabola}}\right)^2 \times 6.75 = \left(\frac{x}{300}\right)^2 \times 6.75$$

For stations 8 + 00 and 12 + 00,

$$\text{Drop} = (^{100}\!/_{300})^2 \times 6.75 = 0.75 \text{ ft}$$

For stations 9 + 00 and 11 + 00,

$$\text{Drop} = (^{200}\!/_{300})^2 \times 6.75 = 3.00 \text{ ft}$$

From the table, it is fairly obvious that the high point of the parabola is at station **11 + 00**. In other cases not so obvious, it is usually a matter of observation on which side of V the high point lies. In this case, knowing the high point is to the right of V,

Elevation of tangent $= 91.00 + 0.03x$

Elevation of curve $= y = 91.00 + 0.03x - \left(\dfrac{x}{300}\right)^2 \times 6.75$

When curve is at maximum elevation, the slope of the tangent is zero. Therefore

$$\frac{dy}{dx} = 0 = +0.03 - \frac{2x \times 6.75}{90{,}000}$$

$$13.5x = 2700 \qquad x = 200$$

or high point is at station $13 + 00 - 2 + 00 = $ **11 + 00.**   **c**

An exact transfer to SI units is not practical because the station would be 30.48 m in length.

MECHANICS (KINETICS)

1.501 Velocity = acceleration $\times$ time, or $v = at$.

$$v = 16 \text{ fps} = a \times 4 \qquad a = 4 \text{ ft/s}^2$$

$$\text{Tension in cable} = Ma = \frac{W}{g}(g + a)$$

$$T = (2000/32.2)(32.2 + 4) = \textbf{2250} \text{ lb } (\textbf{10,008 N}) \qquad \textbf{b}$$

1.502 Final velocity squared = initial velocity squared + twice acceleration $\times$ distance body moves.

$$v^2 = 0 = 16 \times 16 + 2 \times a \times 5 \qquad a = -25.6 \text{ ft/s}^2$$

$$\text{Tension in cable} = Ma = \frac{W}{g}(g + a)$$

$$(2000/32.2)(32.2 - 25.6) = \textbf{410} \text{ lb } (\textbf{1824 N}) \qquad \textbf{c}$$

1.503 Force normal to plane $= 40 \times 0.866$
$\qquad\qquad\qquad\qquad\quad = 34.67$ lb

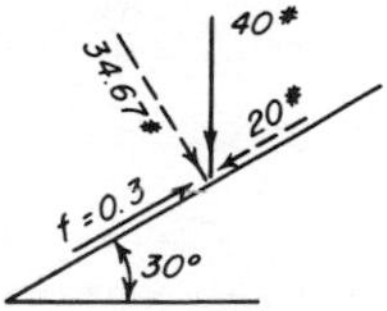

Force parallel to plane $= 40 \times 0.500$
$\qquad\qquad\qquad\qquad = 20.00$ lb
Friction force $= 34.67 \times 0.3 = 10.40$ lb
Resultant force parallel to plane
$\qquad = 20.00 - 10.40 = 9.60$ lb (42.70 N)

Fig. 1.503

Force $=$ mass $\times$ acceleration $\qquad$ or $\qquad 9.60 = \dfrac{40}{32.2} \times a$

$$a = 7.73 \text{ ft } (2.362 \text{ m})/\text{s}^2$$

Distance wt moves = ½ × acceleration × time squared

$$D = \tfrac{1}{2}at^2 \quad \text{and} \quad 60 = \tfrac{1}{2} \times 7.73t^2$$

$$t = \textbf{3.94 s} \quad \textbf{1c}$$

Distance wt moves during third second = $(7.73/2)(3^2 - 2^2)$

$$= \textbf{19.33} \text{ ft } (\textbf{5.89 m}) \quad \textbf{2a}$$

1.504 Initial velocity = $(30 \times 5280)/(60 \times 60) = 44$ fps (13.41 m/s)

Final velocity = initial velocity minus acceleration × time

$$0 = 44 - a \times 5 \quad a = 8.8 \text{ ft/s}^2 \text{ (2.68 m/s}^2)$$

Force = mass × acceleration $\quad F = (6000/32.2) \times 8.8 = 1640$ lb

Taking moments about bottom of rear wheel,

$$\Sigma M_{W_R} = 0 = +10 \times 2 \times W_F - 6000 \times 5 - 1640 \times 2$$
$$W_F = 1500 + 164 = 1664 \text{ lb (7.402 N)}$$

Normal pressure on each front wheel = **1664** lb (**7402** N)   **d**
Normal pressure on each rear wheel = **1336** lb (**5943** N)   **b**

1.505 Weight × distance of free fall + weight × displacement of spring = work of spring = average force of spring × displacement of spring squared.

$$1000 \times 6 + 1000 \times d_s = (2000/2) \times d_s^2$$
$$d_s^2 - d_s - 6 = 0 = (d_s - 3)(d_s + 2)$$
$$d_s = +\textbf{3.0} \text{ in } (\textbf{7.62 cm}) \quad \textbf{a}$$

(The spring does not elongate, hence $d_s = -2$ is meaningless.)

1.506 Work of spring = kinetic energy of car. Let R = resistance of spring, lb/in.

$$\frac{R}{2} \times d_s^2 = \frac{1}{2} Mv^2$$

$$\frac{R}{2} \times 2.5 \times 2.5 = \frac{1}{2} \times \frac{100,000}{32.2} \times 2 \times 2 \times 12$$

$$R = \textbf{23,850} \text{ lb/in } (\textbf{41.77 kN/cm}) \quad \textbf{c}$$

1.507 Block M has a 50-lb component parallel to the 30° plane and an 86.67-lb component normal to the 30° plane.

Friction force block M = $86.67 \times 0.2 = 17.33$ lb (7.86 kg)

Block N has an 86.67-lb component parallel to the 60° plane and a 50-lb component normal to the 60° plane.

Friction force block N = $50 \times 0.2 = 10$ lb (4.54 kg)
Net force block M = $50 + 17.33 = 67.33$ lb (30.54 kg)

Net force block $N = 86.67 - 10 = 76.67$ lb (34.78 kg)

Resultant force $= 76.67 - 67.33 = 9.33$ lb down 60° plane

$$\text{Force} = \text{mass} \times \text{acceleration} \qquad 9.33 = \frac{100 + 100}{32.2} \times a$$

$$a = \mathbf{1.505} \text{ ft } (\mathbf{0.459} \text{ m)/s}^2 \qquad \mathbf{1a}$$

Distance weights move $= \frac{1}{2} \times$ acceleration $\times$ time squared

$$15 = \frac{1}{2} \times 1.505 \times t^2 \qquad t = 4.47 \text{ s} \qquad \mathbf{2b}$$

Tension in connecting cord $= T$

(On weight M side) $T = 67.33 + (100/32.2) \times 1.505$
$$= \mathbf{72.00} \text{ lb}$$

(On weight N side) $T = 76.67 - (100/32.2) \times 1.505$
$$= \mathbf{72.00} \text{ lb } (\mathbf{32.66} \text{ kg}) \qquad \mathbf{3d}$$

1.508 $T_x = T_y$ and $T_x + T_y = T_z = 2T_x = 2T_y$ (where $T =$ tension in cords carrying weights).

Weights y and z will fall, weight x will rise.

Say weight x rose 10 ft and weight y fell 8 ft, then the pulley carrying their joint cord must rise one-half the difference or 1 ft for cord to remain taut. That gives $d_z = (d_x - d_y)/2$.

$$\therefore a_z = \frac{a_x - a_y}{2} = 0.5(a_x - a_y) \tag{1}$$

$$T_x = \frac{10g}{g} + \frac{10}{g}a_x = +10 + 0.31a_x \tag{2}$$

$$T_y = \frac{20g}{g} - \frac{20}{g}a_y = +20 - 0.62a_y \tag{3}$$

$$T_z = \frac{30g}{g} - \frac{30}{g}a_z = +30 - 0.93a_z \tag{4}$$

$$= +20 + 0.62a_x \tag{2'}$$

$$= +40 - 1.24a_y \tag{3'}$$

Equation (5) = Eq. (3') − Eq. (2'):

$$0 = +20 - 0.62a_x - 1.24a_y \tag{5}$$

Equation (6) = Eq. (3') − Eq. (4):

$$0 = +10 - 1.24a_y + 0.93a_z \tag{6}$$

but Eq. (1') $= 0.93 \times$ Eq. (1) $+ 0.93a_z = +0.465a_x - 0.465a_y$

Substituting Eq. (1') in Eq. (6)

$$-0.465a_x + 0.465a_y + 1.24a_y = +10 \tag{6'}$$

Equation $(7) = \frac{4}{3} \times$ Eq. $(6')$

$$-0.62a_x + 2.27a_y = +13.33 \qquad (7)$$
$$\underline{+0.62a_x + 1.24a_y = +20.00} \qquad (5)$$
$$3.51a_y = +33.33 \qquad (8)$$

Equation $(8) =$ Eq. $(5) +$ Eq. (7)
Equation $(8')$ is

$$a_y = +9.5 \text{ ft/s}^2, \ v_y = \textbf{9.51} \text{ fps } \textbf{(2.90 m/s)} \qquad \textbf{1c}$$

From Eq. (5)

$$a_x = +13.3 \text{ ft/s}^2, \ v_x = \textbf{13.3} \text{ fps } \textbf{(4.05 m/s)} \qquad \textbf{1d}$$

From Eq. (1)

$$a_z = +1.9 \text{ ft/s}^2, \ v_z = \textbf{1.9} \text{ fps } \textbf{(0.579 m/s)} \qquad \textbf{1a}$$

Now $d = \frac{1}{2}at^2$ and $v = at$. At the end of 1 s,

$$d_x = \textbf{6.70} \text{ ft } \textbf{(2.04} \text{ m) up} \qquad \textbf{2c}$$
$$d_y = \textbf{4.75} \text{ ft } \textbf{(1.45} \text{ m) down} \qquad \textbf{2b}$$
$$d_z = \textbf{0.95} \text{ ft } \textbf{(0.289} \text{ m) down} \qquad \textbf{2a}$$

1.509   Positive work $-$ negative work $=$ final kinetic energy.

Positive work $=$ ft·lb of drop

$$= \frac{1 \times 100 \text{ ft}}{100} \times 120{,}000 \text{ lb} = 120{,}000 \text{ ft·lb}$$

Negative work $=$ friction loss
$$= 10 \text{ lb/ton} \times 60^T \times 100 = -60{,}000 \text{ ft·lb}$$

Final kinetic energy $= \dfrac{1}{2} \times \dfrac{120{,}000}{32.2} \times v^2 = 60{,}000 \text{ ft·lb}$

$$v^2 = 32.2$$

Striking velocity $= \textbf{5.67}$ fps **(1.73 m/s)**   **1c**

Final kinetic energy $= 60{,}000 \times 12 = \dfrac{100{,}000}{2} \times d_s^2$

Spring will shorten **3.795** in **(9.639** cm)   **2d**

1.510   Initial kinetic energy $=$ uphill work $+$ friction

$$30 \text{ mph} = \frac{30 \times 5280}{60 \times 60} = 44 \text{ fps } (13.41 \text{ m/s})$$

$$\frac{1}{2} \times \frac{80{,}000}{32.2} \times (44)^2 = \frac{80{,}000}{2000} \times 10 \times d_1 + \frac{2 \times d_1}{100} \times 80{,}000 \qquad (1)$$

Divide Eq. (1) by 800:

$$3010 = 0.5d_1 + 2d_1 = 2.5d_1$$

or

$$d_1 = \textbf{1204} \text{ ft } (\textbf{367.0} \text{ m}) \text{ on slope} \qquad \textbf{1d}$$

Downhill work − friction = kinetic energy

$$+ \frac{2 \times 1204}{100} \times 80,000 - 10 \times 1204 \times \frac{80,000}{2000} = \frac{80,000 \times v^2}{64.4}$$

$$+24.08 - 6.02 = 18.06 = \frac{v^2}{64.4}$$

$$v = \textbf{34} \text{ fps } (\textbf{10.36} \text{ m/s}) \qquad \textbf{2a}$$

Kinetic energy = friction

$$80,000 \times \frac{v^2}{64.4} = \frac{80,000}{2000} \times 10 \times d_2$$

$$18.06 = \frac{d_2}{200}$$

$$d_2 = \textbf{3612} \text{ ft } (\textbf{1100.9} \text{ m}) \text{ on level track} \qquad \textbf{3c}$$

1.511 $\dfrac{50}{r} = \sin 5° = 0.0872, r = 574 \text{ ft } (174.96 \text{ m})$

$$45 \text{ mph} = 45 \times 5280/3600 = 66 \text{ fps } (20.12 \text{ m/s})$$

Gage of standard railroad track = 4 ft 8½ in or 4.9 ft (1.49 m) center-to-center rails.

   1. Superelevation for equal wheel pressures is

$$\frac{G \times v^2}{g \times r} = \frac{4.9 \times 66 \times 66}{32.2 \times 574} = \textbf{1.15} \text{ ft } (\textbf{0.351} \text{ m}) \qquad \textbf{1b}$$

   2. Horizontal force 5 ft above rail for 60-mph speed is

$$\frac{Wv^2}{g \times r} = \frac{100,000 \times 88 \times 88}{32.2 \times 574} = 41,950 \text{ lb } (20,287 \text{ kg})$$

Superelevation = 8 in in 4.9 ft or 0.67/4.9, and position of vertical force (weight) off center at top of rails is

$$x = 5 \times 0.67/4.9 = 0.683 \text{ ft } (0.208 \text{ m})$$

Distant vertical force from center inner rail is

$$4.9/2 - 0.683 = 1.767 \text{ ft } (0.539 \text{ m})$$

$\Sigma M_{\text{inner rail}}$

$$0 = +5 \times 41{,}950 + 1.767 \times 100{,}000 - 4.9 \times R_{\text{outer}}$$

Normal force outer rail $= R_{\text{outer}} = 42{,}750 + 36{,}000$
$$= \mathbf{78{,}750} \text{ lb } (\mathbf{35{,}721} \text{ kg}) \qquad \mathbf{2a}$$

1.512 60 mph $= 88$ fps $(26.82$ m/s$)$
  a. $\tan A = 0.15$
  $\qquad A = 8.53° =$ angle of superelevation
  b. Now

$$\tan (A + B) = \frac{88 \times 88}{32.2 \times 500} = 0.482$$

$$A + B = 25.73°$$

$$A = \mathbf{8.53°} \text{ (angle of superelevation)}$$

$$B = 17.20° \qquad \text{and} \qquad \tan B = 0.310$$

Skidding will impend if f is below **0.310.**   **c**

1.513 Vertical is shorter than 100 ft by $(5 \times 5)/200 = 0.125$ ft $= h$.

$$\text{Tension in cable} = \frac{100}{99.875} \times 20{,}000 = 20{,}025 \text{ lb } (9083.3 \text{ kg})$$

$$\text{Horizontal pull} = \frac{5}{99.875} \times 20{,}000 = \mathbf{1001} \text{ lb } (\mathbf{454.1} \text{ kg}) \qquad \mathbf{1b}$$

$$v^2 = 2gh = 2 \times 32.2 \times 0.125 \qquad \text{or} \qquad 0.25g$$

Tension of cable when swinging through vertical

$$= 20{,}000 + \frac{20{,}000 \times 0.25g}{g \times 100} = \mathbf{20{,}050} \text{ lb } (\mathbf{9094.7} \text{ kg}) \qquad \mathbf{2c}$$

1.514 $\sin 45° = \cos 45° = 0.707$
  Vertical velocity at high point of path $= 0 =$ initial velocity $\times \sin 45° - gt$:

$$300 \times 0.707 = 32.2 \times t \qquad t = 6.6 \text{ s}$$

Maximum height projectile will reach is

$$h_{\max} = 300 \times t \times \sin 45° - \tfrac{1}{2}gt^2$$
$$= 300 \times 6.6 \times 0.707 - \tfrac{1}{2} \times 32.2 \times 6.6 \times 6.6$$
$$= 1400 - 700 = \mathbf{700} \text{ ft } (\mathbf{213.4} \text{ m}) \qquad \mathbf{1a}$$

[Also for 45° angle, $h_{\max} =$ initial velocity squared divided by $4g$ or $(300 \times 300)/4g = 700$ ft.]

Range = twice initial velocity × cos 45° × t
  = 2 × 300 × 0.707 × 6.6 = 2800 ft (**853.44** m)   **2c**

[Also max range = v^2/g = (300 × 300)/g = 2800 ft.]

1.515 (See the special formulas in answer 1.514.)

$$\text{Max range} = 40 \times 5280 = \frac{v^2}{32.2} \qquad v^2 = 6,800,000$$

$$v = \textbf{2600} \text{ fps } (\textbf{192.48} \text{ m/s}) \qquad \textbf{1b}$$

Max height = $\dfrac{v^2}{4g}$ = ¼ × max range = 40/4 = **10** mi (**16.093** km)   **2d**

1.516 18 in = 1.5 ft (0.457 m)

sin 60° = 0.866 × 1.50 = 1.30 ft (0.396 m) (radius)
cos 60° = 0.500 × 1.50 = 0.75 ft (0.229 m) (center ball below top end of
                                                                                rod)

$$\tan 60° = 1.732 = \frac{v^2}{gr} \qquad v^2 = 1.732 \times 32.2 \times 1.30$$

$$v = 8.52 \text{ fps } (2.60 \text{ m/s})$$

and

$$60 \times \frac{8.52}{2\pi 1.30} = \textbf{62.6} \text{ rpm} \qquad \textbf{1a}$$

Wt 3-in cast-iron ball = ⅙ × π × (¼)³ × 450
                    = 3.68 lb (1.67 kg)

Moment (about point where arm connects to vertical shaft) of horizontal
effective force and weight (vertical) is

$$\Sigma M = 0 = -0.75 \frac{3.68 \times v^2}{gr} + 1.30 \times 3.68$$

Check: 0.75 × 1.732 = 1.30

Tension in arm = 3.68 $\sqrt{(1)^2 + (1.732)^2}$
          = **7.36** lb (**32.7** N)   **2b**

1.517 60 mph = 88 fps (26.82 m/s)
Radius = 1000 ft (304.8 m) very nearly.
Let A = angle cord makes with vertical:

$$\tan A = \frac{v^2}{gr} = \frac{88 \times 88}{32.2 \times 1000} = 0.2735 \qquad A = \textbf{15.3°} \qquad \textbf{1c}$$

Horiz. displacement of wt $= 6 \times \sin 15.3°$
$$= 6 \times 0.264 = \textbf{1.58} \text{ ft } (\textbf{0.482} \text{ m})\qquad \textbf{2d}$$

$$\text{Tension in cord} = 25 \sqrt{(1)^2 + (0.274)^2}$$
$$= 25 \times 1.036$$
$$= \textbf{25.9} \text{ lb } (\textbf{115.2} \text{ N})\qquad \textbf{3a}$$

1.518 Distance projectile must travel $= (\text{init. velocity})^2 \dfrac{\sin 2A}{g}$

$$= 2(v \cos A)\, \frac{v \sin A}{g}$$

$$10 \times 5280 = \frac{1500 \times 1500}{32.2} \times \sin 2A$$

$$\sin 2A = \frac{322 \times 0.528}{225} = 0.756 \qquad 2A = 49.1°$$

$$A = \textbf{24°33'}\qquad \textbf{c}$$

1.519 $I = \frac{1}{2}Mr^2$ for solid cylinder about geometric axis

$$M(\text{24-in diam. cylinder}) = \pi \left(\frac{12}{12}\right)^2 \times \frac{4}{3} \times \frac{450}{32.2} = 58.5$$

$$M(\text{22-in diam. cylinder}) = \pi \left(\frac{11}{12}\right)^2 \times \frac{4}{3} \times \frac{450}{32.2} = 49.2$$

$$I(\text{24-in cylinder}) = \frac{1}{2} \times 58.5 \times 1^2 \quad\ = 29.3$$
$$I(\text{22-in cylinder}) = \frac{1}{2} \times 49.2 \times (11/12)^2 = 20.6$$

$$I(\text{net}) \; (\text{lb}\cdot\text{ft}\cdot\text{s}^2) = \textbf{8.7} \; (\textbf{11.80} \text{ N}\cdot\text{m}\cdot\text{s}^2)\qquad \textbf{d}$$

1.520 $M = (490/32.2) \times \pi \times r^2 \times L = 47.8 \times r^2 \times L$

$$I = \frac{1}{2}Mr^2 = 23.9 \times r^4 \times L$$
$$I(\text{about axis of 40-in disk}) = I_{cg} + M \times d^2$$

where $d = $ distance between axes

$$I \,(40 \text{ in}) = 23.9 \times \frac{1}{3} \times (20/12)^4 = +61.50$$
$$I \,(10 \text{ in}) = \ \ 7.97 \times (5/12)^4 \qquad\quad = -\ 0.24$$
$$I \,(12 \text{ in}) = \ \ 7.97 \times (6/12)^4 + 15.93 \times (6/12)^2 \times (12/12)^2$$
$$= -0.50 - 3.98 \qquad = -4.48$$
$$I(\text{net}) = \ +61.50 - 4.72$$
$$= \textbf{56.78} \text{ lb}\cdot\text{ft}\cdot\text{s} \; (\textbf{76.99} \text{ N}\cdot\text{m}\cdot\text{s}^2)\qquad \textbf{c}$$

1.521 Angular velocity $= 1500$ rpm $= 1500 \times \dfrac{2\pi}{60} = 157$ rad/s

$$157 = \alpha t = \alpha 60 \qquad \alpha = \mathbf{2.617} \text{ rad/s}^2 \qquad \mathbf{1d}$$

No. rev made by wheel in 1 min $=$ mean angular velocity $\times$ time

$$= \frac{1500 + 0}{2} \times 1 \text{ (min)}$$

$$= \mathbf{750} \text{ rev} \qquad \mathbf{2a}$$

Angular velocity at end 40 s $= 157 - 2.617 \times 40$
$$= 52.3 \text{ rad/s} = \mathbf{500} \text{ rpm} \qquad \mathbf{3b}$$

1.522 Angular velocity $= 60$ mph $= 88$ fps $(26.82$ m/s$)$

$$88 \text{ fps} = \frac{88}{3\pi} \text{ r/s} = \frac{88}{3\pi} \times 2\pi = 58.7 \text{ rad/s}$$

$$\text{No. rev to stop wheel} = \frac{500}{3\pi} \text{ or } \left(\frac{500}{3\pi}\right) 2\pi$$

$$= \mathbf{333} \text{ rad} \qquad \mathbf{1c}$$

Angular velocity squared
$\qquad\qquad =$ twice total no. rev to stop times angular acceleration

$$(58.7)^2 = 2 \times 333 \times \alpha \qquad \alpha = \mathbf{5.17} \text{ rad/s}^2 \qquad \mathbf{2b}$$

1.523 M (hemisphere) $= \dfrac{490}{32.2} \times \dfrac{1}{6} \times \pi \times \left(\dfrac{6}{12}\right)^3 \times \dfrac{1}{2}$

$$= 0.498 \text{ lb·s}^2/\text{ft} \ (7.27 \text{ N·s}^2/\text{m})$$

Center of gravity hemisphere $= \frac{3}{8}r$
$$= \frac{3}{8} \times 3$$
$$= \mathbf{1.125} \text{ in } (\mathbf{2.858} \text{ cm}) \text{ above diametral}$$
$$\text{plane} \qquad \mathbf{1d}$$

Angular velocity $= 300$ rpm $= 31.4$ rad/s
Outward (horizontal) force (ohf) $=$ mass $\times$ radius distance (axis to
$\qquad\qquad$ center of gravity of weight) $\times$ angular velocity squared

$$\text{ohf} = 0.498 \times {}^{12}\!/_{12} \times 31.4 \times 31.4 = 493 \text{ lb}$$

$$\text{Moment in rod} = \frac{1.125 - 0.500}{12} \, 493 = \mathbf{26.5} \text{ ft·lb } (\mathbf{34.7} \text{ N·m}) \qquad \mathbf{2a}$$

For zero moment in rod, place rod at center of gravity of weight or 1.125
in (2.858 cm) above diametral plane.

1.524

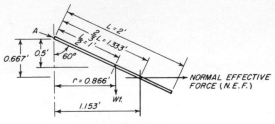

Fig. 1.524

$$\Sigma M_A = 0 = -0.667 \times \text{nef} + 0.866 \times \text{wt}$$

$$0 = -0.667 \times M \times 0.866 \times V_a^2 + 0.866 \times 32.2M$$

$$V_a^2 = \frac{32.2}{0.667} = 47.6$$

V_a (angular velocity) = 6.9 rad/s = **66** rpm   **c**

1.525 Angular velocity = V_a = 240 rpm = 25.1 rad/s

$$2T = \frac{M}{2} r_{cg} \times V_a^2 \qquad \text{where } r_{cg} = \frac{2}{\pi} \times \text{mean radius of rim}$$

$$\frac{M}{2} = \frac{1}{2} \times \frac{450}{32.2} \times \frac{18}{12} \times \pi \left[\left(\frac{72}{12}\right)^2 - \left(\frac{68}{12}\right)^2 \right]$$

$$= 125$$

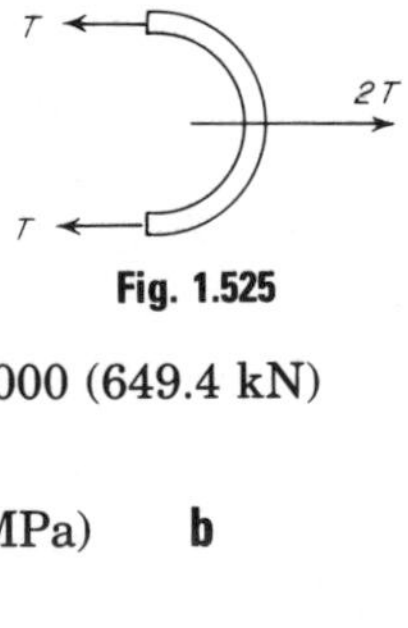

Fig. 1.525

$$T = \frac{125}{2} \times \frac{2}{\pi} \left(\frac{72 + 68}{2 \times 12} \right) \times 25.1 \times 25.1 = 146,000 \ (649.4 \text{ kN})$$

Tension in rim = $\dfrac{146,000}{4 \times 18}$ = **2032** psi (**14.01** MPa)   **b**

MECHANICS (STATICS)

1.601 $\Sigma V = +10 - 10 - 10 + 10 = 0$

ΣM (about F_1) = $+5 \times 10 + 9 \times 10 - 12 \times 10$

$= +$**20** ft·lb (**27.1** N·m) (clockwise)   **1a**

Resultant is equal to a clockwise couple of 20 ft·lb (27.1 N·m). Had the right-hand force (F_4) acted downward, then

$$\Sigma V = +10 - 10 - 10 - 10$$
$$= -20 \text{ lb } (88.96 \text{ N}) \text{ acting downward}$$

and

$$\Sigma M(F_1) = +5 \times 10 + 9 \times 10 + 12 \times 10$$
$$= +260 \text{ ft·lb (352.6 N·m)}$$

Resultant would be a 20-lb force acting downward and $260/20 =$ **13** ft (**3.96** m) to right of F_1.   **2c**

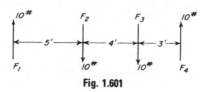

Fig. 1.601

1.602   Find center of gravity of pile group.

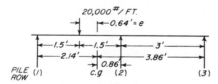

Fig. 1.602

M About Pile Row 1

Pile row	Pile spacing center to center	Area piles/ft	Distance from row 1	Area × distance	$x =$ distance center of gravity pile group to pile	$I =$ area $\times x^2$
1	2.5 ft	0.400 ft²	0	0	+2.14 ft	1.835
2	4.0 ft	0.250	3 ft	0.750	−0.86	0.185
3	6.0 ft	0.167	6 ft	1.000	−3.86	2.485
Sum		0.817	(2.14) ft	1.750		4.505

Center of gravity of pile group lies 1.750/0.817 or 2.14 ft to right of row 1. The 20,000-lb/ft load P is 0.64 ft eccentric e to center of gravity of piles.

$$\text{Vertical load on any pile} = \frac{P}{A} + \frac{Pex}{I}$$

and

$$\frac{P}{A} = \frac{20,000}{0.817} = 24,500 \text{ lb } (109.0 \text{ kN})$$

$$\frac{Pex}{I} = \frac{20,000(0.64)x}{4.505} = 2840x$$

Pile row	$\dfrac{P}{A}$	$\dfrac{Pex}{I}$	Stress in pile
1.	+ 24,500	+ 6,080	= 30,580 lb/pile
2.	+ 24,500	− 2,440	= 22,060 lb/pile
3.	+ 24,500	− 10,960	= 13,540 lb/pile

Check: 1. 30,580/2.5 = **12,210** lb/ft (**54.31** kN/0.305 m)   **a**
2. 22,060/4.0 = 5,515 lb/ft
3. 13,540/6.0 = 2,258 lb/ft
19,983 lb/ft (say OK)

1.603 $\Sigma M_{R,L} = 0 = +4 \times 1000 \times 2 + 10,000 - 12R_R$
$R_R = +1500 \text{ lb } (6.672 \text{ kN})$
$\Sigma M_{R,R} = 0 = +12R_L + 10,000 - 4 \times 1000 \times 10$
$R_L = +2500 \text{ lb } (11.12 \text{ kN})$

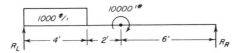

Fig. 1.603a Forces

Check:

$$V = 0 = +R_L + R_R - 4 \times 1000 = +2500 + 1500 - 4000$$

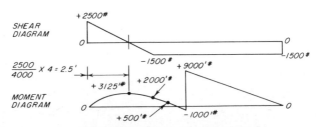

Fig. 1.603b Shear and moment.

M at point 2.5 ft to right of R_L
$$= +2500 \times 2.5 - 1000 \times 2.5 \times 2.5/2 = +3125 \text{ ft·lb}$$
M at point 4.0 ft to right of R_L
$$= +2500 \times 4 - 1000 \times 4 \times 2 = +2000 \text{ ft·lb}$$
M at point 5.0 ft to right of R_L
$$= +2500 \times 5 - 1000 \times 4 \times 3 = +500 \text{ ft·lb}$$
M at point 6.0 ft to right of R_L but to left of couple
$$= +2500 \times 6 - 1000 \times 4 \times 4 = -1000 \text{ ft·lb}$$
M at point 6.0 ft to right of R_L but to right of couple $= 9000 \text{ ft·lb}$

Multiply ft·lb by 1.356 to get newton-meters (N·m).

1.604 $\Sigma M_{R,L} = 0$
$$= +5 \times 2000 + 10 \times 4000 + 15 \times 6000 + 20 \times 1000$$
$$\times 10 - 20R_R$$

$$R_R = 17{,}000 \text{ lb (75.62 kN)}$$

$\Sigma M_{R,R} = 0$
$$= -5 \times 6000 - 10 \times 4000 - 15 \times 2000 - 20 \times 1000 \times 10$$
$$+ 20R_L$$

$$R_L = 15{,}000 \text{ lb (66.72 kN)}$$

Fig. 1.604a   Forces.

Check:

$$\Sigma V = 0 = +15{,}000 - 2000 - 4000 - 6000 - 20{,}000 + 17{,}000$$

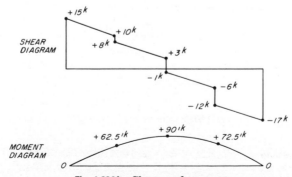

Fig. 1.604b   Shear and moment.

Forces to let point of moments:

M at left quarter point $= +15 \times 5 - 1 \times 5 \times \frac{5}{2} = 62.5$ ft·kips
M at centerline $= +15 \times 10 - 2 \times 5 - 1 \times 10 \times \frac{10}{2} = 90$ ft·kips
M at right quarter point

$$= +15 \times 15 - 2 \times 10 - 4 \times 5 - 1 \times \frac{15 \times 15}{2} = 72.5 \text{ ft·kips}$$

Check: M at right quarter point, forces to right of point of
moments $= +17 \times 5 - 1 \times 5 \times \frac{5}{2} = +72.5$ ft·kips

Multiply ft·kips by 1.356 to get kilonewton-meters (kN·m).

1.605

$\Sigma M_{RL} = 0$
$\quad = +2 \times 20 \times 6 - 14R_R, \quad R_R = 17.14$ kips (76.24 kN)
$\Sigma M_{R,R} = 0$
$\quad = -2 \times 20 \times 8 + 14R_L, \quad R_L = 22.86$ kips (101.68 kN)
Check: $\qquad\qquad\qquad\qquad \Sigma V = 0 = + 40.00$ kips $-(2 \times 20)$

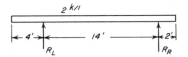

Fig. 1.605a   Forces.

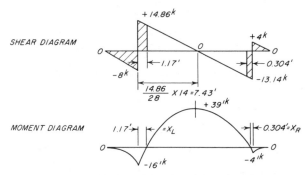

Fig. 1.605b   Shear and moment.

M at $R_L = -2 \times 4 \times 2 = -16$ ft·kips (21.69 kN·m)
M at $R_R = -2 \times 2 \times 1 = -4$ ft·kips (5.42 kN·m)

$$M \text{ at } 11.43 \text{ ft } (M \max) = -2 \times 11.43 \times \frac{11.43}{2} + 22.86 \times 7.43$$

$$= +39 \text{ ft} \cdot \text{kips } (51.88 \text{ kN} \cdot \text{m})$$

$$M = 0 \text{ when } x_L = 1.17, \text{ and } \frac{4 \times 8}{2} = 16 = 14.86x - \frac{2x^2}{2}$$

Similarly,

$$M = 0 \text{ when } x_R = 0.304 \text{ ft } (0.0926 \text{ m})$$

1.606

$\Sigma M_{R,L} = 0$

$\quad = +1 \times 24 \times 6 + 15 \times 9 - 18R_R,$

$$R_R = 15.5 \text{ kips } (68.94 \text{ kN})$$

$\Sigma M_{R,R} = 0$

$\quad = -1 \times 24 \times 12 - 15 \times 9 + 18R_L,$

$$\underline{R_L = 23.5 \text{ kips } (104.53 \text{ kN})}$$

Check: $\qquad\qquad \Sigma V = 0 = -15 - 24 + \overline{39.0 \text{ kips } (173.47 \text{ kN})}$

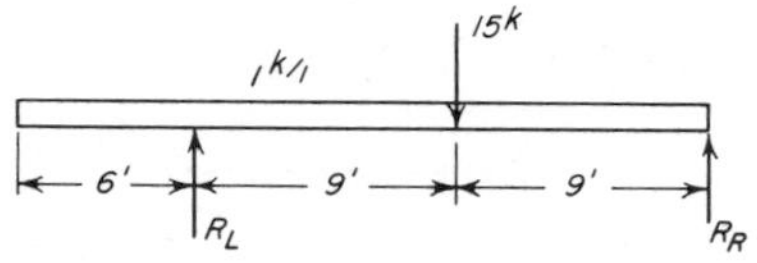

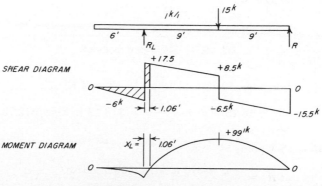

Fig. 1.606a Forces.

Fig. 1.606b Shear and moment.

$$M \text{ at } R_L = -1 \times 6 \times 3 = -18 \text{ ft·kips (24.41 kN·m)}$$
$$M_{\max} \text{ at 15-kips load} = -1 \times 15 \times {}^{15}\!/_2 + 23.5 \times 9$$
$$= 99 \text{ ft·kips (134.24 kN·m)}$$

$$M = 0 \text{ when } x_L = 1.06 \text{ ft (0.323 m), and } 18 = 17.5x - \frac{x^2}{2}$$

1.607

$$\Sigma M_{R,L} = 0$$
$$= +4 \times 15 \times {}^{15}\!/_2 + 30 \times 20 - 30 \times R_R,$$
$$R_R = +35 \text{ kips (155.68 kN)}$$

$$\Sigma M_{R,R} = 0$$
$$= +30R_L - 4 \times 15 \times 22.5 - 30 \times 10,$$
$$\underline{R_L = +55 \text{ kips (244.64 kN)}}$$

Check: $\quad\quad\quad\quad\quad \Sigma V = 0 = -60 - 30 + 90 \text{ kips (480.32 kN)}$

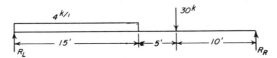

Fig. 1.607a   Forces.

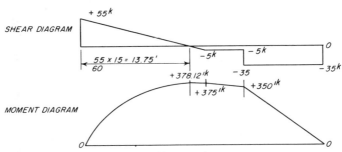

Fig. 1.607b   Shear and moment.

Max moment at point 13.75 ft (4.19 m) to right of R_L

$$= +55 \times 13.75 - 4 \times \frac{13.75^2}{2} = 378.12 \text{ ft·kips (512.73 kN·m)}$$

$$M \text{ at 15-ft point} = +55 \times 15 - 4 \times \frac{15^2}{2}$$
$$= +375 \text{ ft·kips (508.5 kN·m)}$$

$$M \text{ at 20-ft point} = +35 \times 10 = 350 \text{ ft·kips (474.6 kN·m)}$$

1.608   Given that $M_L = PL/16 = 16P/16 = P$ ft·lb.

$$\Sigma M_L = 0 = -P - 16R_R + 10P, \; R_R = \frac{9P}{16} \text{ lb}$$

$$M_P = 6 \times \frac{9P}{16} = \frac{54P}{16} \text{ ft·lb} > M_L$$

$$I_{xx} = \frac{bh^3}{12} = 4 \times \frac{16^3}{12} = 1365 \text{ in}^4 \; (56.82 \text{ kcm}^4)$$

$$f = \frac{Mc}{I} = \frac{12 \times 54P \times 8}{16 \times 1365} = 1350$$

and

$$P = \mathbf{5687} \text{ lb } (\mathbf{25.30} \text{ kN}) \qquad \mathbf{1d}$$

For deflection Y_P, use tables for the deflection of each item of loading and add the deflections. Referring to Merritt, "Structural Steel Designers' Handbook" pp. 3–433 and 3–435, McGraw-Hill, 1972, find the deflection at load P for the load P on a simple beam and the deflection at load P for the moment P on the same simple beam.

$$Y_{P_1} \text{ (due to load } P) = \frac{+Pa^2b^2}{3EIL}$$

$$= \frac{+(5687)(100)(36)(1728)}{(3)(16)(10)^5(1365)(16)}$$

$$= 0.338 \text{ in } (0.859 \text{ cm}) \text{ (down)}$$

Y_{P_2} (due to negative moment P at the left end)

$$= \frac{-P}{6EIL}(2L^2a - 3La^2 + a^3)$$

$$= \frac{-(10)(5687)(1728)}{(6)(16)(10)^5(1365)(16)}(2 \times 256 - 3 \times 16 \times 10 + 100)$$

$$= \frac{-4.72(132)}{10^4} = -0.062 \text{ in } (0.158 \text{ cm}) \text{ (up)}$$

Net $Y_P = +0.338 - 0.062 = \mathbf{0.276}$ in (down) (**0.701** cm)   **2a**

If tables for deflections are not available, a double integration of $EI \; d^2y/dx^2$ may take too much time on an examination and a graphical integration using $Mm \; dx/EI$ is usually faster. Referring to Fig. 1.608a

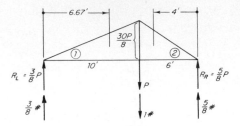

Fig. 1.608a

Area no.	Area	Dummy force	Arm	Product
1	$\dfrac{+10(30P)}{2(8)} = \dfrac{+300P}{16}$	$\dfrac{+3}{8}$ lb	6.67	$+46.9P$
2	$\dfrac{+6(30P)}{2(8)} = \dfrac{+180P}{16}$	$\dfrac{+5}{8}$ lb	4.00	$+28.1P$
				$+75.0P$

$$Y_{P_1} = + \frac{(75)(5687)(1728)}{(16)(10)^5(1365)} = +0.338 \text{ in } (0.859 \text{ cm}) \textbf{ (down)}$$

Referring to Fig. 1.608b

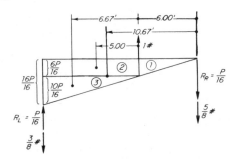

Fig. 1.608b

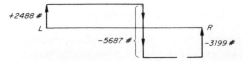

Fig. 1.608c Shear diagram.

Area	Area	Dummy force	Arm	+ Prod.	− Prod.
1 + 2 + 3	$\dfrac{-P(16)}{2} = -8P$	$\dfrac{-5}{8}$ lb	10.67	+53.33P	
2	$\dfrac{-6P(10)}{16} = \dfrac{-60P}{16}$	+1 lb	5.00		−18.75P
3	$\dfrac{-10P(10)}{16(2)} = \dfrac{-100P}{16}$	+1 lb	6.67		−20.83P
					−39.58P
				−39.58P	
				+13.75P	

$$Y_{P,2} = \frac{+(1375)(5687)(1728)}{(16)(10)^5(1365)} = 0.062 \text{ in } (0.158 \text{ cm}) \text{ (up)}$$

$$\text{Net } Y_P = +0.338 - 0.062 = +0.276 \text{ in } (0.701 \text{ cm}) \text{ (down)}$$

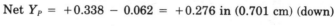

Fig. 1.608d Moment diagram.

1.609 *Using moment distribution:*

Fixed end moments (clockwise $\curvearrowleft = +$, counterclockwise $\curvearrowright = -$)

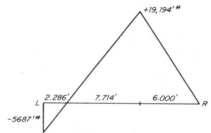

Fig. 1.609

Uniform load:

To left of $R_a = 1 \times 6 \times 3 = +18$ ft·kips (24.41 kN·m)

$$\text{To right of } R_a \text{ (and elsewhere)} = \frac{wl^2}{12}$$

$$= \frac{1 \times 20^2}{12}$$

$$= \pm 33.33 \text{ ft} \cdot \text{kips} \ (\pm 45.20 \text{ kN} \cdot \text{m})$$

Concentrated load:

$$\text{To right of } R_a = \frac{Pab^2}{l^2} = \frac{12 \times 8 \times 12 \times 12}{20 \times 20}$$

$$= -34.56 \text{ ft} \cdot \text{kips} \ (-46.86 \text{ kN} \cdot \text{m})$$

$$\text{To left of } R_b = \frac{Pa^2b}{l^2} = \frac{12 \times 8 \times 8 \times 12}{20 \times 20}$$

$$= +23.04 \text{ ft} \cdot \text{kips} \ (+31.24 \text{ kN} \cdot \text{m})$$

Distribution factors	1.0	0.5	0.5	0
Fixed end +18 moments	−34.56 −33.33	+23.04 +33.33	 −33.33	 +33.33
Net fixed end M +18	−67.89	+56.37	−33.33	+33.33
	+49.89 →	+24.94		
	−11.99 ←	−23.99	−23.99 →	−11.99
	+11.99 →	+ 6.00		
	− 1.50 ←	− 3.00	− 3.00 →	− 1.50
	+ 1.50 →	+ 0.75		
	− 0.19 ←	− 0.38	− 0.38 →	− 0.19
	+ 0.19 →	+ 0.10		
		− 0.05	− 0.05 →	− 0.02
Final moment +18	−18 ft·kips	+60.75	−60.75 ft·kips	+19.63 ft·kips

(To left) M at R_b:

$$0 = -1 \times 26 \times 13 - 12 \times 12 + 20R_a + 60.75$$
$$R_a = +21.06 \text{ kips} \ (93.67 \text{ kN})$$

(Left and right) M at R_b:

$$0 = -1 \times 46 \times 3 - 12 \times 12 + 20 \times 21.06 + 19.63 - 20R_c$$
$$R_c = +7.94 \text{ kips} \ (35.32 \text{ kN})$$

M at R_c:

$$0 = -1 \times 46 \times 23 - 12 \times 32 + 40 \times 21.06 + 20R_b + 19.63$$
$$R_b = +29.00 \text{ kips (128.99 kN)}$$

Check: $\Sigma V = 0 = +21.06 + 29.00 + 7.94 - 46 - 12$

$\qquad M$ at 12-kips load $= -1 \times 14 \times 7 + 21.06 \times 8$
$\qquad\qquad\qquad\qquad = \mathbf{70.48}$ ft·kips **(95.57 kN·m)**　　**1a**

Using three moment equations (signs follow fiber stress convention):

$$M_1 l_1 + 2M_2(l_1 + l_2) + M_3 l_2 = -\tfrac{1}{4}w_1 l_1^3 - \tfrac{1}{4}w_2 l_2^3 - P_1 l_1^2(k_1 - k_1^3)$$

1. Take two spans, R_a to R_c and $k_1 = {}^{8}\!/_{20} = 0.4$

$$-18 \times 20 + 2M_b(20 + 20) + M_c \times 20$$
$$= -\frac{1 \times \overline{20}^3}{4} - \frac{1 \times \overline{20}^3}{4} - 12 \times (20^2)(0.4 - 0.064)$$

$$-360 + 80M_b + 20M_c = -2000 - 2000 - 1614.4$$
$$+4M_b + M_c = -262.70 \qquad\qquad\qquad (a)$$

2. Take two spans R_b to R_d, where $R_d = 0$ ft to right of R_a

$$M_b(20) \;+\; 2M_c(20 + 0) + M_d(0) = -2000$$
$$M_b + 2M_c = -100 \qquad\qquad (b)$$
$$4M_b + 8M_c = -400 \qquad\qquad (b')$$

subtracting Eq. (a) from Eq. (b')

$$+7M_c = -137.30 \qquad M_c = -19.62 \text{ ft·kips } (-26.60 \text{ kN·m})$$

and the minus sign indicates tension in top fiber over R_c.
　　Substituting in Eq. (b),

$$M_b = -100 + 39.25 = \mathbf{-60.75} \text{ ft·kips } (\mathbf{-82.38} \text{ kN·m})　　\textbf{2c}$$

Reactions are found as before.

1.610　The first step is to cut pulley ropes and place rope forces at axles of pulleys or on beam.

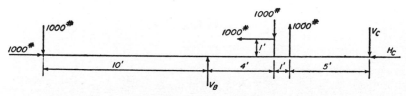

Fig. 1.610a　Forces.

$$\Sigma H = 0 = +1000 - 1000 - H_c \qquad H_c = 0$$
$$M_B = 0$$
$$= -10 \times 1000 - 1 \times 1000 - 5 \times 1000 + 4 \times 1000 + 10V_c$$
$$V_c = 1200 \text{ lb } (5.34 \text{ kN}) \downarrow$$
$$M_C = 0$$
$$= -20 \times 1000 - 1 \times 1000 - 6 \times 1000 + 5 \times 1000 + 10V_B$$
$$V_B = 2200 \text{ lb } (9.79 \text{ kN}) \uparrow$$

Check: $\Sigma V = 0 = -1000 + 2200 - 1000 + 1000 - 1200$

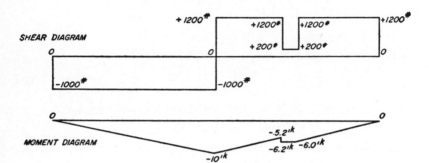

Fig. 1.610b Shear and moment.

Multiply ft·kips by 1.356 to get kN·m.

1.611 $0.75H + 1.333H = 2.083H = 1000 \text{ lb } (454 \text{ kg})$
$$H = 480 \text{ lb and } 360 + 640 = 1000$$

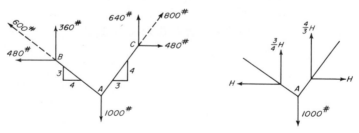

Fig. 1.611

Check:

$$\Sigma M_B = 0 = +4 \times 1000 + 1 \times 480 - 7 \times 640 = +4480 - 4480$$

Multiply lb by 4.448 to get N.

1.612 ΣM(left rope) $= 0 = -200x + 150(10 - x)$

$$x = \textbf{4.29} \text{ ft } (\textbf{1.31} \text{ m}) \qquad \textbf{b}$$

and

$$10 - x = \textbf{5.71} \text{ ft } (\textbf{1.74} \text{ m}) \qquad \textbf{d}$$

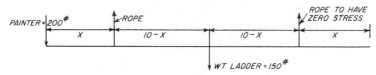

Fig. 1.612

Check: Σ_M(right rope)

$$0 = -200 \times 15.71 - 150 \times 5.71 + \text{rope}_L \times 11.42$$
$$\text{Rope}_L = 350 \text{ lb}, \Sigma V = 0 = +350 - 200 - 150$$

1.613 $h = \sqrt{400 - 64} = 18.32 \text{ ft}$

$$\Sigma M_a = 0 = +4 \times 150 - 18.32H$$

$$H = \textbf{32.75} \text{ lb } (\textbf{145.7} \text{ N}) \qquad \textbf{b}$$

Frictional force at A must at least $= H$

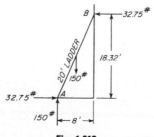

Fig. 1.613

1.614 Length is 1 ft perpendicular to cross section (p. 224).

$$W = 1 \times 6 \times 15 \times 150 = 13,500 \text{ lb (6124 kg)}$$

$$\text{Overturning couple} = P \times \frac{h}{3} = 10.417h^3$$

$$\text{Righting couple} = W(3-2) = 13,500 \text{ lb} \cdot \text{ft } (18.31 \text{ kN} \cdot \text{m})$$

$$h^3 = \frac{13,500}{10.417} = 1296, h = \textbf{10.9} \text{ ft } (\textbf{3.32} \text{ m}) \qquad \textbf{a}$$

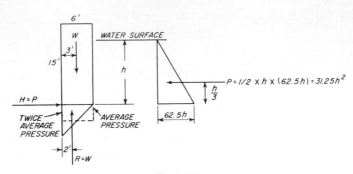

Fig. 1.614

1.615 Length is 1 ft along wall.

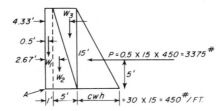

Fig. 1.615a Vertical and horizontal forces.

$$W_1 = 1 \times 1 \times 15 \times 150 = 2,250 \text{ lb}$$
$$W_2 = 1 \times \tfrac{5}{2} \times 15 \times 150 = 5,625 \text{ lb}$$
$$W_3 = 1 \times \tfrac{5}{2} \times 15 \times 100 = \underline{3,750 \text{ lb}}$$
$$\Sigma W = R = 11,625 \text{ lb (51.71 kN)}$$

At the moment of overturning, R would act at A.

For M (About A)

	F		$\times$	a	$=$	M
$W_1 =$	2,250	lb $\times$	0.5	ft $=$	$+$	1,125 ft·lb
$W_2 =$	5,625	$\times$	2.67	$=$	$+15,000$	
$W_3 =$	3,750	$\times$	4.33	$=$	$+16,250$	
$(P =$	3,375)	$\times$	5.00	$=$	$-16,875$	
$R = \Sigma W =$	11,625	$\times$	(1.33)	$=$	$+15,500$	(21.02 kN·m)

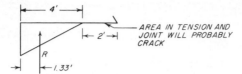

Fig. 1.615b   Force on base.

The wall is safe against overturning.

$$\Sigma W \times f = 3375 \qquad f = \frac{3375}{11,625} = \mathbf{0.29} \qquad \mathbf{c}$$

The wall is safe against sliding.

1.616   To locate R between front and rear axles,

$$\Sigma M(\text{16-ton axle}) = 0 = +20 \times d - 14 \times 4$$

$$d = 2.8 \text{ ft}, \frac{d}{2} = 1.4 \text{ ft } (0.427 \text{ m})$$

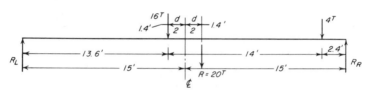

Fig. 1.616

To place a truck on a span so as to produce maximum moment, the rear axle is placed as far to one side of the centerline as the resultant of the front and rear axles lays on the other side of the centerline. If the span is too short to get both front and rear axles on the span, place rear axle at the centerline.

$$R_L = \frac{13.6}{30} \times 20 = 9.067 \text{ tons} \qquad R_R = \frac{16.4}{30} \times 20 = 10.933 \text{ tons}$$

Max moment (about 16-ton axle) $= 9.067 \times 13.6$
$$= \mathbf{123.3} \text{ ft·tons } (\mathbf{334.4} \text{ KN·m}) \qquad \mathbf{1d}$$

If 16-ton axle is placed at the centerline of the span,

$$R_L = \frac{12.2}{30} \times 20 = 8.133 \text{ tons} \qquad M = 8.133 \times 15 = \mathbf{122} \text{ ft·tons}$$

Max shear occurs when 16-ton axle is just to right of R_L

$$= \frac{27.2}{30} \times 20 = \textbf{18.13} \text{ tons } (\textbf{161.3 kN}) \qquad \textbf{2a}$$

1.617 First, read the answer to question 1.616. Then locate the resultant of the tractor and semitrailer axle loads by taking moments about the 8-kip (35.58-kN) axle.

$$14 \times 32 + 28 \times 32 = x(8 + 32 + 32) \qquad x = \frac{42 \times 32}{72} = 18.67 \text{ ft}$$

$$d = 18.67 - 14 = 4.67 \text{ ft} \qquad \frac{d}{2} = 2.33 \text{ ft } (0.71 \text{ m})$$

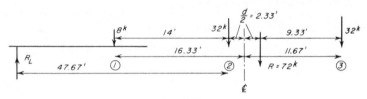

Fig. 1.617

1. The 8-kip axle (①) is $50 - 16.33$ ft or **33.67** ft (**10.26** m) to right of R_L. **1c**
2. Maximum moment (about axle ②) is

$$R_L = \frac{47.67}{100} \times 72 = 34.4 \text{ kips}$$

$$\begin{aligned}
\text{Max } M &= R_L \times 47.67 - 8 \times 14 \\
&= 34.4 \times 47.67 - 112 \\
&= 1640 - 112 = \textbf{1528} \text{ ft} \cdot \text{kips } (\textbf{2072 kN} \cdot \textbf{m}) \qquad \textbf{2b}
\end{aligned}$$

3. Maximum shear is R_R with truck and trailer backed up so that axle ③ is at R_R.

$$R_R = \frac{100 - 9.33}{100} \times 72 = \textbf{65.3} \text{ kips } (\textbf{290.5 kN}) \text{ (max shear)} \qquad \textbf{3a}$$

1.618 M(about line TR)
$$0 = 100 \times 7.07 - V_s \times 17.07, \qquad V_s = 41.4 \text{ kips}$$
M(about line ESF)
$$0 = -100 \times 10 + 2V_T \times 17.07, \qquad \begin{array}{l} V_T = 29.3 \text{ kips} \\ \underline{V_R = 29.3 \text{ kips}} \\ 100.0 \text{ kips} \end{array}$$

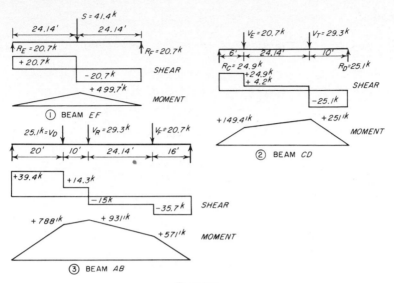

Fig. 1.618

Check:

$$\Sigma V = 0 = -100 + V_R + V_S + V_T$$
$$\text{kips (forces or shears)} \times 4.448 = \text{kN}$$
$$\text{ft·kips (kN)} \times 1.356 = \text{kN·m}$$

1.619 By symmetry and inspection, $R_L = R_R = 5000$ lb (22.24 kN).

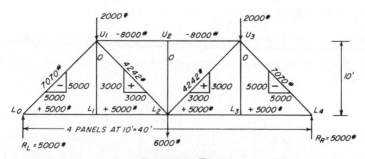

Fig. 1.619a Truss.

By analyzing joints L_1, U_2, and L_3, the stress in the verticals U_1L_1, U_2L_2, and U_3L_3, must be zero because ΣV must equal zero and there is

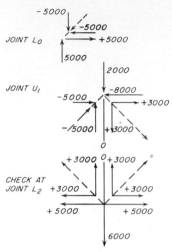

Fig. 1.619b Joints.

no outside vertical force or other vertical member or member having a vertical component to counteract any stress in these three members.

Use the "method of joints" ($\Sigma H = 0$ and $\Sigma V = 0$ for all stresses or components of stresses of all members meeting at a joint), letting $+$ indicate tension and $-$ indicate compression, but noting that for any joint the sum of the forces acting in one direction must equal the sum of the forces acting in a 180° direction regardless of signs.

Also note that no joint can be analyzed if the stresses in more than two members meeting at that joint are unknown.

1.620 To solve for the stress in L_0L_1, pass the section 1-1, cutting mem-

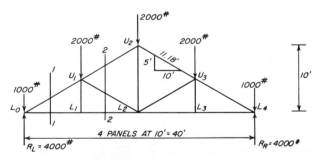

Fig. 1.620a Forces on truss.

bers L_0U_1 and L_0L_1. Take moments about L_0U_1 extended to U_1. Assume stress of L_0L_1 acts out of L_0.

$$\Sigma M_{U1} = 0 = +(4000 - 1000) \times 10 - 5 \times S_{L0L1}$$
$$S_{L0L1} = +\textbf{6000} \text{ lb } (\textbf{26.69 kN}) \text{ (tension)}$$

To solve for the stress in U_1U_2 and U_1L_2, pass the section 2-2, cutting members U_1U_2, U_1L_2, and L_1L_2.

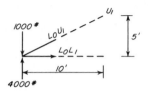

Fig. 1.620b   Stress in L_0L_1.

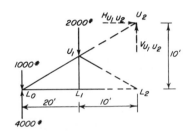

Fig. 1.620c   Stress in U_1U_2.

Take moments at L_2 and assume U_1U_2 to be a tension member (acting out of joint U_1) with components at U_2.

$$\Sigma M_{L2} = 0 = +20 \times 3000 - 10 \times 2000 + 10H_{U1U2}$$
$$H_{U1U2} = -\textbf{4000} \text{ lb } (\textbf{17.79 kN}) \text{ (compression)}$$

(− sign indicates compression)

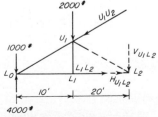

Fig. 1.620d   Stress in U_1L_2.

and $\qquad S_{U1U2} = \dfrac{11.18}{10} \times (-4000)$

$$= -\textbf{4472} \text{ lb } (\textbf{18.89} \text{ kN}) \text{ (compression)}$$

Now take moments at L_0 and assume U_1L_2 to be taken with components acting at L_2. Forces U_1U_2 and L_1L_2 and H_{U1L2} or their lines of action pass through L_0.

$$\Sigma M_{L0} = 0 = +10 \times 2000 + 20 \times V_{U1L2}$$

$$V_{U1L2} = -1000 \text{ lb } (4.45 \text{ kN}) \text{ (compression)}$$

$$S_{U1L2} = \dfrac{11.18}{5} \times 1000 = -\textbf{2236} \text{ lb } (\textbf{9.95} \text{ kN}) \text{ (compression)}$$

1.621

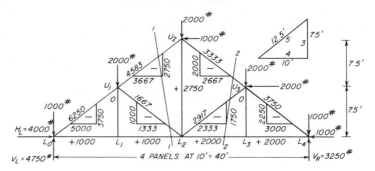

Fig. 1.621

$$\Sigma M_{L0} = 0 = +10 \times 2000 + 20 \times 2000 + 30 \times 2000$$
$$+40 \times 1000 - 7.5 \times 2000 - 15 \times 1000 - 40V_R$$
$$V_R = \textbf{3250} \text{ lb } (\textbf{14.456} \text{ kN})$$
$$\Sigma M_{L4} = 0 = -10 \times 2000 - 20 \times 2000 - 30 \times 2000$$
$$-40 \times 1000 - 7.5 \times 2000 - 15 \times 1000 + 40V_L$$
$$V_L = \textbf{4750} \text{ lb } (\textbf{21.128} \text{ kN})$$

Check:

$$\Sigma V = 0 = -1000 - 2000 - 2000 - 2000 - 1000$$
$$+ 8000 \text{ lb } (35.584 \text{ kN})$$

Pass vertical sections through second and third panels and find H_{U1U2}, V_{U1L2}, H_{U2U3}, and V_{L2U3} as in the answer to question 1.620.

Section 1-1:

$$\Sigma M_{L2} = 0 = +(4750 - 1000) \times 20 - 2000 \times 10 + H_{U1U2} \times 15$$
$$H_{U1U2} = -\textbf{3667} \text{ lb } (\textbf{16.311} \text{ kN})$$
$$\Sigma M_{L0} = 0 = +2000 \times 10 + V_{U1L2} \times 20$$
$$V_{U1L2} = -\textbf{1000} \text{ lb } (\textbf{4.448} \text{ kN})$$

Section 2-2:

$$\Sigma M_{L4} = 0 = -2000 \times 10 - 2000 \times 7.5 - V_{L2U3} \times 20$$
$$V_{L2U3} = -\textbf{1750} \text{ lb } (\textbf{7.784} \text{ kN})$$
$$\Sigma M_{L2} = 0 = +2000 \times 10 - 2000 \times 7.5$$
$$- (3250 - 1000)20 - H_{U2U3} \times 15$$
$$H_{U2U3} = -\textbf{2667} \text{ lb } (\textbf{11.863} \text{ kN})$$

Stresses or components of stress in other members and checks are obtained by applying $\Sigma H = 0$ and $\Sigma V = 0$ at each joint.

1.622   Section 1-1 and truss to right of section (assume U_2U_3 as tension):

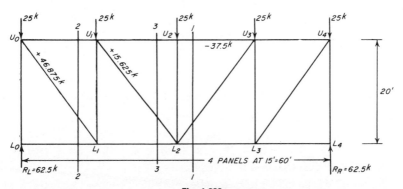

Fig. 1.622

$$\Sigma M_{L2} = 0 = +15 \times 25 + 30 \times 25 - 30 \times 62.5 - 20S_{U2U3}$$
$$S_{U2U3} = -\textbf{37.5} \text{ kips } (\textbf{166.8} \text{ kN}) \text{ (compression)}$$

Section 2-2 and truss to left of section (assume U_0L_1 as tension):

$$\Sigma V = 0 = +62.5 - 25 - V_{U0L1}, V_{U0L1} = +37.5 \text{ kips}$$
$$S_{U1L2} = +\textbf{46.875} \text{ kips } (\textbf{208.5} \text{ kN}) \text{ (tension)}$$

Section 3-3 and truss to left of section (assume U_1L_2 as tension):

$$\Sigma V = 0 = +62.5 - 25 - 25 - V_{U1L2}, V_{U1L2} = +12.5 \text{ kips}$$
$$S_{U1L2} = +\textbf{15.625} \text{ kips } (\textbf{69.5} \text{ kN}) \text{ (tension)}$$

1.623

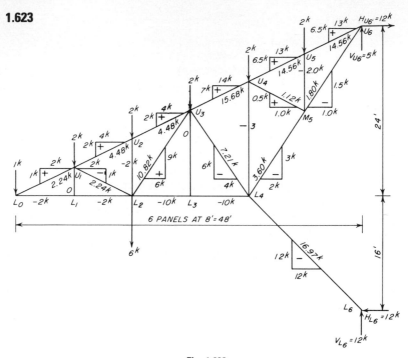

Fig. 1.623

$\Sigma M_{L6} = 0$

$\quad = -1 \times 48 - 2(40 + 32 + 24 + 16 + 8) - 6 \times 32 + 40H_{U6}$

$H_{U6} = +12$ kips (53.38 kN) (to right)

$H_{L6} = -12$ kips (53.38 kN) (to left), $V_{L6} = +12$ kips (53.38 kN) (up)

$\Sigma V = 0 = -17 + 12 + V_{U6}$, $V_{U6} = +5$ kips (22.24 kN) (up)

a. Vertical section in second panel from left end and components of U_1U_2 acting at U_2

$$\Sigma M_{L3} = 0 = -1 \times 16 - 2 \times 8 + 8H_{U1U2}$$
$$H_{U1U2} = +4 \text{ kips (17.79 kN) (tension)}$$

b. Vertical section in fourth panel from left end and components of U_3U_4 acting at U_4

$$\Sigma M_{L4} = 0 = -1 \times 32 - 2 \times 24 - 8 \times 16 - 2 \times 8 + 16H_{U3U4}$$
$$H_{U3U4} = +14 \text{ kips (62.27 kN)}$$

c. Vertical section in sixth panel from left end and components of $U_5 U_6$ acting at U_4

$\Sigma M_{L4} = 0$

$\quad = -1 \times 32 - 2 \times 24 - 8 \times 16 - 2 \times 8 + 2 \times 8 + 16 H_{U5U6}$

$H_{U5U6} = +13$ kips (57.82 kN)

d. The remaining components and the necessary checks are easily obtained by applying $\Sigma H = 0$ and $\Sigma V = 0$ at each joint. Having one component of stress, the other component of stress and the actual stress in the members are found from simple proportion.

1.624   Use moments to find H_{U1U2}, $L_2 L_3$, and H_{U4U5}

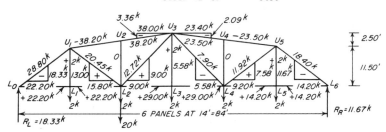

Fig. 1.624

$\Sigma M_{L2} = 0$ (truss to left of L_2), section through second panel

$\quad\quad 0 = +18.33 \times 28 - 2 \times 14 + 12.75 \times H_{U1U2}$

$\quad\quad H_{U1U2} = \mathbf{-38.00}$ kips (**169.02** kN) (compression)

$\Sigma M_{L4} = 0$ (truss to right of L_4), section through fifth panel

$\quad\quad 0 = -11.67 \times 28 + 2 \times 14 - 12.75 \times H_{U4U5}$

$\quad\quad H_{U4U5} = \mathbf{-23.40}$ (**104.08** kN) kips

$\Sigma M_{U3} = 0$ (truss to right of U_3), section through fourth panel

$\quad\quad 0 = -11.67 \times 42 + 2 \times 28 + 2 \times 14 + 14 \times L_3 L_4$

$\quad\quad L_3 L_4 = \mathbf{+29.0}$ kips (**128.99** kN) (tension)

The remaining components and the necessary check are easily obtained by applying $\Sigma H = 0$ and $\Sigma V = 0$ at each joint. Having one component of stress, the other component of stress and the actual stress in the members are found from simple proportion.

1.625 $\Sigma M_{(B-C)} = 0 = +40 \times 2400 + 30 V_{AD}$

$\quad\quad\quad\quad V_{AD} = 3200$ lb (14.234 kN)

$\quad\quad\quad\quad H_{AD} = -1600$ lb (7.117 kN)

$\quad\quad\quad\quad S_{AD} = \mathbf{-3580}$ lb (**15.925** kN) (compression)

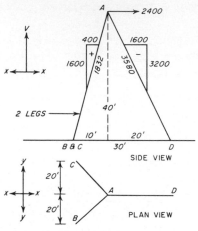

Fig. 1.625

In both legs AB and AC there are horizontal components parallel to BC equal to $^{20}\!/_{10} \times 400 = 800$ lb. Thus each leg, AC and AB, has three components:

$$x = 400 \qquad y = 800 \qquad V = 1600$$
$$\textit{Stress in AB and AC} = 400\sqrt{1^2 + 2^2 + 4^2} = 400/\sqrt{21}$$
$$= +1832 \text{ lb } (8.149 \text{ kN}) \text{ (tension)}$$

THERMODYNAMICS

1.701 *a. Latent heat* (latent heat of evaporation): Quantity of heat required to evaporate 1 lb saturated liquid.

b. Specific heat: Amount of transferred heat required to change the temperature of one unit weight of a substance one degree unit of temperature.

c. Heat of fusion: Quantity of heat required to change the state of a substance from solid to liquid without change of temperature.

d. Btu: $^{1}\!/_{180}$ the quantity of heat required to raise the temperature of 1 lb pure water from 32 to 212°F (0 to 100°C) under standard atmospheric pressure.

e. Absolute temperature: A measure of the ability to transfer heat to other bodies based upon a reference temperature (absolute zero) where a body has given up all the thermal energy it possibly can. Absolute zero $= -273.16°C = -459.69°F$.

f. Calorie: Quantity of heat required to raise the temperature of 1 g water from 14.5 to 15.5°C.

g. Watt: Power expended when 1 A flows between two points having a potential difference of 1 V.

$$1 \text{ W} = 1 \text{ J/s} = 10^7 \text{ erg/s} = 10^7 \text{ dyn·cm/s}$$

1.702 T = final temperature, °F (°C)
T_I = initial temperature ice, °F (°C)
T_w = initial temperature water, °F (°C)
W_I = weight ice, lb (kg)
W_W = weight water, lb (kg)
HF = heat of fusion = 144 Btu/lb (335.2 kJ/kg)
Heat lost by water = heat gained by ice

$$W_W(T_w - T) = W_I(\text{HF}) + W_I(T - T_I)$$
$$100(100 - T) = 30(144) + 30(T - 32)$$
$$10,000 - 100T = 4320 + 30T - 960$$
$$130T = 6640$$
$$T = \mathbf{51}°\text{F } (\mathbf{10.6}°\text{C}) \qquad \mathbf{d}$$

1.703   Initial temperature cold water = T_C, °F
Initial temperature hot water = T_H, °F
Volume cold water = V_C, ft³
Volume hot water = V_H, ft³
Total volume = V_T, ft³
Weight cold water = W_C, lb
Weight hot water = W_H, lb
Total weight = W_T, lb
Specific weight = w, lb/ft³
Heat lost by hot water = heat gained by cold water

$$W_H(T_H - T) = W_C(T - T_C)$$
$$V_H(w)(T_H - T) = V_C(w)(T - T_C)$$
$$V_H(62.5)(160 - 70) = V_C(62.5)(70 - 40)$$
$$90V_H = 30V_C \text{ and } V_C = 3V_H$$
$$V_C + V_H = V_T = (25)(75)(5) = 9375 = 3V_H + V_H = 4V_H$$
$$V_H = \mathbf{2344} \text{ ft}^3 \ (\mathbf{66.38} \text{ m}^3)$$
$$V_C = V_T - V_H = 9375 - 2344 = \mathbf{7031} \text{ ft}^3 \ (\mathbf{199.12} \text{ m}^3) \qquad \mathbf{c}$$

1.704   Weight water = W_w, lb (kg)
Weight ice = W_I, lb (kg)
Weight steam = W_s, lb (kg)

HV = heat of vaporization = 970 Btu/lb (2256.2 kJ/kg)

HF = heat of fusion = 144 Btu/lb (334.9 kJ/kg)

Heat lost by steam in condensing = $(HV)(W_s)$

$$= 970W_s, \text{ Btu}$$

Heat lost by condensed steam = $(T_s - T)W_s$

$$= (212 - 80)W_s$$

$$= 132W_s, \text{ Btu}$$

Heat gained by ice in melting = $(HF)W_1 = 144 \times 100$

$$= 14,400 \text{ Btu}$$

Heat gained by ice and cold water = $(T - T_w)(W_I + W_w)$

$$= (80 - 32)(100 + 300)$$

$$= 19,200 \text{ Btu}$$

Heat lost by steam = heat gained by ice and water

$$970W_s + 132W_s = 1102W_s = 14,400 + 19,200 = 33,600$$

$$W_s = \frac{33,600}{1102} = \textbf{30.5} \text{ lb } (\textbf{13.83} \text{ kg}) \qquad \textbf{b}$$

1.705 V_1 = initial vol. $\quad p_1$ = initial pressure

V_2 = final vol. $\quad p_2$ = final pressure

T_1 = initial temp.

T_2 = final temp.

$$V_1 = 12V_2 \qquad K = 1.4 \text{ for air}$$

For adiabatic process

$$\frac{p_2}{p_1} = \left(\frac{V_1}{V_2}\right)^k \quad \text{and} \quad \frac{p_2}{13} = \left(\frac{12V_2}{V_2}\right)^{1.4}$$

$$p_2 = 13(12)^{1.4} = 13 \times 32.4 = \textbf{421} \text{ psi } (\textbf{2903} \text{ kPa}) \qquad \textbf{1c}$$

$$\frac{T_2}{T_1} = \left(\frac{V_1}{V_2}\right)^{k-1} \quad \text{and} \quad \frac{T_2}{580} = \left(\frac{12V_2}{V_2}\right)^{1.4-1} = (12)^{0.4}$$

(°R = 120°F + 460° = 580°R)

$$T_2 = 580(12)^{0.4} = 580 \times 2.71 = \textbf{1570}°\text{R} = \textbf{1110}°\text{F } (\textbf{599}°\text{C}) \qquad \textbf{2b}$$

1.706 $\dfrac{p_2}{p_1} = \dfrac{T_2}{T_1}$ (constant-volume process)

$$p_1 = 32 + 14.7 = 46.7 \text{ psia } (322.0 \text{ kPa abs})$$

$$T_2 = 75 + 460 = 535°\text{R}$$

$$T_1 = 50 + 460 = 510°\text{R}$$

$$p_2 = 46.7 \times {}^{535}\!/_{510} = 49.1 \text{ psia } (338.5 \text{ kPa abs})$$

$$= 49.1 - 14.7 = \textbf{34.4} \text{ psig } (\textbf{237.2} \text{ kPa gage}) \qquad \textbf{b}$$

1.707 $R = \dfrac{pv}{T} = \dfrac{p_1V_1}{WT_1} = \dfrac{p_2V_2}{WT_2}$, where $v = \dfrac{V}{W}$

$\qquad\qquad v =$ specific vol. $R =$ gas constant,
$\qquad\qquad V =$ total vol. $W =$ total wt
$\quad 0°F = 460°R \qquad$ and $\qquad 200°F = 660°R$

$V_1 = 100V_2, \dfrac{14.7 \times 100V_2}{460} = \dfrac{p_2V_2}{660}$

$p_2 = {}^{660}\!/_{460} \times 14.7 \times 100 = \mathbf{2110}$ psia (**14.6** MPa abs)   **d**

1.708 *a.* The heating value of a fuel is the amount of heat that must be removed from the products of complete combustion of a unit amount of fuel to cool the products down to the temperature of the original air-fuel mixture. Most common fuels contain hydrogen, which when burned forms water vapor or steam. The higher heating value is obtained when all the water vapor of combustion condenses, thus giving up its latent heat of evaporation. The lower heating value is obtained when sufficient excess air is used to prevent any condensation of water vapor.

b. If the final volume of the products of a constant-pressure combustion exceeds the original volume, part of the fuel energy must be used to push the atmosphere out of the way. In this case, the heating value at constant pressure is less than that at constant volume by a quantity equal to the work done in pushing the atmosphere out of the way. If the final volume of the products is less than the original volume, the constant-pressure heating value exceeds the constant-volume value by a quantity equal to the work done on the gas by the atmosphere during the volume contraction.

1.709 When a gas or a gaseous mixture remains in contact with a liquid surface, it acquires vapor from the liquid until the partial pressure due to the water vapor in the air is equal to the saturated vapor pressure of the water at the existing temperature. When the vapor concentration reaches the equilibrium value, the gas is said to be saturated with vapor. It is not possible for the gas to contain a greater concentration of vapor, because as soon as the vapor pressure of the liquid is exceeded by the partial pressure of the vapor, condensation takes place.

Vapor pressure of the pond water at 70°F (21.1°C) = 0.363 psia [2.50 kPa (abs)]. Partial pressure of air at 90°F (32.2°C) dry bulb and 65°F (18.3°C) wet bulb = 0.170 psia (1.17 kPa abs).

Since 0.363 is greater than 0.170, water will be evaporated into the air stream.

1.710 *a.* Near the floor. Since the specific weight of the hot air is less than that of the cold air, the warm air will rise, thus setting up convection currents.

b. *e* = emissivity factor, expressing the degree to which the source surface approaches an "ideal blackbody." An ideal blackbody is one that could absorb (or emit) all the radiant energy that falls on it.

Most nonmetallic substances regardless of color have *e* equal to or greater than 0.8.

Aluminum paint has an *e* = 0.3 to 0.5. Therefore, to heat a room most efficiently, the radiators should be painted with black rather than aluminum paint.

1.711
$$1 \text{ kW} = 3413 \text{ Btu/h (3604 kJ/h)}$$
$$50,000 \text{ kW} = 170,650,000 \text{ Btu/h (180,200 MJ/h)}$$

$$\frac{\text{Btu/h}}{\text{lb/h}} = \frac{\text{Btu}}{\text{lb}} = \frac{170,650,000 \ (180,200 \text{ MJ})}{569,000 \ (258,098 \text{ kg})}$$

$$= h_1 - h_2' = 300 \text{ Btu/lb (698.4 kJ/kg)}$$

$$\eta_e = \text{engine efficiency}$$

$$h_1 - h_2 = \frac{h_1 - h_2'}{\eta_e} = \frac{300}{0.75} = 400 \text{ Btu/lb (931 kJ/kg)}$$

$$h_1 = 950 + 300 = 1250 \text{ Btu/lb (2910 kJ/kg)}$$
$$h_2 = 1250 - 400 = 850 \text{ Btu/lb (1979 kJ/kg)}$$

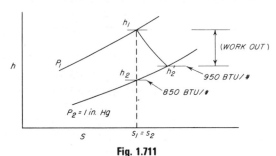

Fig. 1.711

Using a Mollier chart, find s_2 by following constant-pressure line (1 inHg) down to h_2 = 850 Btu/lb and then project vertically up to h_1 = 1250 Btu/lb.

$$\text{Condition at } h_1 = T_1 = \textbf{476}°\text{F (\textbf{246.7}°C)} \quad \textbf{1c}$$
$$P_1 = \textbf{243} \text{ psia (\textbf{1675.5} kPa)} \quad \textbf{2a}$$

1.712   Total wt snow = vol × specific wt
$$= (10 \times 50) \times \tfrac{1}{2} \times 10$$
$$= 2500 \text{ lb (1134 kg)}$$

$$\text{Btu to melt snow} = \frac{\text{wt} \times \text{heat of fusion}}{\text{efficiency}}$$

$$= \frac{2500 \times 144}{0.50}$$

$$= 720{,}000 \text{ Btu (759,600 kJ)}$$

$$3415 \text{ Btu} = 1 \text{ kWh, and kWh needed} = \frac{720{,}000}{3415} = 211$$

$$\text{Cost} = 0.08 \times 211 = \textbf{\$16.88} \qquad \textbf{a}$$

1.713 W_{bo} = lb steam/h = 885,000/4 = 221,250 lb (100,359 kg)/h

h_1 = enthalpy of water (280°F at 400 psia) (137.8°C at 2758 kPa abs) reaching boiler
$\quad$ = 248.4 Btu/lb (578.3 kJ/kg) (see steam tables)
h_2 = enthalpy of steam delivered by boiler (700°F at 400 psia)
$\quad$ = (371.1°C at 2758 kPa abs)
$\quad$ = 1362.7 Btu/lb (3172.4 kJ/kg)

$$\text{Boiler output/h} = W_{bo} \times (h_2 - h_1)$$
$$= 221{,}250 \, (1362.7 - 248.4)$$
$$= 247 \times 10^6 \text{ Btu/h (260,800 MJ/h)}$$

$$\text{Pounds coal necessary/h} = \frac{\text{Btu/h output}}{(\text{heating value/lb})(\text{efficiency})}$$

$$= \frac{247 \times 10^6}{13{,}850 \times 0.825}$$

$$= \textbf{21,600} \text{ lb } (\textbf{9798} \text{ kg})/\text{h} \qquad \textbf{d}$$

1.714

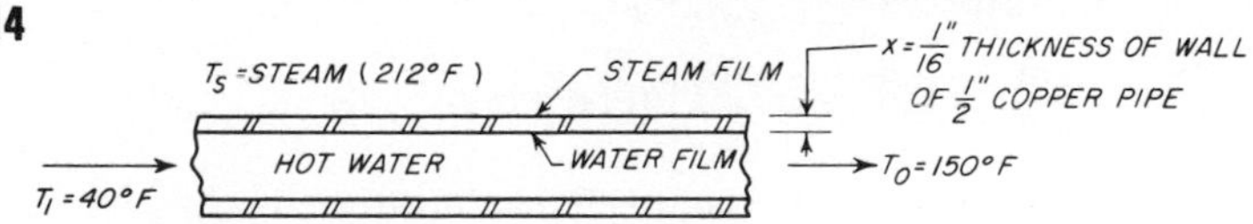

Fig. 1.714

$$Q = UA \Delta t_m,$$

where Q = heat transferred, Btu/h
$\quad U$ = transmittance, Btu/(ft^2) (°F)(h)
$\quad \Delta t_m$ = logarithmic mean temperature difference/°F

$$\text{Surface area} = \pi \frac{D}{12} \times L = \pi \frac{0.5}{12} \times 20 = 2.62 \text{ ft}^2 \ (0.260 \text{ m}^2)$$

$$\Delta t_m = \frac{\Delta t_a - \Delta t_B}{\log_e (\Delta t_a / \Delta t_B)}$$

where $\Delta t_a = T_s - T_1 = 212 - 40 = 172°\text{F} \ (77.8°\text{C})$
$\Delta t_B = T_s - T_0 = 212 - 150 = 62°\text{F} \ (16.7°\text{C})$

$$\Delta t_m = \frac{172 - 62}{\log_e (^{172}/_{62})} = \frac{110}{\log_e 2.77} = \frac{110}{1.02} = 108°\text{F} \ (42.2°\text{C})$$

$$\frac{1}{U} = \frac{1}{h_w} + \frac{x}{K} + \frac{1}{h_s}$$

where K = thermal conductivity copper, $\dfrac{\text{Btu/in thickness}}{\text{h/(ft)(°F)}}$

h_w = surface coefficient heat transfer (water film),
Btu/(h)(ft^2)(°F) [20.44 kJ/(h)(m^2)(°C)]
h_s = surface coefficient heat transfer (steam film),
Btu/(h)(ft^2)(°F) [20.44 kJ/(h)(m^2)(°C)]

$$\frac{1}{U} = \frac{1}{600} + \frac{0.0625}{2100} + \frac{1}{1000}$$

$= 0.00167 + 0.00003 + 0.001$
$= 0.00270$
$U = 370 \text{ Btu/(ft}^2)(°\text{F})(\text{h})$
$Q = 370 \times 2.62 \times 108$
$= 105,000 \text{ Btu } (110,800 \text{ kJ})/\text{h}$
w = wt water, lb/h
c_p = mean specific heat coefficient = 1
$Q = W_{c_p}(T_0 - T_1)$ $105,000 = W(1)(150 - 40) = W(110)$

$$W = \frac{105,000}{110} = \textbf{956} \text{ lb } (\textbf{433.6} \text{ kg})/\text{h}$$

$$\frac{W}{500} = \frac{956}{500} = \textbf{1.91} \text{ gal } (\textbf{7.23} \text{ liters}) \text{ water} \qquad \textbf{b}$$

1.715 Btu output/lb coal = (heat combustion/lb coal)(eff.)
$= (14,000)(0.50)$
$= 7000 \text{ Btu/lb } (16296 \text{ kJ/kg})$
h_2 = enthalpy steam (212°F and 50 percent quality)
$h_2 = h_f + x_2 h_{fg}$ (from vapor tables)
$= 180 + 0.50 \times 970 = 665 \text{ Btu/lb } (1548 \text{ kJ/kg})$
h_{fg} = enthalpy fluid and vapor (gas)

$$h_1 = \text{enthalpy water (70°F)} = 38 \text{ Btu/lb (88.5 kJ/kg)}$$
$$\text{Btu output/lb coal} = w_{bo}(h_2 - h_1)$$

where $w_{bo} = \text{lb (kg) steam /lb (kg) coal}$

$$7000 = w_{bo}(665 - 38) = 627w_{bo}$$

$$w_{bo} = \frac{7000}{627} = \textbf{11.2} \text{ lb (kg) steam/lb (kg) coal} \qquad \textbf{a}$$

1.716 Indicated hp $= I_{\text{hp}} = \dfrac{P_m LAN}{33,000}$

where $P_m = \text{mean effective pressure, psia}$
$L = \text{stroke, ft}$
$A = \text{net area of piston, in}^2$
$N = \text{power cycles/min per cylinder end}$
$D = \text{piston diameter, in}$
$D_R = \text{piston rod diameter, in}$

$$\text{Area head end piston} = \frac{\pi D^2}{4} = \frac{\pi}{4} \times (6)^2$$

$$= 28.4 \text{ in}^2 \text{ (183.2 cm}^2)$$

$$\text{Area crank end piston} = \frac{\pi}{4}(D^2 - D_R^2)$$

$$= \frac{\pi}{4}(36 - 1.56)$$

$$= 27.0 \text{ in}^2 \text{ (174.2 cm}^2)$$

I_{hp}, head end $= 62 \times 0.67 \times 28.4 \times 300 \div 33,000 = 10.65$ hp
I_{hp}, crank end $= 62 \times 0.67 \times 27 \times 300 \div 33,000 = 10.15$ hp
Total $I_{\text{hp}} = 10.65 + 10.15 = 20.8$ hp
Brake hp $= (\text{total } I_{\text{hp}})(\text{mean eff.})$
$= 20.8 \times 0.83 = 17.25$ hp
1 hp $= 0.746$ kW
Generator output $= (\text{brake hp})(0.746)(\text{generator eff.})$
$= 17.25 \times 0.746 \times 0.92 = \textbf{11.85}$ kW \qquad **c**

1.717 $I_{\text{hp}} = \dfrac{P_m LAN}{33,000}$ (see answer to question 1.716)

$$\text{Work/revolution} = 2P_m LA = 2 \times 120 \times 1 \times \frac{\pi}{4}(10)$$

$$= \textbf{18,850} \text{ ft·lb (\textbf{25,561} N·m)}$$
$$I_{\text{hp}} = 18,850 \times 300 \div 33,000 = \textbf{171.5} \text{ hp (\textbf{128} kW)} \qquad \textbf{d}$$

1.718 $\dfrac{P_1V_1}{T_1} = \dfrac{P_2V_2}{T_2}$ and $\dfrac{740 \times 400}{291} = \dfrac{760 \times V_2}{273}$

where T (°abs) = °C + 273

$$V_2 = {}^{273}\!/_{291} \times {}^{740}\!/_{760} \times 400 = \mathbf{366}\ \text{cm}^3 \qquad \mathbf{c}$$

1.719 1 liter water = 1000 cm³ = 1000 g

$$Q = WC(T_2 - T_1)$$

where Q = heat added, cal
W = wt water, g

$$C = \text{thermal capacity, } \frac{\text{cal}}{\text{°C/g}}$$

$$500 = 2000\ (1)(T_2 - 30)$$
$$T_2 - 30 = 0.25$$
$$T_2 = \mathbf{30.25}\text{°C} \qquad \mathbf{c}$$

1.720 For a Carnot cycle the $T - S$ diagram is a rectangle.

$$T_1 = 273 + 157 = 430\ \text{K}$$
$$T_2 = 273 + 100 = 373\ \text{K}$$

$$\text{Thermal efficiency} = \frac{T_1 - T_2}{T_1} \times 100$$

$$= \frac{430 - 373}{430} \times 100 = \mathbf{13.25}\ \text{percent} \qquad \mathbf{b}$$

1.721 1 W = 5.689×10^{-2} Btu/min
Wt 1 gal water = 62.5/7.5 = 8.33 lb
Btu/min effective in heating water = (W input)(eff.)(5.689×10^{-2})
$$= (300)(0.75)(5.689)(10)^{-2}$$
$$= 12.8\ \text{Btu/min}$$

Btu/gal necessary to heat water

$$= (\text{wt water, lb}) \times (\text{temp change in °F})$$
$$= 8.33(120 - 60) = 500\ \text{Btu/gal}$$

$$\text{Time to heat 1 gal water} = \frac{\text{Btu/gal}}{\text{Btu/min}} = \frac{500}{12.8}$$

$$= \mathbf{39.1}\ \text{min/gal or 0.65 h/gal}$$

$$\text{Cost to heat 2 gal water} = \frac{\text{cost}}{\text{kWh}} \times \text{kW} \times \frac{\text{h}}{\text{gal}}$$

$$= 0.08 \times 0.300 \times 0.65 \times 2$$

$$= \mathbf{\$0.031}\ \text{or } \mathbf{3}\ \text{cents} \qquad \mathbf{c}$$

1.722 Saving, Btu $= \dfrac{\text{Btu saved}}{(\text{h})(\text{°F})(\text{area})} \times (\text{h})(\Delta T)(\text{area, ft}^2)$

$\qquad\qquad = \dfrac{0.40 - 0.18}{1 \times \text{°F} \times 1} \times 5000 \times 35 \times 10{,}000$

$\qquad\qquad = 385 \times 10^6 \text{ Btu}$

lb coal saved $= \dfrac{\text{saving, Btu}}{(\text{Btu/lb coal}) \times \text{eff.}}$

$\qquad\qquad = \dfrac{385 \times 10^6}{13 \times 10^3 \times 0.60} = 49{,}360 \text{ lb}$

Saving in \$ $=$ (cost/ton)(tons saved) $= 40 \times 49{,}360/2000$
$\qquad\qquad = $ **\$987**   **d**

1.723 psia $= 14.7 + \text{psig} = 14.7 + 20 = 34.7$ (239.3 kPa abs)
$\qquad\qquad h_2 = h_f + x h_{fg} \text{ and } h_1 = h_f$
where $x =$ quality steam
$\quad h_1 =$ enthalpy condensed water
$\quad h_2 =$ enthalpy original mixture/lb
$\quad h_f =$ enthalpy 1 lb (0.454 kg) water
$\quad h_{fg} =$ change of enthalpy during vaporization

$$\Delta(\text{Btu/lb}) = h_2 - h_1 = h_f + x h_{fg} - h_f = x h_{fg}$$

$$x = \frac{\Delta(\text{Btu/lb})}{h_{fg}} = \frac{475}{939.5} = 0.507 = \textbf{50.7} \text{ percent} \qquad \textbf{a}$$

1.724 (See answer to question 1.716 for symbols.)

$$I_{hp} = \frac{P_m LAN}{33{,}000}$$

Area piston, head end $= \dfrac{\pi}{4} \times 144 = 113.1 \text{ in}^2$

Area piston, crank end $= \dfrac{\pi}{4}(12^2 - 2^2) = 110.0 \text{ in}^2$

I_{hp} head end $= \dfrac{70 \times 1.5 \times 113.1 \times 350}{33{,}000} = 126.0$

I_{hp} crank end $= \dfrac{70 \times 1.5 \times 110.0 \times 350}{33{,}000} = 122.5$

$\qquad\qquad\qquad\qquad\qquad I_{hp(tot)} = \overline{248.5} \text{ hp}$
Brake hp $= (I_{hp})$(mechanical eff.) $= 248.5 \times 0.92$
$\qquad\qquad = $ **228.5** hp (**170.5** kW)   **a**

1.725   1 hp = 42.41 Btu/min

    Btu/min dissipated as heat = 42.41 × hp = 42.41 × 300
$$= 12,720$$
Btu/min necessary to heat water from 80 to 180°F
$$= 12,720 = \text{wt water/min} \times 100$$
    wt water/min = 127.2 lb/min

$$= \frac{127.2 \times 7.5}{62.5}$$

$$= \textbf{15.3} \text{ gal (57.9 liters)/min} \qquad \textbf{b}$$

GENERAL

1.801   Assuming a current exists in the direction as shown in Fig. 1.801 of the question, Kirchhoff's voltage law states that the sum of the E's is zero. Therefore, the vector sum

$$\mathbf{E}_{CB} + \mathbf{E}_{BA} + \mathbf{E}_{AD} + \mathbf{E}_{DC} = 0$$

Fig. 1.801

    1. Assuming $\mathbf{E}_{CB}$ as reference, the voltage vector diagram is shown in Fig. 1.801. For a 45° angle

$$|IR| = |IX_{c,c}| \qquad \text{and} \qquad R = X_{c,c}$$
$$R = \frac{1}{2\pi f C_c} = \frac{1}{2\pi 60 \times 10^{-6}} = \textbf{2650} \text{ ohms} \qquad \textbf{1c}$$

    2. $|E_{CB}| = |E_{BA}| = |E| = |E_{CA}/2|$
       $|E_{CA}| = 1.414|E_{AD}|$
and $|E_{AD}| = 2|E|/1.414 = \textbf{1.414}|E| \qquad \textbf{2a}$

1.802   1. At resonant frequency in a series RLC circuit, the supply voltage equals the voltage across the resistance.

    Therefore, $\qquad P = \dfrac{E_R^2}{R} \qquad \text{or} \qquad R = \dfrac{E^2}{P}$

and $$R = \frac{100^2}{10} = \textbf{1000} \text{ ohms} \quad \textbf{1a}$$

2. The Q at the resonant frequency equals the resonant frequency divided by the bandwidth.

$$Q_0 = \frac{\omega_0}{\Delta\omega} = \frac{2\pi \times 10^6}{0.1(2\pi \times 10^6)} = 10$$

$$Q_0 = \frac{\omega_0 L}{R} \quad \text{or} \quad L = \frac{Q_0 R}{\omega_0}$$

Therefore,

$$L = \frac{(10)(1000)}{2\pi \times 10^6} = \textbf{1.59} \times 10^{-3} \text{ H} \quad \textbf{2b}$$

3. $Q_0 = \dfrac{1}{\omega_0 RC}$ or $C = \dfrac{1}{Q_0\omega_0 R}$

Therefore,

$$C = \frac{1}{(10)(2\pi \times 10^6)(1000)} = \textbf{15.9} \times 10^{-12} \text{ F} \quad \textbf{3d}$$

Check:

$$\omega_0 = \frac{1}{\sqrt{LC}} = \frac{1}{\sqrt{1.59 \times 10^{-3} \times 15.9 \times 10^{-12}}} = 2\pi \times 10^6 \text{ rad/s}$$

1.803   1. Induction motor kVA load $= \dfrac{80}{0.8} \underline{/\cos^{-1} 0.8}$
$$= 100\underline{/36.8°} \text{ kVA}$$

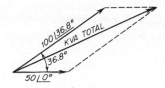

Fig. 1.803

Total kVA load is the vector sum of the loads as shown in Fig. 1.803.

$$\text{KVA}_{tot} = 100\underline{/36.8°} + 50\underline{/0°} = 80 + j60 + 50$$
$$= 130 + j60 = \textbf{143}\underline{/\textbf{24.7°}} \text{ kVA} \quad \textbf{1d}$$

2. Power factor $= \cos 24.7° = \textbf{0.910} \quad \textbf{2d}$

3. Current per phase = kVA × 1000/(1.732 × voltage)
$$= 143 \times 1000/(1.732 \times 2200)$$
$$= \textbf{37.6 A} \qquad \textbf{3c}$$

1.804 1. If $R = 0$ and $G = 0$, the resonant angular frequency is $\omega_0 = 1/\sqrt{LC} = 10^6$ rad/s. By definition of the Q of a coil, $Q = \omega L/R$. The frequency at which the Q was measured should have been stated in the question. Since it was not, assume that it was measured at ω_0 and that the resonant frequency will not differ greatly from ω_0. Then, approximately,

$$R = \frac{\omega_0 L}{Q} = \frac{10^6 \times 10^{-3}}{2} = \textbf{500 ohms} \qquad \textbf{1a}$$

From the LC product

$$C = \frac{LC}{L} = \frac{10^{-12}}{10^{-3}} = \textbf{1} \times \textbf{10}^{-9} \textbf{ F} \qquad \textbf{1a}$$

From the formula for the time constant of the series GC circuit comprising the capacitor alone

$$T = \frac{C}{G} \qquad G = \frac{C}{T} = \frac{10^{-9}}{2 \times 10^{-6}}$$
$$= \textbf{5} \times \textbf{10}^{-4} \textbf{ mho} \qquad \textbf{1a}$$

2. The equivalent impedance of the given circuit is given by

$$Z_e = R + j\omega L + \frac{1}{G + j\omega C}$$

Series resonance may be defined as either of two conditions as ω is varied: (1) the imaginary (reactive) part of Z_e equal to zero, or (2) the magnitude of Z_e equal to a maximum. In general, for low-Q circuits, these two definitions will yield slightly differing values of the resonant frequency. Choosing here to use the first definition,

$$Z_e = R_e + jX_e = R + j\omega L + \frac{1}{G + j\omega C}$$

Multiplying both sides by $G + j\omega C$

$$GR_e - \omega C X_e + j(X_e G + \omega C R_e) = RG - \omega^2 LC + j(\omega LG + \omega CR) + 1$$

Equating the real terms,

$$GR_e - \omega C X_e = RG - \omega^2 LC + 1$$

Equating the imaginary terms,

$$X_e G + \omega C R_e = \omega LG + \omega CR$$

From the latter, if $X_e = 0$,

$$R_e = \frac{LG}{C} + R = \frac{10^{-3} \times 5 \times 10^{-4}}{10^{-9}} + 500 = 1000 \text{ ohms}$$

From the former, if $X_e = 0$, $GR_e = RG - \omega^2 LC + 1$.

Whence

$$\omega = \sqrt{\frac{1 + G(R - R_e)}{LC}} = \omega_0 \sqrt{1 - G(R_e - R)}$$

$$= \omega_0 \sqrt{1 - \frac{LG^2}{C}}$$

$$= \omega_0 \sqrt{1 - \frac{10^{-3} \times 25 \times 10^{-8}}{10^{-9}}} = \omega_0 \sqrt{1 - 0.25}$$

$$= 0.865\omega_0 = \mathbf{0.865 \times 10^6} \text{ rad/s} \qquad \mathbf{2c}$$

Thus the series-resonant frequency is lower than the LC resonance by the factor 0.865. At this frequency, the value of R is $0.865 \times 500 = 433$ ohms, but this correction does not change the resonant frequency.

3. $R_e = \dfrac{LG}{C} + R = 500 + 433 = \mathbf{933}$ ohms $\qquad \mathbf{3d}$

1.805 The kVA per phase is

$$\text{kVA}_\phi = \frac{\text{kW}}{3 \times \text{pf}} = \frac{1000}{3 \times 0.85} = 392$$

The voltage per phase is

$$|E_R| = \text{kVA}_\phi \times \frac{10^3}{I_\phi} = 392 \times \frac{10^3}{56.4} = 6960 \text{ V to neutral}$$

If I_ϕ is chosen as reference, then for $\cos\theta = 0.850$ or $\theta = 31.6°$ and $\sin\theta = 0.524$,

$$I_\phi = 56.4 \underline{/0°} \text{ A/wire}$$
$$E_R = 6960 \underline{/31.6°} \text{ V to neutral}$$

In each phase the line impedance, $Z_\phi = 1 + j10$, and for a balanced load there is no current, and hence no voltage drop, in the neutral. Hence, the sending-end voltage is simply $E_s = E_R + T_\phi Z_\phi = 5930 + j3650 + 56.4 + j564 = 5990 + j4214 = 7344 \underline{/35.3°}$ V to neutral.

1. The line voltage required at the sending end will be

$$7344 \times 1.732 = \mathbf{12,700} \text{ V line to line} \qquad \mathbf{1c}$$

2. The regulation is

$$\frac{|E_s|}{|E_R|} - 1 = \frac{7344}{6960} - 1 = 0.052 = \textbf{5.2 percent} \qquad \textbf{2b}$$

1.806 $R \times \tan 24° = 0.445R$

$(0.445R - 10)^2 + (R - 300)^2 = (R - 150)^2$ Fig. 1.806
$0.198R^2 - 8.9R + 100 + R^2 - 600R + 90,000 = R^2 - 300R$
$\qquad\qquad\qquad\qquad\qquad\qquad\qquad\qquad\qquad\qquad + 22,500$

$$0.198R^2 - 308.9R = -67,600$$
$$R^2 - 1560R = -341,400$$
$$(R - 780)^2 = (-780)^2 - 341,400 = 608,400 - 341,000$$
$$= 267,000 = (\pm517)^2$$
$$R = +780 + 517 = \textbf{1297} \text{ ft } (\textbf{395.3 m}) \qquad \textbf{d}$$

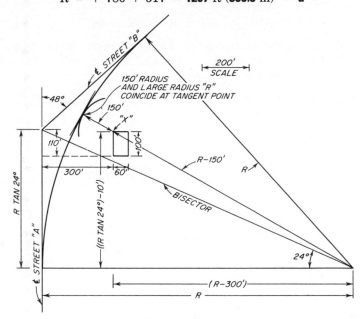

Fig. 1.806

Check: $0.445 \times 1297 = 577$, and $577 - 10 = 567, 1297 - 300 = 997$,
$1297 - 150 = 1147$.

$$567^2 = 321,500$$
$$997^2 = 994,000$$
$$1147^2 = \overline{1,315,500}$$

1.807 $1100 \times \sin 5° = 1100 \times 0.0872 = 95.9$
$1100 \times \cos 5° = 1100 \times 0.996 = 1095.8$

Coordinates of the radius center are (Fig. 1.807)

$$1570.0 - 1095.8 = 474.2\text{N}$$
$$420.0 + 95.9 = 515.9\text{E}$$

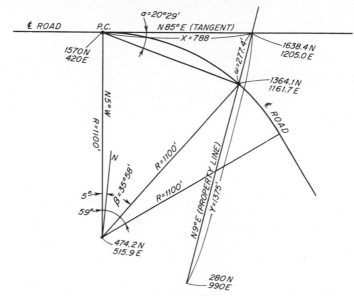

Fig. 1.807

To find coordinates of centerline of road N85°E (tangent) and intersection of property line bearing N9°E through coordinates 280 N and 990 E:

$$420 + 0.996X = 990 + 0.1564Y$$
$$0.996X - 0.1564Y = 570 \tag{1}$$
$$1570 + 0.0872X = 280 + 0.988Y$$

$$-0.0872X + 0.988Y = 1290 \tag{2}$$

Eq. (3) = 6.32 × Eq. (1)

$$+6.3000X - 0.988Y = 3600 \tag{3}$$

Eqs. (2) + (3): $+6.213X = 4890$
$X = 788$ ft

From Eq. (2) $0.988Y = 1290 + 0.0872 \times 788$
$= 1290 + 68.6 = 1358.6$
$Y = 1375$ ft

Check coordinates:

$$1570 + 788 \times 0.0872 = 1570 + 68.6 = 1638.4N$$
$$280 + 1375 \times 0.998 = 280 + 1358.6 = 1638.4N$$
$$420 + 788 \times 0.996 = 420 + 785 = 1205E$$
$$990 + 1375 \times 0.1564 = 990 + 215 = 1205E$$

To find coordinates of centerline of road on the curve and the intersection of line bearing N9°E, solve for angle and distance.

$$1100 \sin \beta + 0.1564W = 1205.0 - 515.9 = 689.1 \tag{4}$$
$$1100 \cos \beta + 0.988W = 1638.4 - 474.2 = 1164.2 \tag{5}$$

Eq. (6) = 6.32 × Eq. (4):

$$6952 \sin \beta + 0.988W = 4355.1 \tag{6}$$

Eq. (7) = Eq. (6) − Eq. (5):

$$6952 \sin \beta - 1100 \cos \beta = 3190.9 \tag{7}$$

Eq. (7a) = Eq. (7) ÷ 1100:

$$6.32 \sin \beta = \cos \beta + 2.90 \tag{7a}$$

$$39.94 \sin^2 \beta = 39.94 - 39.94 \cos^2 \beta = \cos^2 \beta + 5.80 \cos \beta + 8.41$$
$$40.94 \cos^2 \beta + 5.80 \cos \beta = 31.53$$
$$\cos^2 \beta + 0.1416 \cos \beta = 0.770$$
$$(\cos \beta + 0.0708)^2 = 0.770 + 0.005 = 0.775 = \pm 0.8800^2$$
$$\cos \beta = +0.8800 - 0.0708 = 0.8092$$
$$= 35°58'$$
$$\alpha = 0.5 \, (5° + 35°58') = \mathbf{20°29'} \quad \mathbf{b}$$
$$1100 \sin \beta = 1100 \times 0.587 = 645.8$$
$$1100 \cos \beta = 1100 \times 0.809 = 889.9$$

from Eq. (5)
$$889.9 + 0.988W = 1164.2$$
$$0.988W = 274.3$$
$$W = 277.6$$

Check coordinates:

$$0.1564W = 0.1564 \times 277.6 = 43.3$$
$$474.2 + 889.9 = 1364.1N + 274.3 = 1638.4N$$
$$515.9 + 645.8 = 1161.7E + 43.3 = 1205.0E$$

1.808 $\mu = 0.005/32$ (lb/ft per ft/s^2) and $F = 25$ lb

$$V_{\text{transverse}} = \sqrt{F/\mu} = \sqrt{25 \times 32/0.005} = \sqrt{160,000}$$
$$= \mathbf{400} \text{ fps } (\mathbf{121.92} \text{ m/s}) \quad \mathbf{d}$$

1.809 $V = \sqrt{\gamma RT/M}$ (ideal gas)

where γ = ratio heat capacity at constant pressure to heat capacity at constant volume

R = universal gas constant

T = temperature, K

M = mean molecular wt air

$V = \sqrt{1.40 \times 8.3 \times 10^7 \times {}^{300}\!/_{29}}$

$= 34{,}670$ cm/s $=$ **1137** fps (**346.7** m/s)　　**b**

1.810　Assume the speed of sound in air to be 1130 fps (344.4 m/s).

1. $\lambda = (1130 - 88)/500 =$ **2.09** ft (**0.637** m)　　**1c**
2. $\lambda = (1130 + 88)/500 =$ **2.44** ft (**0.744** m)　　**2d**
3. $\int = \int_0 \times V/(V - v_w) = 500 \times 1130/(1130 - 88)$
   $=$ **5.42** cps　　**3b**
4. $\int = \int_0 \times V/(V + v_w) - 500 \times 1130/(1130 + 88)$
   $=$ **4.64** cps　　**4c**
5. $f = f_0(V + v_L)/(V - v_w)$
   $= 500(1130 + 44)/(1130 - 88)$
   $=$ **5.62** cps　　**5d**
6. $f = f_0(V - v_L)/(V + v_w)$
   $= 500(1130 - 44)/(1130 + 88)$
   $=$ **4.46** cps　　**6a**

1.811　1. $B_2 - B_1 = 10 = 10 \log I_2/I_1$
   $\log I_2/I_1 = 1 = \log 10$
   $\therefore I_2/I_1 =$ **10**　　**1a**
2. $B_2 - B_1 = 20 = 10 \log I_2/I_1$
   $\log I_2/I_1 = 2 = \log 100$
   $\therefore I_2/I_1 =$ **100**　　**2d**
3. $B_2 - B_1 = 10 \log I_2/I_1 = 10 \log 4$
   $= 10 \times 0.6 =$ **6** dB　　**3b**
4. $B_2 - B_1 = 20 \log p_2/p_1 = 20 \log 6 = 20 \times 0.778$
   $=$ **15.56** dB　　**4c**

1.812　1. Potential horsepower available is

$$300 \times 62.5 \times {}^{125}\!/_{550} = 4260 \text{ hp } (3178 \text{ kW})$$

Efficiency of the plant is

$$100 \times 3500/4260 = \textbf{82.2} \text{ percent}　\textbf{1c}$$

2. Area of pipe $= 0.785 \times 25 = 19.63$ ft^2 (1.82 m^2)
   Velocity of water in pipe $= 300/19.63 = 15.3$ fps
   Friction factor for pipe (from table) $= f = 0.01276$

and $\qquad f \times l/d = 0.01276 \times 1000/5 = 2.552$

Head loss in pipe $= (1.03 + 2.552) \times 15.3^2/64.4 = 13$ ft
Head available for turbine $= 125 - 13 = 112$ ft (34.14 m)

Potential horsepower of the turbine is

$$300 \times 62.5 \times {}^{112}\!/_{550} = 3820 \text{ hp } (2850 \text{ kW})$$

Efficiency of the turbine is

$$100 \times 3500/3820 = \textbf{91.5} \text{ percent} \qquad \textbf{2d}$$

1.813   Area of 12-in-diameter pipe $= 0.785$ ft^2
2000 gpm $= 2000/(60 \times 7.48) = 4.45$ ft^2/s
$v = Q/A = 4.45/0.785 = 5.68$ fps (1.73 m/s)
6.5 psi $= 6.5/0.434 = 15$ ft of head
Pump develops a net head of $1000 - 15 = 985$ ft
$\mu = $ coefficient of viscosity (lb·s/ft^2)
$\rho = $ mass per unit volume
$\nu = $ kinematic viscosity $= \mu/\rho$
API specific gravity relative to water at 60°F
$\qquad = 141.5/(131.5 + \deg \text{API}) = 141.5/(131.5 + 40)$
$\qquad = 141.5/171.5 = 0.825$
$\mu = (0.0022 \times 400 - 1.30/400) \times 0.825$
$\qquad = (0.88 - 0.0032) \times 0.825 = 0.724 \text{ P}$
$\qquad = 0.724/478.69 = 0.00151 \text{ lb·s/ft}^2$
$\rho = 62.4 \times 0.825/32.2 = 1.60 \text{ slugs/ft}^3$
$\nu = \mu/\rho = 0.00151/1.60 = 0.000944 \text{ ft}^2/\text{s}$
Re $= vd/\nu = 15.3 \times 1.0/0.000944 = 16{,}200$ (turbulent flow)

From a table of friction factors versus Reynolds numbers

$f_{\text{max}} = 0.039$
$985 = (f \times l \times v^2)/(d \times 2g) = 0.039 \times l \times 5.68^2/64.4$
$l = (985 \times 64.4)/(0.039 \times 32.2) = \textbf{50,500}$ ft (**15,392** m)   **b**

1.814   Area 4-in pipe $= 0.0872$ ft^2 $= 12.56$ in^2
μ (viscosity) at 100°F $= 0.397 \times 10^{-6}$ (from tables)
$w_1 = p/RT$
$\qquad = [(100 + 14.7) \times 144]/[53.3 \times (100 + 459.4)]$
$\qquad = 0.554 \text{ lb/ft}^3$
$v_1 = W/(0.554 \times 0.0872) = 20.7W$
$\nu = \mu/\rho = 0.397 \times 10^{-6} \times 32.2/0.554 = 23.1 \times 10^{-6}$
Re $= v_1 d/\nu = (20.7W \times 0.33)/(23.1 \times 10^{-6}) = 295{,}000W$

$$(p_1^2 - p_2^2) = \frac{W^2 RT(f \times l/d)}{ga^2}$$

$$W^2 f = \frac{[(114.7)^2 - (74.7)^2] \times (12.56)^2 \times 32.2}{53.3 \times 559.4 \times 3000}$$

$$= \frac{(13,140 - 5580) \times 5070}{89,400 \times 1000} = \frac{7560 \times 5.07}{89,400} = 0.428$$

Assume a value for W, as 5.00 lb; then Re $= 1,475,000$ and $f = 0.0151$:

$$W^2 = 0.428/0.0151 = 28.4$$
$$W = 5.33 \text{ lb}$$

Then for $W = 5.33$, Re $= 1,570,000$, and $f = 0.0149$

$$W^2 = 0.428/0.0149 = 28.7$$
$$W_s = \textbf{5.36} \text{ lb } (\textbf{2.43} \text{ kg}) \text{ (say OK)} \qquad \textbf{d}$$

1.815 $\dfrac{T_2}{T_1} = e^{f\theta} = 2.718^{0.12 \times 2\pi n}$

where θ = angle, rad
$\quad\quad n$ = no. turns about capstan

$$4000/50 = 80 = 2.718^{0.755n}$$
$$\log 80 = n \times 0.755 \log 2.718$$
$$1.9031 = n \times 0.755 \times 0.4343 = n \times 0.327$$
$$n = 1.9031/0.327 = 5.81, \text{ or use } \textbf{6} \text{ turns} \qquad \textbf{a}$$

1.816 $P = \dfrac{2\pi nT}{33,000 \times 12} \text{ hp}$

$$\text{Torque} = \frac{33,000 \times 12 \times 80}{6.28 \times 300} = 16,800 \text{ in·lb}$$

$$\text{Moment} = 30 \times 1000 = 30,000 \text{ in·lb}$$

$$\text{Shear due to torque} = s = \frac{Tr}{\pi r^4/2} = \frac{16,800}{\pi r^3/2}$$

$$\text{Fiber stress due to moment} = f = \frac{30,000r}{\pi r^4/4} = \frac{60,000}{\pi r^3/2}$$

Shear stress due to torque and moment $= s' = (\frac{1}{2}) \sqrt{4s^2 + f^2}$

$$10,000 = \frac{1 \times 2}{2 \times \pi r^3} \sqrt{4(16,800)^2 + 60,000^2}$$

$$\pi r^3 = \frac{1000}{10,000} \sqrt{1130 + 3600} = 6.88$$
$$r^3 = 2.196$$
$$r = 1.30 \text{ in}$$
$$d = 2.60 \text{ in, or use } \textbf{2.625} \text{ in } (\textbf{6.67} \text{ cm}) \qquad \textbf{d}$$

Check bending stress due to torque and moment
$$f = 17,442 \qquad f/2 = 8721 \qquad s = 4884$$
$$f_{max} = (f/2) + \sqrt{(f/2)^2 + s^2} = 8721 + \sqrt{(8721)^2 + (4884)^2}$$
$$= 8721 + 9995 = 18,716 < 20,000$$
$$s_{max} = 9995 < 10,000$$

1.817 Neglect fillets. Then
$$\text{Stem of } T = 0.885 \times 6.5 \quad = \quad 5.75 \text{ in}^2$$
$$\text{Net flange of } T = 1.570 \times 15.71 = 24.70 \text{ in}^2$$
$$\text{Area of } T = 30.45 \text{ in}^2$$

Center-of-gravity distance from back of flange is
$$x = \frac{3.25 \times 5.75 + 0.785 \times 24.70}{30.45} = \frac{18.70 + 19.40}{30.45} = 1.25 \text{ in}$$

I of T:
$$\text{Stem } 5.75 \times \frac{6.5^2}{12} + 5.75 \times 2^2 = 20.20 + 23.00 = 43.20$$
$$\text{Flange } 24.70 \times \frac{1.57^2}{12} + 24.70 \times 0.465^2 = \quad 5.07 + \quad 5.35 = 10.42$$
$$\overline{ 53.62}$$

Use $I = 54 \text{ in}^4$:
$$I_2 = 2(54 + 30.45 \times 28.75^2) = 2(54 + 25,200) = 50,500$$
$$S_2 = 50,500/30 = 1683 \text{ in}^3$$
$$S_2/S_1 = 1683/1031 = \textbf{1.63} \text{ (times as strong)} \qquad \textbf{c}$$

1.818 1. $E = PL/A\Delta = (10,000 \times 600)/(1 \times 0.222)$
$$= \textbf{27} \times \textbf{10}^6 \text{ psi } (\textbf{186.2} \text{ GPa}) \qquad \textbf{1a}$$
  2. $\Delta = (3.4 \times 600 \times 0.5 \times 600)/27 \times 10^6$
$$= \textbf{0.0226} \text{ ft } (\textbf{0.689} \text{ cm}) \qquad \textbf{2b}$$

1.819 1. $\Delta = \frac{3}{32} = 36 \times 65 \times 10^{-7}t$

$$t = \frac{3 \times 10,000,000}{32 \times 36 \times 65} = \textbf{400}°\text{F } (\textbf{204.4}°\text{C}) \qquad \textbf{1d}$$

  2. Tension in ring $= \Delta \times A \times E/L = \frac{1}{32} \times \frac{1}{4} \times 30,000,000 \times \frac{1}{36}$
$$= \textbf{26,000} \text{ lb } (\textbf{115.6} \text{ kN}) \qquad \textbf{2c}$$
  3. $2T = 4 \times 36 \times p$
$$p = \frac{2 \times 26,000}{4 \times 36} = \textbf{361} \text{ psi } (\textbf{2.49} \text{ MPa}) \qquad \textbf{3a}$$

1.820 $\Delta_s = (P_s \times L)/1 \times 30 \times 10^6 = \Delta_c = (P_c \times L)/1 \times 17 \times 10^6$

$$17P_s = 30P_c$$
$$17P_s - 30P_c = 0$$
$$P_s - 1.765P_c = 0 \quad\quad (1)$$
$$P_s + P_c = 30,000 \quad\quad (2)$$

Eq. (2) − Eq. (1): $+2.765P_c = 30,000$
$$P_c = 10,860 \text{ lb}$$
$$P_s = \mathbf{19{,}170} \text{ lb } (\mathbf{85.3} \text{ kN}) \quad\quad \mathbf{1c}$$

$$\Delta_s = \Delta_c + 60 \times (93 - 65) \times 10^{-7} \times L$$

$$\frac{P_s \times L \times 10^{-6}}{1 \times 30} = \frac{P_c \times L \times 10^{-6}}{1 \times 17} + \frac{6 \times 28 \times 10^{-6} \times L}{1}$$

$$17P_s = 30P_c + 17 \times 30 \times 168$$
$$P_s - 1.765P_c = 5040 \quad\quad (3)$$
$$P_s + P_c = 30,000 \quad\quad (2)$$

Eq. (2) − Eq. (3): $+2.765P_c = 24,960$
$$P_c = \mathbf{9030} \text{ lb } (\mathbf{40.2} \text{ kN}) \quad\quad \mathbf{2b}$$
$$P_s = 20,970 \text{ lb } (93.3 \text{ kN})$$

1.821

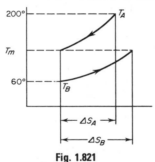

Fig. 1.821

$$\text{lb water} \times \text{temp} = \text{product}$$
$$4 \times 200.0 = 800$$
$$\underline{2 \times 60.0 = 120}$$
$$6 \times (153.3) = 920$$

$T_m = 153.3°\text{F} = 153.3 + 460 = 613.3°\text{R} (340.7 \text{ K}); T_A = 660°\text{R} (366.7 \text{ K}); T_B = 520°\text{R} (288.9 \text{ K})$. By Fig. 1.821

$$\Delta S_A = wc_p \log_e (T_m/T_A) = (4)(1) \log_e (613.3/660)$$
$$= 4 \log_e 0.93 = -0.290 \text{ units (drop)}$$
$$\Delta S_B = wc_p \log_e (T_m/T_B) = (2)(1) \log_e (613.3/520)$$
$$= 2 \log_e 1.18 = +0.331 \text{ units (rise)}$$
$$\Delta S_{A-B} = 0.331 - 0.290 = \mathbf{0.041} \text{ units} \quad\quad \mathbf{a}$$

1.822 1. For ideal turbine

$$W = \frac{KwRT_1}{1 - K}\left[\left(\frac{p_2}{p_1}\right)^{\frac{K-1}{K}} - 1\right]$$

$$= \frac{(1.4)(1)(53.3)(1660)}{1 - 1.4}\left[\left(\frac{15}{60}\right)^{\frac{1.4-1}{1.4}} - 1\right]$$

$$= 103,800 \text{ ft·lb/s}$$

$$= 103,800/550 = \textbf{188.5} \text{ hp } (\textbf{140.6} \text{ kW}) \qquad \textbf{1b}$$

2. For turbine developing 150 hp (111.9 kW):

$$\text{Temp water entering} = 50°\text{F } (10.0°\text{C})$$
$$\text{Temp water leaving} = 100°\text{F } (37.8°\text{C})$$
$$\text{Each lb water picks up } (100 - 50) \times 1 = 50 \text{ Btu}$$
$$= (52.8 \text{ kJ})$$

General expression for steady flow (neglecting change in kinetic energy) is

$$H_1 + Q = H_2 + W$$

$$Q = W + (H_2 - H_1) = W + wc_p \, \Delta T$$
$$= (150 \times 550)/778 + (1)(0.24)(300 - 1200)$$
$$= 106 - 216 = -110 \text{ Btu/s (to be carried away by cooling water)}$$
$$(^{110}\!/_{50})(60/8.34) = \textbf{15.85} \text{ gal } (\textbf{59.99} \text{ liters})/\text{min} \qquad \textbf{2c}$$

1.823

Steam	Ice	Calorimeter	Water
$w_s = 7$ lb	$w_i = 8$ lb	$w_c = 5$ lb	$w_w = 50$ lb
$t_s = 242°$F	$t_i = 25°$F	$t_c = 60°$F	$t_w = 60°$F
$c_s = 0.48$	$c_i = 0.50$	$c_c = 0.093$	
$p_s = 14.7$ psia			

Let t_m = resulting temperature of mixture

Heat given up by steam = heat gained by ice + heat gained by calorimeter + heat gained by water

$$w_s c_s(t_s - 212) + 970w_s + w_s(212 - t_m)$$
$$= [w_i c_i(32 - 25) + 144w_i + w_i(t_m - 32)]$$
$$+ [w_c c_c(t_m - 60)] + [w_w(t_m - 60)]$$

$$7 \times 0.48(242 - 212) + 970 \times 7 + 7(212 - t_m)$$
$$= [8 \times 0.50 \times 7 + 144 \times 8 + 8(t_m - 32)]$$
$$+ [5 \times 0.093(t_m - 60)] + [50(t_m - 60)]$$
$$101 + 6790 + 1484 - 7t_m = (28 + 1152 + 8t_m - 256)$$
$$+ (0.465t_m - 27.9) + (50t_m - 3000)$$
$$65.465t_m = +11,659 - 1180 = 10,479$$
$$t_m = \mathbf{160}°\text{F} \ (\mathbf{71.1}°\text{C}) \qquad \mathbf{d}$$

1.824 Heat loss without insulation:

$$Q = U_1 A \ \Delta t = 0.45 \times 12,000 \times 40 = 216,000 \ \text{Btu/h}$$

Heat loss with insulation:

$$Q = U_2 A \ \Delta t = 0.25 \times 12,000 \times 40 = 120,000 \ \text{Btu/h}$$
$$\text{Saving} = \ 96,000 \ \text{Btu/h}$$

Total saving for heating season:

$$Q = 96,000 \times 6000 = 596$$
$$\text{Coal saved} = (596 \times 10^6)/(14,000 \times 0.65 \times 2000)$$
$$= 32.75 \ \text{tons}$$

$$\text{Dollar saving} = \$35 \times 32.75 = \$1146 \qquad \mathbf{b}$$

1.825 1. $Q_A = 10$ tons $= 10 \times 200$ Btu/min per ton
$$= 2000 \ \text{Btu/min}$$
$$\text{Input to compressor} = 15 \times (2545/60) = 637 \ \text{Btu/min}$$

Condensing water must carry away $2000 + 637 = 2637$ Btu/min

$$Q_{\text{water}} = wc_p \ \Delta t$$
$$2637 = w \times 1 \times (70 - 50)$$
$$w = 2637/20 = 131.85 \ \text{lb/min} = 131.85/8.34$$
$$= \mathbf{15.8} \ \text{gal} \ (\mathbf{59.8} \ \text{liters})/\text{min} \qquad \mathbf{1a}$$

2. C.O.P. $= \dfrac{T_1}{T_2 - T_1} = \dfrac{0 + 460}{(70 + 460) - (0 + 460)} = \dfrac{460}{70}$

$$= \mathbf{6.57} \qquad \mathbf{2a}$$

Chemical Engineering

Reference is made a number of times in this section to R. H. Perry, "Chemical Engineers' Handbook," 3d ed., McGraw-Hill, 1950, simply by stating "Perry, p.—."

2.01

Coal analysis (dry), %	Refuse	Flue gas (volatile), %
C = 75 Ash = 8 $\left.\begin{array}{l} H_2 \\ O_2 \end{array}\right\} = 17$	C = 0	CO_2 = 12.6 CO = 1.0 O_2 = 6.2 N_2 = 80.2
100		= 100.0

Basis will be 100 lb fuel.

100 lb fuel contains 75 lb carbon, or

$$^{75}\!/_{12_{(at\ wt)}} = 6.25\ lb \cdot atoms$$

1 lb·mol flue gas contains 0.136 lb·atoms carbon; that is,

$$\begin{array}{l} CO_2 = 0.126\ lb \cdot mol \\ CO = \underline{0.010}\ lb \cdot mol \\ 0.136\ lb \cdot mol \end{array}$$

Therefore
$$\frac{6.25}{0.136} = \frac{46 \text{ lb·mol dry stack gas}}{100 \text{ lb fuel}}$$

The stack gas (dry) contains $\dfrac{0.062 \text{ lb·mol oxygen}}{\text{lb·mol stack gas}}$ and for each 100 lb fuel, there is $46 \times 0.062 = 2.86$ lb·mol oxygen in the stack gas.

Since this oxygen came from the air (21 percent oxygen and 79 percent nitrogen), there is $2.86 \times (0.79/0.21)$ or 10.75 lb·mol N_2 in stack gas from excess air.

Nitrogen in the stack gas that comes from the air supplies for combustion is

$$46 - (5.79 + 0.46 + 2.86 + 10.75) = 26 \text{ lb·mol } N_2$$

where

$$\frac{0.126}{0.136} \times 6.25 = 5.79 \text{ lb·mol } CO_2$$

and

$$\frac{0.010}{0.136} \times 6.25 = 0.46 \text{ lb·mol } CO$$

Oxygen supplied for combustion by the air is

$$26 \times \frac{0.21}{0.79} = 6.90 \text{ lb·mol}$$

Oxygen required for combustion of carbon is

$$C + O_2 \rightarrow CO_2 \text{ or } \frac{0.126}{0.136} \times 6.25 = 5.79 \text{ lb·mol}$$

$$C + \tfrac{1}{2} \times O_2 \rightarrow CO \text{ or } \tfrac{1}{2} \times \frac{0.010}{0.136} \times 6.25 = \underline{0.23 \text{ lb·mol}}$$
$$6.02 \text{ lb·mol}$$

Oxygen supplied by air for combustion of H in fuel is $6.90 - 6.02 = 0.88$ lb·mol/100 lb fuel.

Available O_2 and H_2 in fuel per 100 lb fuel is

$$H_2 + \tfrac{1}{2} \times O_2 \rightarrow H_2O$$
$$2X + (\tfrac{1}{2}X - 0.88) \, 32 = 17 \text{ (from fuel analysis)}$$
$$X = 2.55 \text{ lb·mol hydrogen or } 5.10 \text{ lb/100 lb fuel}$$
$$\tfrac{1}{2}X = 1.275 \text{ lb·mol total oxygen required}$$
$$\tfrac{1}{2}X - 0.88 = 0.395 \text{ lb·mol available oxygen in fuel}$$
$$= 12.65 \text{ lb/100 lb fuel}$$

1. *Percent oxygen*
lb·mol O_2 for combustion of 100 lb fuel

$$\begin{array}{l} CO_2 = 5.79 \text{ lb·mol} \\ CO = 0.23 \text{ lb·mol} \\ H_2 = \underline{1.27} \text{ lb·mol } (= \tfrac{1}{2}X) \\ 7.29 \text{ lb·mol} \end{array}$$

But there are 2.86 lb·mol oxygen in stack gas.

$$\therefore \text{ percent oxygen } = \frac{2.86}{7.29 - 0.39} \times 100 = \textbf{41.5} \text{ percent} \qquad \textbf{1a}$$

2. *Complete analysis of fuel*

$$\begin{array}{l} C = \textbf{75.0} \text{ percent} \qquad \textbf{2b} \\ Ash = 8.0 \text{ percent} \\ H_2 = 5.1 \text{ percent } (\times \text{ above}) \\ O_2 = 11.9 \text{ percent } (17.0 - 5.1) \end{array}$$

3. *Ft³ stack gas/lb fuel* (at 680°F)
Basis—100 lb fuel

H_2O in air (80 percent relative humidity and 36 mm vapor pressure)

$$= \frac{36 \times 0.80}{760 - 36 \times 0.80} = 0.0393 \text{ mol } H_2O/\text{mol air}$$

$$\therefore \frac{46 \times 0.802}{0.79} = 46.75 \text{ lb·mol air (dry) and } 46.75 \times 0.0393 = 1.84$$

lb·mol water vapor in air. H_2O from combustion of $H = 5.1/2 = 2.55$
lb·mol

$$\text{Total lb·mol stack gas/100 lb fuel} = 46 + 1.84 + 2.55$$
$$= 50.39 \text{ lb·mol}$$

$$\text{Total lb·mol stack gas/1 lb fuel} = 0.504 \text{ lb·mol} = n$$

$$\therefore V = \frac{nRT}{P} = \frac{0.504 \times 10.71 \times (460 + 680)}{14.7}$$
$$= \textbf{418.0} \text{ ft}^3/\text{lb } (\textbf{26.1} \text{ m}^3/\text{kg}) \text{ fuel} \qquad \textbf{3c}$$

4. *Ft³/lb fuel* (at 90°F)

$$n = 0.4675 + 0.0184 = 0.486 \text{ lb·mol/lb fuel}$$

$$V = \frac{0.486 \times 10.71 \times (460 + 90)}{14.7}$$
$$= \textbf{195.0} \text{ ft}^3/\text{lb } (\textbf{12.2} \text{ m}^3/\text{kg}) \text{ fuel} \qquad \textbf{4a}$$

2.02   At top, 740 mm, 20°C, 0.13 percent NH_3 by volume. At base, 745 mm, 40°C, 4.90 percent NH_3 by volume 100 ft³/min gas.

$$100 \text{ percent } - 4.90 \text{ percent } = 95.1 \text{ percent } = 0.951$$

$$\text{Air, ft}^3/\text{min} = 100 \times 0.951 = 95.1$$

$$mV_1 = mV_0 \times \frac{T_1}{T_0} \times \frac{P_0}{P_1}$$

$$= \text{mol} \cdot \text{vol at other than standard condition}$$

$$mV_1 \text{ air at 745 mm and 40°C} = 359 \times \frac{273 + 40}{273} \times \frac{760}{745}$$

$$= 421 \text{ ft}^3/\text{mol}$$

$$\frac{95.1}{421.0} = 0.226 \text{ mol air flowing/min}$$

$$\frac{\text{mol } NH_3}{\text{mol air}} \text{ at entrance} = \frac{0.049}{0.951} = 0.0515$$

$$\frac{\text{mol } NH_3}{\text{mol air}} \text{ at exit} = \frac{0.0013}{0.9987} = 0.0013$$

$$\text{mol ammonia absorbed/mol air flowing} = 0.0515 - 0.0013$$
$$= 0.0502$$

1. Rate of flow of gas at exit, ft³/min

$$V = 0.226 \times 359 \times \frac{273 + 20}{273} \times \frac{760}{740}$$
$$= \textbf{89.5} \text{ ft}^3/\text{min } (\textbf{2.53} \text{ m}^3/\text{min}) \qquad \textbf{1b}$$

2. Wt ammonia absorbed/min $= 0.226 \times 0.0502 \times 17$
$$= \textbf{0.193} \text{ lb } NH_3$$
$$(\textbf{0.088} \text{ kg } NH_3)/\text{min} \qquad \textbf{2a}$$

where atomic weight $NH_3 = 17$.

2.03   1. Look up xy data (Perry, p. 574)
   2. Construct xy diagram.
   3. Locate top product and feed, and bottom product on xy diagram. xy data indicate that a top product boiling at 78.41°C would have a vapor composition of 0.7815 mol fraction of ethanol.
   4. Material balance per 44 mol feed
   a. $0.17F = 0.7815D + 0.01W$
   or $17 \times 44 = 78.15D + W$

    b. $44 = D + W$
$$77.15D = 16 \times 44$$
$$D = 9.12 \text{ mol/h}$$
$$W = 34.88 \text{ mol/h}$$

5. Intercept of operating line $= \dfrac{9.12}{9.12 + 31} \times 0.7815 = 0.177$.

6. Construct operating line for enriching section.

7. Since the feed is a saturated liquid, q (where $c_p\, \Delta t = 0$) $= \dfrac{c_p\, \Delta t + \lambda}{\lambda} = 1$ and slope of q line $= \dfrac{q}{q-1} = \dfrac{1}{1-1} = \dfrac{1}{0} = \infty$. Construct q line.

8. Construct operating line in stripping section.

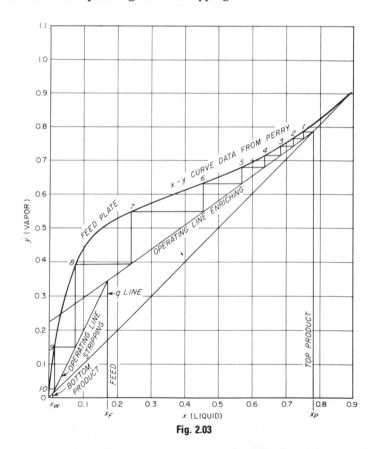

Fig. 2.03

9. Construct McCabe-Thiele stepwise representation of theoretical plates.

1. Ten theoretical plates are required. Therefore 10/0.60 or **17** actual plates are required. (If reboiler is considered to be 100 percent efficient, then 9/0.60 + 1 or 16 plates are required.)   **1c**

2. Condenser must remove the latent heat from 9.12 + 31 or 40.12 mol top product per hour. Latent heat of a vapor (78.15 mol percent ethanol) is 420 Btu/lb. Hence

$$40.12 \times 39.84 \times 420 = \textbf{672,000} \text{ Btu } (\textbf{709} \text{ MJ})/h \qquad \textbf{2b}$$

where

$$0.7815 \times 46 + 0.2185 \times 18 = 39.84$$

3. Reboiler duty is calculated as follows:

$$\hat{V} = F(q - 1) + V$$

where $F = 44$ mol/h

$$q - 1 = 1 - 1 = 0$$
$$\hat{V} = V = 40.12 \text{ mol/h}$$
$$\hat{V} = \text{vapor volume in stripping section}$$
$$V = \text{vapor volume in enriching section}$$

The vapor in equilibrium with a 1 mol percent ethanol solution contains approximately 0.10 mol percent ethanol and its latent heat of vaporization is 850 Btu/lb. Therefore

$$q = 40.12 \times 18.28 \times 850$$
$$= \textbf{623,000} \text{ Btu } (\textbf{657} \text{ MJ})/h \qquad \textbf{3a}$$

(where $0.01 \times 46 + 0.99 \times 18 = 18.28$.)

Note: Feed-plate location can be obtained from the xy diagram.

2.04   This problem could be solved directly by using the graph in Perry, p. 682, where ΔP is shown as a function of L and G/ϕ for a variety of packing material.

Solution: 5 gal water/min is equivalent to

$$\frac{5}{7.5} \times 62.5 \times 60 = 2500 \text{ lb/h}$$

$$L = 2500 \div \frac{\pi}{4} \times 1 \times 1 = 3180 \text{ lb/(h)(ft}^2)$$

$$\frac{G}{\phi} = (1.5 \times 3600)\left(\frac{\pi}{4} \times 1^2\right)\left(\frac{492}{530} \times 0.0807\right)$$

$$= 318 \text{ lb/h}$$

$$= (\text{fps} \times \text{s/h}) \times \text{ft}^2 \times (\text{wt air, lb/ft}^3 \text{ at } 70°\text{F})$$

From graph, for ½-in Berl saddles, $\Delta p = 0.20$ in/ft

$$\therefore 7 \times 0.20 = \mathbf{1.40} \text{ in } (\mathbf{35.6} \text{ mm}) \text{ water}$$
$$= \text{drop of pressure through tower} \quad \mathbf{c}$$

2.05 Perry, p. 674.
 a. For 25 g NH_3/100 g H_2O, partial pressure of $NH_3 = 227$ mm.
 b. For 2.04 g NH_3/100 g H_2O, partial pressure of $HN_3 = 12.2$ mm.
Assume that the total pressure of the exit gas is 760 mm and the gas leaves saturated with water at 20°C.

p (water) $= 17.5$ mm (from steam tables)
p (air $+ NH_3$) $= 760 - 17.5 = 742.5$ mm
p (NH_3) $= 0.10 \times 742.5 = 74.25$ mm
Top: $Pai - Pag = 227 - 74.2 = 152.8$ mm
Bottom: $Pai - Pag = 12.2 - 0 = 12.2$ mm

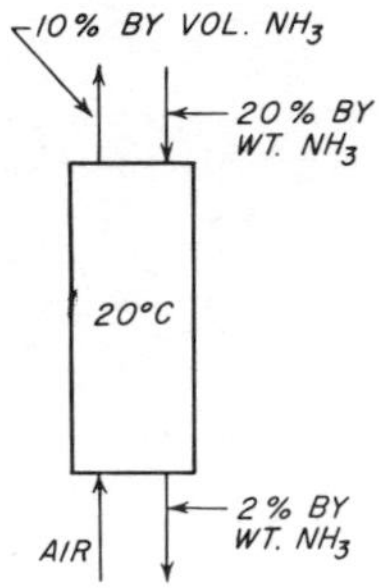

$$(\Delta p)_{lm} = \frac{152.8 - 12.2}{2.3 \times \log_{10}(152.8/12.2)} = 55.6 \text{ mm}$$

$$= 0.0732 \text{ atm}$$

$N = \text{lb·mol } NH_3 \text{ stripper per hour} = pV/RT$

$$= \frac{74.25 \times 8000}{760 \times 0.730 \times 293 \times 1.8} = 2.03$$

Fig. 2.05

1. $V = \dfrac{N}{K_g a(\Delta p)_{lm}} = \dfrac{2.03}{7 \times 0.0732} = \mathbf{3.96} \text{ ft}^3 \ (\mathbf{0.112} \text{ m}^3)$ **1d**

When gas rate is decreased 25 percent,

2. $V_2 = (\frac{4}{3})^{0.8} \times V_1 = 1.259 \times 3.96 = \mathbf{5.00} \text{ ft}^3 \ (\mathbf{0.142} \text{ m}^3)$ **2c**

2.06 It is assumed that the alcohol to be cooled is 95 percent alcohol. The arithmetic mean temperature of each fluid will be used in evaluating fluid properties:

$$T_{alc} = \frac{172 + 70}{2} = 121°F \ (49.4°C)$$

$$T_{water} = \frac{80 + 50}{2} = 65°F \ (18.3°C)$$

Volume of alcohol to be handled is assumed to have been measured at room temperature. Hence

Alcohol, lb/h $= 200 \times 8.33 \times 0.804 = 1340$ lb (608 kg)/h
Specific heat 95 percent alcohol $= 0.742$

Heat to be transferred from alcohol to water $= q$
$$= 1340 \times 0.742 \times (172 - 70) = 101{,}400 \text{ Btu } (107 \text{ MJ})/h$$

Calculation of film coefficient h_i for alcohol:

$$\text{Mass velocity} = G = \frac{1340 \times 4 \times (16)^2 \times (12)^2}{60 \times 60 \times \pi \times (19)^2}$$

$$= 48.4 \text{ lb}/(s)(ft^2) \ [(2.04 \text{ kg}/(s)(m^2)]$$

$$\text{Re} = \frac{DG}{\mu} = \frac{19}{16 \times 12} (48.4) \frac{1}{(0.77)(0.000672)}$$

$$= 9250 \text{ (no units)}$$

where $\mu = \text{lb}/(s)(ft) = \text{cP} \times 0.000672$

$k = 0.087 \text{ Btu}/(h)(ft^2)(°F) \ [1.78 \text{ kJ}/(h)(m^2)(°C)]$

$$h_i = 0.0225 \frac{k}{D} (\text{Re})^{0.8} \left(\frac{C\mu}{k}\right)^{0.3}$$

$$= 0.0225 \frac{(0.087)(16)(12)}{19} (9250)^{0.8} \left(\frac{0.742 \times 0.77 \times 2.42}{0.087}\right)^{0.3}$$

$$= 67.5 \text{ Btu}/(h)(ft^2)(°F) \ [1.38 \text{ MJ}/(h)(m^2)(°C)]$$

where $2.42 = 0.000672 \times 3600$

Calculation of h_0 for water:

$$\text{Rate of flow} = \frac{101{,}400}{(80 - 50) \times 3600} = 0.939 \text{ lb } (0.426 \text{ kg})$$

$$\text{Area of annulus} = \frac{\pi}{4 \times 144} \left[(2.067)^2 - \left(\frac{21}{16}\right)^2\right]$$

$$= 0.01395 \text{ ft}^2 \ (0.013 \text{ m}^2)$$

$$\text{Mass velocity} = G = \frac{0.939}{0.01395}$$

$$= 67.3 \text{ lb}/(s)(ft^2) \ [(2.84 \text{ kg}/(s)(m^2)]$$

$$\text{Equivalent diameter } D_e = \frac{4 \times \text{area annulus}}{\text{area heating surface}}$$

$$D_e = \frac{4 \times 0.01395}{\pi \times {}^{21}/_{16} \times {}^{1}/_{12}} = 0.162 \text{ ft } (0.0494 \text{ m})$$

$$\text{Re} = \frac{D_e G}{\mu} = \frac{0.162 \times 67.3}{1.052 \times 0.000672} = 15{,}420 \text{ (no units)}$$

By McAdams' equation (modified D and B equation):

$$h_0 = 160(1 + 0.01 \times 65)\left(\frac{67.3}{62.3}\right)^{0.8}\frac{1}{(0.162 \times 12)^{0.2}}$$

$$= 246 \text{ Btu/(h)(ft}^2)(°F) \text{ } [5.03 \text{ MJ/(h)(m}^2)(°C)]$$

Overall coefficient V_0, based upon outside surface of copper tube, is given by:

$$\frac{1}{U_0} = \frac{1}{h_0} + \frac{LA_0}{KA_{av}} + \frac{A_0}{h_iA_i}$$

$$= \frac{1}{246} + \frac{(\frac{1}{16} \times \frac{1}{12})(\pi^{21}\!/_{16}) \times 1}{222(\pi^{20}\!/_{16}) \times 1} + \frac{(\pi^{21}\!/_{16}) \times 1}{67.5(\pi^{19}\!/_{16}) \times 1}$$

$$= 0.02047 \text{ and } U_{0_c} = 48.9 = \text{clean heat-transfer coeff.}$$

Fouling resistances of at least 0.001 should be assumed on the inside and outside of the tubing.

$$\frac{1}{U_{0_D}} = 0.02047 + 0.002 = 0.02247$$

$$U_{0_D} = 44.5 = \text{heat-transfer design coefficient}$$

Logarithmic mean temperature difference from alcohol to water:

$$(\Delta t)_{lm} = [(172 - 80) - (70 - 50)] \div 2.3 \log_{10}\frac{172 - 80}{70 - 50}$$

$$= 47.2°F \text{ } (8.4°C)$$

Required heat-transfer area:

$$A = \frac{q}{U_0(\Delta t)_{lm}} = \frac{101,400}{44.5 \times 47.2} = 48.3 \text{ ft}^2 \text{ } (4.49 \text{ m}^2)$$

Area per pass, based upon outside surface of copper tube:

$$a = (\pi^{21}\!/_{16}) \times (\frac{1}{12}) \times 24 = 8.25 \text{ ft}^2 \text{ } (0.77 \text{ m}^2)$$

$$\text{No. pipes in series} = \frac{48.3}{8.25} = 5.8$$

Therefore, use **6** pipes.   **a**

2.07 *XY* data from Perry, p. 574.

Boiling point 50 mol percent solution = 73.1°C

$\Delta t = 73.1 - 20 = 53.1°C$

c_p = specific heat of feed = 0.81

λ = latent heat of feed = 16,300 Btu/lb·mol

$$q = \frac{\lambda + c_p \times \Delta t}{\lambda} = \frac{16{,}300 + 0.81 \times 25 \times 53.1 \times 1.8}{16{,}300} = 1.12$$

where 25 = average mol weight feed (lb/lb·mol)

$$\text{Slope of } q \text{ line} = \frac{q}{q-1} = \frac{1.12}{0.12} = 9.33$$

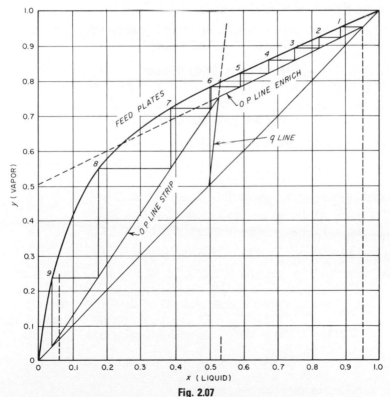

Fig. 2.07

1. $R_{\min} = \dfrac{0.95 - 0.79}{0.79 - 0.53} = \mathbf{0.615}$ **1a**

2. $R = \dfrac{100 + 40}{100} \times 0.615 = 0.861$

$$Y \text{ intercept} = \frac{X_D}{R + 1} = \frac{0.95}{1.861} = 0.51$$

From the diagram, number of perfect plates is 8.85. Therefore, use **9** plates.   **2c**

3. 2000 lb/h feed per 25 = 80 lb·mol/h. For 50% solution,

$$40 \ (0.50 \times 80) \ \text{lb·mol/h} = 40 \ \text{lb·mol/h}$$
$$\text{Product} = 0.95 \times 40 \times 32 + 0.05 \times 40 \times 18$$
$$= \mathbf{1252} \ \text{lb} \ (\mathbf{568} \ \text{kg}) \quad \mathbf{3a}$$

4. Residue = 40 lb·mol/h.

$$0.95 \times 40 \times 18 + 0.05 \times 40 \times 32 = \mathbf{748} \ \text{lb} \ (\mathbf{339} \ \text{kg}) \quad \mathbf{4c}$$
$$\text{Total} = \mathbf{2000} \ \text{lb} \ (\mathbf{907} \ \text{kg})$$

$$V = RD = 0.861 \times 40 = 34.44 \ \text{lb·mol/h}$$
$$\bar{V} = F(q - 1) + V = 80(1.12 - 1) + 34.44 = 44.0 \ \text{lb·mol/h}$$

5. lb (kg) steam/h = $\dfrac{970}{960} \times 18 \times 44.0$

$$= \mathbf{800} \ \text{lb/h} \ (\mathbf{363} \ \text{kg/h}) \quad \mathbf{5a}$$

2.08   Heat of this reaction at 25°C is:

$$\Delta H = 369.4 - 356.2 + 10.5 = 23.7 \ \text{kcal/g·mol}$$
$$= 23{,}700 \ \text{cal/g·mol}$$

Standard entropy change for this reaction at 25°C is:

$$\Delta S = 35.0 - 42.0 + 45.1 = 38.1 \ \text{cal/(g·mol)(K)}$$
$$\Delta c_p = c_p(\text{H}_2\text{O}) = 6.89 + 3.283 \times 10^{-3}T - 0.343 \times 10^{-6}T^2$$

At 500 K:

$$\Delta H = 23{,}700 + \int_{298}^{500} (6.89 + 3.283 \times 10^{-3}T - 0.343 \times 10^{-6}T^2) \, dT$$

$$= 23{,}700 + 6.89(500 - 298) + 1.641 \times 10^{-3}(500^2 - 298^2)$$
$$- 0.114(500^3 - 298^3) \times 10^{-6}$$

$$= 23{,}700 + 1393 + 264 - 11 = 25{,}346 \ \text{cal/g·mol}$$

$$\Delta S = 38.1 + \int_{298}^{500} \left(\frac{6.89}{T} + 3.283 \times 10^{-3} + 0.343 \times 10^{-6}T \right) dT$$

$$= 38.1 + 6.89 \times 2.3 \log_{10} (500/298)$$
$$+ 3.283 \times 10^{-3}(500 - 298) - 0.172 \times 10^{-6}(500^2 - 298^2)$$

$$= 38.1 + 3.57 + 0.66 - 0.03$$
$$= 42.3 \text{ cal/(g·mol)(K)}$$

$$\Delta G° = 25{,}346 - 500 \times 42.3 = +4196 \text{ cal/(g·mol)}$$

$$\log K = \log P_{(H_2O)} = -\frac{\Delta G°}{2.3RT} = -\frac{4196}{2.3 \times 1.987 \times 500}$$

$$= -1.835 = \frac{1}{1.835}$$

$$P_{(H_2O)} = \frac{1}{68.4} = 0.0146 \text{ atm} = 11.1 \text{ mmHg (1.48 kPa)}$$

At 700 K:

$$\Delta H = 23{,}700 + \int_{298}^{700} (6.89 + 3.283 \times 10^{-3}T - 0.343 \times 10^{-6}T^2)\, dT$$

$$= 27{,}093 \text{ cal/g·mol}$$
$$\Delta S = 38.1 + \int_{298}^{700} (6.89/T + 3.283 \times 10^{-3} - 0.343 \times 10^{-6}T)\, dT$$

$$= 45.2 \text{ cal/(g·mol)(K)}$$
$$\Delta G° = 27{,}093 - 700 \times 45.2 = -4547 \text{ cal/g·mol}$$

$$\log P_{(H_2O)} = +\frac{4547}{2.3 \times 1.98 \times 700} = 1.420$$

$$P_{(H_2O)} = 26.3 \text{ atm} = 386 \text{ psi (\textbf{2.66 MPa})}$$

Thermal dehydration of the acid appears to be **commercially feasible** in the temperature range of 500 to 700 K **without** the use of any powerful desiccating agents.   **a**

2.09 *Reference:* Hougen, Watson, and Ragatz, "Chemical Process Principles," part 3, pp. 834–835, Wiley, 1936.

$$\text{Eq. } (b) = \frac{Vr}{F}$$

$$= \frac{n_0^2}{K\pi^2}\left[Hx_A + J\frac{x_A}{n_{A0}(n_{A0} - x_A)} + M \ln \frac{n_{A0}}{n_{A0} - x_A} + N(\cdots) \right]$$

where V_r = volume of reactor = 5 ft³ (0.14 m³)

      F = feed rate = 1 lb·mol (0.454 kg)/h

      $n_0 = 1$ (assumed) $n_{A0} = 1$ $w = -\frac{1}{2}$

      $x_A = \frac{1}{2}$ $\pi = 1$ atm (101.3 kPa)

      $H = \omega^2 = \frac{1}{4}$ $J = 1 + \omega n_{A0})^2 = \frac{1}{4}$

$$M = -2\omega(1 + \omega n_{A0}) = \tfrac{1}{2}$$
$$N = 0$$

$$\frac{5}{1} = \frac{(1)^2}{K \times (1)^2}\left[\left(\frac{1}{4}\right)\left(\frac{1}{2}\right) + \left(\frac{1}{4}\right)\frac{\tfrac{1}{2}}{1(1 - \tfrac{1}{2})} + \frac{1}{2}\ln\frac{1}{1 - \tfrac{1}{2}}\right]$$

$$K = \frac{1}{5}\left(\frac{1}{8} + \frac{1}{4} + \frac{2.3}{2}\log_{10} 2\right) = 0.1442 \text{ lb}\cdot\text{mol}/(\text{ft}^3)(\text{h})(\text{atm}^2)$$

Let rate constant at 300°C $= K'$

$$\log\frac{K}{K'} = \frac{30,000}{4.576}\frac{773 - 573}{773 \times 573} = 2.97$$

$$K' = \frac{0.1442}{932} = 1.55 \times 10^{-4} \text{ lb}\cdot\text{mol}/(\text{ft}^3)(\text{h})(\text{atm}^2)$$

$$\text{Feed rate} = \frac{PV}{RT} = \frac{5 \times 10,000}{0.730 \times 573 \times 1.8} = 66.5 \text{ lb}\cdot\text{mol}/\text{h}$$

$H = J = \tfrac{1}{4}, M = \tfrac{1}{2}, N = 0$ as before, but $x_A = \tfrac{1}{4}$ and $\pi = 5$ atm (507 kPa):

$$\frac{V_r}{66.5} = \frac{1}{(1.55 \times 10^{-4})(5)^2}\left[\left(\frac{1}{4}\right)\left(\frac{1}{4}\right) + \left(\frac{1}{4}\right)\frac{\tfrac{1}{4}}{1(1 - \tfrac{1}{4})} + \frac{1}{2}\ln\frac{1}{1 - \tfrac{1}{4}}\right]$$

$$V_r = \frac{66.5 \times 0.2886}{(1.55 \times 10^{-4})(25)} = \textbf{4960} \text{ ft}^3\ (\textbf{140.5} \text{ m}^3) = \text{volume of reactor} \qquad \textbf{a}$$

2.10 Since the vapor pressures of the two pure components are given, this question will be solved on the assumption that Raoult's and Dalton's laws hold.

$$\text{Total pressure} = P = 0.50 \times 3.0 + 0.50 \times 0.9$$
$$= \textbf{1.95} \text{ atm } (\textbf{198} \text{ kPa}) \qquad \textbf{1c}$$

Composition of vapor phase:

$$Yn\text{-butane} = 1.50/1.95 = \textbf{0.769} \text{ mol fraction}$$
$$Yn\text{-pentane} = 0.45/1.95 = \textbf{0.231} \text{ mol fraction} \qquad \textbf{2a}$$

This question may also be solved using K constants (thermodynamics books give px data).

Assume $P = 2$ atm:

	x	K	K_x	$Y = K_x/\Sigma K_x$
n-Butane	0.50	1.8	0.90	0.773
n-Pentane	0.50	0.53	0.265	0.227
			1.165	

Assume $P = 3$ atm:

n-Butane	0.50	1.2	0.60	0.769
n-Pentane	0.50	0.36	0.18	0.231
			0.78	

Interpolating for $\Sigma K_x = 1.00$

$$P = 2 + (3 - 2) \frac{1.165 - 1.000}{1.165 - 0.780} = 2.43$$

Yn-butane $= \mathbf{0.771}$ mol fraction
Yn-pentane $= \mathbf{0.229}$ mol fraction   **2a**

2.11 *Reference:* Hougen and Watson, "Chemical Process Principles," 2d ed., part 3, p. 824, Wiley, 1936.

$$kt = \ln \frac{n_{A0}}{n_{A0} - x_A} = 2.3 \log_{10} \frac{100}{100 - 99} = 4.606$$

$$t = 4.606/0.0098 = \mathbf{470} \text{ min} \quad \mathbf{c}$$

2.12 *Reference:* Perry, pp. 1870–1872.

It is assumed that the plant has been properly designed so as to include the proper materials of construction, venting and ventilating systems, fire equipment, etc.

(1) An active safety- and fire-protection organization should exist in the plant. (2) Special safety-protection equipment should be available and used. (3) Special fire-protection equipment should be available and its proper use should be common knowledge in the plant. (4) Preventive maintenance should be practiced continuously.

2.13 *a.* Fats are glycerol esters of saturated and unsaturated fatty acids. Hydrolysis or splitting of the fat may be accomplished in either alkaline or acid media. The alkaline media method is described here.

$$C_xH_yCOO-CH_2 \qquad\qquad\qquad CH_2OH$$
$$C_xH_yCOO-CH + 3NaOH \longrightarrow 3C_xH_yCOONa + CHOH$$
$$C_xH_yCOO-CH_2 \qquad\qquad\qquad CH_2OH$$

Fat   Soap   Glycerol

The fat is heated to approximately 80°C with live steam, and sodium hydroxide is added. The removal of the acid radical in the fat is a stepwise procedure. The partially hydrolyzed fat is "salted out" with sodium chloride and the procedure is repeated until the fat is completely hydrolyzed, the soap appearing in the top layer and the glycerol in the bottom layer.

b. Sodium chloride is added to salt out the soap.

Reference: Riegel, "Industrial Chemistry," 4th ed., Reinhold, 1942, or Groggins, "Unit Processes in Organic Synthesis," 4th ed., McGraw-Hill, 1952, for a description of acid hydrolysis.

2.14 $\dfrac{50{,}000 \text{ units} \times 0.020/100}{1{,}000{,}000 \text{ units}} = \dfrac{10 \text{ units}}{1{,}000{,}000 \text{ units}}$

Use 50/1000 or **5** percent of 0.020 wt percent manganese.   **b**

2.15 (*a*) Formic acid. (*b*) Alcohol. (*c*) Acid. (*d*) Glycerol and salts of higher fatty acids. (*e*) Glucose and fructose.

2.16 When resistance offered by press and cloth is negligible, $V^2 = K\theta$, where V = volume of filtrate collected in time θ and K = constant for constant pressure filtration. Volume of filtrate to be collected is directly proportional to thickness of the cake. Hence, new time of filtration is $\theta_2 = \theta_1(V_2/V_1)^2 = 2(\frac{3}{2})^2 = 4.50$ h. The rates of washing are in each case inversely proportional to the final volumes of filtrate. Hence, the wash rate $R_2 = R_1(V_1/V_2) = R_1(\frac{2}{3})$. The time for washing is proportional to the volume of wash water and inversely proportional to the wash rate. Hence, $(\theta_{wash})_2 = (\theta_{wash})_1(V_2/V_1)(R_1/R_2) = (30)(\frac{3}{2})(\frac{3}{2}) = 67.5$ min $= 1.125$ h. The new total time per cycle $= 4.50 + 1.125 + 0.500 = 6.125$ h. The new gal/cycle $= \frac{3}{2}$(old gal/cycle). Therefore,

$$\frac{\text{New output press, gal/h}}{\text{Old output press, gal/h}} = \frac{1.5(3.00)}{1(6.125)} = 0.735$$

Hence, average output from press is **reduced** by **26.5** percent.   **a**

2.17

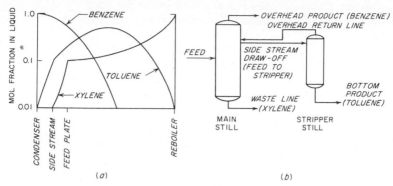

Fig. 2.17

2.18 *a.* Equation that expresses the liquid level Y as a function of the controller set point R is

$$Y(s) = \frac{GG_vG_c}{1 + GG_vG_cH} R(s) = \frac{(1/4\pi s)(3)(6)}{1 + (1/4\pi s)(3)(6)(0.5)} R(s)$$

$$= \frac{9/2\pi}{s + (9/4\pi)} R(s)$$

b. For this case the above equation becomes

$$Y = \frac{9/2\pi}{s + (9/4\pi)} \frac{\gamma_0}{s} = \frac{9}{2\pi} \gamma_0 \frac{1}{s[s + (9/4\pi)]}$$

$$= \gamma_0 \left[\frac{1}{s} - \frac{1}{s + (9/4\pi)} \right]$$

Time domain response $y(t) = 2\gamma_0[1 - e^{-(9/4\pi)t}]$. If $t \to \infty$, $y(t) = 2\gamma_0 = 6$ or $\gamma_0 = $ **3** psi (**20.7** kPa).

2.19 Both columns operate with overhead vapors at the azeotrope; condensation yielding two liquid phases, one butanol-rich (upper layer) and one water-rich phase (lower layer).

Heterogeneous azeotrope $\pi = 760$ mmHg (**101.3** kPa)

Vapor: 25 percent *n*-butanol/75 percent water (mol basis)

Liquid butanol-rich: 40 percent butanol/60 percent water (mol basis)

Water-rich: 2.5 percent butanol/97.5 percent water (mol basis), liquid is at ~200°F no subcooling.

Operation of the recovery plant depends upon a steady condition existing. Therefore, check separation in the decanter.

Butanol-rich layer; essentially temperature-independent:

	Mol	mW	Wt	Wt percent
Butanol	40	74	2960	73.27
Water	60	18	1080	26.73
			4040	

Butanol-rich storage-tank composition will vary with time.

Water-rich layer; temperature-dependent:

	Mol	mW	Wt	Wt percent
Butanol	2.5	74	185	9.54
Water	97.5	18	1755	90.46
			1940	

Water-rich storage-tank composition will vary with time.

At 90°F (~110° subcooling), the 92.3 wt percent water is a possibility. Since indicated flow sketch is inconsistent with separation achieved in the decanter, the entire flow arrangement needs to be modified. Suggested modification is to return decanted streams directly to the respective columns onto the top stage along with the fresh feeds from the storage/holdup tanks.

2.20 The reactants in the Fischer-Tropsch synthesis of motor fuels are carbon monoxide and hydrogen. This synthesis gas may be prepared from coal, coke, natural gas (that is, CH_4), etc.

The principle reactions, which are a function of

a. Temperature (200–300°C)

b. Pressure (1–25 atm)

c. Type of catalyst (CO, Ni, Fe, or Cu with alkali oxide promotors)

d. Carbon monoxide–hydrogen ratio

e. Purity of synthesis gas

are of two general types:

$$aCO + \frac{(2a + b)H_2}{2} \rightarrow C_aH_b + aH_2O$$

$$2aCO + \frac{bH_2}{2} \rightarrow C_aH_b + aCO_2$$

The products of the reaction vary in chain length from 1 to 40 or more carbon atoms and a variety of saturated and unsaturated products. The principle by-products are alcohols and ketones.

2.21 *a.* The weight of sulfur conveyed from the storage pile to the melting tank may be controlled automatically by a Harding weigh feeder, a Schaeffer belt system, or a Jeffrey vibrating feeder if the process is continuous. An automatic hopper scale or conveyor-type batching scale may be used in a batch operation. (See Perry, p. 1293.)

    b. A two-control valve system for the heating tank is described in Perry, p. 1330. A control system for a continuous process is described in Perry, p. 1334.

    c. A ratio flow-control system is described in Perry, pp. 1333 and 1334.

2.22 *a.* Iron and aluminum removal is costly because the precipitate of the iron and aluminum ion drags down, by adsorption, appreciable quantities of phosphate that must be reclaimed. Finely ground limestone is added to the phosphoric acid. Soluble CO_2HPO_4, CaH_2PO_4, and a precipitate of iron and aluminum phosphate and calcium fluoride are formed. The precipitate must be washed thoroughly to remove the soluble phosphates. The wash water is added to the acid solution of CaH_2PO_4 and pure sulfuric acid is added with the resulting precipitate of $CaSO_4$. After the $CaSO_4$ has been removed, the acid solution is concentrated in a vacuum evaporator or a spray chamber.

    b. There are three general methods of defluorination of phosphate rock to produce tricalcium phosphate, which is used as an ingredient in animal food products. The chief difference in the two methods that employ SiO_2 and steam is in the heat treatment. One process involves fusion and the other does not. The principle reaction is $Ca_{10}F_2(PO_4)_6 + H_2O + SiO_2 \rightarrow 3Ca_3(PO_4)_2 + CaSiO_3 + 2HF$. For a complete discussion, refer to Waggaman, "Phosphoric Acid, Phosphates, and Phosphate Fertilizers," p. 390, Reinhold.

    c. $Na_3PO_4 \cdot 12(H_2O)$ is used as an internal "softening agent" in cases where the boiler acidity and/or the natural hardness of the water is relatively low. Phosphates, when added internally, produce a flocculant precipitate of the scale-producing ions, which is easily removed in boiler blowdown. The principle reactions for the removal of scale-producing ions are:

$$3Ca(HCO_3)_2 + 3Na_3PO_4 \rightarrow Ca_3(PO_4)_2 \downarrow + 6NaHCO_3$$
$$3MgSO_4 + 2Na_3PO_4 \rightarrow Mg_3(PO_4)_2 \downarrow + 3Na_2SO_4$$
$$FeCl_3 + Na_3PO_4 \rightarrow FePO_4 \downarrow + 3NaCl$$

The phosphates have an added advantage in that they control the pH of the boiler water.

2.23 24,000 gal (90.84 m³) is equivalent to 24,000 × (62.4/7.5) × (1/2000) or 100 tons (90.72 Mg) water.

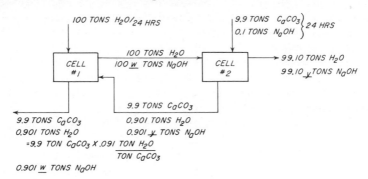

Fig. 2.23

Let w = equilibrium concentration NaOH in cell #1 in tons
    NaOH/ton H_2O
  y = equilibrium concentration NaOH in cell #2 in tons
    NaOH/ton H_2O
Material balance NaOH in cell #1 is

$$0.901w + 100w = 100.901w = 0.901y$$
$$100w = 0.89y \qquad (1)$$

Material balance NaOH in cell #2 is

$$0.901y + 99.10y = 100.001y = 0.1 + 100w$$
$$= 0.1 + 0.89y \qquad (2)$$
$$99.111y = 0.1 \quad y = 1.01 \times 10^{-3} \text{ tons NaOH/ton } H_2O \qquad (3)$$

Therefore, $w = 9 \times 10^{-6}$ tons NaOH/ton H_2O.

Therefore, since each ton $CaCO_3$ carries 0.091 ton water and each ton water carries 9×10^{-6} ton NaOH, the product will contain per ton $CaCO_3$

1 (ton $CaCO_3$) × 0.091 (ton H_2O/ton $CaCO_3$) × 9 × 10⁻⁶ ton
    NaOH/ton H_2O = 8.19 × 10⁻⁷ ton (0.0743 g) NaOH

2.24 Following symbols will be used for solution of the problem. (*Reference:* William H. McAdams, "Heat Transmission," 3d ed., chap. 9, McGraw-Hill, 1954.

S = cross section stream in tube, ft^2 (m^2)

G = mass velocity, $\text{lb}/(\text{h})(\text{ft}^2)$ $[\text{kg}/(\text{h})(\text{m}^2)]$

q = heat-transfer rate, Btu/h (kJ/h)

U_0 = overall coefficient of heat transfer between two streams based on outside area, $\text{Btu}/(\text{h})(\text{ft}^2)$ $[\text{kJ}/(\text{h})(\text{m}^2)]$

A_0 = area for heat transfer at outside radius

N_{Re} = Reynolds number = DG/μ, μ = velocity of fluid

k = thermal conductivity of fluid

h = coefficient of heat transfer between fluid and surface, $\text{Btu}/(\text{h})(\text{ft}^2)$ $[\text{kJ}/(\text{h})(\text{m}^2)]$

T = absolute temperature

t_1, t_2 = temperature at stations 1 and 2, respectively

t_w = temperature of wall

i = as subscript means isothermal

w = mass rate of flow per tube, lb (kg) fluid/h

c_p = specific heat at constant pressure, $\text{Btu}/(\text{lb})(°\text{F})$ $[\text{kJ}/(\text{kg})(°\text{C})]$

$S = 0.00413(13) = 0.0537 \ \text{ft}^2$

$G = 60{,}000/0.0537 = 1.118 \times 10^6 \ \text{lb}/(\text{h})(\text{ft}^2)$ $[5459 \ \text{Mg}/(\text{h})(\text{m}^2)]$

$q = wc_p(t_2 - t_1) = UA \, \Delta T_L \qquad A_0 = 0.2618(52)(6) = 81.6 \ \text{ft}^2$

$$U_0 = \cfrac{1}{\cfrac{1}{2000} + \cfrac{1}{0.935}\cfrac{0.065/12}{220} + \cfrac{1}{0.87h_i}} = \cfrac{1}{0.000526 + \cfrac{1.15}{h_i}}$$

$$N_{\text{Re,min}} = \frac{(0.87/12)(1.118 \times 10^6)}{2.42} = 33{,}500 \qquad \text{so turbulent}$$

Assume $t_2 = 100$, $t_{\text{av}} = 80°$, $\mu = 0.862$ cP, $k = 0.35$. Neglecting the μ/μ_w term,

$$h_{i(\text{approx})} = \frac{0.023(0.35)}{0.87/12}\left(\frac{33{,}500}{0.862}\right)^{0.8}\left(\frac{2.42 \times 0.862}{0.35}\right)^{0.33} = 944$$

Therefore $U_0 = 573$.

$$q = 60{,}000(t_2 - 60) = 573(81.6)\frac{t_2 - 60}{\ln[(227 - 60)/(227 - t_2)]}$$

$$t_2 = 151° \qquad \text{so no good}$$

Try $t_2 = 160°$, $t_{\text{av}} = 110°$, $\mu = 0.62$, $k = 0.367$. Now include $(\mu/\mu_w)^{0.14}$; $t_w \approx 192°\text{F}$, $\mu_w = 0.32$.

$$h_i = \frac{0.023(0.367)}{0.87/12}\left(\frac{33{,}500}{0.62}\right)^{0.8}\left(\frac{2.42 \times 0.62}{0.367}\right)^{1/3}\left(\frac{0.62}{0.32}\right)^{0.14} = 1267$$

$$U_0 = 697 \qquad q = 60,000(t_2 - 60)$$

$$= 697(81.6) \frac{t_2 - 60}{\ln[(227 - 60)/(227 - t_2)]}$$

Above equations yield $t_2 = $ **162°F** (**72.2°C**), approx   **c**

2.25 Following symbols will be used for solution of the problem. (*Reference:* William H. McAdams, "Heat Transmission," 3d ed., chap. 6, McGraw-Hill, 1954.)

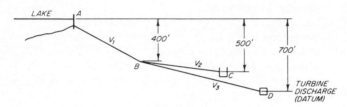

Fig. 2.25

P = absolute pressure (intensity of), lbf/ft² (kPa)
V = average velocity, fps (m/s)
f = friction factor, dimensionless, in Fanning equation
N_{Re} = Reynolds number, dimensionless
ρ = absolute viscosity of fluid, lb/(s)(ft) [kg/(s)(m)]
H = skin friction
$P_D = (95.3)(144) = 13,720$ lbf/ft² (657 kPa)
$P_A = $ atm $= 2115$ lbf/ft² (101 kPa)
$V_1 = V_2 + V_3$

$A \rightarrow B$: $\qquad \dfrac{2115 - P_B}{62.3} = (300 - 700) + H_1$

$B \rightarrow C$: $\qquad \dfrac{P_B - 2115}{62.3} = (200 - 300) + H_2$

$B \rightarrow D$: $\qquad \dfrac{P_B - 13,720}{62.3} = (0 - 300) + H_3$

$$H_1 = \frac{(4f_1)(1000)(V_1^2)}{(3.826/12)64.4} \qquad H_2 = \frac{(4f_2)(8000)(V_2^2)}{(3.826/12)64.4}$$

$$H_3 = \frac{(4f_3)(3000)(V_3^2)}{(3.826/12)64.4}$$

f values are functions of N_{Re}, $N_{Re} = \dfrac{DV\rho}{\mu}$. Solve by trial and error using tables for values of f.

$$N_{Re} = \frac{(3.826/12)(62.3)V}{0.000672} = 29{,}500V$$

Solution gives $V_1 = 18$ fps (5.49 m/s), $V_2 = V_3 = 9$ fps (2.74 m/s).

$N_{Re1} = 511{,}000$ $f_1 = 0.004$ $P_B = 9000$ lbf/ft² (431 kPa)

$N_{Re2} = 268{,}500$ $f_2 = 0.0045 = f_3$

gal (liters)/min $= 18 \left(\dfrac{\pi}{4}\right)\left(\dfrac{3.826}{12}\right)^2 (7.48)(60) = \textbf{645 (2441)}$ **c**

2.26   Following symbols will be used for solution of the problem. (*Reference:* William H. McAdams, "Heat Transmission," 3d ed., chaps. 6 and 15, McGraw-Hill, 1954.)

a_0 = constant, dimensionless

B_0 = correction factor

D_i, D_o = inside and outside diameters of tube, ft

y_o = clearance, ft, between outer surfaces of tubes in bundle, taken to correspond to minimum free area

μ_o = viscosity for fluid outside tubes

m = exponent, dimensionless, in dimensionless equations

K = dimensional term

g_c = conversion factor in Newton's law equal to 4.17×10^8 (lb fluid)(ft)/(h)(lbf)

ρ_o = density of fluid for fluid outside tubes, lb/ft³

G_{oo} = mass velocity for optimum velocity outside tubes, lb/(h)(ft² of cross section)

G = mass velocity, same units as G_{oo}

t = bulk temperature of fluid, °F

S = cross section normal to flow of fluid, ft²

q = quantity of heat transfer, Btu/h

A = area of heat-transfer surface, ft²

$$a_0 = 0.25 + \frac{0.1175}{1.85^{1.08}} = 0.311$$

$$K_0' = K_0 \frac{D_i}{D_0} = \frac{2B_0 a_0 y_0 \mu_0^m}{\pi g_c \rho_0^2 D_0^{m+1}}$$

$$= \frac{(2)(1)(0.311)(1.95/12)(0.0455)^{0.15}}{(3.14)(4.17 \times 10^8)(0.072)^2(1.05/12)^{1.15}}$$

$$= 1.542 \times 10^{-7}$$

$$G_{oo} = \left[\frac{9 \times 10^{-5}}{(7.55 \times 10^{-9})(1.542 \times 10^{-7})} \times \frac{1}{3.75} \right]^{0.351}$$

$$= 4225 \ \text{lb/(h)(ft}^2\text{)} \ [20.63 \ \text{Mg/(h)(m}^2\text{)}]$$

$$\frac{\Delta t_m}{\Delta t_0} = 1.0$$

$Check:$ $\dfrac{D_0 G_{\max}}{\mu} = \dfrac{(1.05/12)(4225)}{0.0455} = 8120 \quad \text{OK}$

$$S_{\min} = \frac{w}{4225} = \frac{64,000}{4225} = 15.15 \ \text{ft}^2 \ (1.41 \ \text{m}^2)$$

If N = number of tubes in one row, then

$$S_{\min} = \frac{(3N + 2.55) - 1.05N}{12} \ 10 = 15.15$$

and $$N = \textbf{8} \qquad \textbf{1c}$$

$$h_0 = \frac{(0.33/1.25)(0.25)(4225)}{(8120)^{0.4}(0.74)^{2/3}} = 9.35 = U_0$$

$$\Delta t_L = \frac{100}{\ln (175/75)} = 118.3°F \ (47.9°C)$$

$$q = 64,000(0.25)(100) = 9.35A(118.3),$$

and $$A = 1440 \ \text{ft}^2 \ (133.8 \ \text{m}^2)$$

Let N' equal total number of tubes.

$$1440 = \pi \left(\frac{1.05}{12} \right) N'(10) \qquad N' = 526$$

$$\frac{526}{8} = \textbf{66} \ \text{rows} \qquad \textbf{2d}$$

2.27 *a.* Solvent recovery by adsorption of solvent vapors on activated adsorbing agents operates efficiently and safely for the recovery of solvent vapors over a wide range of concentrations of solvent in solvent-laden air.

The solvent-laden air is removed from the drying operation by a suction blower. The air is usually passed through a dust-removal device and then to adsorbers.

The solvent-laden air then passes to an adsorber, where the solvent is selectively adsorbed. When the adsorbing agent becomes "saturated," the solvent-laden gas is directed to a recently activated adsorber. Steam is directed to the "saturated adsorber," and the solvent and water are sent to decanters and stills.

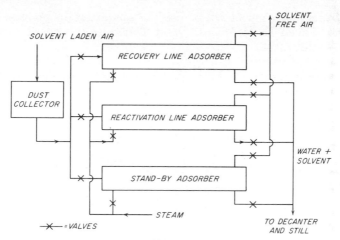

Fig. 2.27

b. Other methods of solvent recovery to be considered are:
1. Scrubbers
2. Compression and cooler

References: Perry, "Chemical Engineers' Handbook," 3d ed., McGraw-Hill, 1950; Robinson, "The Recovery of Vapors," Reinhold, 1942.

2.28 1. Assume adiabatic operation.

$$Q = 0 = \Delta H \qquad \text{energy balance around condenser}$$

Let t = temp. leaving condenser. Ref. 60°F.

$$
\begin{aligned}
0 = {} & (100)(0.5)(t - 60) + 27.5(0.909)(0.4)(200 - 60) \\
& + 27.5(0.091)(1)(200 - 60) - 27.5(0.909)(0.3)(300 - 260) \\
& + 100 + 0.4(260 - 60) - 27.5(0.091)(0.5)(300 - 212) \\
& \hspace{5cm} + 970 + 1.0(212 - 60)
\end{aligned}
$$

Above equation yields $t =$ **180°F** (**82.2°C**)   **1b**

2. Energy balance around cooler. Ref. 65°F. Let w = lb/h water. Stream entering cooler is

$$48,000 \text{ lb/day} \frac{27.5}{100} \frac{1}{24} = 550 \text{ lb/h}$$

$$Q = 0 = \Delta H$$

$$
\begin{aligned}
0 = {} & 550(0.909)(0.4)(100 - 65) + 550(0.091)(1)(100 - 65) \\
& + w(1)(130 - 65) - 550(0.909)(0.4)(200 - 65) \\
& \hspace{4cm} - 550(0.091)(1)(200 - 65)
\end{aligned}
$$

Above equation yields $w =$ **385** lb (**174.6** kg)/h   **2d**

2.29 Following symbols will be used for solution of the problem. (*Reference:* William H. McAdams, "Heat Transmission," 3d ed., chap. 6, McGraw-Hill, 1954.)

D_p = diameter of packing, average
G_o = superficial mass velocity based upon cross section of empty tube
ϵ = fraction voids, dimensionless
V = average velocity, fps
ρ = density, lb fluid/ft^3
f' = friction factor, dimensionless

$$D_p = \frac{6}{76} = 0.079 \text{ ft } (0.024 \text{ m})$$

$$V_s = \frac{1000 \text{ ft}^3/\text{min}}{(60 \text{ s/min})(0.785)(4)^2} = 1.333 \text{ fps}$$

$$\rho_m = \frac{29}{359}\frac{492}{530} = 0.075 \text{ lb/ft}^3 \qquad \mu = 0.018 \text{ cP}$$

$G_o = V_s\rho_m$; substitute the above values in

$$\frac{D_pG_o}{\mu(1 - \epsilon)} = \frac{(0.079)(1.333)(0.075)}{(0.018)(0.000672)(1 - 0.69)} = 2110$$

$$f' = 1.75 + \frac{150}{2110} = 1.82$$

$$\Delta P_f = \frac{f'LG_0^2(1 - \epsilon)}{\epsilon^3 g_c D_p \rho_m} = \frac{1.82(30)[(1.333)(0.075)]^2(1 - 0.69)}{(0.69)^3(32.2)(0.079)(0.075)}$$

$$= 27.1$$

$$\frac{27.1 \text{ lbf/ft}^2}{62.35 \text{ lbf/ft}^3} \times 12 = \mathbf{5.21} \text{ in } (\mathbf{13.23} \text{ cm}) \text{ H}_2\text{O} \qquad \mathbf{a}$$

2.30 Flow rate, lb/h $= \dfrac{70,000 \times 60}{359}\dfrac{492}{710}\dfrac{1.1}{1.0} 29.1$

$$= 260,000 \text{ lb/h}$$

$$G_s = \text{lb/(h)(ft}^2) = \frac{260,000}{(\pi D^2)/4} = \frac{331,000}{D^2}$$

where D = diameter, ft.

$$\text{Tower height} = (0.32G_s^{0.18})(15) \text{ ft}$$

$$\text{Tower volume} = (\text{height})(\text{cross-sectional area}) = (\text{height})\left(\frac{\pi}{4}D^2\right)$$

$$= 0.32 G_s^{0.18}(15)\,\frac{260,000}{G_s}\;\text{ft}^3$$

Tower cost $/h (fixed cost) $= (\text{volume})(\$1.0)(0.20)\,\dfrac{1}{8000}$

$$= \frac{0.32 \times 15 \times 260,000}{G_s^{0.82}}\left[(1)(0.2)\,\frac{1}{8000}\right] = \frac{31.2}{G_s^{0.82}}$$

Total cost/h = operation cost/h + fixed cost/h

$$= \left(1.8 G_s^2 \times 10^{-8} + \frac{81}{G_s} + \frac{4.8}{G_s^{0.8}}\right) + \frac{31.2}{G_s^{0.82}}$$

$$\frac{d}{dG_s}\,(\text{total cost}) = 0 = 3.6 \times 10^{-8} G_s - \frac{81}{G_s^2} - \frac{3.84}{G_s^{1.80}} - \frac{25.6}{G_s^{1.82}}$$

$$= 3.6 \times 10^{-8} G_s^3 - 3.84 G_s^{0.20} - 25.6 G_s^{0.18} = 81$$

By trial and error, $G_s = 1760$ lb/(h)(ft²). Therefore, the desired values are

$$\text{Diameter} = \sqrt{\frac{331,000}{1760}} = \textbf{13.8 ft (4.21 m)}\qquad \textbf{1d}$$

$$\text{Height} = (0.32)(1760^{0.18})(15) = \textbf{18.5 ft (5.64 m)}\qquad \textbf{2a}$$

2.31 This is a trial-and-error solution. As a first approximation, assume that the theoretical deliverabilities of the parallel lines are in the ratio to their diameters:

$$\frac{Q_{22}}{Q_{20}} = \left(\frac{22}{20}\right)^{2.67} = 1.29$$

The actual deliverabilities will be in the ratio:

$$\frac{Q'_{22}}{Q'_{20}} = 1.29\,\frac{0.88}{0.92} = 1.233$$

This gives an estimated actual flow of

$$Q'_{22} = (1.233/2.233) \times 6 \times 10^6/\text{h} = 3,310,000\ \text{ft}^3\ (93.8\ \text{km}^3)/\text{h}$$

for 22-in line and

$$Q'_{20} = (1.000/2.233) \times 6 \times 10^6/\text{h} = 2,690,000\ \text{ft}^3\ (76.2\ \text{km}^3)$$

for 20-in (50.8-cm) line. Theoretical flow, assuming smooth pipes, is then computed as

$$Q_{22} = 3,310,000/0.88 = 3,760,000\ \text{ft}^3\ (106.5\ \text{km}^3)/\text{h}$$
$$Q_{20} = 2,690,000/0.92 = 2,930,000\ \text{ft}^3\ (83.0\ \text{km}^3)/\text{h}$$

Weymouth equation is

$$(P_1)^2 - (P_2)^2 = \frac{fLW^2zRT}{Dg_cMA^2}$$

where P_1 and P_2 = inlet and outlet pressures, psf (abs)
  f = friction factor for smooth pipe
  L = length of pipeline, ft
  W = theoretical rate of flow, lb/s
  z = compressibility factor
  R = universal gas constant
  T = absolute temperature
  D = internal diameter of pipe, ft
  g_c = 32.17 ft/s^2
  M = molecular weight of gas, lb·s/ft^3 (this case = 0.65 × 29)
  A = cross section of pipe, ft^2

The Weymouth equation is used to compute the pressure drop in each line. If the calculated values of P_2 for the two lines agree within reason, an average P_2 value may be taken as the answer. If there is no close agreement of the calculated values of P_2, the assumed ratio of Q_{22}/Q_{20} is readjusted and the calculations repeated.

2.32 Approximate calculations for the amount of heat transfer: Temp. steam = 228°F (108.9°C), and temp. in condenser = 101°F (38.3°C). Assume forward feed and boiling point levels of 3°F (-16.1°C) in the first effect and 6°F (-14.4°C) in the second effect, giving temperatures estimated to be 170°F (76.7°C) for boiling point in first effect and 107°F (41.7°C) for boiling point in second effect.

  lb H_2O evaporated/lb NaOH = $^{90}/_{10} - ^{84}/_{16}$ = 3.75 (1.70 kg/kg)

Heat transferred for vaporization in two effects

$$= 3.75 \times \frac{997 + 1033}{2}$$

$$= 3800 \text{ Btu/lb (8846 kJ/kg) NaOH}$$

Heat required to preheat feed if admitted to first effect
$$= 9(170 - 70) = 900 \text{ Btu/lb (2095 kJ/kg) NaOH}$$

It is seen that approximately 24 percent of the heat-transfer requirement is for preheating the feed to its boiling point. Accordingly, it is recommended that for a 10 percent increase in production, available heat exchangers be used to preheat the feed to about 160°F (71.1°C) by steam bled from the vapor line leaving the first effect. The evaporators are to be operated as a double effect with forward feed. It is estimated that the resultant increase in capacity will be well over 10 percent and that there will be no loss in steam economy as compared with normal operation with backward feed.

To handle a temporary 50 percent overload, it will be necessary to operate the two evaporators in parallel, using the regular condenser on one of the effects and the spare condenser on the other one. High-pressure steam (5 psi) (34.5 kPa) will be admitted to the steam chest of each evaporator, and the feed will enter both effects in parallel. If the capacity of the spare condenser is insufficient to give a 50 percent increase in production, then available heat exchangers may be used to preheat the feed, thus taking some load off the evaporators. High-pressure steam will have to be used for preheating the feed in the heat exchangers. Since high-pressure steam will be the heating medium in both evaporators and in the heat exchangers, there will be a resultant loss in steam economy that can be tolerated on a temporary overload basis.

2.33 *Reference:* Walker, Lewis, McAdams, and Gilliland, "Principles of Chemical Engineering," pp. 415–416, McGraw-Hill, 1937.

$$\frac{1}{V^2} = a + b\theta \qquad a = \frac{1}{(225)^2} \qquad \theta = 160$$

$$\frac{1}{(50)^2} = \frac{1}{(225)^2} + 160b \qquad b = 2.38 \times 10^{-6}$$

$$V_{av}\theta = \frac{Q}{A\,\Delta t} = \frac{2(\sqrt{a + b\theta} - \sqrt{a})}{b}$$

a. It is assumed that the evaporator is operated for 160 h and cleaned for 8 h, both each week. It follows that on substituting values for a and b,

$$V_{av} = \frac{Q}{A\,\Delta t(\theta - \theta_c)} = 1.27 \times \frac{10^4}{168}$$

$$= 77.7 \text{ Btu/(h)(ft}^3)(°F) \ [1590 \text{ kJ/(h)(m}^2)(°C)]$$

b. Let time of operation $= \theta$
            time of cleaning $= \theta_c$

It is required that $Q/(\theta + \theta_c)$ be a maximum. Hence

$$\frac{d}{d\theta}\left(\frac{Q}{\theta - \theta_c}\right) = \frac{d}{d\theta}\left[2A\,\Delta t\frac{\sqrt{a + b\theta} - \sqrt{a}}{b(\theta + \theta_c)}\right] = 0$$

When $\dfrac{d}{d\theta}\left(\dfrac{\sqrt{a + b\theta} - \sqrt{a}}{\theta + \theta_c}\right) = 0$, then $\theta = \theta_c + 2\sqrt{\dfrac{a\theta_c}{b}}$.

Substituting values of a, b, and θ_c, $\theta = 24.3$ h. This time corresponds to a value for $V_{av} = \dfrac{2\sqrt{b\theta_c}}{b(\theta + \theta_c)} = 113$ Btu/(h)(ft^2)(°F) as compared with $V_{av} = 77.7$ in (a).

Accordingly, for maximum capacity, the evaporator should be cleaned after 24.3 h of operation, corresponding to a total time for the cycle of 32.3 h. It is suggested that the evaporator be operated during three shifts and cleaned during the fourth shift, corresponding to a 32-h cycle.

2.34 a. Plate and frame presses are suitable for batch operation with both compressible and noncompressible sludges. This press is the cheapest per unit filtering surface and requires the smallest floor space but involves a large amount of hand labor. This press is best adapted to sludges that contain small amounts of solids or in which the solids are highly valuable, as in the manufacture of dyes.

b. Rotary continuous filters are used where the process is continuous, large amounts of solids must be handled, and labor costs must be kept low. These filters are well-adapted to homogeneous sludges of high solid concentrations and low compressibility, which require little washing and do not require high pressures.

c. The Sweetland press is a pressure leaf filter that involves batch operation and has high washing efficiency, the wash water displacing the filtrate when admitted into the press. This filter is suitable for handling slurries when the filtrate is valuable and dilution with wash water is to be held to a minimum.

d. The Moore filter is an intermittent vacuum filter adapted to batch operation. Vacuum is applied to the inside of the leaves during filtration, while compressed air is applied to discharging the cake. The filter basket is moved from filtering tank to washing tank and dump tank with a traveling crane. This filter is suitable for sludges that do not easily settle out and that form an adherent cake. It may be used with compressible sludges provided a high rate of filtration is not required.

2.35 The problem is the application of the phase rule in metallurgy. Precipitation hardening is applied to alloy systems in which there is a decrease of solid solubility in solid solutions with decreasing temperature.

If an alloy whose composition is represented by A' is heated to a temperature T', a single phase of solid solution results. If this solution is cooled suddenly to T''', the solid solution will persist but be unstable with respect to precipitation of component B (i.e., the solid solution is supersaturated with respect to component B). The precipitation of component B distorts the space lattice and interferes with the normal slip of the

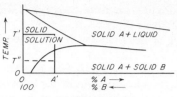

Fig. 2.35

alloy; this in turn hardens and strengthens the alloy. There is an optimum value in this process, beyond which the strength and hardness of the alloy are decreased. If this process takes place at room temperature or is the result of cold working, the process is known as age hardening.

The two terms—precipitation hardening and age hardening—have currently become synonymous. The "Metals Handbook," ASM, however, still gives the following definitions: 1. "Precipitation hardening is a process which increases the hardness of an alloy by the controlled precipitation of a constituent from a supersaturated solid solution"; and 2. "Age hardening is a process that increases hardness and strength and usually decreases ductility. The process usually involves rapid cooling or cold work."

2.36 See Perry, p. 844.
Initial mixture in lb alcohol/lb air

$$0.05 \times 46 = 2.30 \text{ lb alcohol}$$
$$0.95 \times 29 = 27.60 \text{ lb air}$$

$$\therefore H = \frac{2.3}{27.6} = 0.0833$$

Hs for final air:

$$Hs = \frac{60}{[(114.7/14.7) \times 760] - 60} \times \frac{46}{29} = 0.0162 \frac{\text{lb alcohol}}{\text{lb air}}$$

and

$$\frac{0.0833 - 0.0162}{0.0833} \times 100 = \textbf{80.6} \text{ percent recovery of ethanol} \qquad \textbf{c}$$

2.37
$$6000 \times 500 \times \frac{144}{163} \times \frac{1}{33,000 \times 60 \times 0.25} = 5.4 \text{ hp}$$

$$6000 \times (4200 - 500) \times \frac{144}{163} \times \frac{1}{33,000 \times 60 \times 0.30} = 33.0 \text{ hp}$$

$$\text{Total power} = \textbf{38.4} \text{ hp } (\textbf{28.6} \text{ kW}) \qquad \textbf{a}$$

2.38 *a.* Use aluminum or chromium alloys (over 18 percent Cr) for cold acid where strength is over 68 percent.

b. Use iron or steel, or use monel or nickel when contamination must be avoided.

c. Use iron or steel when the benzol is water and acid-free.

d. Use ordinary cast iron when slight contamination is permissible. Use monel metal or nickel when contamination must be avoided.

e. Use iron or steel when there is less than 20 percent water and when the H_2SO_4 is greater than 15 percent. Use high-silica cast iron when the water content is between 20 and 30 percent. Use lead when the water content of the mixture is greater than 30 percent.

2.39 Basis of solution: 1 ft^2 and capacity of 1-in cake on leaf.

1.
$$K'\theta = \frac{V^2}{A^2} = K' \times 5 = \frac{(3.8)^2}{(1)^2}$$
$$K' = 2.89$$

2. Differentiating Eq. (1), the rate of filtration at a given time is obtained

$$\frac{dV}{d\theta} = \frac{2.89A^2}{2V}$$

3. Cake thickness will be described in terms of t. When $t = 2$, then the cake thickness will be $1/t$ or one-half the thickness of the original cake and the volume of the filtrate delivered will be one-half the value delivered at $t = 1$.

Time Schedule for Leaf Press

Cycles, no.	Cake thickness, in	Filter cycle	Wash cycle	Clean	Total time	Time (n), cycles
		Time, h				
1	1	5	5	0.5	10.5	10.5
2	½	1.25	1.25	0.5	3	6
3	⅓	0.549	0.549	0.5	1.6	4.8
4	¼	0.312	0.319	0.5	1.12	4.48
5	⅕	0.200	0.200	0.5	0.9	4.5
6	⅙	0.139	0.139	0.5	0.78	4.68

Sample calculations:

1 cycle [$V = 3.8 \text{ ft}^3 \ (0.108 \text{ m}^3)$]:

$$\text{Wash rate} = \frac{2.89 \times 1^2}{2 \times 3.8} = 0.38 \text{ ft}^3 \ (0.0108 \text{ m}^3)/\text{h}$$

$$\text{Wash time} = \frac{3.8/2}{0.38} = 5 \text{ h}$$

2 cycles ($V = 1.9 \text{ ft}^3$):

$$\text{Filter time} = \theta = \frac{(1.9)^2}{1^2 \times 2.89} = 1.25 \text{ h}$$

$$\text{Wash rate} = \frac{dv}{d\theta} = \frac{2.89 \times 1^2}{2 \times 1.9} = 0.76 \text{ ft}^3 \ (0.0215 \text{ m}^3)/\text{h}$$

$$\text{Wash time}, \theta = \frac{1.9/2}{0.76} = 1.25 \text{ h}$$

2.40 $H_w - H = \dfrac{0.26}{\lambda} (t_g - t_w)$

where $t_g = 100°\text{F}$
$\quad\quad t_w = 75°\text{F}$
$\quad\quad H_w = 0.019$
$\quad\quad \lambda = 1050$

$$\textit{Air in:} \quad H = 0.019 - \frac{0.26}{1050} (100 - 75)$$

$$= 0.019 - 0.006 = 0.013 \frac{\text{lb water}}{\text{lb air}}$$

Air out: $H = 0.059$ (psychometric chart)
$\therefore 0.059 - 0.019 = 0.040$ lb water removed/lb air

and $\quad\quad \dfrac{1}{0.04} = 25$ lb air circulated/lb water removed

$$q = \underset{\text{(lb air)}}{25} \times \underset{(c_p \text{ air})}{0238} \times \underset{(\Delta t)}{20} + \underset{\text{(lb water)}}{0.013} \times 25 \times \underset{(c_p)}{0.48} \times \underset{(\Delta t)}{20} + \underset{(\lambda)}{1032}$$
$$= \textbf{1154} \text{ Btu/lb } (\textbf{2684} \text{ kJ/kg}) \text{ water removed} \quad \textbf{b}$$

See Perry for experimental data, *TH* charts, and steam tables.

2.41   Use the symbols given in the problem.

a. Tank 1: $VsC \dfrac{d\hat{T}_1(t)}{dt} = WC[\hat{T}_i(t) - \hat{T}_1(t)] + \hat{q}(t)$

or
$$0.5s \frac{d\hat{T}_1(s)}{dt} = WC[\hat{T}_i(t) - \hat{T}_1(t)] + \frac{1}{400} \hat{q}(t)$$

$$0.5s\hat{T}_1(s) = \hat{T}_i(s) - \hat{T}_1(s) + \frac{1}{400} \hat{Q}(s)$$

or
$$\hat{T}_1(s) = \frac{1}{0.5s + 1} \left[\frac{1}{400} \hat{Q}(s) + \hat{T}_i(s) \right]$$

Tank 2: $VsC \dfrac{dT(t)}{dt} = WC[T_1(t) - T(t)]$

or
$$0.5s \frac{d\hat{T}}{dt} = \hat{T}_1(t) - \hat{T}(t)$$

$$0.5s\hat{T}(s) = \hat{T}_1(s) + \hat{T}(s)$$

or
$$\hat{T}(s) = \frac{1}{0.5s + 1} \hat{T}_1(s) = \left(\frac{1}{0.5s + 1}\right)^2 \left[\frac{1}{400} \hat{Q}(s) + \hat{T}_i(s) \right]$$

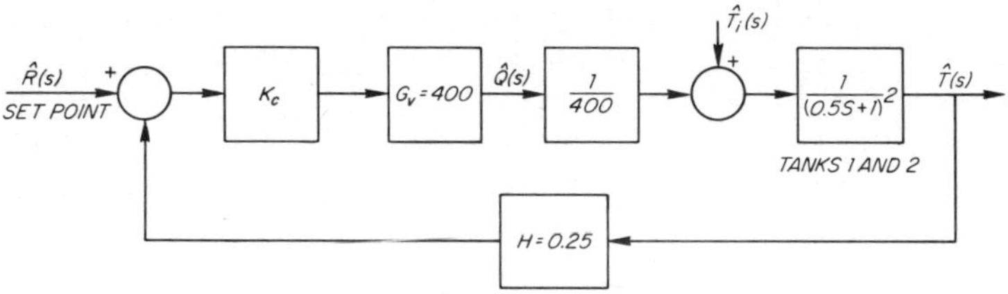

Fig. 2.41a

b. $\hat{T}(s) = \dfrac{[1/(0.5s + 1)^2](1/400) \, 400K_c}{1 + [1/(0.5s + 1)^2](1/400) \, 400(0.25K_c)} \hat{R}(s)$

$\qquad = \dfrac{K_c}{(0.5s + 1)^2 + 0.25K_c} \hat{R}(s)$

2.42 *a.* Packed volume is a function of the column diameter and the height of the column required to give the desired separation. Column diameter is determined on the basis of permissible vapor velocity. This information must be determined for individual types of packing and is usually presented under flooding data.

The height of the column is a function of:

1. The degree of removal of solute
2. The vapor-liquid equilibrium and solubility data for the solvent chosen

3. Height-of-transfer-unit $K_g a$, or heat-equivalent-transfer-plate data for the system involved

b. 1. 50 to 75 percent of the flooding velocity is chosen as vapor velocity.

2. An economic balance of cost of solute waste versus cost of column is the ultimate aim of most design calculations. To calculate the column height, the following steps are used:

(a) Plot equation data.

(b) Choose liquid-gas ratio.

(c) (1) Determine the number of transfer units by calculation or graphically, and the height equals the number of transfer units × HTU.

(2) Evaluate $dx/(x^* - x)$(liquid phase controlling) or $dy/(y^* - x)$ (gas phase controlling) by graphical integration and solve $dx/(y^* - x) = K_l aS \ dL/L$ or $dy/(y^* - x) = K_y aS \ dL/L$ when $K_y a$ or $K_y a$ are known and where y^* and x^* are equilibrium values.

c. The advantage of the HTU over the HETP is that the HTU is based upon a true differential countercurrent process rather than a stepwise countercurrent process. The HTU method is essentially the same as capacity of $K_g a$ and $K_y a$ methods, but the HTU method has an advantage in that it involves only one dimension and does not vary appreciably with gas velocity. When the operating and equilibrium lines are parallel, HTU, HETP, and capacity methods are identical.

2.43 A multiple-effect, probably a triple-effect, evaporator would be used. If necessary, an economic balance could be carried out to determine the number of effects for which the sum of the fixed charges and operating costs is a minimum. To design the evaporator, it would be necessary to know the temperature of the steam available for heating in the first effect and the temperature of the cooling water for the condenser in the last effect. The boiling point elevations of caustic solutions of various strengths between 8 and 50 percent would be required.

For making heat balances around the effects, it would be desirable to have an enthalpy concentration chart for caustic solutions or at least data on heats of mixing and specific heats of caustic solutions. The temperature of the feed to the evaporator would determine whether forward feed (for hot feed) or backward feed (for cold feed) should be used. The three effects would normally be designed for equal areas. In carrying out the calculations, it would first be necessary to assume the concentrations of the solutions in the three effects. An approximation would be to assume equal evaporations. The overall coefficients of heat transfer in each effect would then be estimated from some available correlation. The temperatures of the vapors leaving each effect would then have to be estimated, assuming either equal temperature drops across the heat-transfer surface in each effect or the temperature drops inversely proportional to the overall coefficients. Allowance must be made for the boiling point elevations of the solutions. A heat balance is then made around each effect, and the amounts evaporated in each effect are calculated from these heat balances. The calculated amounts of heat transferred may now be used to readjust the assumed temperature drops across the heat-transfer surfaces; the calculated evaporations can be used to correct the assumed concentrations and overall coefficients in the effects. The calculations are then repeated until the calculated evaporations and temperature drops agree with the assumed ones.

Materials of construction will be determined by the degree of corrosiveness of the solutions, which will depend upon the concentrations and the temperatures. In the case of caustic solutions up to concentrations of 50 percent and temperatures of 200°F, mild steel can be used. With forward feed, the effect containing the most concentrated solutions (50 percent) could be operated below 200°F so that mild steel could be used for all effects. With backward feed, however, the first effect would contain the 50 percent solution at some temperature above 200°F (93.3°C) so that stainless steel or nickel-clad steel or monel metal would then have to be used, at least in the first effect, and for any lines, valves, or pumps leading out of this effect. The higher cost of these materials may be a deciding factor in choosing between backward feed and forward feed.

2.44

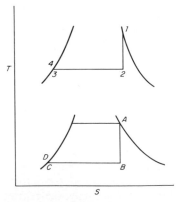

Fig. 2.44

1. $-W_{Hg} = h_1 - h_2 = 148.6 - h_2$

$S_1 = S_2 = 0.1277 = 0.02323 + x(0.1385)$

or $\qquad\qquad x = 0.754$

$$h_2 = 15.85 + 0.754(126.95) = 111.45$$
$$-W_{Hg} = 148.6 - 111.45$$
$$= \textbf{37.15} \text{ Btu/lb } (\textbf{86.41} \text{ kJ/kg}) \qquad \textbf{1c}$$

2. $-W_{steam} = h_A - h_B = 1204.5 - h_B$
$S_A = S_B = 1.4946 = 0.1151 + x(1.8867)$

or $\qquad\qquad x = 0.731$

$$h_B = 59.99 + 0.731(1040.9) = 820.99$$
$$-W_{steam} = 1204.5 - 821$$
$$= \textbf{383.5} \text{ Btu/lb } (\textbf{892.02} \text{ kJ/kg}) \qquad \textbf{2b}$$

3. For heat balance around Hg condenser, let $y = $ lb (kg)Hg/lb (kg) steam

$$y(h_2 - h_3) = h_A - h_D$$
$$y(111.45 - 15.85) = 1204.5 - 59.99$$

$$y = \textbf{11.98} \qquad \textbf{3a}$$

4. Efficiency $= \dfrac{W_{net}}{Q} = \dfrac{11.98(37.15) + 383.5}{11.98(148.6 - 15.83)}$

$$= 0.521 \text{ or } \textbf{52.1} \text{ percent} \qquad \textbf{4d}$$

2.45 $\dfrac{V}{v_0} = \dfrac{C_{A0} - C_A}{kC_A}$ solving for v_0:

$$v_0 = \frac{kC_A}{C_{A0} - C_A} V$$

$$Q = (24)(60) \text{ min/day}$$

Profit/day $= P_D = Q[(v_0 C_{A0} x)2.5 - v_0(10)] - 10$

$$= Q\frac{kC_A}{C_{A0} - C_A} V(C_{A0}x)(2.5) - Q\frac{kC_A}{C_{A0} - C_A} V(10) - 10$$

Substitute $C_{A0}(1 - x)$ for C_A:

$$P_D = Q\frac{kC_{A0}(1 - x)}{C_{A0} - C_{A0}(1 - x)} V(C_{A0}x)(2.5)$$

$$- Q\frac{kC_{A0}(1 - x)}{C_{A0} - C_{A0}(1 - x)} V(10) - 10$$

$$= 2.5kQVC_{A0}(1 - x) - \frac{10kQV(1 - x)}{x} - 10$$

$$\frac{\partial P_d}{\partial x} = -2.5kQVC_{A0} - 10kQV\frac{-1}{x} - \frac{10kQV(1 - x)(-1)}{x^2} = 0$$

with $C_{A0} = 5$, the above becomes

$$-12.5x^2 + 10x + 10 - 10x = 0$$
$$x = \textbf{0.89} \text{ percent} \quad \textbf{c}$$

2.46 *a.* There are three general methods of demineralizing or desalting. All methods involve replacing metal cation by hydrogen ion and non-metal anion by hydroxyl ion.

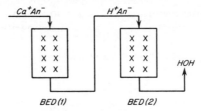

Fig. 2.46a Method I.

Method I: Double-bed operation with the removal of cation and then anion, Fig. 2.46, method I. Bed (1) is charged with cation resin (H^+) form and bed (2) is charged with weakly basic anion resin (OH^-) form. The reaction in bed (1) is

$$RSO_3H + Ca^+An^- \rightleftharpoons RSO_3Ca + H^+An^-$$

where $Ca^+ \backsim Na^+$, K^+, Ca^{2+}, Mg^{2+}, Fe^{3+}, etc. The reaction in bed (2) is

$$RNH_3OH + H^+An^- \rightleftharpoons RNH_3An + HOH$$

Note that in this type of treatment the solution is acidic after treatment in bed (1) and until treatment in bed (2) is completed.

Method II: Double-bed operation with the removal of anion and then cation, Fig. 2.53, method II. Bed (1) is charged with a strong base anion exchanger and the bed reaction may be represented by the equation

$$R_4NOH + Ca^+An^- \rightarrow R_4NAn + Ca^+OH^-$$

The solution in transit between beds (1) and (2) will be basic. The reaction in bed (2) is

$$RCOOH + Ca(OH)_2 \rightarrow RCOOCa + HOH$$

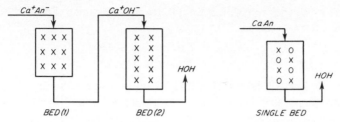

Fig. 2.46b   Method II.   **Fig. 2.46c**   Method III.

Method III: Simultaneous removal of cation and anion (mixed bed), Fig. 2.53, method III. The simultaneous reactions in the mixed bed are

$$Ca^+An^- + RSO_3H + R_4NOH \rightarrow RSO_3Ca + R_4NAn + HOH$$

and conditions of pH deviating from 7 are localized.

  b. The important factors to consider in comparing the methods described in (*a*) are:

1. Possible increased space requirements for methods I and II due to multiple beds and increased rinse requirements in regeneration
2. Influence of pH (less or greater than 7 in methods I and II) where other materials present in the stream are susceptible to changes in pH
3. The problem of "leakage" is not a major consideration in method III
4. Effluent qualities in method III are more nearly independent of regenerant levels.
5. All methods may be operated to give high-quality effluent (0.1 to 1.0 ppm iron in the case of water).

2.47   Perry, p. 569, gives data on $K = Y/X$ as a function of temperature at 115 psia for hydrocarbons:

	1 Mol fraction in feed, $L = 1$	2 K	3 X_L, when $L = 0.07$	4 Y_γ, when $\gamma = 0.93$
Methane	0.20	27	0.008	0.213
Ethane	0.20	4.7	0.045	0.212
Propane	0.30	1.6	0.192	0.307
n-Butane	0.10	0.51	0.184	0.094
Isobutane	0.10	0.71	0.137	0.097
n-Pentane	0.10	0.17	0.438	0.074
			1.004	0.997

Values of L_n (the remaining liquid) are now assumed. From a material balance for 1 mol feed

$$X_m L_n + Y_m(1 - L_n) = 1X_{fm}$$
$$X_e L_n + Y_e(1 - L_n) = 1X_{fe}$$

where X_m and X_e refer to equilibrium values of mol fraction of methane and ethane, respectively. The Y values apply to equilibrium conditions in vapor and X_f values represent mol fraction of component in feed. A similar equation can be written for each component in the feed. Since $K_m = Y_m/X_m$, dividing both sides of the material balance by X_m and substituting for Y_m/X_m, the following equation is obtained:

$$L_n + K_m(1 - L_n) = \frac{1X_{fm}}{X_m} \text{ and } X_m = \frac{X_{fm}}{L_n + K_m(1 - L_n)}$$

The same type of equation can be written for each component. The proper solution is reached when

$$X_m + X_e + X_p + X_{n-b} + X_{i-b} + X_{n-p} = 1$$

The solution is trial and error. The values shown in the table are values of X (mol fraction in remaining liquid) when L (mol liquid remaining) $= 0.07$ for 1 mol feed. The vapor is therefore 0.93 mol and

$$X_m = \frac{0.20}{0.07 + 27(0.93)} = \frac{0.20}{25.2} = 0.008$$

$$X_e = \frac{0.20}{0.07 + 4.7(0.93)} = \frac{0.20}{4.44} = 0.045$$

$$X_p = \frac{0.30}{0.07 + 1.6(0.93)} = \frac{0.30}{1.56} = 0.192$$

$$X_{n-b} = \frac{0.10}{0.07 + 0.51(0.93)} = \frac{0.10}{0.544} = 0.184$$

$$X_{i-b} = \frac{0.10}{0.07 + 0.71(0.93)} = \frac{0.10}{0.730} = 0.137$$

$$X_{n-p} = \frac{0.10}{0.07 + 0.17(0.93)} = \frac{0.10}{0.228} = 0.438$$

Summation X $= \overline{1.004}$

The vapor compositions are now calculated from $Y = KX$

and $$Y_m = 27 \times 0.008 = 0.216$$

The composition of the remaining liquid (0.07 mol) is shown in column 3 of the table and that of the vapor (0.93 mol) in column 4.

2.48 *a.* Figure 2.48 is a schematic flow diagram.

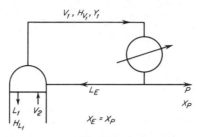

Fig. 2.48

$$R_E = \frac{L_E}{P} = \text{external reflux ratio}$$

$V_1 = L_E + P$
     $= $ material balance top of column and reflux splitter
     $= R_E P + P = P(R_E + 1)$
     $L_E + V_2 = L_1 + V_1 = $ material balance top plate
$L_E H_{L,E} + V_2 H_{V2} = L_1 H_{L1} + V_1 H_{V1} = $ enthalpy balance

Assuming that the enthalpies of vapors V_1 and V_2 are equal,

$$L_E H_{L,E} + V_2 H_{L1} = L_1 H_{L1} + V_1 H_{V1}$$
$$H_{V1}(V_2 - V_1) = L_1 H_{L1} - L_E H_{L,E}$$

and therefore

$$(L_1 - L_E)(H_{V1}) = L_1 H_{L1} - L_E H_{L,E}$$
$$L_1 H_{V1} - L_1 H_{L1} = L_E H_{V1} - L_E H_{L,E}$$
$$L_1 = L_E \frac{H_{V1} - H_{L,E}}{H_{V1} - H_{L1}}$$

The external reflux ratio is defined as $R_E/(R_E + 1) = (L_E/V_1)$, and the internal reflux ration is defined as $R_I/(R_I + 1) = (L_1/V_2)$.

$$R_E V_1 = R_E L_E + L_E$$
$$V_1 = \frac{R_E L_E + L_E}{R_E}$$

and
$$\frac{R_I}{R_I + 1} = \frac{L_1}{V_2} = \frac{L_1}{L_1 + V_1 - L_E}$$
$$R_I L_1 + R_I V_1 - R_I L_E = R_I L_1 + L_1$$

$$V_1 = \frac{R_I L_E + L_1}{R_I}$$

and therefore

$$\frac{R_E L_E + L_E}{Re} = \frac{R_I L_E + L_1}{R_I}$$

$$R_I L_E = R_E L_I$$

$$R_I = \frac{R_E L_E}{L_E} \frac{H_{V1} - H_{L,E}}{H_{V1} - H_{L1}} = R_E \frac{H_{V1} - H_{L,E}}{H_{V1} - H_{L1}}$$

For a saturated vapor 0.8 mass fraction ethanol,

$$H_{V1} = 595 \text{ Btu/lb } (1385 \text{ kJ/kg})$$

For a saturated liquid 0.8 mass fraction ethanol

$$H_{L,E} = 103 \text{ Btu/lb } (240 \text{ kJ/kg})$$
$$H_{L,E} = 58 \text{ Btu/lb } (135 \text{ kJ/kg}) \text{ at } 120°F \ (48.9°C)$$

Composition of liquid in equilibrium with 0.8 mass fraction vapor is 0.62 and

$$H_{L1} = 115 \text{ Btu/lb } (267.7 \text{ kJ/kg})$$

$$R_E = 2 \frac{595 - 58}{595 - 115} = 2 \frac{537}{480} = 2.236$$

Slope of the operating line is 0.69.

b. For 1 lb (0.4536 kg) feed, the amounts of top and bottom product are obtained by a material balance.

$$0.35 = 0.8V + 0.05B = 0.75V + (0.05V + 0.05B)$$
$$1.0 = V + B \text{ and } 0.05 = 0.05V + 0.05B$$

therefore

$$0.35 = 0.75V + 0.05$$

and $0.75V = 0.30$, whence $V = 0.40$ lb (0.18 kg) and $B = 0.60$ lb (6.27 kg). For every pound (0.4536 kg) feed, 0.40 lb (0.18 kg) of top product are withdrawn and 0.80 lb (0.36 kg) of reflux are returned to the top plate. When the reflux is fed to the top plate as a saturated liquid, only the latent heat of vaporization that is removed in the condenser must be supplied in the reboiler, or

$$0.8 \times (595 - 103) = 393.6 \text{ Btu/lb } (916 \text{ kJ/kg})$$

When the reflux is returned at 120°F (48.9°C) but with a composition of 0.8 mass fraction ethanol, its enthalpy is 58 Btu/lb (135 kJ/kg). Therefore, $0.8 \times (595 - 58) = 429.6$ Btu/lb (1000 kJ/kg) feed must be supplied. When the "cold" liquid stream enters the top plate, it removes latent heat from vapors on the top plate and internal or operating reflux is greater than external reflux. Data on enthalpies of ethanol-water mixtures are in Brown and Associates, "Unit Operations," pp. 327 and 582, Wiley, 1950. Discussion on influence of cold reflux is in Treybal, "Mass-Transfer Operations," McGraw-Hill, p. 312, 1955.

2.49   At the point of incipient fluidization, the P due to the gas flow will be such that the forces tending to raise the particles of the catalyst will be equal to the weight of the particles.

     a. $-\Delta P_f = L(1 - X)(\rho_s - P_F)$

$$\frac{-\Delta P_f}{L} = (1 - 0.4)(152 - 0.5) = 0.6(151.5) = 90.90 \text{ psf}$$

     b.   It is now necessary to determine the gas velocity that will give a pressure drop equal to that calculated in (*a*) using the method of Brownell and Katz explained in Brown and Associates, "Unit Operations," Wiley, 1950.

$$\text{Re} = \frac{D_P F_{\text{Re}} V P_F}{\mu} = \frac{0.174 \times 44 \times V \times 0.5}{12 \times 0.03 \times 6.72 \times 10^{-4}} = 1.59 \times 10^4 \times V \text{ fps}$$

$$f = \frac{2 g_c D_P(-\Delta P)}{F_f L V^2 P_F}$$

$$\frac{\Delta P_f}{L} = \frac{f F_f V^2}{2 g_c D_P} = \frac{f V^2 \times 1400 \times 0.5 \times 12}{64.4 \times 0.174} = f V^2 \times 749$$

Assume $V = 2.6$ fps:

$$\text{Re} = 1.59 \times 10^4 \times 2.6 = 4.13 \times 10^4$$

$$f = 0.0225$$

$$\frac{\Delta P_f}{L} = 0.0225 \times (2.6)^2 \times 749 = 112.35 \text{ psf}$$

Assume $V = 2.3$ fps:

$$\text{Re} = 1.59 \times 10^4 \times 2.3 = 3.65 \times 10^4$$

$$f = 0.023$$

$$\frac{\Delta P_f}{L} = 0.023 \times (2.3)^2 \times 749 = 89.88 \text{ psf}$$

This velocity could be estimated as suggested by Leva in *Chemical Engineering*, November 1957.

$$G_{mf} = 688D_P^{1.82} \frac{[\rho_f(\rho_s - \rho_F)]^{0.94}}{\mu^{0.88}}$$

$$= \frac{688(0.174)^{1.82}[0.5(151.5)]^{0.94}}{(0.03)^{0.88}} = 9.91 \text{ lb/ft}^2 \cdot \text{s}$$

$$\text{Re} = \frac{D_P G_{mf}}{\mu} = \frac{0.0145 \times 9.91}{0.03 \times 6.72 \times 10^{-4}} = 0.71 \times 10^4 = 7100$$

Estimate the correction factor = 0.2 and

$$G_{mf} = \frac{9.91 \times 0.2}{0.5} = 4 \text{ fps } \textbf{(1.22 m/s)} \qquad \textbf{1c}$$

c. Now assuming the porosity = 1, when the velocity of the gas is sufficient to elutriate particles whose diameter is 0.174 in, determine the velocity of the fall of this particle. From Brown and Associates, "Unit Operations," p. 79, Wiley, 1950,

$$f_D = \frac{4g(\rho_s - P_F)D_P^3 P_F}{3\mu^2} = \frac{128.8(151.5)(0.0145)^3(0.5)}{3(9 \times 10^{-4})(6.72 \times 10^{-4})} = 2.45 \times 10^7$$

When Re = 1, $f_D = 2.45 \times 10^7$. When $f_D = 1$, Re = 4950. Therefore, Re = 7900 and $V = 22.2$ fps. Richardson and Laki in *Transactions of the Institute of Chemical Engineers*, London, vol. 32, p. 35, 1954, have found that a straight line results when the logarithm of the porosity is plotted as a function of the logarithm of the Reynolds number.

When porosity = 0.4, Re = 816
When porosity = 1.0, Re = 7900

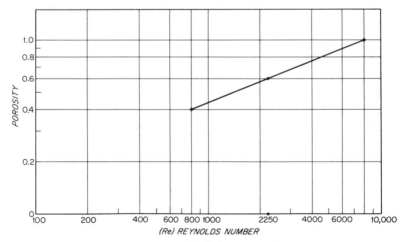

Fig. 2.49

Plot on loglog paper and draw a straight line as in Fig. 2.49. From this plot, when the porosity = 0.6, Re = 2250 and V = 6.35 fps. There will be 10 × 0.5 = 5 ft available to accommodate the expanded bed. Each foot of static bed will occupy 1.25 ft in the fluidized state. Therefore, 5/1.25 = **4** ft (**1.22** m) of static bed that can be accommodated.   **2b**

2.50   1. Volumetric displacement of the piston is

$$\frac{\pi}{4} \times (0.667)^2 \left(\frac{10}{12}\right) (200) = 58.2 \text{ ft}^3 \ (1.65 \text{ m}^3)/\text{min}$$

Since the mechanical efficiency is only 88 percent, the operating volumetric displacement is

$$0.88 \times 58.2 = 51.2 \text{ ft}^3 \ (1.45 \text{ m}^3)/\text{min}$$

The specific volume of saturated ammonia vapor at 0°F is 9.116 ft³/lb (0.569 m³/kg) (Perry, p. 250). And

$$51.2 \ \frac{\text{ft}^3}{\text{min}} \ \frac{1 \ (\text{lb ammonia})}{9.116 \ (\text{ft}^3)} = \textbf{5.61} \text{ lb } (\textbf{2.54} \text{ kg}) \text{ ammonia/min} \qquad \textbf{1b}$$

2. Enthalpy of ammonia vapor at 0°F (-17.8°C) = 611.8 Btu/lb (1424 kJ/kg). Entropy of ammonia vapor at 0°F (-17.8°C) = 1.3352 Btu/(lb)(°F) [5.60 kJ/(kg)(°C)]. Enthalpy of ammonia at 140 psia (965.3 kPa abs) at h_B and S = 1.3352.

T	h	S
190°	704.0	1.3328
	h_B	1.3352
200°	709.9	1.3418

(Table 204, Perry, p. 253).

$$h_B = 704 + 5.9 \times \frac{0.0024}{0.0090} = 705.6 \text{ Btu/lb } (1643 \text{ kJ/kg})$$

Isentropic shaft work = 705.6 − 611.8 = 93.8 Btu/lb

Actual shaft work = 1.2 × 93.8
                  = 112.6 Btu/lb (262 kJ/kg)

$$\text{hp input} = \frac{\text{actual shaft work} \times \text{lb/min}}{\text{efficiency} \times \text{Btu/hp}}$$

$$= \frac{112.6 \times 5.61}{0.85 \times 42.4} = \textbf{17.5} \text{ hp } (\textbf{13.05} \text{ kW}) \qquad \textbf{2a}$$

3. Enthalpy of ammonia after compression = 611.8 + 112.6 = 724.4 Btu/lb, and as a saturated liquid at 70°F = 120.5 Btu/lb (Table 203, Perry, p. 250). Heat removed = (724.4 − 120.5)5.61 = **3380** Btu (**356** kJ)/min.   **3b**

4. Enthalpy of ammonia (sat. liquid) at 70°F (21.1°C) = 120.5

       Enthalpy of ammonia (liquid) at 30.42 psia = 42.9
       Enthalpy of ammonia (vapor) at 30.42 psia = 611.8

Therefore, by a heat balance

$$120.5 = 42.9 + X(611.8 - 42.9)$$

$$X = \frac{120.5 - 42.9}{611.8 - 42.9} = \frac{77.6}{568.9} = 0.1364$$

or **13.64** percent is vaporized.   **4a**

2.51   1. Total weight water in the feed stock is $10 \times 2000 \times 0.1 = 2000$ lb/day. Weight of salt is 20,000 − 2000 = 18,000 lb/day. Weight of salt leaving is 18,000/0.99 = 18,181 lb = **9.09** tons (**8246** kg)/day.   **1a**

2. Weight of water removed is 2000 − 181 = 1819 lb/day, or 1819/24 = **75.8** lb (**34.38** kg) water/h.   **2c**

3. Humidity of entering air is determined from wet-bulb dry-bulb reading.

$$H_w - H_g = \frac{H_g}{29\lambda_w k_g} (T_g - T_w)$$

$$H_w - H_g = \frac{0.26}{\lambda_w} (T_g - T_w)$$

$$0.0595 - H_g = \frac{0.26}{962} (225 - 110) = 0.0310$$

Humidity of air entering = H_g = 0.0595 − 0.0310

$$= \mathbf{0.0285} \; \frac{\text{lb (kg) water}}{\text{lb (kg) dry air}} \quad \textbf{3a}$$

4. When all the heat required for vaporizing water is supplied by air or other gases and the heat transfer by conduction through boundaries and by radiation is negligible, the drying operation may be termed adiabatic. The condition of the air with respect to humidity would therefore follow an adiabatic cooling line on the humidity chart.

$$\text{Humidity of exit air} = \mathbf{0.047} \; \frac{\text{lb (kg) water}}{\text{lb (kg) dry air}} \quad \textbf{4c}$$

2.52 Weight of copper is $3 \times 100 \times 0.5$, or 150 lb. Current requirements per pound of material produced are found in tables of electrochemical equivalents. See Mantell, "Industrial Electrochemistry," 3d ed., p. 741, McGraw-Hill, 1950. Cu^{2+} requires 382.5 Ah/lb. Since the current efficiency is 85 percent, the actual current requirement will be $382.5/0.85 = 450$ Ah/lb. Therefore, 150 lb copper requires 450×150 or 67,500 Ah. Power requirement will be $67,500 \times 2.5/1000$, or 168.75 kWh. Since conversion is 90 percent and transformer loss is 3 percent, actual power will be $168.75/(0.90 \times 0.97)$ or 193.3 kWh, and the cost will be $193.3 \times 0.05 =$ **\$9.65.** **c**

2.53 A 20,000-A cell with a current efficiency of 92 percent would effectively deliver 18,400 Ah. The table of electrochemical equivalents indicates that 342.9 Ah is required to produce 1 lb chlorine when the change in valence is 1. Therefore, $18,400 \times 24/342.9 =$ **1286** lb **(583 kg)**/day chlorine. (Current efficiency includes undesirable anode products, current leaks, heat losses, etc.)   **c**

2.54 1. Work of an adiabatic compression $(-w)$ is

$$-w = +\Delta E = +C_v\, dT$$

Since $C_p - C_v = R$, then

$$C_v = C_p - R$$

and

$$-w = +\Delta E = \int (C_p - R)\, dT$$

where

$$R = \frac{T_2}{T_1} = \frac{460 + 600}{460 + 80} = \frac{1060}{540} = 1.987.$$

Therefore

$$-w = \frac{100}{28} \int \left(9.46 - \frac{3.29 \times 10^3}{T} + \frac{1.07 \times 10^6}{T^2} - 1.987\right) dT$$

$$= 3.57 \int \left(7.47 - \frac{3.29 \times 10^3}{T} + \frac{1.07 \times 10^6}{T^2}\right) dT$$

$$= 3.57[(7.47)(1060 - 540) - 3.29 \times 10^3 \ln {}^{1060}\!/_{540}$$
$$+ 1.07 \times 10^6({}^1\!/_{540} - {}^1\!/_{1060})]$$

$$= 3.57[7.47 \times 520 - 3290(6.97 - 6.29) + 1070(1.85 - 0.95)]$$

$$= 3.57(3880 - 2240 + 960)$$

$$= 3.57 \times 2600 = \textbf{9300 Btu (9811 kJ)}\qquad \textbf{1b}$$

2. General expression for the change in entropy of a system undergoing polytropic change is

$$\Delta S = R \ln \frac{T_2}{T_1} + R \ln \frac{p_1}{p_2} + \int_{T_1}^{T_2} (C_p - R) \frac{dT}{T}$$

$$= R \ln \frac{p_1}{p_2} + R \ln \frac{T_2}{T_1} - R \ln \frac{T_2}{T_1} + \int_{T_1}^{T_2} (C_p) \frac{dt}{T}$$

$$= R \ln \frac{p_1}{p_2} + \int_{T_1}^{T_2} (C_p) \frac{dT}{T}$$

Since the process is reversible and adiabatic, $\Delta S = 0$. Therefore,

$$0 = -R \ln \frac{p_2}{p_1} + \int_{T_1}^{T_2} (C_p) \frac{dT}{T}$$

or $\quad R \ln \dfrac{p_2}{p_1} = \displaystyle\int_{T_1}^{T_2} (C_p) \dfrac{dT}{T}$

$$= \int_{540}^{1060} \left(\frac{9.46}{T} - \frac{3.29 \times 10^3}{T^2} + \frac{1.07 \times 10^6}{T^3} \right) dT$$

$$= 9.46 \ln \frac{1060}{540} - 3.29 \times 10^3 \left(\frac{1}{540} - \frac{1}{1060} \right)$$

$$+ 0.535 \times 10^6 \left[\frac{1}{(540)^2} - \frac{1}{(1060)^2} \right]$$

$$= 9.46 \times (6.97 - 6.29) - 3.29 \times (1.85 - 0.95)$$

$$+ 0.535 \times (3.42 - 0.89)$$

$$= +6.43 - 2.96 + 1.35 = 4.82$$

$$\log \frac{p_2}{p_1} = \frac{4.82}{2.3 \times 1.987} = 1.056$$

$$\log p_2 - \log p_1 = 1.056$$

$$\log p_2 = 1.056 + 1.175 = 2.232$$

$$p_2 = \textbf{170.6 } \text{psia [(\textbf{1176} kPa (abs)]} \qquad \textbf{2c}$$

2.55 A schematic diagram of the process is shown in Fig. 2.55a. A line connecting the feed and solvent is constructed on the phase diagram in Fig. 2.55b. The gross composition of the mixture is calculated from the material balance.

$$1200 X_{cm1} = 1000 \times 0.5 \qquad \text{component } C$$

$$X_{cm1} = 500/1200$$

$$= 0.416 \text{ wt fraction or 41.6 percent } C$$
$$1200 X_{bm1} = 200 \qquad \text{component } B$$
$$X_{bm1} = 200/1200$$
$$= 0.166 \text{ wt fraction or 16.6 percent } B$$

This falls on a tie line parallel to and a little above the tie line for point ①. A tie line is drawn through M_1 and the composition of the first raffinate (R_1) and extract (E_1) is determined.

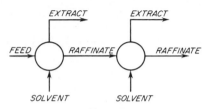

Fig. 2.55a

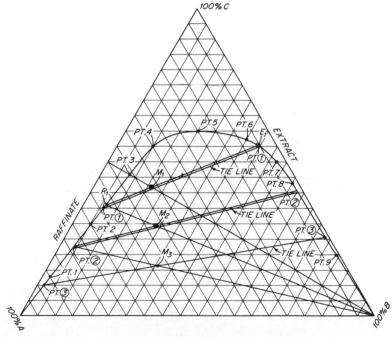

Fig. 2.55b

$$R_1 \text{ wt percent } A \ 58.0 \text{ and } E_1 \text{ wt percent } A \ 5.5$$
$$B \ 6.2 \qquad\qquad B \ 39.4$$
$$C \ 35.8 \qquad\qquad C \ 55.1$$

The weights of R_1 and E_1 are now determined by a material balance.

Component C:
$$1200(0.416) = 0.358R_1 + 0.551E_1$$
$$1200 = R_1 + E_1 \text{ and } E_1 = 1200 - R_1$$
$$R_1(0.551 - 0.358) = 1200(0.551 - 0.416)$$
$$R_1 = 840 \text{ lb } (381 \text{ kg}) \text{ and } E_1 = 360 \text{ lb } (163.3 \text{ kg})$$

Since the operation is to remove $500 \times 0.8 = 400$ lb (181.4 kg), only $500 - 400 = 100$ lb (45.4 kg) is to remain in raffinate. One extraction with 200 lb (90.7 kg) of C leaves $840 \times 0.358 = 300.7$ lb (136.4 kg) of component C. Hence one extraction is not sufficient. A new line is constructed from R_1 through C and a new mixture point M_2 is located.

$$1040X_{cm2} = 840(0.358) \qquad \text{component } C$$
$$X_{cm2} = 301/1040 = 0.29$$
$$1040X_{am2} = 840(0.58) \qquad \text{component } A$$
$$X_{am2} = 488/1040 = 0.469$$

This falls approximately on the tie-line data for point ②.

$$R_2 \text{ wt percent } A \ 73.7 \text{ and } E_2 \text{ wt percent } A \ 1.8$$
$$B \ 3.8 \qquad\qquad B \ 57.7$$
$$C \ 22.5 \qquad\qquad C \ 40.5$$

Weight of R_2 is now determined from a material balance.

$$840(0.29) = R_2(0.225) + E_2(0.405)$$
$$840 = R_2 + E_2 \text{ and } E_2 = 840 - R_2$$
$$840(0.29) = R_2(0.225) + 840(0.405) - R_2(0.405)$$
$$R_2(0.405 - 0.225) = 840(0.405 - 0.290)$$
$$R_2 = 840(0.115)/(0.180) = 537 \text{ lb } (243.6 \text{ kg})$$
$$537(0.225) = 120.8 \text{ lb } (54.8 \text{ kg}) \ C \text{ left in raffinate}$$

Therefore, a third extraction is necessary if at least 80 percent of the component C is to be removed. The new mixture point will be at:

$$737X_{am3} = 537(0.737) \qquad \text{component } A$$
$$X_{am3} = 0.537$$
$$737X_{bm3} = 537(0.038) + 200 \qquad \text{component } B$$
$$X_{bm3} = 220.4/737 = 0.304$$

This point is on tie-line data for point ③.

$$737(0.537) = R_3(0.984) + E_3(0.013) \qquad \text{component } A$$
$$737 = R_3 + E_3 \text{ and } E_3 = 737 - R_3$$
$$737(0.537) = R_3(0.984) + 737(0.013) - R_3(0.013)$$
$$R_3 = 737(0.537 - 0.013)/(0.984 - 0.013)$$
$$= 398.16 \text{ lb } (180.5 \text{ kg})$$

Weight of component C in raffinate $= 398(0.10) = 39.8$ lb (18.1 kg)

Civil Engineering

3.01 Normal duration critical path requires 19 days $(A\text{–}H\text{–}I\text{–}J\text{–}K\text{–}M)$. Reduce A to 2 days, D to 2 days, H to 3 days, and L to 1 day. Normal cost = \$12,750. Added cost = A + \$100, D + \$50, H + \$600, and L + \$500 for a sum of \$1250.

Total cost = **\$14,000.**   **1a**

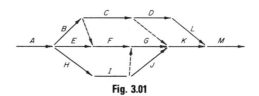

Fig. 3.01

$A + H + I + J + K + M =$ **15** days   **2c**
$A + B + C + D + L + M = 15$ days

3.02 a, 11; b, 3; c, 8; d, 5; e, 13; f, 4; g, 6; h, 12; i, 20; j, 19.

3.03 1. For the necessary thickness of pavement and base courses, see U.S. Corps of Engineers "Engineering Manual," part XII, chap. 2, p. 40, fig. 3, 1951, or similar charts in other publications.

Compacted subgrade CBR = 10, wheel load = 60,000 lb (266.9 kN), tire pressure = 200 psi (1379 kPa), combined thickness = 25.5 in, say, 26 in (66.04 cm). Necessary pavement thickness ("Engineering Manual," p. 31), tire pressure = 200 psi (1379 kPa), wheel load = 60,000 lb (266.9 kN), thickness = 4 in (10.16 cm) (for CBR in upper base = 80).

Thickness necessary above material (d) = 20 in (50.8 cm) ("Engineering Manual," p. 40). Make material (d) 6 in (15.24 cm) thick.

Thickness necessary above material (c) = 11 in (27.9 cm). Make material (c) = 8 in (20.32 cm) thick.

Thickness necessary above material (b) = 10 in (25.4 cm). Thickness = 1 in. Omit material (b).

Make material (a) = 8 in (20.32 cm) and pavement = 4 in (10.16 cm).

Relative cost: Pavement	2 layers at 20.0 =	40.0
Material (a)	2 layers at 7.0 =	14.0
Material (c)	2 layers at 4.5 =	9.0
Material (d)	2 layers at 2.0 =	4.0
Subgrade	2 layers at 1.0 =	2.0
	Total =	69.0

2. The pavement shall consist of a binder course 2.5 in (6.35 cm) thick and a surface course 1.5 in (3.81 cm) thick. The binder course shall have 5.5 percent asphaltic cement and the surface 7 percent asphaltic cement. The aggregate gradation for the surface course shall have the following characteristics:

Screen	Percent passing	Opening in	mm
1½	100	1.500	38.1
1	84–100	1.000	25.4
½	66–84	0.500	12.7
No. 4	42–66	0.187	0.76
No. 10	31–55	0.0787	2.00
No. 40	16–34	0.0165	0.420
No. 80	10–22	0.0072	0.182
No. 200	4–8	0.0029	0.074

The binder course shall have the following aggregate gradation limits:

Screen	Percent passing	Opening	
		in	mm
2	100	2.000	50.8
1½	86–100	1.500	38.1
1	70–88	1.000	25.4
½	52–72	0.500	12.7
No. 4	32–52	0.187	4.76
No. 10	20–40	0.0787	2.00
No. 40	10–22	0.0165	0.420
No. 80	6–15	0.0072	0.182
No. 200	3–7	0.0029	0.074

A prime coat of cutback asphalt, grade MC-O at a rate of ½ gal/yd^2 (2.26 liters/m^2) will be applied to the base before the binder course is applied.

A heated tack coat of RC-3 should be applied to the binder course at the rate of 0.2 gal/yd^2 (0.91 liter/m^2) before placing the surface course.

A seal coat of 0.2 gal/yd^2 (0.91 liter/m^2) of hot cutback asphalt should be applied to the surface, followed by 8 to 15 lb (3.63 to 6.80 kg) of sand.

3.04 Elevation of low point must equal 98.42 ft.

$$\text{Station (L.P.)} = \frac{1.5L}{1.5 - (-2)} = 0.429L \text{ from PVT}$$

Tangent offset at PVI is T.O.

$$\frac{(0.429L)^2}{(0.5L)^2} (\text{T.O.}) = 98.42 - (95.00 + 0.015 \times 0.071L)$$

$$\text{T.O.} = \frac{+3.42 - 0.00106L}{0.736} \tag{1}$$

$$2(\text{T.O.}) = \frac{(95.00 + 0.02 \times 0.5L) + (95.00 + 0.015 \times 0.5L)}{2} - 95.00$$

$$\text{T.O.} = \frac{0.0100L + 0.0075L}{4} \tag{2}$$

By equating Eqs. (1) and (2),

$$L = 3.42/0.00428 = \textbf{800} \text{ ft (\textbf{243.8} m)} \quad \textbf{c}$$

and $$\text{T.O.} = 3.50 \text{ ft } (1.07 \text{ m})$$

$$\text{Station (L.P.)} = 100 + (0.500 - 0.429) \times 800 = 100 + 56.8$$
$$= (3065.3 \text{ m})$$

As an alternate, the length of curve could have been assumed as 800 ft (243.8 m). The T.O. is 3.50 ft (1.07 m). The elevation at the low point can be expressed as Y and the distance in stations from the PVT as X. Then

$$Y = 101.00 - 1.5X + \frac{X^2}{4^2} \ (3.50)$$

At the low point the tangent will be horizontal and

$$\frac{dY}{dX} = -1.5 + \frac{2X}{16} \ (3.50) = 0$$

$$7X = 24$$

and $X = 3.43$ stations from PVT or $X = 100 + 57 = 3065.3$ m.

3.05 a, false; b, true; c, true; d, true; e, false; f, true; g, false; h, false; i, true; j, false.

3.06   Using test data,

$$d = \frac{V^2}{30(f - 0.05)} = 45 \text{ ft } (13.72 \text{ m})$$

and $$f = \frac{25^2}{30 \times 45} + \frac{1.5}{30} = 0.51$$

For accident car:

$$V^2 = 130 \times 30(0.51 - 0.05) + 5^2 = \textbf{42.7} \text{ mph } (\textbf{68.7} \text{ km/h}) \qquad \textbf{b}$$

3.07 $p = \frac{1}{12} = 0.08333$, $q = 1 - p = 0.91667$, $n = 10$.

1. $p(1) = \dfrac{10!}{1! \ 9!} \ (0.08333)^1 (0.9167)^9$

$$= (10)(0.08333)(0.458) = 0.381 \ (\textbf{38.1} \text{ percent}) \qquad \textbf{1d}$$

2. $p(0) = \dfrac{10!}{0! \ 10!} \ (0.08333)^0 (0.9167)^{10}$

$$= (1)(1)(0.419) = 0.419 \ (\textbf{41.9} \text{ percent}) \qquad \textbf{2a}$$

3.08   1, d; 2, f; 3, l; 4, a; 5, u; 6, b; 7, j; 8, s; 9, t; 10, h.

3.09   It may be inferred from (3) that a distribution reservoir is to be provided, rather than pumping directly from a small, clear well to the mains. Therefore filter capacity can be designed for normal rather than maximum flow.

1. Daily consumption = 14,000 × 100 = 1.4 mgd (5299 kl/day)

Water filtered, including washwater = 1.4/0.96
$$= 1.46 \text{ mgd (5526 kl/day)}$$
$$\text{Required filter area} = (1.46/125) \times 43,560$$
$$= 509 \text{ ft}^2 \text{ (47.28 m}^2\text{)}$$

For a plant as small as this, the filters can be operated on one 8-h shift.

$$\text{Area} = 3 \times 509 \times (8/7.5) = 1628 \text{ ft}^2 \text{ (151.24 m}^2\text{)}$$
Provide 3 units 20 × 27.5 ft in plan = 1650 ft² (153.29 m²)

If one unit is out of service for extensive repairs, the plant can be operated on a double shift to provide the required flow. An alternative would be to install 4 units.

2. Alum = 4.5 gr/gal

$$1 \text{ gr/gal} = 142.9 \text{ lb/million gal (17.13 kg/Ml)}$$
$$4.5 \times 142.9 \times 1.46 = \textbf{939 lb (425.9 kg)} \text{ alum/day}$$

3. National Board of Fire Underwriters

$$Q = 1020 \sqrt{P} \, (1 - 0.01 \sqrt{P})$$

When P = population in thousands,

$$Q = 1020 \times \sqrt{14} \, (1 - 0.01 \sqrt{14})$$
$$= 1020 \times 3.74(1 - 0.037)$$
$$= 1020 \times 3.74 \times 0.963 = \textbf{3674} \text{ gal (13,906 kl)/min}$$

National Board of Fire Underwriters recommends providing for a 10-h fire in towns exceeding 2500 in population. Therefore required storage = 3674 × 60 × 10 = 2,204,000 gal (8342 kl).

Peak of daily flow may occur during fire. Therefore domestic storage = 2.25 × 1.46 mgd = 3.29 million gal (12,450 kl). Total storage, fire + domestic use = 2.204 + 3.29 = 5.494 million gal (20,792 kl)/day.

3.10 *a.* Detention period **30** to **60** s. Most states recommend 60 s.

b. Maximum rate of flow **0.5** to **1** ft (**0.152** to **0.305** m)/s. The higher figure is generally used.

c,d,e. $wd = Q/V, \, wL = A$

For medium-size grit particles, the recommended rate of flow is 50,000 gal/(day)(ft²) [2037 kl/(day)(m²)] of surface area. For 1 mgd (3785 kl/day), the required surface area would be 20 ft² (1.86 m²). Assume $w = \textbf{1}$ ft (**0.305** m) **d**, $L = 20$ ft (6.10 m) **c**, and $d = Q/wV = 1.547/(1$

$\times$ 1) = 1.547 ft (0.472 m). Since 1 mgd = 1.547 ft³/s to provide grit storage capacity (g), make depth = **2** ft (**0.61** m)   **e.**

f. Use 3 units in parallel to take care of variations in flow.

g. Grit will vary in amount from 0.5 to 10 ft³ (0.014 to 0.282 m³)/million gal (3785 kl). Increasing depth to 2 ft (0.61 m) will provide storage for 1 $\times$ 20 $\times$ 0.453 = **9**+ ft³ (**0.257** m³).

h, i. Clean **daily, manually,** for so small a plant.

3.11 1. For raw sewage containing 250 ppm suspended solids, 1.5-h sedimentation should effect a reduction of $\pm$45 percent or $\pm$**138** ppm in the effluent.   **1a**

2. V = 1.5 $\times$ 500,000/24 = **31,250** gal (**118.3** kl)   **2b**

V = 31,250/7.48 = 4180 ft³ (118.3 m³)

3. Assume $\pm$1000 gal/ft² (40.7 kl/m²) per day loading. Then wl = 500,000/1000 = 500 ft² (46.45 m²). Take $t = 4w$, then $4w^2 = 500$, $w^2 = 125$, $w = 11.2$; using width = 12 ft; length = 48 ft; surface area = 576 ft² (53.51 m²); depth = V/A = 4180/576 = 7+.

Use 8 ft (2.44 m).

w = **12** ft (**3.61** m), L = **48** ft (**14.63** m), D = **8** ft (**2.44** m)   **3a**

4. On basis of 45 percent removal, we have 112 ppm settled out. This equals (112 $\times$ 8.33)/2 = 468 lb (212.3 kg)/day.

468 lb solids + 8900 lb water = 9368 lb (4249.3 kg)

9368 lb at 63 lb/ft³ = 148+ ft³, say **150** ft³ (**4.25** m³)   **4c**

3.12 *a.* The separate sludge-digestion process comprises the collection of sewage solids in primary sedimentation tanks (and also in the final tanks of certain secondary treatment processes) and the passing of these solids, by pumping or by gravity flow, to one or more tanks where digestion of the sludge occurs. Digestors are generally provided with floating covers, devices for the collection and utilization of sludge gases, and arrangements for heating and controlling the reaction of the sludge to reduce the time required for digestion. Many plants use two-stage digestion, the primary stage of about 1 week taking place in one digestor and the sludge then being transferred to a second digestor for about 3 weeks.

b. Sludge may be dewatered on open or covered drying beds or on vacuum filters. Sand beds require considerable acreage and may be high in first cost, particularly when covered. Drying is slow but does not require skilled attendance.

Vacuum filters require little space and dry the sludge quickly. They are not low in either first cost or maintenance. Skilled attendance is necessary since conditioning of the sludge may be required to obtain a satisfactory filter cake.

3.13   1*b*; 2*b*; 3*d*; 4*d*; 5*d*; 6*b*; 7*b*; 8*a*; 9*d*; 10*b*.

3.14   Recommended capacity of septic tank for 10 persons = 900 gal (3.41 kl). Recommended dimensions are 3.5 ft (1.07 m) wide, 8.5 ft (2.59 m) long, 4.5 ft (1.37 m) liquid depth, 5.5 ft (1.68 m) total depth. If a domestic garbage grinder is to be installed, the capacity should be increased by 50 percent. Capacity shown above provides for 2 years' accumulation of sludge plus 50 gal (189 liters) per capita sewage flow per day.

For fairly porous soil, allow 450 ft^2 (41.81 m^2) of absorption area in bottom of disposal trenches. For 18-in (48.72-cm) width of trench, length will equal 450/1.5 $\mp$ 300 ft (91.44 m).

Sewage from house will flow into septic tank shown in Fig. 3.14*a*. Clarified effluent will flow out to distribution box and thence to 300 ft (91.44 m) of 4-in (10.16-cm) open-joint tile drain (Fig. 3.14 *b*). Top of tile should be surrounded with gravel, coarse sand, or cinders for a distance

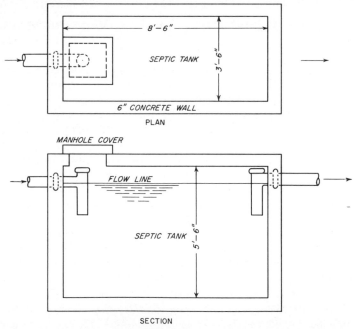

PLAN

SECTION

Fig. 3.14*a*

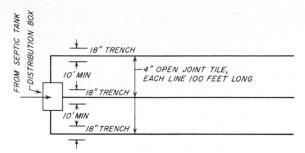

Fig. 3.14b

of at least 6 in (15.24 cm) on sides and bottom and at least 2 in (5.08 cm) above top of pipe. Open joints should be covered by tar paper. A more effective septic tank than that shown in Fig. 3.14b would involve a small secondary chamber following the main septic tank. This would be provided with an automatic siphon to permit dosing the entire tile field periodically instead of permitting the effluent to trickle slowly through the first few joints, as is likely to occur when no siphon is provided.

3.15 Malaria is transmitted by various mosquitoes of the *Anopheles* group. *Anopheles quadrimaculatus* is the most important vector of the disease in the eastern United States.

The mosquito must bite persons whose blood contains the malaria parasites. Certain stages in the life cycle of the parasite must then take place in the mosquito before the insect can transmit the disease by biting others. Obviously, if there are no cases in the area, there can be no transmission of the disease even though the density of *quadrimaculatus* infestation may be heavy.

Where there are cases in the area, a number of control measures are possible: (1) Infected persons should receive proper medical treatment to reduce or eliminate the parasites from their blood. (2) Infected persons should be protected from the bites of the insects. This is not difficult with proper screening since *quadrimaculatus* is predominantly a dusk and night biter. (3) The raising of cattle in rural areas reduces the incidence of the disease since the insect prefers cattle to people for a blood meal. (4) Standard mosquito-control measures aimed principally at the insect in the larval stage are effective in reducing mosquito density. (5) Blood tests of all persons in a heavily infected area should be carried out so that all infected persons can receive proper medical treatment. (6) DDT can be used as a residual spray in inhabited buildings to kill adult mosquitoes before they can transmit the disease. (7) Educational campaigns are effective in informing the public concerning the steps necessary to eliminate the disease or to reduce its incidence.

3.16 1. Case rate = $^{2891}\!/_{2000}$ = **1.446**/1000   **1a**

2. Death rate = $^{1304}\!/_{2000}$ = **0.652**/1000   **2c**

3. Fatality rate = $^{1304}\!/_{2891}$ × 100 = **45** percent   **3c**

4. Morbidity is the number of cases of a disease occurring during a year in terms of the population in hundred thousands. Although generally computed for the entire population, it may also be determined for certain age groups or classes.

3.17 *a.* Many diseases are transmitted by milk, but tuberculosis and undulant fever are those most prevented by protecting the dairy herd. The tuberculin test is applied periodically, usually annually or semiannually, to detect tubercular cows in the herd. Such animals are destroyed. Undulant fever can be controlled by removing infected animals from the herd and by pasteurizing the milk.

b. Measures to protect the milk line are designed to assure healthy cows, clean and healthy workers, clean and dustless barns, prompt cooling of the milk in a separate milk house, proper utensils and the effective cleaning and sterilization thereof, and pasteurization of the milk.

c. The above steps will minimize or eliminate the transmission of typhoid, scarlet fever, diphtheria, undulant fever, and other diseases by milk.

d. Milk is an ideal medium for the development of many bacteria, and bacterial growth is rapid in dirty milk. Pasteurization is not a sterilization process, and hence a high-count raw milk still has considerable numbers of bacteria in it after pasteurization. The most effective measures to eliminate this danger are providing sanitary conditions in the dairy and properly pasteurizing and bottling the milk.

e. Milk may be pasteurized at 143°F (61.7°C) for 30 min or flash pasteurized at 160°F (71.1°C) for 15 s. Control of early flash pasteurizers was uncertain and the method was consequently little used until recently. Modern flash pasteurizers provide excellent control and are as safe as the holder types. They require much less space and are consequently increasing in use.

3.18 *a.* Bubonic plague is transmitted to people by the bite of fleas from an infected rat.

b. Murine typhus fever (endemic) is transmitted by the bite of fleas or mites from an infected rat.

c. Malaria is carried by various mosquitoes of the *Anopheles* genus.

d. The common vector of dengue is the *Aedes aegypti* mosquito.

e. Smallpox is commonly communicated by discharges from the nose and throat of persons suffering from the disease. It may also be transmitted through the agency of material from pustules or skin lesions.

f. Undulant fever may be transmitted by unpasteurized milk, contact with infected slaughtered animals, contact with the dust of pens containing infected animals, and by contact with these animals when they abort since the disease causes infectious abortion in cows, sows, and goats.

g. Tularemia is usually communicated to people in skinning or plucking rabbits, quail, and other game. It may also be transmitted by certain biting flies.

h. The common vector of yellow fever is the *Aedes aegypti* mosquito. In jungle areas, other mosquitoes are involved, including *Sabethini* and *Haemagogus*.

i. Rocky Mountain spotted fever is transmitted from various small wild animals to humans by ticks.

3.19 The steps to be taken and data to be collected to determine the source of the typhoid outbreak would include:

1. Attending physicians promptly reporting to the local health department all cases.

2. Filling in questionnaires relative to the sources of milk, water, shellfish, and other foods consumed by those persons affected by the disease.

3. Preparing spot maps showing the areas within the community where there are cases of the disease.

4. The prompt bacteriological analysis of supplies of water, milk, shellfish, and other foods.

5. Checking restaurants and other establishments where food had been served or sold to victims of the disease for possible typhoid carriers.

6. Checking private well supplies and cross connections between potable and industrial water supplies.

3.20 Two methods suitable for the disposal of garbage are sanitary fill and incineration.

Sanitary fill differs from open dumping in that the garbage is covered with about 2 ft (0.61 m) of earth at the end of each working day. It is suitable where there are low areas, abandoned borrow pits, and other suitable land within or near the municipality and where suitable earth, preferably sandy soil, is available for covering. The method is generally cheaper than incineration and permits the eventual utilization of tracts of land for parks, housing developments, and other uses.

Incinerators provide for the complete stabilization of organic materials and a great reduction in volume of the wastes collected. First costs and operating costs are high. They are particularly useful where extensive tracts of land for sanitary fill are not locally available.

3.21 There are several methods for determining the safe yield of a watershed from the storage that can be developed economically at a given reservoir site or, conversely, the storage that will be necessary to furnish a desired yield. One of the most satisfactory procedures is to use a mass curve of modified runoff.

Actual records of runoff, usually monthly, are obtained for a stream-gaging station on the stream relatively close to the site of the proposed dam. The longer the period of years for which records are available, the greater the likelihood of the record including a period of extreme drought that will control the yield or the required storage.

If the records are given in cubic feet per second at the gage, they are converted to inches of depth on the watershed area above the gage; these figures are then used for the slightly different watershed area above the site of the proposed dam.

When the proposed reservoir will have a large surface area, the actual runoff records can be modified downward to account for the expected increased evaporation losses from the reservoir. The amount of water to be released downstream for the use of riparian owners below the reservoir site is also deducted from the monthly values of actual runoff. The remainder will be water available for storage, power or water-supply use, and waste over the spillway at times when the reservoir is full.

The accumulated totals of modified runoff are then added month by month and plotted as a mass curve (Fig. 3.21). Suppose a yield of 100 mgd (378.5 ml/day) is desired. This value is expressed in inches depth per month on the watershed area and plotted as the sloping line *AB*. Lines *CD* and *EJ* are drawn parallel to *AB* from points *C* and *E* where the runoff starts to decline. It is obvious that the worst period of drought

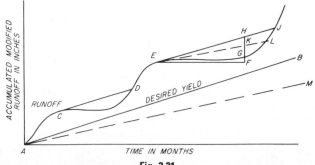

Fig. 3.21

has occurred between E and F. During this period, the total inflow is represented by the ordinate FG. The total required yield is represented by FH. Therefore water in the amount GH must be released from storage to supply the demand. This storage, shown here as inches depth on the watershed, can be converted into acre feet, billions of gallons, or other convenient units.

It may be uneconomical to develop this volume of storage at the site. If the storage that can be developed is GK, we draw the dotted line EKL. The safe yield is then reduced to that represented by AM, drawn parallel to EJ.

3.22 *a.* The reduction of losses through leakage and waste in a water-distribution system is becoming ever more important as the cost of supplying water to a community increases year after year. Water pressures will also improve as waste is minimized. In some cities, the actual flow of water in the system may be from 1½ to 2 times the necessary consumption.

b. The complete metering of a city is most effective for reducing wastes. Leakage from broken mains may be detected by visual inspection; the Cole pitometer is widely used to detect leaks in mains. Audio devices to amplify the sound of escaping water are also used. Periodic wastewater surveys and careful maintenance of the distribution system may reduce the unaccounted-for water to perhaps 2 percent of the delivered water.

c. When water shortages exist, educational campaigns have been quite effective in reducing actual consumption and minimizing the actual waste of water.

3.23 A rapid-sand-water-filtration plant normally includes the following units or items: (*a*) screens for removing leaves and other large objects; (*b*) a chemical storage house for storing chlorine, ammonia, lime, alum, activated carbon, and soda ash; (*c*) a mixing device, such as a flash, mixer, baffled mixing chamber, hydraulic jump, a low-lift pump suction to provide intimate mixing of activated carbon, lime, and alum, or other coagulant with the raw water; (*d*) one or more flocculators in which gentle stirring action permits satisfactory floc formation; (*e*) one or more sedimentation basins in which the coagulant can settle out, removing turbidity, color, and many bacteria; (*f*) one or more filter units as described in 3.17*b*; (*g*) a clear well for storing filtered water; (*h*) provisions for final chlorination and pH adjustment in the clear well; (*i*) washwater pumps and a washwater tank to permit backwashing the filters; (*j*) air compressors for air washing filters in some cases; (*k*) a sewer connection for the dirty washwater; (*l*) suitable meters and gages to show the quantity of water being processed at any time; and (*m*) a well-equipped laboratory, adequately staffed, for analyzing the raw, finished, and delivered water. Low- and high-lift pumps and power-generation equipment will be required in some cases.

3.24 *a.* The zeolite softening process is essentially one of base exchange. The zeolite, through which the water passes, consists of a chemical compound so loosely bound that it will readily exchange its sodium radical for the calcium and magnesium radicals in the water to be treated. The process is also reversible, depending upon the concentration of dissolved chemicals. Zeolites may be natural, such as glanconite or greensand, or synthetic, such as Permutit.

In passing the raw water through the zeolite, the reactions are:

$$NaZ + \frac{Ca}{Mg}(HCO_3)_2 \rightarrow \frac{Ca}{Mg}(Z) + ZNaHCO_3$$

| Sodium zeolite | Calcium or magnesium bicarbonate hardness | Calcium or magnesium zeolite | Sodium bicarbonate |

$$NaZ + \frac{Ca}{Mg}(SO_4) \rightarrow \frac{Ca}{Mg}(Z) + Na_2SO_4$$

| Sodium zeolite | Sulfate hardness | Calcium or magnesium zeolite | Sodium sulfate |

The end products of these reactions are soluble and do not produce hardness in the water.

When the softening capacity of the zeolite is exhausted, the raw-water flow is cut off and the zeolite is backwashed with a brine solution. The reaction of the salt with the calcium zeolite is:

$$CaZ + NaCl \rightarrow NaZ + CaCl$$

| Calcium zeolite | | Sodium zeolite | Calcium chloride |

After such backwashing, the restored zeolite can again be used to soften water.

b. The effect of H_2S in raw water to be chlorinated is to increase the required dosage of chlorine to provide a disinfecting residual in the water since a part of the chlorine added will react with the hydrogen sulfide.

c. Lime and soda ash are added to water to remove the bicarbonates and sulfates of calcium and magnesium that cause hardness. The chemical reactions with the bicarbonates and sulfates of calcium are:

$$Ca(HCO_3)_2 + Ca(OH)_2 \rightarrow 2CaCO_3 + 2H_2O$$

| Calcium bicarbonate | Lime | Calcium carbonate | Water |

$$CaSO_4 + Na_2CO_3 \rightarrow CaCO_3 + Na_2SO_4$$

| Calcium sulfate | Soda ash | Calcium carbonate | Sodium sulfate |

The reactions with magnesium compounds are similar.

The addition of the lime and soda ash results in insoluble compounds that will settle out or in soluble compounds which do not cause hardness.

3.25 *a.* Chloramines are used in water treatment for (1) disinfection, (2) taste and odor control, (3) as algicides, (4) weed control in reservoirs.

b. Sodium hexametaphosphate (Calgon) is used to control the solubility equilibrium of calcium carbonate in water-softening plants, in the processing of industrial water supplies, and in the "threshold treatment" of water in an attempt to inhibit corrosion.

c. Sodium bisulfite is a reducing agent. As such, it may be used in the deaeration of water to minimize its corrosiveness. It may also be used in dechlorination, to avoid chlorinous tastes after heavy applications of chlorine. Because careful chemical control is necessary, its use has not become widespread in water treatment.

d. Sulfur dioxide in aqueous solution has been used to clean rapid-sand filters that have become coated with iron or manganese. This process is patented. Sulfur dioxide is also used in dechlorination.

e. Activated carbon is chiefly useful in water treatment to control tastes and odors.

f. Aluminum sulfate (alum) has a variety of uses in water treatment, chiefly as a coagulant to remove color, turbidity, and bacteria in connection with rapid-sand filtration.

g. Hydrated lime is sometimes used before filtration in an attempt to reduce the bacterial load on filters; it is frequently used after filtration to combat the corrosive quality of the water resulting from the presence of oxygen and carbon dioxide and also to adjust the pH value; it is widely used in water softening.

h. Sodium chloride is most widely used in water treatment in the regeneration of zeolite water-softening tanks.

i. Sodium carbonate is added to waters of low alkalinity to permit the use of alum as a coagulant; it is used in cleaning filters; and it is widely used with lime in water softening.

3.26 *a.* (1) *A single-stage centrifugal pump* is one having a single impeller. It is not generally economical to use when the operating head exceeds 300 ft (91.44 m) (see question *b*).

(2) In a *centrifugal* pump a rotating impeller imparts velocity to the water entering from the suction line and forces it to flow under pressure into the discharge line.

(3) For higher heads (*c*), a *multistage* pump is used. Here two or more impellers are connected in series, the discharge from the first impeller passing to the suction of the second, etc. (see question *c*).

(4) Both the reciprocating and the rotary pumps are *positive-displacement* pumps.

(5) *Reciprocating pumps* are provided with a piston or plunger moved by reciprocating action in a cylinder provided with the necessary inlet and outlet valves. The movement of the piston alternately draws water into the cylinder from the suction line and forces it out under pressure into the discharge line.

(6) In the single-acting reciprocating pump, the flow is intermittent and unsteady. This condition is largely corrected by using a *double-acting plunger pump* in which there are two sets of inlet and outlet valves so that there is continuous discharge of water.

(7) A *rotary pump* has cams or gears that rotate in a casing, thus forcing the incoming water around the casing and out the discharge line on each revolution.

3.27 *a. Bacillus coli* and other bacteria of *coli-aerogenes* group are normally found in the intestinal tracts of humans and animals. The presence of these organisms in water supplies indicates sewage pollution. Although these organisms are not in themselves pathogenic, their presence indicates possible pollution of the water by pathogenic intestinal bacteria such as *B. typhosus*, and hence the need to purify the supply.

The *E. coli* index for a sample of water is the reciprocal of the smallest quantity of the sample, in cubic centimeters, that gives a positive test for coliform organisms. The determination of the "most probable number" is in greater use today for determining the numbers of *E. coli* present in a sample.

b. Anaerobic bacteria thrive in the presence of suitable nutrients and in the absence of free oxygen.

Aerobic bacteria require free oxygen for their development.

c. Hardness in water is commonly expressed as parts per million of $CaCO_3$ even though the total hardness may be made up of the carbonates and sulfates of calcium and magnesium. A carbonate hardness of 100 ppm indicates a water supply about on the border between slightly and moderately hard. Such a supply would not, as a rule, be softened for municipal use.

3.28 The factors that control the accuracy of computation of the lengths of the sides of a triangulation network include: (1) precision of measurement of baseline, (2) precision of measurement of angles, (3) size of "distance angles" involved in the computations, (4) number of side and angle equations provided by the geometrical figures used in the network, (5) number of stations occupied by the transit party, and (6) frequency with which check baselines are measured.

The effects of items (1) and (2) are obvious. Since the lengths of triangle sides are computed by the law of sines, the angles used in the computations (distance angles) should be of such size that their sines are changing slowly so that errors of angular measurement will have a minimum effect upon the computations. Thus, the closer the angles used approach 90°, rather than 0° or 180°, the less effect errors in angular measurement will have upon the computed lengths of network sides. The effect of the size of the angles is indicated by

$$\delta A^2 + \delta A \delta B + \delta B^2$$

in the strength of figure equation. δA equals the tabular difference for 1 s in the log sine of angle A in the sixth decimal place. δB is the corresponding value for log sine of angle B.

Items (4) and (5) reflect the number of geometrical checks on the computed lengths permitted by the network. They are included in the $(D - C)/D$ term in the strength of figure equation. D represents the number of new directions observed; C is the number of geometrical conditions (side and angle equations) in the network figure.

If a triangulation network is extended indefinitely, its strength gradually becomes less. Specifications accordingly provide for the measurement of a check base periodically in extending a network.

3.29 Watershed control should include the right to approve or disapprove sewage-disposal facilities in the resort area around the lakes; constant vigilance with respect to records of the quality of the effluent from the sewage-treatment plants; provision for the application of copper sulfate for algae control in the lakes; and the establishment of many sampling stations so that the source of any unusual pollution or contamination can be quickly determined and steps taken to adjust the situation. A chlorinator at the outlet of the lowest of the lakes may reduce the load on the plant.

A boom should protect the influent canal from the entrance of large floating objects. Mechanically cleaned screens should be provided to remove leaves and other objects and to protect the pumps. Gates are desirable for controlling the water intake.

Although a flash mixer can be used for mixing various chemicals with the raw water, satisfactory mixing should be provided for by applying the chemicals to the suction line of the low-lift pumps. Therefore, the low-lift pumphouse should have, in addition to space for pumps and motors, adequate space for chemical storage, for solution- or dry-feed machines, and for measuring devices so that the quantity of chemicals added can be proportionate to the flow. At this point, we may prechlorinate, apply activated carbon, add alum or other coagulant, and apply soda ash when the alkalinity of the raw water is low. Rail or truck delivery of chemicals should be provided. A storage yard for chlorine containers is necessary. Provision should be made for fork lifting pallets of bagged chemicals and transportation, via elevator, to overhead storage from which the chemicals can be fed via hoppers by gravity to the proportionate feeders.

The discharge from the low-lift pumps will pass to a flocculation tank and thence to a sedimentation basin. This basin can be cleaned periodically or continuously, depending upon the volume of sludge that can be anticipated. The sludge can be returned to the river below the plant, lagooned, or dried and used as fill, as dictated by local conditions.

The settled effluent will pass to rapid-sand filters of the usual type and then, through rate controllers, to the clear well. Pumps to raise filtered water to the washwater tank shall be provided. Customary provisions for backwashing the filters shall be provided. The backwash water shall pass to the sewer or to the river below the plant. Provision may be made for surface and air wash as well. In the latter case, air-compression equipment must be installed in the control house.

Postchlorination should be provided in the clear well. Lime will probably be necessary for pH adjustment and for CO_2 removal, although aeration is sometimes used for the latter purpose. Necessary facilities for chemical storage must be provided near the clear well.

The high-lift pumps will raise the water from the clear well to the distribution reservoir. Both low- and high-lift pumps will normally be powered by electric motors. Standby power units, either gasoline or diesel engines, must be provided in the event of primary power failure.

Within the plant buildings, space must be provided for an administrative office, a chemical and bacteriological laboratory, heating equipment, change room, lavatories, garage space for maintenance trucks, storage space not only for chemicals but for maintenance supplies, fuel storage, motor and pump rooms, etc.

3.30 The inflow hydrograph will have the form shown in Fig. 3.30a. Determine the values of inflow in acre-feet (1 acre-ft = 1233.5 m^3) for each time interval. Accumulate these values and plot the accumulated inflow in acre-ft as shown in Fig. 3.30b. On a separate sheet, plot the pool elevation-discharge curve and the pool elevation-storage curve as shown in Fig. 3.30c. The first is determined by assuming various pool elevations and computing the corresponding discharge from the equation $Q = CbH^{1.5}$. The latter is determined from a contour map of the area flooded at various elevations of flow above the spillway.

Now assume a storage increment in acre-ft small enough so that the

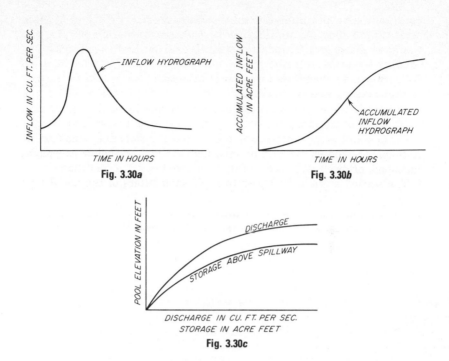

Fig. 3.30*a*

Fig. 3.30*b*

Fig. 3.30*c*

discharge curve during the period of accumulation of this storage is practically a straight line. This amount of storage (from Fig. 3.30*c*) will give a certain spillway discharge at the end of the period (from the same figure). Take the mean discharge for the period and convert it to acre-ft for a time period of say 12 h. Plot this at the right of Fig. 3.30*b* (as shown in Fig. 3.30*d*) and, from the assumed inflow storage in acre-ft, draw a

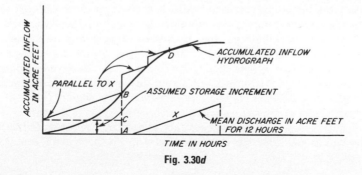

Fig. 3.30*d*

parallel line until it intersects the curve of accumulated inflow. At this point (Fig. 3.30*d*), the total inflow is represented by *AB*, the total discharge is shown by *CB*, and the volume stored above the spillway *AC* is equal to the assumed value.

Continue adding storage increments, of greater and then lesser values, and follow the same procedure until the discharge line becomes tangent to the curve of accumulated inflow at *D*. At this point the inflow and outflow will be equal and the flood will have reached its maximum height above the spillway. The process may be continued, using negative increments of storage, to determine the form of the discharge hydrograph.

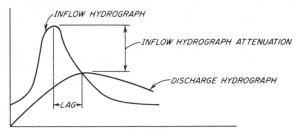

Fig. 3.30e

As shown in Fig. 3.30*e*, this may be compared with the inflow hydrograph to indicate the time lag in cresting and the attenuation, or reduction in the crest height, resulting from reservoir storage.

3.31   *a*, false; *b*, false; *c*, true; *d*, true; *e*, false; *f*, true; *g*, false; *h*, false; *i*, true; *j*, false; *k*, true; *l*, true; *m*, true; *n*, false; *o*, true; *p*, false; *q*, true; *r*, false; *s*, true; *t*, false; *u*, true; *v*, false; *w*, false; *x*, false; *y*, true.

3.32   *a*. Precipitation in northern New Jersey results from cyclonic storms of continental origin during the fall, winter, and spring months. Much of the summer precipitation results from local convective air movements accompanied by thunderstorms. In the late summer and early fall, hurricanes originating in the equatorial regions of the Atlantic may bring copious rainfall to this area. The net result is to produce reasonably uniform precipitation throughout the months of the year, although dry and wet periods are observed, depending upon the effectiveness of each factor. Some increase in orographic precipitation is observed in the northern highlands, but this may be counterbalanced by the proximity of the coastal regions to hurricane paths.

   b. The Thiessen method weights each station in proportion to the part of the watershed closest to it and is useful when there are few stations on or near the watershed and conditions in the area are reasonably homogeneous. When many stations are located on or near the watershed, isohyetal lines can be drawn on the map and the mean precipitation determined from these. The mean value of all Weather Bureau records in the neighborhood is the least satisfactory.

   c. Precipitation in the San Francisco area occurs chiefly during the winter months, when the prevailing winds from the Pacific are cooled to below the dew point in crossing the cooler land areas.

   d. In the Upper Mississippi Valley, cold fronts from Canada provide relatively dry air during the winter months and hence precipitation is light. During the summer, warm moisture-laden air from the Gulf moves up the valley. Because of high temperatures, its moisture is not dropped in the lower valley. With increased altitude and latitude, it is cooled. If cooled below the dew point, precipitation will occur during the crop-growing season. When the summer temperatures in this region are above normal, precipitation will not occur and droughts and dust storms have their inevitable effect.

3.33 *a.* 1. Biochemical reactions are accelerated at high temperatures so that oxidation and reduction are likely to occur in the sewer system rather than being delayed until the changes can occur under controlled conditions at the disposal plant. Also, the corrosion of iron, steel, and concrete is accelerated at high temperatures.
   2. The appearance of the stream will suffer. Spawning beds for fish may be destroyed. Reaeration will be retarded in the stream.
   3. The water will become turbid and unsightly. Sludge deposits will form, which may interfere with bottom life or navigation. Odors are likely to develop. Fish may be killed by clogging of their gills.
   4. The water will become turbid. Light penetration will be reduced, thus reducing photosynthesis. Fine organic solids will increase the BOD loading and cause a reduction in dissolved oxygen. Fish may be killed if the waste is toxic or if it injures the gills.
   5. Fire and explosion hazards may result. The waste will interfere with many phases of treatment-plant operation.
   b. 1. Heat-recovery systems may be shown to be of economic value to the company discharging the wastes. Otherwise the wastes may be cooled in holding basins or diluted with cold water.

2. Sawdust may be burned with waste lumber to provide steam power. It may be removed by fine screens or settling basins.
3. Lagoons may be used. Vacuum filters may be used for drying the sludge. The dried sludge may be burned, with lime recovered for reuse.
4. Usually chemical or biological treatment is required. Coagulation followed by flocculation and sedimentation will remove much of the solid matter. Treatment of the effluent in trickling filters, contact aerators, or activated-sludge tanks will result in further removal.
5. Oil and grease traps should be placed in waste lines. Additives may break up emulsions chemically. Coagulants and filter aids may be used.

3.34 *a.* Physical processes: dispersion, sedimentation, aeration, action of sunlight.

Chemical processes: oxidation, reduction, disinfection, coagulation.

Biological processes: aerobic and anaerobic bacteria oxidize and reduce the organic compounds; algae, fungi, protozoa, rotifera, crustacea, and larger forms of plant and animal life also play a part.

b. Hygienic considerations: contamination of water supplies, waters used for bathing, shellfish, ice; pollution that affects the public health in any way.

Economic considerations involve depreciation of the value of abutting property, damage to fish life, industrial water supplies, livestock that are watered from the polluted stream, river and harbor improvements and navigation, impairment of recreational facilities.

Legal considerations involve the right of other riparian owners to receive the water in a reasonable condition and to make reasonable use of it.

Aesthetic considerations are involved in the appearance and odor of the receiving stream.

Whenever any of the above is seriously affected by the disposal of raw sewage by dilution, either primary or primary and secondary treatment must be provided before dilution, the extent of the treatment depending upon the pollutional load and the ability of the receiving water to handle this load satisfactorily.

3.35 *a.* Racks, coarse and fine screens, grit chambers, preliminary and secondary plain sedimentation basins, grease traps, skimming tanks, chemical precipitation tanks

b. Intermittent sand filters, standard and high-rate trickling filters, contact aerators, the activated-sludge process and modifications thereof

   c. Chlorination

   d. Separate sludge-digestion tanks (preferably heated), Imhoff tanks, septic tanks (for use in rural areas)

   e. Sludge-drying beds (either covered or uncovered), sludge elutriation and thickening, vacuum filters, centrifuges, filter presses, kiln dryers, incinerators; disposal at sea, burial, lagooning; use of dried digested sludge as fill or as a soil conditioner

   f. Dilution, lagooning, irrigation

3.36 1. 1*b.*

   2. Ferric chloride, a coagulant used to precipitate and condition sludge.

   3. The sewer is being cleaned.

   4. To kill tree roots that have entered the sewer through cracks or poorly constructed joints.

   5. That a test was being made for the presence or absence of adequate oxygen in the sewer after it had been first determined that no explosive gases were present.

   6. Two feet per second.

   7. Upgrade.

   8. Manholes should be placed at every change in alignment, grade, size of pipe, and at intervals not to exceed 500 ft (152.4 m).

   9. The velocity in a circular sewer is a maximum at slightly above 0.8 depth since it varies directly with the hydraulic radius. At greater depths, the wetted perimeter increases more rapidly than does the area, and hence the hydraulic radius decreases. Thus the maximum discharge, the product of velocity and area, is obtained at about 0.93 depth, when both velocity and area are somewhat below their maximum values but when their product is a maximum.

   10. In the septic tank, sedimentation and sludge digestion take place in the same chamber. Rising masses of sludge, buoyed up by entrapped gas, interfere with the sedimentation process. The required detention period is therefore much longer than when the sedimentation and digestion processes are separated. Septic tanks are therefore uneconomical for municipal use. They are useful in rural disposal systems where there are no provisions for skilled operating personnel.

3.37 1. $\dfrac{5 \times 10^6}{4} \dfrac{6}{24} \dfrac{1.25}{7.48} = 52{,}500$ ft^3 (1487 m^3)

The BOD of the entering raw sewage at aeration tanks is equal to

$(5 \times 8.33 \times 300)0.70 = 8750$ lb (3969 kg)/day. Using four tanks,

$$\frac{8750/4}{38/1000} = \textbf{55,750} \text{ ft}^3 \ (\textbf{1579} \text{ m}^3) \qquad \textbf{1d}$$

$$\text{Area (using depth of say 15 ft) (4.57 m)} = \frac{55,750}{15}$$

$$= 3750 \text{ ft}^2 \ (348.4 \text{ m}^2)$$

Use 30×125 ft $(9.14 \times 38.10$ m) tanks.

2. Aeration capacity $= \dfrac{1.5 \times 5 \times 10^6}{24 \times 60}$

$\qquad\qquad\qquad\quad = \textbf{5200}$ ft^3 ($\textbf{147.3}$ m^3)/min $\qquad$ **2c**

3. Diffuser capacity $= \dfrac{8750 \times 1.5 \times 1000}{4(24 \times 60)}$

$\qquad\qquad\qquad\quad = \textbf{2280}$ ft^3 ($\textbf{64.6}$ m^3)/min $\qquad$ **3a**

3.38 Streeter-Phelps equation:

$$400 \text{ mgd} \times 1.55 = 620 \text{ ft}^3 \ (17.6 \text{ m}^3)/\text{s}$$

$$\text{Weighted BOD (5 day at 20°C)} = \frac{620 \times 150 + 2,000 \times 0.05}{2620}$$

$$= 35.9 \text{ ppm}$$

$$\text{BOD} = LA(1 - 10^{-k_1 t})$$
or $\qquad\qquad 35.9 = LA[1 - 10^{-(0.5)(5)}]$

Solving LA at 20°C $= 52.5$ ppm.

$$LA_{T°C} = LA \text{ at } 20°C(0.02T°C + 0.6)$$
hence $\qquad LA$ at 15°C $= 52.5(0.02 \times 15 + 0.6) = 47.25$ ppm

At 15°C,

Dissolved oxygen (D.O.) saturation concentration $= 10.15$ ppm

$$\text{Weighted D.O.} = \frac{2000 \times 10.15 + 0 \times 620}{2620} = 7.75 \text{ ppm}$$

Initial deficit $= 10.15 - 7.75 = 2.40$ ppm.

$$\text{Time in stretch} = \frac{80}{20} = 4 \text{ days flow downstream}$$

$$\text{Deficit (at 4 days flow)} = \frac{k_1 A}{k_2 - k_1}(10^{-k_1 t} - 10^{-k_2 t}) + \text{D.O.}(10^{-k_2 t})$$

or

$$\text{Deficit} = \frac{0.08(47.25)}{0.28 - 0.08}(10^{(-0.08)(4)} - 10^{(-0.28)(4)}) + 2.4(10^{(-0.28)(4)})$$

$$= \mathbf{7.80} \text{ ppm} \quad \mathbf{b}$$

3.39 $1a$; $2a$; $3b$; $4a$; $5a$; $6d$; $7d$; $8b$; $9b$; $10b$

3.40 Total hardness due to Ca^{2+}, Mg^{2+}, Sr^{2+} is:

Cation	Equivalent weight	Hardness (as mg/liter $CaCO_3$)
Ca^{2+}	20.0	$20(50/20) = 50.0$
Mg^{2+}	12.2	$15(50/12.2) = 61.5$
Sr^{2+}	43.8	$5(50/43.8) = 5.8$
		117.3

Alkalinity is:

Anion	Equivalent weight	Hardness (as mg/liter $CaCO_3$)
HCO_3	61.0	$5(50/61) = 4.1$
CO_3	30.0	$10(50/30) = 16.7$
		20.8

Carbonate hardness = **20.8** mg/liter as $CaCO_3$ ($20 + 30 = 50$). Noncarbonate hardness = $117.3 - 20.8 = $ **96.5** mg/liter as $CaCO_3$.

3.41 *a.* *Atterberg limits* comprise the liquid, plastic, and shrinkage limits. The liquid limit is the water contents at which the soil has such a small shearing strength that it flows to close a preformed groove of standard width when jarred in a specified manner. The plastic limit is the water content at which a soil begins to crumble when rolled into threads of a specified diameter. The shrinkage limit is the water content that is just sufficient to fill the pores when the soil is at its minimum volume attained by drying. These limits furnish a sound basis for identifying and classifying soils and predict qualitatively their general behavior when they are correlated with soil tests of a more specific nature. The information furnished by these tests may be misleading since, for the most part, they are performed on remolded soil.

For determining the liquid limit, a special standard-cup device is used. All the dimensions of the device are fixed, and the jarring is accomplished by turning the crank at the rate of 2 revolutions/s. The moisture content when the groove is closed by 25 blows of the cup is defined as the liquid limit.

To find the plastic limit, a thread of soil is rolled out on a glass plate to a diameter of ⅛ in (3.18 mm). When the moisture content has been reduced to the point where crumbling starts, the moisture content is measured.

The shrinkage limit is found by measuring the volume (by displacement of mercury) of a soil pat that has been dried in a desiccator. The volume is then used to determine volume of the voids remaining in the soil pat.

b. The *permeability test* determines the coefficient of permeability in Darcy's equation, $q = kiA$, for a given soil. In the constant-head permeameter, the net head of water on the soil sample of a known length and area is kept constant, and the rate of flow is measured. The permeameter tube or container holds the sample in place by screens or fitted porous stones. Before the test is run, it is necessary to rid the system of any entrapped air.

c. The *direct-shear test* measures the cohesion and angle of internal friction of a given soil under somewhat arbitrary conditions of loading rates, seepage conditions, and internal stresses.

The test is performed on a soil sample that is either an undisturbed sample or representative of the true soil condition being investigated. The equipment for the test consists principally of the following: (1) a split rectangular box for holding the specimen constructed so that the upper half of the box can move horizontally with respect to the lower half; (2) a jacking arrangement by which a given vertical pressure can be applied to the specimen; (3) a means of pulling the upper half of the box with respect to the lower half; (4) a means of measuring the shearing resistance of the soil to the horizontal shearing displacement; and (5) a means of measuring the change of thickness of the sample during the test. At the start of a particular run, the vertical pressure is applied and maintained constant. As shearing displacement is applied, the shearing resistance is measured by a proving ring dial and the vertical displacement by an Ames dial.

In cases where seepage takes place under actual conditions, the test is performed with porous stones or plates top and bottom of the specimen and the load applied slowly or quickly, depending upon whether consolidation takes place.

d. The *consolidation test* measures the time rate of consolidation for an undisturbed soil sample. From these measurements, the coefficient of consolidation, Cv, in $T = Cvt/H^2$ can be found.

A carefully prepared undisturbed soil sample is cut to fit into the consolidometer. The consolidometer is equipped with porous stones to allow seepage of the pore water. A constant vertical pressure is applied to the specimen, during which time its change in thickness is observed at various intervals. Several such runs are made with different soil pressures to include the pressures to be encountered in the actual soil.

3.42 *a*. Pour cinders into the water along the outside face of sheet piling. The flow of water will carry the cinders into the interlocks in the sheet piling and greatly reduce these leaks. The sump pumps will take care of the remainder of the leakage.

b. Using a mix of 1:1 sand and cement, "grout" the river bottom for about 10 ft (3.05 m) around the outside of the sheet piling. The mix is forced through a pipe by air pressure and the pipe is "air jetted" down through the ooze of the river bottom. Bags filled with sand and bags containing cinders should be stacked at convenient points around the cofferdam to be available if and when needed.

c. A pile should be completely driven before stopping for lunch, particularly when driving in a clay soil. If stoppage is permitted, redriving to completion may be very difficult because of the additional shearing resistance that the soil has acquired during the lunch interval (frictional take-up on the pile).

d. Fairly heavy boulders if the velocity of the stream is apt to be high.

e. Prefabricating timber that is to be pressure-creosoted allows every part of the completed member to be protected. The exposed timber in holes bored subsequent to creosoting could be treated by plugging one side of the hole and applying creosote from a pressure gun at the other side of the hole.

3.43 *a*. A relatively shallow trench dug in hardpan in dry weather would not require sheeting. In the case of a deep trench that might stand exposed for some time before backfilling, some bracing would be required because of the possibility of rain and water entering the trench, with the subsequent loss of cohesion in the soil.

b. Paint the steel with a coat of oil. Lead paints under the heat of the torch produce a gaseous vapor that may be poisonous to the welder.

c. Oiling the forms is beneficial to concrete since it prevents the surface from drying out and renders the concrete more impervious.

d. Gunite is a rich cement mortar applied by spraying under high air pressure.

e. The major advantage of ac over dc welding is in the reduction of arc blow. Alternating current usually increases welding speeds and produces a higher-quality weld. Where single-phase current is available, the ac current is more economical.

3.44

Weight of container plus dry soil	582.04 g
Weight of container	194.40 g
Weight of dry soil	387.64 g
Weight of dish plus dry soil	175.50 g
Weight of dish	118.77 g
Weight of dry soil	56.73 g
Weight of bottle plus water plus soil	732.22 g
Weight of bottle plus water	687.39 g
Submerged weight of soil	35.83 g
Weight of dry soil	56.73 g
Weight of submerged soil	35.83 g
Weight of displaced water	20.90 g

$$G_s = 56.73/20.90 = 2.72$$

Initial height of specimen $= H_i$
$$= 1.235 + (0.4200 - (0.4010)$$
$$= 1.254 \text{ in } (31.9 \text{ mm})$$

Cross-sectional area of specimen $= 0.7854 \times 4.23^2$
$$= 14.1 \text{ in}^2 \ (9059 \text{ mm}^2)$$

$$\text{Volume of solids} = \frac{387.6 \times 2.2 \times 1.728}{1000 \times 2.72 \times 62.4}$$

$$= 8.69 \text{ in}^3 \ (142{,}403 \text{ mm})$$

Height of solids $= H_s = 8.69/14.1 = 0.615$ in (15.6 mm)
Height of voids $= H_v = 1.254 - 0.615 = 0.639$ in (16.2 mm)
Initial voids ratio $= 0.639/0.615 = 1.04$

Pressure, kg/cm²	ΔH	$\Delta e = \Delta H/H_s$	Voids ratio e
0			1.040
¼	0.0069	− 0.011	1.029
½	0.0044	− 0.008	1.021
1	0.0077	− 0.012	1.009
2	0.0182	− 0.030	0.979
4	0.0401	− 0.065	0.914
8	0.0513	− 0.0845	0.830

(See Fig. 3.44.)

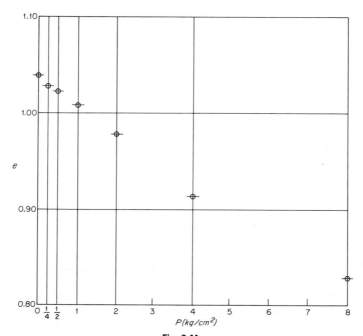

Fig. 3.44

3.45 For horizontally layered soil, use the Westergaard theory, and refer to D. W. Taylor, "Fundamentals of Soil Mechanics," Wiley, 1948, chart p. 260.

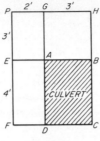

Fig. 3.45

Rectangle *PHCF* (Fig. 3.45)
  $M = \frac{7}{9} = 0.777$
  $N = \frac{5}{9} = 0.555$
  $K = 0.075$ (from chart)

Rectangle *PGDF*
  $M = \frac{7}{9} = 0.777$
  $N = \frac{2}{9} = 0.222$
  $K = 0.036$

Rectangle *PHBE*
  $M = \frac{5}{9} = 0.555$
  $N = \frac{3}{9} = 0.333$
  $K = 0.043$

Rectangle *PGAE*
  $M = \frac{3}{9} = 0.333$
  $N = \frac{2}{9} = 0.222$
  $K = 0.021$

$$K_{ABCD} = K_{PHCF} - K_{PHBE} - K_{PGDF} + K_{PGAE}$$
$$= 0.075 - 0.043 - 0.036 + 0.021 = 0.017$$

Load on culvert $= 7000$ lb $\times 0.017 = $ **119** lb (**540** kg)   **c**

3.46 *a*. The foundations may consist of *two unreinforced-concrete ped-estals* running down to bedrock. The top of the pedestal should have an area approximately 2 to 3 times as great as that of the column cross section. The compressive stress in the pedestal should be somewhat less than $0.25 f_c'$. The pedestal should be approximately twice as deep as the amount of flare or taper at each side of the column.

Dowels, whose area equals the area of the main steel in the column, should run down into the pedestal a distance sufficient to develop their full strength in bond. These dowels should not be hooked. A mat of rods should be used across the top of the pedestal. Tie rods should be used around the outside perimeter. Vertical rods should be used parallel to the four faces of the pedestal. This system of reinforcement prevents the upper corners of the pedestal from cracking loose.

Investigate punching shear around the perimeter of the column, using the full depth of the pedestal and an allowable stress equal to $0.05 f_c'$.

The shear force should be computed as that on the bottom of the pedestal outside the perimeter of the column.

Bending stress need not be investigated since the pedestal sits on a roughened rock surface and cannot elongate laterally. Hence no tensile stress may develop.

b. The foundations may consist of *two separate reinforced-concrete footings.* The footings can be placed directly under the column, or a pedestal can be used above the footing proper. Sometimes the use of a pedestal reduces the necessary size of footing enough to make the cost of the combination less than the cost of a footing alone.

The necessary bearing area of the spread footing, which is preferably made square, is determined by considering the soil's allowable bearing pressure and the weight of the footing, the column, and the live load minus the soil pressure that originally existed at that depth before construction.

Steel to be used in the bottom of the footing [3-in (76.2-mm) cover] may be computed on the basis of moment on a section through the face of the column or pedestal, multiplying the computed moment by 0.85 and using $M = A_s f_s jd$.

Investigation for punching shear seems unnecessary since failures from this cause in reinforced-concrete spread footings are virtually unknown.

Diagonal tension (shear) is computed at a section a distance d away from the face of the column or the pedestal toward the edge of the footing, where d is the effective depth of the footing. The shear V on this section is taken as the total pressure on the bottom of the footing acting between the critical section, the outside face of the footing, and two 45° lines drawn out from the center of the column. $v = V/bjd$, where b is the width of the critical section between 45° lines. v should not exceed $0.03f'_c$, assuming that the main reinforcement is hooked as it should be.

Bond should be checked, using the shear as the total pressure acting on an area bounded by the outside of the footing, the face of the column, or pedestal, and 45° lines from the column center. Bond stress should not exceed $0.05f'_c$.

c. A *combined footing* may be used. It should be so proportioned that the center of gravity of the loads and the center of gravity of the footing area coincide. This can be accomplished by making the footing rectangular and extending it beyond the heavier of the two columns for a greater distance than beyond the lighter or by making the footing trapezoidal in plan with the wider portion near the heavier column. The footing is designed in its length as an overhanging beam with two supports (the columns). Appropriate shear and moment diagrams are drawn, and reinforcing is placed according to the principles of reinforced-concrete beam design.

The short direction is designed as is an ordinary spread footing, taking critical sections as described in (b).

In the long direction, shears for computing diagonal tension are taken from the shear diagram.

Bottoms of all footings should be below the expected depth of frost penetration.

The variable depth to bedrock beneath the two columns may cause unequal settlement. Thus there would be a tendency for the combined footing to tilt. This should be considered when choosing among the three alternate possibilities.

3.47 a. *Wet unit weight:*

$$\text{Volume of sample} = V_t = 298.1/62.4 = 4.78 \text{ ft}^3 \ (0.14 \text{ m}^3)$$
$$\gamma_w = 587.4/4.78 = 122.8 \text{ lb/ft}^3 \ (1967 \text{ kg/m}^3)$$

b. *Dry unit weight:*

$$\text{Dry wt of sample} = \frac{103.6 \times 587.4}{112.4} = 540.5 \text{ lb} \ (245 \text{ kg})$$

$$\gamma_d = 540.5/4.78 = 113.2 \text{ lb/ft}^3 \ (1813 \text{ kg/m}^3)$$

c. *Moisture content:*

$$w = \frac{112.4 - 103.6}{103.6} \times 100 = 8.49 \text{ percent}$$

d. *Voids ratio:*

$$v_s = \frac{540.5}{2.67 \times 62.4} = 3.25 \text{ ft}^3 \ (0.0920 \text{ m}^3)$$

$$\text{Volume of voids} = v_v = v_t - v_s = 4.78 - 3.25$$
$$= 1.53 \text{ ft}^3 \ (0.0433 \text{ m}^3)$$

$$e = \frac{v_v}{v_s} = \frac{1.53}{3.25} = 0.471$$

e. *Porosity:*

$$n = \frac{v_v}{v_t} 100 = 1.53 \frac{100}{4.78} = 32.0 \text{ percent}$$

f. Saturation:

Wt water in original sample = $587.4 \times 8.8/112.4$
$$= 46 \text{ lb } (20.9 \text{ kg})$$

$$s = \frac{46 \times 100}{62.4 \times 1.53} = 48.2 \text{ percent}$$

g. Relative density:

$$D_R = \frac{(1/\gamma_{min}) - (1/\gamma_d)}{(1/\gamma_{min}) - (1/\gamma_{max})} 100 = \frac{(1/103.9) - (1/113.2)}{(1/103.9) - (1/117.4)} 100$$

$$= \frac{0.00963 - 0.00883}{0.00963 - 0.00852} 100 = 72 \text{ percent}$$

3.48 *a.* Interbedded fine silts and coarse sands with possibly gravels at depth. High groundwater possible. General absence of coarse gravels but plenty of clean sands and fine gravels. Bridges will require deep and expensive foundations.

b. Subsoils often wet and plastic. Beach ridges will be sandy and gravelly in texture. Transition may require special care because of differential settlement. Adequate drainage will be a major problem. Material is difficult to handle when wet. High volume changes will occur as the moisture content changes.

c. The low areas probably will have poor internal drainage. Better on the higher slopes. Soil probably clay with high organic content. Excavation will be easy. Borrow is probably available. Frost action can be expected. Limestone honeycombing possible. Till will reflect presence of limestone base.

d. Generally quartz sand in dunes. Excavation is easy but excessively rough on equipment. Probably dunes will have to be stabilized.

e. Excavation will be easy. Excellent source of borrow. Highly stable material and makes good road base and foundation material.

3.49 Assume $\phi = 35°$. From design charts obtain $D/h = 0.71$. Then, depth of sheet pile penetration = $0.71 \times 15 = \textbf{10.7}$ ft (**3.26** m).
1a Also

$$\frac{M_{max}}{M'} = 1.58 \quad \text{or} \quad M_{max} = 1.58M'$$

$$p_a = K_a\gamma$$

where $K_a = \dfrac{1 - \sin \phi}{1 + \sin \phi} = 0.27$

Using $\gamma = 125$ lb/ft^3 (2002 kg/m^3), $p_a = (0.27)125 = 34$.

$$M_{\max} = 1.58 \times 5 \times 0.5 \times 15^2 \times 34$$
$$= \mathbf{30{,}218} \text{ ft·lb } \mathbf{(40{,}976 \text{ N·m})} \qquad \mathbf{2a}$$

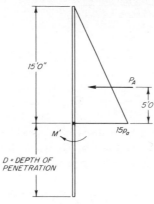

Fig. 3.49

This is a specialty problem. If a "USS Steel Sheet Piling Design Manual" or the book "Foundation Design," Wayne C. Teng (Prentice-Hall), is available, the reader will find the problem worked out in detail.

3.50 *a.* A balanced design is one in which both the concrete and the steel are so proportioned as to work to their full working stresses when the member carries its full allowable load.

b. The transformed section is one in which the flexure steel is conceived to be replaced by a large area of imaginary concrete that can take tension. This gives a homogeneous section of concrete to which ordinary beam analyses may be applied.

 c. 1. In a new design the neutral axis is located by the allowable working stresses in the steel and the concrete.

  2. In an investigation of an existing beam, the neutral axis is located at the centroid of the existing transformed area.

d. The fiber stress in a beam is assumed to be proportional to the distance of the fiber from the neutral axis.

e. Fireproofing in columns should consist of at least 1½ in (38.1 mm) of concrete or a thickness of concrete equal to 1½ times the maximum size of the coarse aggregate. Where the concrete is to be exposed to weather or be in contact with the ground, the fireproofing shall be not less than 1½ in (38.1 mm) for bars ⅝ in (15.9 mm) or less in diameter.

f. In a T beam, the shear is assumed to be carried by an area whose width is equal to the width of the stem and whose depth is equal to the depth of the T beam.

g. The stirrups prevent diagonal tension cracks.

h. Stirrups are not needed in slabs because the depth of the slab is so proportioned that the intensity of shearing stress is kept below the critical limit, which indicates that diagonal tension would require stirrups.

i. Diagonal tension is computed by a formula for the shearing stress, but the failure to be guarded against is one of diagonal tension and not shear. Concrete is rarely used in a manner that would subject it to shearing failure, and thus shear is only used as a measure of tensile stress.

j. Three in (76.2 mm) of concrete should intervene between the reinforcement and the ground surface of the footing.

3.51 *List 1*   *List 2*

1	b
2	e
3	a
4	c
5	d
6	f

3.52 *a.* A sole plate is a steel plate attached to the lower flange of the girder below the end stiffeners and distributes the girder reaction to the masonry bearing plate. It shall have a thickness of not less than ¾ in (19.1 mm) and not less than the thickness of the flange angles plus ⅛ in (3.2 mm). Preferably it shall not be longer than 18 in (457 mm).

b. 1. Stiffeners shall be placed at all points of end bearing and bearing of concentrated loads. According to the American Railway Engineering Association, "if the depth of the web between the flanges or side plates of a plate girder exceeds 60 times its thickness, it shall be stiffened by pairs of angles riveted to the web."

2. The clear distance between stiffeners shall not exceed 72 in (182.9 cm).

3. The clear distance between stiffeners shall not exceed $\dfrac{10{,}500t}{\sqrt{S}}$

where t = thickness of web, in
S = unit shearing stress, gross section, in web at point considered

Other specifications have similar requirements.

c. Stiffener angles at points of concentrated load shall have milled ends.

d. One-eighth of the web may be included in the net area of the flange, and one-sixth of the web may be used for the gross area.

e. Anchor bolts shall be not less than 1¼ in (31.8 mm) in diameter. Anchor bolts not carrying uplift shall extend 12 in (30.48 cm) into the masonry. Those carrying uplift shall be designed to engage a mass of masonry whose weight is 1½ times the uplift.

f. Combined stresses include dead-load stresses, live-load stresses, impact stresses, wind stresses, snow stresses, ice stresses, etc. in such a combination as to give both maximum negative and maximum positive effect.

g. Alternate stresses are those used to design a member that goes through a reversal of stress. The alternates would be the maximum positive and the maximum negative stresses.

h. A gusset plate connects the various members of a truss that come together at a joint. It is made large enough to include the necessary number of connection rivets.

i. A filler plate is used at truss joints to compensate for any difference in the depth or thickness of the members being joined so that connection can be made to a flat gusset plate.

j. Where the open side of truss members are joined by latticing, stay plates are used as near the end of the member as possible and at intermediate points where the lacing is interrupted. The stay plates lend a little additional rigidity to the member.

3.53 a. Fills may be compacted by: (1) jetting, which involves forcing water into the fill under pressure using nozzles on high-pressure hoses and relying on the downward pressure of the water to align the grains in a dense condition; (2) tamping, which may be accomplished using hand tampers or pneumatically or gasoline-driven vibratory tampers and compacting in layers; and (3) rolling in layers using a sheep's-foot roller and a number of passes for each layer. Jetting's advantage is that it can be used for relatively thick layers and accomplished rather quickly. The disadvantage is that it introduces large quantities of water into the embankment, which greatly decreases the embankment's strength and stability. Hand tamping when properly executed is every effective, but it is expensive and lengthy. Rolling using a sheep's-foot roller is probably the most effective method for getting adequate compaction on jobs of fairly large extent. Its disadvantage is that its effectiveness is not quite as great with noncohesive as with cohesive soils.

b. Shrinkage means that a cubic foot of soil in its natural state will not yield a cubic foot as artificially compacted in the embankment because of greater relative density in the embankment and incidental losses as the material is transported from one place to the next. The amount of shrinkage is determined by the comparative relative densities from the natural state to the embankment and the amount of material lost in transit. A shrinkage factor must be used in balancing cuts and fills or there will not be sufficient material to complete the embankment. To compensate for the shrinkage, all fill quantities are multiplied by an amount equal to $1/(1 - S)$, where S = shrinkage factor expressed as a ratio.

c. The temporary pavement should be a bituminous macadam. Lay a 6- to 8-in (15.24- to 20.32-cm) foundation of crushed stone ranging from 4 to 1 in (10.16 to 2.54 cm) in size. Roll with a 10-ton (9072-kg) three-wheel roller, moving longitudinally with subsequent passes from the edge of the pavement toward the center. Then sweep in choke stone and continue to roll, wetting down the surface as you go. The wearing course should consist of broken stone ranging in size from 2½ to 1 in (6.35 to 2.54 cm). The depth of the wearing course is about 4 in (10.16 cm). This course is rolled just as the base course was rolled. High-penetration asphalt is then applied at the rate of 2 gal/yd² (9.06 liters/m²). Choke stone is then spread immediately on the surface and broomed and rolled in. Then a seal coat of asphalt is applied at the rate of ¾ to 1 gal/yd² (3.40 to 4.53 liters/m²). Sand or cover stone is then applied and rolled into the surface. All excess is swept off, and the surface is ready for use. The strength and durability of this pavement depend partly upon the subgrade and partly upon the waterbound macadam base course used. This type of pavement was recommended because it is relatively inexpensive, does not require a curing period before it can be used, and is flexible so that it can withstand a certain amount of settlement in the newly compacted approach embankment.

3.54   Referring to Fig. 3.54, a base AB is taped along one bank, the quadrilateral $ABCD$ is triangulated and tied into the bridge centerline EF. CD is measured as a check base and the control net is adjusted by least squares.

With the design stationing of E, F, and the centerlines of piers 1 to 5 known, we may set range flags on shore at G, H, J, K, L, M, N, and O, drive batter frames upstream from each pier as shown at pier 2, and compute the necessary angles to be laid off from the bases on either shore to set P and Q on these batter frames on the centerline of pier 2. Then

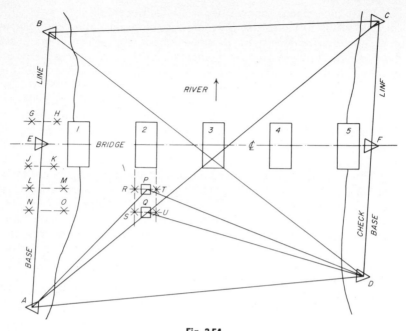

Fig. 3.54

range flags can be set at R, S, T, and U, to permit driving the sheet piles for the long side of the pier-2 cofferdam. A similar procedure is applied to all other piers.

With the cofferdams in place, the system of triangulation is applied to the center of each cofferdam in turn to permit the establishment of points controlling the placing of forms, stone facing, and arch hinges.

3.55 Such a traverse should be executed using taping bucks or stools since stakes cannot be driven. A minimum of three bucks is required. Tape measurements, each substantially the full length of the tape where possible, should be made from the end monument to mark on the first buck and then carried successively from buck to buck. A tension balance and thermometers should be provided. Where a long tape (200 ft or 50 m) is used, a center support, adjusted to grade by eye, may be necessary.

A transit party consisting of an instrument operator and one helper is required to align the bucks as they are successively moved from the rear to the head of the line. A level operator and rod person must establish the differences in elevation between each pair of bucks to permit computation of the inclination corrections. The taping party will normally consist of front and rear chain operators, front and rear stretcher operators, and a recorder who may also hold the center support when needed. Necessary corrections include standardization, temperature, slope, and sag.

4.01–4.55

Electrical Engineering

4.01 *References:* "I.E.S. Lighting Handbook," 1947.

Krachenbuehl, "Electric Illumination," Wiley, 1951.

Sharp, "Introduction to Lighting," Prentice-Hall, 1951.

For parking lots, recommended illumination is 1 to 2 fc (10.764 to 21.528 lm/m²). For a downtown area assume 2 fc (21.528 lm/m²). A broad-beam (30° or more) general-service type of floodlight is indicated for this area of 200 × 300 = 60,000 ft² (60.96 × 91.44 = 5574.2 m²). Select a 500-W floodlight unit having initial beam lumens = 3400. Assume a depreciation factor = 0.7.

$$\text{No. of units} = \frac{\text{ft (lm/m}^2) \times \text{area}}{0.7 \times \text{initial beam lumens}}$$

$$= \frac{2 \times 60,000 \ (\text{or } 21.528 \times 5574.2)}{0.7 \times 3400} = 50$$

Refer to table 8-8, "I.E.S. Lighting Handbook," 1947, p. 8–27. For a 25-ft (7.62-m) mounting height = D, assume Z = 40 ft (12.19 m) for a 40° beam. The coverage of one unit = 2300 ft² (213.7 m²). 60,000 (5574.2 m²)/50 = 1200 ft² (111.5 m²), minimum coverage required, so the 50 (500-W) units will give ample uniformity. The 50 units could be mounted five to a pole on 10 poles spaced 60 ft (18.3 m) along the two long sides of the lot. A more efficient but less diversified arrangement could be made by using 1000-W general-service lamps in 18- to 24-in (45.72- to 60.96-cm) broad-beam reflectors which would give 8900 initial beam lumens.

$$\text{No. of 1000-W units} = \frac{2 \times 60{,}000 \ (\text{or } 21.528 \times 5574.2)}{0.7 \times 8900}$$

$$= 20$$

Area to be covered by one unit = 60,000/20 = 3000 ft² (278.7 m²). Refer to table 8-8, "I.E.S. Lighting Handbook," p. 8–27. For D = 25 ft (7.62 m) and Z = 50 ft (15.24 m), a 35° beam would cover 3270 ft² (303.8 m²). The 1000-W units could be mounted in groups of two on 10 poles spaced 60 ft (18.29 m) along each long side of the lot.

4.02 *Reference:* "I.E.S. Lighting Handbook," 1947, pp. 8–38 to 8–40.
The formula for this point-by-point calculation is

$$E_h = (I/h^2) \cos^3 \theta_h$$

where E_h = the horizontal illumination in footcandles (lm/m²) at a point P in a plane h ft (m) below a point source of intensity I candlepower (lm) in the direction of P, and θ_h = the angle between a vertical line at P and the line connecting P and the source.

1. At the center, the contribution of one lamp is

$$E_{h_1} = (800/10^2) \ (\text{or } 800/3.048^2) \cos^3 21.8°$$

(where $\tan \theta_h = {}^4\!/_{10}$ and θ_h = 21.8°)

$$= 6.4 \text{ fc } (68.9 \text{ lm/m}^2)$$

The contribution of the other lamp is the same, so at the center of the table (both ways) E_h = 2 × 6.4 = **12.8** fc (**137.8** lm/m²).   **1a**

2. At the center of either end of the table from the nearer lamp, E_h = 6.4 fc (as in part 1), but from the farther lamp, E_h = (800/10²) cos³ 50.2° = 2.1 fc. The resultant E_h = 6.4 + 2.1 = **8.5** fc (**91.5** lm/m²).   **2b**

3. At any corner, E_h = 8 cos³ 26.5° + 8 cos³ 51.2° = 5.6 + 2.0 = **7.6** fc (**81.8** lm/m²).   **3c**

4.03 *Reference:* "I.E.S. Lighting Handbook," p. 6–10.
Take efficiency of a 750-W PS Mazda lamp as 20 lm/W. Assume the same luminaire efficiency and depreciation factor for incandescent and fluorescent lamps:

The power ratio of fluorescent to incandescent for 20 fc = ${}^{20}\!/_{42}$. If the illumination is raised to 30 fc, ${}^{30}\!/_{20}$ more lumens will be required. The ratio of power for a change would be ${}^{20}\!/_{42} \times {}^{30}\!/_{20}$ = 0.715, and the power-cost saving would be (1 − 0.715)100, or **28.5** percent.   **d**

4.04 No. "The National Electric Code," vol. V, 1949, table 4, p. 326, specifies that a 1-in (25.4-mm) conduit be used for two or three #6 (5.11-mm diam.) wires.

4.05 *a.* $P_{12} = \dfrac{1}{R^2 + X^2} (R|\tilde{V}_1|^2 - R|\tilde{V}_1| \, |\tilde{V}_2|\cos \delta + X|\tilde{V}_1| \, |\tilde{V}_2|\sin \delta)$

$\qquad = \dfrac{1}{5^2 + 40^2} [5(345)^2 - 5(345)(360)\cos 10°$

$$+ 40(345)(360)\sin 10°]$$

$\qquad = 520.8 \text{ MW}$

$\quad Q_{12} = \dfrac{1}{R^2 + X^2} (X|V_1|^2 - X|\tilde{V}_1| \, |\tilde{V}_2|\cos \delta - R|\tilde{V}_1| \, |\tilde{V}_2|\sin \delta)$

$\qquad = \dfrac{1}{5^2 + 40^2} [40(345)^2 - 40(345)(360)\cos 10°$

$$- 5(345)(360)\sin 10°]$$

$\qquad = -147.3 \text{ Mvar}$

$\quad P_{21} = \dfrac{1}{5^2 + 40^2} [5(360)^2 - 5(124{,}200)\cos 10°$

$$- 40(124{,}200)\sin 10°]$$

$\qquad = -508.5 \text{ MW}$

$\quad Q_{21} = \dfrac{1}{1625} [40(360)^2 - 40(124{,}200)\cos 10°$

$$+ 5(124{,}200)\sin 10°]$$

$\qquad = 245.7 \text{ Mvar}$

b. $P_{\text{loss}} = P_{12} + P_{21} = 520.8 - 508.5 = \mathbf{12.3} \text{ MW}$
$\quad Q_{\text{loss}} = Q_{12} + Q_{21} = -147.3 + 245.7 = \mathbf{98.4} \text{ Mvar}$

4.06 Rated current, flowing before the fault, is

$$I_L = P_L/E_L = 10{,}000/250 = 40 \text{ A}$$

The generated emf under these conditions is

$$E_g = E_L + I_0 R_a = 250 + 40 \times 0.50 = 270 \text{ V}$$

On short-circuit, it is assumed that E_g, the inductance L, and the armature resistance R_a remain constant. The circuit resistance becomes $R = R_a + R_{sc} = 0.5 + 0.2 = 0.7$ ohm. The equivalent circuit from $t = 0$, when the fault occurs, until the breaker opens is 0.02 s.

The current I_0 is the initial current flowing in the inductive circuit at

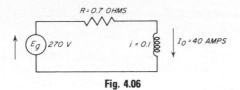

Fig. 4.06

$t = 0$, and the current $I = E_g/R = 270/0.7 = 386$ A is the current which will flow eventually if the breaker never operates. The transient equation for this case is

$$i = I - (I - I_0) \exp(-Rt/L) \qquad \text{for } 0 \le t \le 0.02$$

The maximum current withstood by the armature will occur at $t = 0.02$ s:

$$I_{max} = 386 - (386 - 40) \exp(-0.7 \times 0.02/0.1)$$
$$= 386 - 346 \times 0.869 = 386 - 301 = \mathbf{85} \text{ A} \qquad \mathbf{b}$$

4.07 At the instant the field is disconnected from the line, the circuit will be as in Fig. 4.07, where R_d is the field-discharge resistance, and by Ohm's law, $R_f = 120/8.4 = 14.3$ ohms, is the field resistance. Because of the field inductance L_f, the field current cannot change instantaneously; hence it remains at 8.4 A.

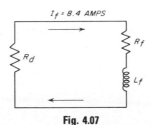

Fig. 4.07

The voltage across the two resistances will be equal to the induced voltage across L_f, and this value must be kept to one-third of the standard high-potential test voltage of 1000 plus twice the voltage rating of the machine. We then have the voltage equation:

$$I_f(R_d + R_f) = \frac{1000 + 2 \times 120}{3} = 413.3$$

whence $R_d = \dfrac{413.3 - I_f R_f}{I_f} = \dfrac{413.3 - 120}{8.4} = \mathbf{34.9}$ ohms   **c**

Note that the values of the field inductance and the field resistance were not required for a solution.

4.08 From the first measurement we have (1) $3C_g = 0.5$ where $C_g =$ the capacitance of each conductor to the sheath. From the second measurement we have (2) $2C_m + C_g = 0.6$ where $C_m =$ the capacitance between any two conductors. From (1), $C_g = 0.5/3 = 0.167$ μF. Substituting (1) in (2), $C_m = (0.6 - 0.167)/2 = 0.217$ μF. We thus have a balanced Y load of three 0.167-μF capacitors and a balanced Δ load of three 0.217-μF capacitors. By a ΔY impedance transformation, the balanced Δ is equivalent to a balanced Y having capacitors of $0.217 \times 3 = 0.651$ μF each.

The charging current is thus due to an equivalent balanced Y load each leg of which consists of a capacitor of $0.167 + 0.651 = 0.818$ μF.

Then the charging current per phase is

$$I_L = I_\phi = E_\phi B_\phi = E_\phi(2\pi fC) = \frac{25{,}000 \times 377 \times 0.818}{\sqrt{3} \times 10^6}$$

$$= \textbf{4.45 A} \qquad \textbf{d}$$

4.09 In the new installation we tolerate total losses in the amount

$$(0.8)(0.5/100)(500) = 2000 \text{ W}$$

New core losses: the voltage is the same but the frequency is only $50/60 = 0.833$ per unit.

$$P_{ec} = K(f\phi)^2 = KE^2 = 625 \text{ W (same as given)}$$

$$P_{\text{hyst.}} = K_1 fB^{1.6} = K_1 \frac{(fB)^{1.6}}{f^{0.6}} = \frac{K_1 E^{1.6}}{f^{0.6}} = \frac{K_2}{f^{0.6}}$$

$$= 625 \text{ W} \frac{1}{0.833^{0.6}} = 700 \text{ W}$$

Permitted new copper loss:

$$2000 - (700 + 625) = 675 \text{ W}$$

The old copper losses were related to the current as follows: $1250 = I^2R$. In the new installation, $675 = I^2R$. Thus

$$I^2 = \frac{675}{1250} = 0.735 \text{ per unit}$$

$$\text{kVA} = (0.735)(500) = 370 \text{ kVA}$$

New rating: 50 Hz, 1000/500 V, **370 kVA**.   **a**

4.10   *a.* $N_{600} = \dfrac{600 \times 200}{2400} = 50$ turns

$N_{240} = \dfrac{240 \times 200}{2400} = 20$ turns

b, c. $I_p = \dfrac{200,000}{2400} = 83.3$ A

d. $I_{600} = \dfrac{100,000}{600} = 166$ A

$I_{240} = \dfrac{100,000}{240} = 416$ A

e. $I_p = A\underline{/\alpha°} + B\underline{/\beta°}$

$-\dfrac{600 \times 166}{2400}\underline{/-45.6°} + \dfrac{416 \times 240}{2400}\underline{/0°}$

$= 29.5 - j29 + 41.6$

$I_p = 76.6$ A

4.11 $I_R = 15,000/2300 = 6.52$ A

$Z_H = \dfrac{V}{I} = \dfrac{65}{6.52} = 9.96$ ohms

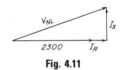

Fig. 4.11

$R_H = \dfrac{P}{I^2} = \dfrac{350}{(6.52)^2} = 8.22$ ohms

$X_H = \sqrt{Z_H^2 - R_H^2} = 5.65$ ohms

1. $Ix = 6.52 \times 5.65 = 36.8$ V
   $IR = 6.52 \times 8.22 = 53.6$ V
   $V_{NL} = \sqrt{(2300 + 54)^2 + (37)^2} = 2354$ V

   Regulation $= \dfrac{V_{NL} - V_{FL}}{V_{FL}} \times 100$

   $= \dfrac{2354 - 2300}{2300} \times 100 =$ **2.34** percent   **1c**

2. Maximum efficiency occurs when $I^2R = CL$
   $I^2R = I^2 \times 8.22 = 245$ W
   $I = 5.45$ A
   Power out (pf $= 1.0$) $= 2300 \times 5.45 = 12.5$ kW (neglecting regulation)

$$\text{Maximum efficiency} = \frac{\text{power out}}{\text{power in}}$$

$$= \frac{12,500}{12,500 + 245 + 245}$$

$$= \frac{12,500}{12,990} = 0.962$$

$$= \mathbf{96.2} \text{ percent} \qquad \mathbf{2b}$$

4.12 Assuming rated voltage out:

$$Z_{\text{rated}} = 2200/45.5 = 48.5 \text{ ohms}$$
$$Z_{A_{\text{po}}} = 72/2200 = 0.0326$$
$$R_{A_{\text{po}}} = 1000/100,000 = 0.01$$
$$\cos \theta_A = 0.01/0.0326 = 0.307 \qquad \theta_A = 72.1°$$
$$Z_{B_{\text{po}}} = 75/2200 = 0.0341$$
$$R_{B_{\text{po}}} = 1100/100,000 = 0.011$$
$$\cos \theta_B = 0.011/0.0341 = 0.322 \qquad \theta_B = 71.1°$$

$$\dot{V} = \dot{E}_1 + \dot{I}_1 Z_1 = \dot{E}_2 + \dot{I}_2 Z_2$$

$$\dot{E}_1 = \dot{E}_2 \text{ and } \dot{I}_1 + \dot{I}_2 = \dot{I}_T = 2\underline{/-37°}$$

$$E_1 + \dot{I}_1(0.0326\underline{/72.1°}) = 1$$

$$E_1 + \dot{I}_2(0.0341\underline{/71.1°}) = 1$$

$$\dot{I}_1 = \frac{0.0341\underline{/71.1°}}{0.0326\underline{/72.1°}} \dot{I}_2 = (1.04\underline{/-1°})\dot{I}_2$$

$$(1.04\underline{/-1°})\dot{I}_2 + \dot{I}_2 = 2\underline{/-37°}$$

$$\dot{I}_2(1.04 - j0.017 + 1) = 1.6 - j1.2$$

$$\dot{I}_2 = \frac{2\underline{/-37°}}{2.04\underline{/-0.5°}} = 0.98\underline{/-36.5°}$$

$$\therefore |I_2| = 0.98PO = \mathbf{98} \text{ percent rated.} \qquad \mathbf{1a}$$

$$\dot{I}_1 = (1.04\underline{/-1°})\dot{I}_2 = (1.04\underline{/-1°})(0.98\underline{/-36.5°}) = 1.02\underline{/-37.5°}$$
$$\therefore |I_1| = 1.02PO = \mathbf{102} \text{ percent rated.} \qquad \mathbf{2c}$$

4.13 Since unit 2 has a smaller per unit Z on its own base, it will be the limiting transformer. Let unit 2 then pass only **30** MVA.　　**a**

$$\text{MVA}_1 = \text{MVA}_2 \frac{Z_2}{Z_1}$$

where $Z_2 = j0.07(^{45}\!/_{30}) = j0.105$ per unit on 45-MVA base.

$$\text{MVA}_1 = 30 \frac{0.105}{0.08} = \textbf{39.4 MVA}　\textbf{b}$$

$$\text{Total MVA} = 30 + 39.4 = \textbf{69.4 MVA}　\textbf{d}$$

4.14 Although the power could be transmitted at 11,000 V, it would be preferable to increase the transmission voltage to about 33 kV. Transformers at each end would then be required. Assuming the use of Y-Y banks, the line voltages and turn ratios are shown in Fig. 4.14, with each bank rated at 3000 kVA.

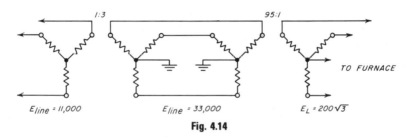

Fig. 4.14

At 33,000 V each wire will have

$$I_{\text{line}} = \frac{VA}{\sqrt{3} \times E_{\text{line}}} = \frac{3000 \times 10^3}{\sqrt{3} \times 33 \times 10^3} = 52.5 \text{ A}$$

The allowable loss for each of the three wires is

$$P_{\text{loss/wire}} = P_{\text{phase}} \times 0.05 = \frac{3000 \times 10^3 \times 0.8}{3} \times 0.05 = 40,000 \text{ W}$$

Hence the maximum allowable resistance per wire per mile (where d is the length of the transmission line in miles) is

$$R_{\text{wire mile}} = \frac{P_{\text{loss/wire}}}{I_L^2 \times d} = \frac{40,000}{(52.5)^2 \times 10}$$

$$= 1.45 \text{ ohms/mi (1.6093 km)}$$

This requirement may be met with selection of #4 (5.19-mm diam) AWG standard annealed copper wire having 1.34 ohms/mi at 25°C. On wooden poles with 250-ft (76.2-m) span, a minimum spacing of 5 ft (1.52 m) between wires should be allowed. The resistance drop in this line is $IR = 52.5 \times 14.5 = 760$ V. Tables for this size wire and spacing give 0.80 ohm reactance per mile per wire at 60 cps. Hence, the reactive drop for 10 mi is $IX = 52.5 \times 8.0 = 420$ V. Therefore, the total drop (neglecting charging current) is

$$IZ = \sqrt{(IR)^2 + (IX)^2} = \sqrt{(760)^2 + (420)^2} = 870 \text{ V}$$

This is only 4.6 percent of $33,000/\sqrt{3}$ V and is completely negligible for a furnace application. Regulation of generator and transformer have not been included.

4.15 Lower transmission voltages result in larger wire size or in high resistance and low efficiency. On the other hand, higher transmission voltages require more expensive tower and protective equipment, but the larger spacing required results in higher inductive reactance. Furthermore, for economical pole spacing, there is a minimum size wire for sufficient mechanical strength. In this problem construction costs are neglected.

$$I_{\text{line}} = \frac{P_{\text{delivered}}}{\text{pf} \times \sqrt{3} \times E_{\text{line}}} = 25,000 \times \frac{10^3}{0.85 \times \sqrt{3} \times E_{\text{line}}}$$

The allowable voltage drop and power loss in each wire are, respectively,

$$IZ_{\text{per wire}} = 0.08 E_{\text{phase}} = 0.08 \times E_{\text{line}}/\sqrt{3}$$
$$P_{\text{loss/wire}} = 0.06 P_{\text{phase}} = 0.06 \times 25,000 \times 10^3/3 = 500 \times 10^3 \text{ W}$$

Based on allowable power loss only, the allowable resistance of wire per mile (1.6093 km) is, where $d = 25$ mi (40.23 km),

$$R_{\text{wire mile}} = \frac{P_{\text{loss/wire}}}{d \times I_{\text{line}}^2} = \frac{500 \times 10^3}{25 \times I_{\text{line}}^2}$$

For the three available transmission voltages, the values are tabulated below per wire:

kV	I_{line}, A	P_{loss}, kW	R_{mile}, ohms	Wire size, AWG
12.5	1360	500	0.0108	
66.0	256	500	0.303	4/0
115.0	143	500	0.970	2

The last column gives the nearest standard copper-wire size, taken from wire tables at 25°C, having a resistance a trifle lower than the allowable ohms per mile (1.6093 km). On this basis, the 12.5 kV is immediately eliminated as requiring impracticable size (large) wire.

In determining IZ drop, as this line is short, neglect capacitance (charging current). The inductive reactance depends on spacing, and minimum spacing for safety and minimal corona loss depend on voltage. Consider the data in the following table (at sea level):

kV	Wire size, AWG (mm diam)	R, ohms/mi	Min. allowance spacing, ft	X, ohms/mi
66	4/0 (11.68)	0.264	9	0.770
115	2 (6.54)	0.840	60	
115	2/0 (9.27)	0.420	20	0.895

None of these choices satisfies the regulation requirement. The 60-ft (18.29-m) spacing (approx.) is too great even to consider the reactance. Choosing a larger wire size, for example, #2/0 as shown in the third line of the table, gives $Z_{\text{wire mile}} = \sqrt{R^2 + X^2} = 0.990$ ohm, and the drop per wire is then $(IZ)_{\text{drop/wire}} = I_{\text{line}} \times Z_{\text{wire mile}} \times d = 143 \times 0.990 \times 25 = 3550$ V which satisfies with ample margin both the loss and regulation requirements.

4.16 If all reactive kVA are supplied by the synchronous condenser, 5000 kW, including the 300-kW loss in the condenser, may be added.

a. This leaves 4700 kW for the new motors. At 80 percent pf, this means 4700/0.18 = **5875** kVA in motors.

b. For a pf of 0.8, the reactive factor is 0.6, and 0.6 × 5875 = 3525 kvar in the new motors. Total kvar to be supplied by the condenser are 8667 + 3525 = **12,192** kVA = condenser rating.

c.

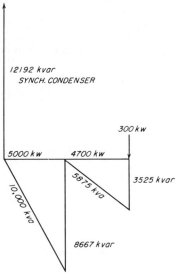

Fig. 4.16

4.17

(1)	(2)	(3)	(4)	(5)	(6)	(7)	(8)	(9)
	Load			Input	Factors		Input	
Rating	hp	kW	Effi-ciency	kW	Power reactance		kVA	kvar
(Given)	(Given)	0.746 × (2)	(Given)	(3)/(4)	(Given) cos θ	sin θ	(5)/(6)	(7) × (8)
1–50 hp	30	22.4	0.68	26.05	0.70	0.715	37.20	26.60
1–100 hp	75	56.0	0.89	63.00	0.80	0.60	78.75	47.20
2–15 hp	15	11.2	0.92	12.20	0.85	0.527	14.35	7.56
1–300 hp	310	231.5	0.92	252.00	0.85	0.527	296.00	156.00
1–33 kW	—	33.0	1.00	33.00	1.00	0	33.00	0
1–500 hp	{ (×0.746) 300	373 224	1.000 0.925	373.00 242.00	0.80	0.60	459.35 466.0	−237.36 +280.00
Total				628.25				42.64

This combination can be adjusted to unity pf where kW = kVA. Otherwise

$$kVA = \sqrt{(kW)^2 + (unbalanced\ kvar)^2}$$

The above figures indicate almost unity power factor so that kVA in this case will be very close to 629.5 kVA.

4.18

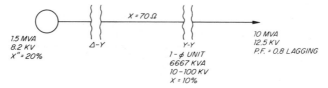

Fig. 4.18a

Base chosen: 10 MVA, 12.5 kV in the load circuit.

$$Z_{base}(\text{for H.T. line}) = \frac{125^2}{10}$$

$$X_{line} = \frac{70}{(125^2/10)} = 0.0448\ \text{per unit}$$

$$X_{\text{transf.}} = 0.1\,\frac{1}{3 \times 6.667}\left(\frac{\sqrt{3} \times 10}{12.5}\right)^2 = 0.0096$$

$$X_{\text{gen.}} = 0.2\,\frac{10}{15}\left(\frac{8.5}{7.22}\right)^2 = 0.1845$$

where $\qquad\qquad \tilde{V}_{\text{base}(G)} = \dfrac{125}{\sqrt{3}}\dfrac{1}{10} = 7.22\ \text{kV}.$

$$I_L = \frac{1}{0.8 + j0.6} = 0.8 - j0.6 = 1\underline{/-36.8°}$$

$$\tilde{V}_t = 1.0 + (0.8 - j0.6)(j0.0448 + 2Xj0.0096)$$
$$= 1.0384 + j0.0512 = 1.04\underline{/2.8°}\ \text{per unit}$$

Generator line-to-line voltage $= 1.04 \times 7.22$ kV per unit
$$= \textbf{7.79}\ \text{kV}\qquad \textbf{d}$$

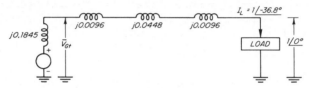

Fig. 4.18*b*

4.19   The propagation constant per loop mile (1.6093 km) for the transmission line with uniformly distributed parameters is

$$\gamma = \alpha + j\beta = \sqrt{(r + j1\omega)(g + jc\omega)}$$

At a frequency of 796 cps, $\omega = 2\pi f = 5000$

$$\gamma = \sqrt{(10.44 + j18.3)(0.300 + j41.9)(10^{-6})}$$
$$= \sqrt{21.1\underline{/60.2°} \times 42.0 \times 10^{-6}\underline{/90°}}$$
$$= 0.0298\underline{/75.1°} = 0.00766 + j0.0287$$

whence

$$\alpha = 0.00766\ \text{neper/loop mile (attenuation)}$$
$$\beta = 0.0287\ \text{rad/loop mile (phase shift)}$$

a. The wavelength is $\lambda = 2\pi/\beta = 6.28/0.0287 = 219$ mi (351 km).

b. The phase velocity is given by $v = \omega/\beta = 5000/0.0287 = 174{,}000$ mi/s or $(174{,}000/186{,}000) \times 100 = 93.5$ percent of the speed of light.

c. The attenuation, if the line is properly matched at the load is $\alpha = 0.00766$ neper/loop mile. There are 8.68 dB in one neper, hence the attenuation is

$$0.00766 \times 8.68 = 0.0665 \text{ dB/loop mile (1.6093 km)}$$

d. For distortionless transmission it is necessary that

$$lg = rc \quad \text{or} \quad l = \frac{rc}{g} = \frac{10.44 \times 0.00838 \times 10^{-6}}{0.300 \times 10^{-6}}$$

$$= 0.290 \text{ H/loop mile (1.6093 km)}$$

As the line has but 0.00366 H/loop mile, it is necessary to use additional loading of $0.290 - 0.00366 = 0.286$ H/loop mile.

4.20 To match properly the generator to the line, the latter must have a characteristic impedance $Z_0 = 100 + j0$ ohms, equal to the generator impedance. Since the stud is to be connected in parallel with the line, it is convenient to work with admittances.

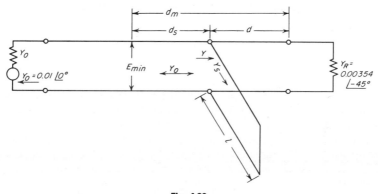

Fig. 4.20

In this case there is a conductive characteristic admittance

$$Y_0 = \frac{l}{Z_0} = G_0 = 0.01 \text{ mho}$$

The stub must be connected to the line at a distance d, preferably as near the load as possible, where the line admittance looking toward the load, $Y = G + jB$, has a real or conductive component $G = G_0$. To determine this location, compute first the receiving-end reflection factor

$$N_R = \frac{(Z_R/Z_0) - 1}{(Z_R/Z_0) + 1} = \frac{2 + j2 - 1}{2 + j2 + 1} = 0.62\underline{/30°}$$

The standing-wave ratio of the line prior to stubbing is

$$\frac{1 + N_R}{1 - N_R} = \frac{1 + 0.62}{1 - 0.62} = 4.27$$

Next compute the location, d_m, at which the standing wave of voltage is at its minimum,

$$d_m = \frac{(\varphi + \pi)\lambda}{4\pi}$$

where φ = angle, rad, of N_R = 0.525

$$\lambda = \text{wavelength, m,} = \frac{3 \times 10^8}{f = 600 \text{ Mc}} = 0.5$$

Consequently, $d_m = (0.525 + \pi) \times 0.5/4\pi = 0.136$ m. Now the stub must be connected a distance d_s either side of the point of voltage minimum, where

$$d_s = \frac{\lambda \times \cos^{-1} N_R}{4\pi} = 0.5 \times \frac{51.6/57.3}{4\pi} = 0.0358 \text{ m}$$

As it is desirable to place the stub as close to the load as possible, locate it at $d = d_m - d_s = 0.136 - 0.036 = 0.1$ m. As mentioned above, a point has been located at which the admittance of the line looking toward the load is $Y = G_0 + jB$. The lossless stub has an admittance $Y_s = 0 + jB_s$. By adjusting the length of the stub so that $B_s = -B$, then the admittance of the line and stub in parallel, Y', will be $Y' = Y + Y_s = G_0 + jB + 0 - jB = G_0$, and the line is matched, as desired, to the left of the stub.

The length of the lossless short-circuited stub required is given in terms of the standing-wave ratio of the line without the stub as

$$l = \frac{\lambda}{2\pi} \times \tan^{-1} \frac{\sqrt{\rho}}{\rho - 1} = \frac{0.5}{6.28} \times \tan^{-1} \frac{\sqrt{4.27}}{4.27 - 1}$$

$$= 0.045 \text{ m}$$

It should be noted that answers to questions of this type are obtained with much greater facility by means of a universal transmission line diagram, such as the Smith chart.

4.21 *a.*

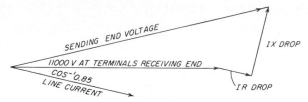

Fig. 4.21

b. Line current $= \dfrac{\text{watts}}{E \cos \theta} = \dfrac{1200 \times 10^3}{11{,}000 \times 0.85} = \textbf{128.4}$ A

$$7.5 \text{ percent of } 1200 \text{ kW} = 90 \text{ kW}$$
$$I^2R = (128.4)^2R \leq 90{,}000 \text{ W}$$

$R \leq \dfrac{90{,}000}{(128.4)^2} = 5.45$ ohms or 0.2725 ohms/mi (1.6093 km) of wire

#4/0 solid copper wire has 0.264 ohms/mi and diameter = 0.4600 in (11.68 mm).

c. Resistance per wire = $0.264 \times 10 = \textbf{2.64}$ ohms

d. Inductance/mi = $0.7411 \times \log_{10}(18/0.23 = 78.2) + 0.080$
$$= 0.7411 \times 1.8932 + 0.080$$
$$= 1.485 \text{ mH/mi (1.6093 km)}$$

$\quad$ Reactance/wire $= 2 \times \pi \times 50 \times 1.485 \times 10^{-3} \times 10$
$$= \textbf{4.66} \text{ ohms}$$

e. Sending-end voltage, where $0.527 = \sin(\cos^{-1} 0.85)$,

$$= 2\sqrt{(5500 \times 0.85 + 128.4 \times 2.64)^2 + (5500 \times 0.527 + 128.4 \times 4.66)^2}$$
$$= 2\sqrt{(4670 \times 339)^2 + (2900 + 596)^2} = \textbf{12,220} \text{ V}$$

f. Regulation, assuming no load, the receiving-end voltage becomes equal to the sending-end voltage, is

$$\frac{12{,}220 - 11{,}000}{11{,}000} = \frac{1220}{11{,}000} = \textbf{11.1} \text{ percent}$$

g. Line loss $= I^2R = (128.4)^2 \times 2.64 \times 2 = 87{,}000$ W

$\quad$ Efficiency $= \dfrac{1{,}200{,}000}{1{,}200{,}000 + 87{,}000} \times 100 = \textbf{93.2}$ percent

4.22 $Z = zl = (0.35 + j0.8)100$

$$= 100 \sqrt{0.1275 + 0.64} \underline{/\tan^{-1} {}^{180}\!/_{35}} = 87.3 \underline{/66.4°}$$

$$Y \frac{\tanh \theta}{\theta} = \frac{Is}{Es} = \frac{j50}{100,000}$$

$$= 5 \times 10^{-4} \underline{/90°} \text{ (from tests with 100,000 V)}$$

Assuming, for the moment, that $\tanh(\theta)/\theta = 1\underline{/0°}$ and $Y = 5 \times 10^{-4}$ $\underline{/90°}$,

$$ZY = (87.3\underline{/66.4°})(5 \times 10^{-4}\underline{/90°}) = 436.5 \times 10^{-4}\underline{/156.4°}$$
$$= 0.04365\underline{/156.4°}$$

From chart in Woodruff, "Principles of Electric Power Transmission," chap. 5,

$$\frac{\tanh \theta}{\theta} = 1.003\underline{/-0.1°} \quad \text{and} \quad Y = \frac{5 \times 10^{-4}\underline{/90°}}{1.003\underline{/-0.1°}}$$

$$Y = 5 \times 10^{-4}\underline{/90.1°}$$

The correction is negligible except in the angle. Then for $ZY = 0.04365\underline{/156.5°}$,

$$\cosh \sqrt{ZY} = \cosh \theta = 0.98\underline{/0.05°}$$

and

$$\frac{\sinh \theta}{\theta} = 0.993\underline{/0.018°}$$

$$\text{Voltage to neutral} = \frac{132,000}{\sqrt{3}} = 76.3 \text{ kV}$$

$$\text{Line current} = \frac{50,000}{3 \times 76.3 \times 0.80} = 273 \text{ A}$$

$$= 273(0.8 - j0.6) = (218.4 - j164) \text{ A}$$

From Woodruff (as above),

$$E_S = E_R \cosh \theta + I_R Z \frac{\sinh \theta}{\theta}$$

$$= 76.3 \times 0.98\underline{/0.05°} + \frac{(273\underline{/-36.9°})(87.3\underline{/66.4°})(0.993\underline{/0.018°})}{1000}$$

$$= 74.8\underline{/0.05°} + 23.68\underline{/29.68°}$$

$$\cos 29.68° = 0.8688$$
$$\sin 29.68° = 0.4952$$

$$E_S = 74.8 + j0 + 23.68(0.8688 + j0.4952)$$
$$= 74.8 + 20.55 + j11.7 = (95.35 + j11.7) \text{ kV}$$

$$I_S = I_R \cosh \theta + E_R Y \frac{\sinh \theta}{\theta}$$

$$= (273\underline{/-36.8°})(0.98\underline{/0.05°})$$

$$+ (76,300\underline{/0°})(5 \times 10^{-4}\underline{/90.1°})(0.993\underline{/0.018°})$$

$$= 267.5\underline{/-36.75°} + 37.85\underline{/90.12°}$$

$$\cos 36.75 = 0.5983$$
$$\sin 36.75 = 0.8012$$

$$I_S = 267.5(0.8012 - j0.5983) + j37.85$$
$$= 214 - j160 + j37.85 = (214 - j122.15) \text{ A}$$

Sending-end power = real part of $(95.35 + j11.7) \times (214 + j122.15)$
$$= 20,400 - 1430 = 18,970 \text{ kW/phase}$$

$$\text{Efficiency} = \frac{\text{receiving-end power}}{\text{sending-end power}} = \frac{50,000}{3 \times 18,970} \times 100$$

$$= \frac{50}{56.9} \times 100 = 87.9 \text{ percent}$$

Check by finding approximate value of line current by assuming half of the charging current supplied at each end.

Charging current at 76.3 kV to neutral is

$$I_c = \frac{76.3}{100} \times 50 = 38.15 \text{ A}$$

With half at each end, current at receiving end is

$$I_R + \frac{I_c}{2} = \text{line current} + \frac{38.5}{2} = 218.4 - j164 + j19.1$$

$$= 218.4 - j144.9 = 262.5\underline{/-33.5°}$$

$$I^2R \text{ per line} = (262.5)^2(0.35 \times 100) = 2890 \text{ kW}$$

$$\text{Efficiency} = \frac{50,000}{50,000 + 3 \times 2890} \times 100 = \mathbf{87.4} \text{ percent}$$

4.23 One phase of the circuit is

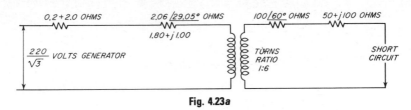

Fig. 4.23a

The equivalent circuit referred to generator voltage is

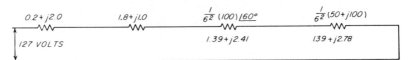

Fig. 4.23b

Simplified, this becomes

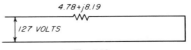

Fig. 4.23c

The current at the generator is

$$I_G = \frac{127}{4.78 + j8.19} = \frac{127}{9.48\underline{/59.7°}} = 13.4\underline{/-59.7°} \text{ A}$$

The current at the short circuit is one-sixth of this, or

$$I_{\text{short circuit}} = \frac{1}{6} \times 13.4 = \mathbf{2.63} \text{ A} \qquad \mathbf{a}$$

4.24 From blocked test

$$R_{\text{eq}} \text{ per phase} = \frac{P_B}{3I^2} = \frac{40,500}{3(170)^2} = 0.466 \text{ ohm}$$

$$Z_{\text{equiv}} = \frac{V}{I} = \frac{440}{\sqrt{3} \times 170} = 1.5 \text{ ohm}$$

$$X_{\text{equiv}} = \sqrt{Z_{\text{equiv}}^2 - R_{\text{equiv}}^2} = \sqrt{2.25 - 0.22} = 1.42 \text{ ohm}$$

$$R_S + R_R = 0.466 \text{ ohm}$$

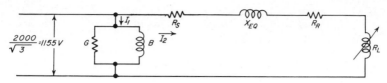

Fig. 4.24

Using dc ratios for split,

$$R_{S_{dc}} = 0.165 \text{ ohm}$$

$$R_{R_{dc}} = 0.0127 \times n^2$$
$$= 0.0127 \times 4^2 = 0.203 \text{ ohm (referred to stator)}$$

$$R_S = \frac{0.165}{0.165 + 0.203}\, 0.466 = 0.21 \text{ ohm}$$

$$R_R = \frac{0.203}{0.165 + 0.203}\, 0.466 = 0.256 \text{ ohm}$$

$$R_L = R_R \frac{1 - S}{S} = 0.256 \frac{1 - 0.015}{0.015} = 16.8 \text{ ohm}$$

From no-load test (assuming F and W inherent in R_L),

$$G = \frac{10{,}100 - 2000}{3 \times (1155)^2} = 0.00202 \text{ mho}$$

$$Y = \frac{I}{V} = \frac{15.3}{1155} = 0.0132 \text{ mho}$$

$$\text{pf} = \frac{p}{\sqrt{3}\, VI} = \frac{10{,}100}{\sqrt{3} \times 2000 \times 15.3} = 0.191$$

$$B = Y \sin\theta = 0.0132 \times 0.9815 = 0.013 \text{ mho}$$

$$I_2 = \frac{V}{Z} = \frac{1155}{0.21 + 0.256 + 16.8 + j1.42} = \frac{1155}{17.27 + j1.42}$$

$$= \frac{1155}{17.4\underline{/4.7^\circ}} = 66.4\underline{/-4.7^\circ} \text{ A} = 66 - j5.5$$

$$I_1 = V(G - jB) = 1{,}155(0.00202 - j0.013)$$
$$= 2.33 - j15 = 15.2\underline{/-81.2^\circ}$$

$a.$ $I_{\text{stator}} = I_1 + I_2 = 68.3 - j20.5 = \mathbf{71.3\underline{/-16.7^\circ}}$ **A**

$\quad I_{\text{rotor}} = nI_2 = 4 \times 66.4 = \mathbf{265.6}$ **A**

b. $P_{\text{out}} = 3I_2^2R_L - F$ and W

$\quad\quad = [3(66.4)^2(16.8) - 2000] = 220$ kW

$\quad$ hp $= 220/0.746 = $ **295** hp

c. $N = \dfrac{120f}{p}(1 - s) = 1000 \times 0.985 = $ **985** rpm

d. Torque developed $= 3I_2^2(R_L + R_R) = 3(66.4)^2(17.1)$

$\quad\quad\quad\quad\quad\quad\quad\quad\quad = 226{,}000$ sync watts

$\quad\quad\quad\quad\quad\quad\quad\quad\quad = \dfrac{226{,}000}{105} = $ **2150** N·m

$\quad\quad$ Net torque $= \dfrac{226{,}000 - 2000}{105} = 2130$ N·m

e. pf $= \cos \underline{/I_{\text{stator}}} = \cos 16.7° = $ **0.958**

f. Efficiency $= \dfrac{P_0}{P_I} \times 100 = \dfrac{220{,}000 \times 100}{\sqrt{3} \times 2000 \times 71.3 \times 0.958}$

$\quad\quad\quad\quad\quad\quad = $ **93.0** percent

4.25 Present excitation $= 480 \times 11 = 5280$ A·turns/pole

$\quad\quad$ Output current $= 35{,}000/125 = 280$ A

$\quad\quad$ Series field turns per pole $= 5280/280 = 18.85$

Therefore, wind **19.5** turns per pole. **b**

4.26 1. At unity pf:

$\quad\quad\quad\quad$ Output $= 2500 \times 1$ kW

$\quad\quad\quad\quad$ Current $= \dfrac{2500}{\sqrt{3} \times 6.6 \times 1} = 219$ A

$\quad\quad$ Armature copper loss $= \dfrac{3 \times (219)^2 \times 0.072}{1000} = \;\;10.4$ kW

$\quad\quad\quad\quad\quad\quad$ Field loss $= \dfrac{(200)^2 \times 0.43}{1000} = \;\;17.2$ kW

$\quad\quad\quad\quad\quad\quad\quad\quad$ Friction loss $= \;\;35.0$ kW

$\quad\quad\quad\quad\quad\quad\quad\quad\quad\quad$ Core loss $= \;\underline{\;47.5}$ kW

$\quad\quad\quad\quad\quad\quad\quad\quad$ Total loss $= 110.1$ kW

$\quad\quad$ Efficiency $= \dfrac{\text{output}}{\text{output} + \text{losses}} = \dfrac{2500 \times 100}{2610.1}$

$\quad\quad\quad\quad\quad\quad\quad\quad = $ **95.8** percent **1c**

2. At 0.8 pf:

$$\text{Output} = 2500 \times 0.8 = 2000 \text{ kW}$$

$$\text{Current} = \frac{2000}{\sqrt{3} \times 6.6 \times 0.8} = 219 \text{ A}$$

$$\text{Armature copper loss} = \frac{3 \times (219)^2 \times 0.072}{1000} = 10.4 \text{ kW}$$

$$\text{Field loss} = \frac{(240)^2 \times 0.43}{1000} = 24.8 \text{ kW}$$

$$\begin{aligned}\text{Friction loss} &= 35.0 \text{ kW} \\ \text{Core loss} &= \underline{47.5} \text{ kW} \\ \text{Total loss} &= 117.7 \text{ kW}\end{aligned}$$

$$\text{Efficiency} = \frac{2000 \times 100}{2117.7} = \textbf{94.4} \text{ percent} \qquad \textbf{2d}$$

4.27 *a.* Reads average. $V_{av} = \dfrac{3V \times 1 \text{ ms}}{3 \text{ ms}} = \textbf{1 V}$

b. Reads rms. $V_{rms} = \sqrt{\dfrac{(3)^2(1)}{3}} = \sqrt{\dfrac{9}{3}} = \sqrt{3} = \textbf{1.732 V}$

c. Peak above average $= 3V - 1V = 2$ V
rms reading $= 2/\sqrt{2} = \textbf{1.414 V}$

d. Leads reversed. Average $= \dfrac{3 \times 2}{3} = 2$ V

Peak above average $= 3 - 2 = 1$ V
rms reading $= 1/\sqrt{2} = \textbf{0.707 V}$

See Fig. 4.27 with question.

4.28 1. Current at start varies directly as the applied voltage. For rated current at start, $V_{start} = 550/4.25 = \textbf{130 V}$. **1a**

2. Starting torque varies as the square of the voltage. At 440 V, T_{start} $= [(440)^2/(550)^2] \times 160$ percent $= \textbf{102}$ percent of rated torque. **2b**

3. $I_{start} = (440/550) \times 425$ percent $= \textbf{340}$ percent of rated current. **3c**

4.29 If 85 percent efficiency is assumed, the rated motor current will be

$$I_m = \frac{746 \times \text{hp}}{E \times \text{efficiency}} = \frac{746 \times 5}{120 \times 0.85} = 36.6 \text{ A}$$

Rated field current is, by Ohm's law,

$$I_f = \frac{E}{R_f} = \frac{120}{50} = 2.4 \text{ A}$$

Whence rated armature current is

$$I_a = I_m - I_f = 36.6 - 2.4 = 34.2 \text{ A}$$

The starter is to limit the armature current to 200 percent of rated value, or 68.4 A. If the armature circuit resistance is R_a and the maximum starter resistance is R_s, then by Ohm's law

$$68.4 = \frac{E}{R_a + R_s}$$

Whence the starter resistance must be

$$R_s = \frac{E}{68.4} - R_a = \frac{120}{68.4} - 0.8$$

$$= 1.755 - 0.800 = \textbf{0.955} \text{ ohm} \qquad \textbf{d}$$

4.30 $I_{28.8} = 100 \text{ A}$

$$= \frac{28,800}{\sqrt{3} \times 208 \times 0.8}$$

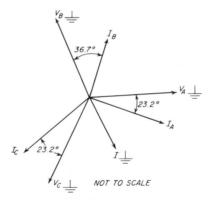

Fig. 4.30

Using $V_{A_{\text{ground}}}$ as reference,

$$I_B = 100\underline{/120° - 36.7°}$$
$$= 100\underline{/83.3°} \text{ A}$$

$$I_A = 60\underline{/0°} + 100\underline{/-36.7°}$$
$$= 60 + 80 - j60$$
$$= 152\underline{/-23.2°} \text{ A}$$
$$I_C = 60\underline{/-120°} + 100\underline{/-120° - 36.7°}$$
$$= 152\underline{/-143.2°} \text{ A}$$
$$I_{\text{ground}} = 60\underline{/0°} + 60\underline{/-120°}$$
$$= 60\underline{/-60°} \text{ A}$$

4.31 Gain with feedback is $A_f = 100$. Gain without feedback is

$$A = (10^2 \times 10^2) \pm 2000 = 10^4 \pm 2000 \text{ approx.}$$

Since $A_f = A/(1 - A\beta)$, from any textbook on electronics,

$$100 = \frac{10^4}{1 - A\beta} \quad \text{or} \quad 1 - A\beta = 100$$

Also

$$\frac{dA_f}{100} = \frac{1}{1 - A\beta}\frac{dA}{A},$$

from any electronics textbook, or

$$\frac{dA_f}{100} = \frac{1}{100} \times \frac{2000}{10^4} = 20 \times 10^{-4}$$

$$= \textbf{0.2} \text{ percent maximum variation in gain} \qquad \textbf{b}$$

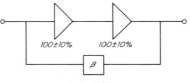

Fig. 4.31

4.32 Motor input $= 62.5 + 30.5 = 93$ kW

$$\text{Tangent of pf angle} = \sqrt{3} \times \frac{62.5 - 30.5}{62.5 + 30.5}$$

$$= \sqrt{3} \times {}^{32}\!/_{93} = 0.596 = \tan 30.8°$$

Power factor $= \cos 30.8° = 0.859$
Power input $= \sqrt{3} \times EI \cos \theta$
$$= \sqrt{3} \times 600 \times I \times 0.859 = 93,000 \text{ W}$$
$I = \textbf{104.3}$ A in each line   **1a**
Power output $= 0.90 \times 93,000/746 = \textbf{112.2}$ hp (**83.67** kW)   **2b**

4.33   Using complex notation,

$$kVA = \sqrt{3} \times kV \times I \cos \theta + j \sqrt{3} \times kV \times I \sin \theta$$

Total load $kVA = 4000 + j0$

Alternator no. 1:

$$kVA_1 = \sqrt{3} \times 6.6 \times 200 \times 0.85 + j \sqrt{3} \times 6.6 \times 200 \times 0.527$$
$$= 1943 + j1205$$

Alternator no. 2 must supply the difference.

$$kVA_2 = kVA - kVA_1 = (4000 - 1943) + j(0 - 1205)$$
$$= 2057 - j1205 = 2308\underline{/-30.4°}$$

Alternator no. 2 supplies **2057** kW, at $\cos(-30.4°) = $ **0.863** pf lagging. $I_2 = 2380/(\sqrt{3} \times 6.6) = $ **208** A.

4.34   1. $I = \dfrac{150/2}{100} = 0.75$ per unit

$$\begin{aligned}
E_{af} &= \bar{V} + jX_dI \\
&= 1.0 + j1.1(0.8 - j0.6)(0.75) \\
&= 1.495 + j0.66 \\
&= 1.63 \underline{/23.85°} \text{ pu} \\
&= \frac{(1.63)(13.8 \text{ kV})}{\sqrt{3}} = \textbf{13} \text{ kV per phase} \qquad \textbf{1a}
\end{aligned}$$

2. $\delta = $ **23.85°**   **2c**

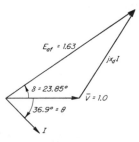

Fig. 4.34

4.35   By definition, $dB = 10 \log P_1/P_2$, and the power delivered to the coil based upon $P_2 = 0.005$ W is

$$P_1 = 0.005 \text{ antilog } {}^{20}\!/_{10} = 0.5 \text{ W}$$

Inasmuch as this power is $P_1 = I^2R$, the rms current in the 10-ohm voice coil must be $I = \sqrt{P_1/R} = \sqrt{0.5/10} = 0.224$ A. If mks units are desired, which are easier and require no conversion factors, the flux density of 10,000 lines/in² (6.45 cm²) is

$$B = 0.155 \text{ Wb/m}^2$$

and for the active length, which is the circumference of the voice coil,

$$l = -d = 3.14 \times 1 \times 0.0254 = 0.0799 \text{ m}$$

Then the rms force exerted is simply $f = BlNI$

$$f = 0.155 \times 0.0799 \times 10 \times 0.224 = 0.0277 \text{ N}$$

(This force is equivalent to 1 oz.) If the current is sinusoidal, the maximum thrust will be $\sqrt{2}$ times the rms value found above.

4.36 First find the Thevenin equivalent to the left of the base.

$$V_1 = 20 \frac{R_b}{R_b + R_a}$$

$$= 20 \frac{10}{100} = 2 \text{ V}$$

$$R_1 = \frac{R_a R_b}{R_a + R_b}$$

$$= \frac{(10 \text{ k}\Omega)(90 \text{ k}\Omega)}{100 \text{ k}\Omega} = 9 \text{ k}\Omega$$

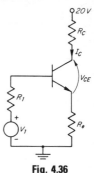

Fig. 4.36

From any textbook on electronics,

$$I_c = \frac{\beta(V_1 - V_{BE})}{R_1 + R_e(1 + \beta)} = \frac{40(2 - 0.7)}{9 \text{ k}\Omega + 1 \text{ k}\Omega(41)} = \mathbf{1.04} \text{ mA} \qquad \mathbf{1d}$$

$$V_{CE} = 20 - I_c(R_c + R_e)$$

$$= 20 - 1.04(5 + 1) = \mathbf{13.76} \text{ V} \qquad \mathbf{2a}$$

4.37 The base current is

$$I_B = \frac{I_C}{\beta} = \frac{2 \text{ mA}}{10} = 200 \text{ }\mu\text{A}$$

The emitter current is

$$I_E = I_C + I_B = 2 \text{ mA} + 0.2 \text{ mA} = 2.2 \text{ mA}$$

The voltage across R_c is

$$V_{R_c} = 20 - V_{CE} - I_E(2 \text{ k}\Omega) = 20 - 10 - 4.4 = 5.6 \text{ V}$$

and

$$R_c = \frac{V_{R_c}}{I_C} = \frac{5.6}{2 \times 10^{-3}} = \mathbf{2.8} \text{ k}\Omega \qquad \mathbf{1c}$$

The voltage across R_a is

$$V_{R_a} = 20 - I_E(2 \text{ k}\Omega) - 0.7 = 20 - 4.4 - 0.7 = 14.9 \text{ V}$$

The current through R_a is

$$I_{R_a} = I_B + \frac{4.4 + 0.7}{10 \text{ k}\Omega} = 0.2 \text{ mA} + 0.51 \text{ mA} = 0.71 \text{ mA}$$

$$R_a = \frac{V_{R_a}}{I_{R_a}} = \frac{14.9}{0.71 \text{ mA}} = \textbf{21.0} \text{ k}\Omega \qquad \textbf{2b}$$

4.38 *a.* $\sqrt{2} \times 120 = 170\text{-V peak-to-neutral-peak inverse}$

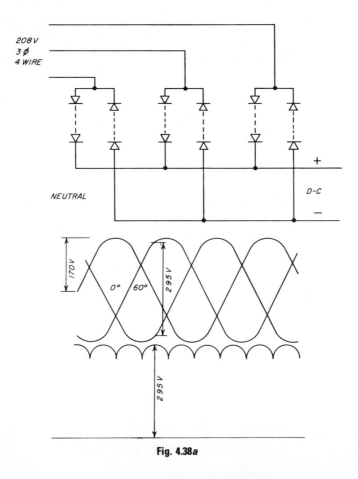

Fig. 4.38a

$$\text{Voltage per stack} = 2 \times \frac{\sqrt{3}}{2} \times 170 = 295 \text{ V}$$

$$\frac{295}{20} \text{ V} = 14.75, \text{ or } 15 \text{ disks}$$

Six single stacks required, two stacks in series always. Current capacity, $3 \times 2 = 6$ A.

Output voltage:

$$E_{av} = \frac{170}{\pi/6} \int_{0°}^{30°} [\cos \theta - \cos(120° + \theta)] \, d\theta$$

$$= \frac{170 \times 6}{\pi} \int_{0°}^{30°} (\cos \theta - \cos 120° \times \cos \theta + \sin 120° \times \sin \theta) \, d\theta$$

$$= 325 \int_{0°}^{30°} \left[(1 + 0.5) \cos \theta + \frac{\sqrt{3}}{2} \sin \theta \right] d\theta$$

$$= 325(1.5 \sin \theta - 0.866 \cos \theta) \Big]_{0°}^{30°}$$

$$= 325(1.5 \times 0.5 - 0 - 0.866 \times 0.866 + 0.866)$$

$$= 325 \times 0.866 = 282 \text{ V no load}$$

or $$0.85 \times 282 = 240 \text{ V full load}$$

$$\text{Output capacity} = 6 \times 240 = 1440 \text{ W}$$

b. Rectifier of part *a* is also the rectifier that gives the smallest voltage ripples.

c. The rectifier that takes the smallest number of disks and still draws a balanced load is one-half of the rectifier of part *a*, which uses voltages to neutral only once. Peak inverse voltage is as in part *a*.

Fifteen disks per stack are required but only three stacks current capacity = $3 \times 2 = 6$ A.

Output voltage:

$$E_{av} = \frac{170}{\pi/3} \int_{0°}^{60°} \cos \theta \, d\theta = 162(\sin \theta) \Big]_{0°}^{60°}$$

$$= 162 \left(\frac{\sqrt{3}}{2} - 0 \right) = 140 \text{ V no load}$$

$$= 0.85 \times 140 = 119 \text{ V full load}$$

$$\text{Voltage ripple in part } a = \frac{295 - 255}{2} = 20 \text{ V}$$

$$\text{Voltage ripple in part } c = \frac{170 - 85}{2} = 42.5 \text{ V}$$

Voltage ripple in part c is $42.5/20 = 2.13$ times voltage ripple in part a.

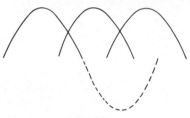

Fig. 4.38c

4.39 Power $= EI \cos \theta$
$$550 = 220 \times 2.5 \cos \theta$$

$$\cos \theta = \frac{550}{220 \times 2.5} = 1$$

$$\theta = 0°$$

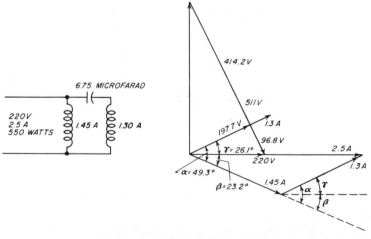

Fig. 4.39

a. The current in the main winding, 1.45 A, lags the line voltage. The condenser current, 1.3 A, must lead the other current if their sum, 2.5 A, is to be in phase with the line voltage.

By the cosine law,

$$(2.5)^2 = (1.45)^2 + (1.3)^2 + 2 \times 1.45 \times 1.3 \cos \alpha$$

$$6.25 = 2.10 + 1.69 + 3.77 \cos \alpha$$

$$\cos \alpha = \frac{6.25 - 3.79}{3.77} = 0.652$$

$$\alpha = 49.3°$$

$$\sin \alpha = 0.758$$

By the sine law,

$$\frac{2.5}{\sin(\pi - \alpha)} = \frac{1.3}{\sin \beta}$$

$$\sin \beta = \frac{1.3}{2.5} \times \sin \alpha$$

$$= 0.52 \times 0.758 = 0.394$$

$$\beta = 23.2°$$

$$\cos \beta = 0.918$$

Let $\gamma = \alpha - \beta = 49.3° - 23.2° = 26.1°$, $\cos \gamma = 0.898$. The components of the line voltage, 220, with respect to the condenser current, 1.3 A, are $220 \cos \gamma = 220 \times 0.898 = 197.7$ V.

$$220 \sin \gamma = 220 \times 0.440 = 96.8 \text{ V}$$

The voltage drop across the condenser is

$$E_c = IX_c = \frac{1.3}{377 \times 6.75 \times 10^{-6}}$$

$$= \frac{1300}{2.545} = 511 \text{ V}$$

Apparent impedance drop in condenser branch

$$220 = IZ = IR_w + j(IX_w - IX_c) = 197.7 - j96.8$$
$$IX_w - 511 = -96.8 \text{ V} \quad \text{and} \quad IX_w = 511 - 96.8 = 414.2 \text{ V}$$
$$\text{Apparent } IZ_w = IR_w + jIX_w = 197.7 + j414.2$$
$$\text{Apparent } Z_w = 197.7/1.3 + j414.2/1.3 = 152 + j319$$

$$= \sqrt{(152)^2 + (319)^2} = 353 \text{ ohms}$$

Main winding apparent $Z = 220/1.45 = 152$ ohms

b. Power to main winding $= 220 \times 1.45 \times \cos 23.2°$
$$= 293 \text{ W}$$
Power to auxiliary winding $= 197.7 \times 1.3 = \textbf{257 W}$

4.40 1.

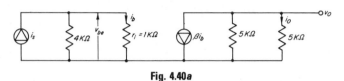

Fig. 4.40a

2. The base current is
$$i_b = \tfrac{4}{5} i_s$$

The output current is
$$i_0 = -\tfrac{1}{2}\,\beta i_b$$

Dividing these two equations, we have

$$\frac{i_0}{i_s} = -\frac{0.5\beta i_b}{\tfrac{4}{5} i_b} = -\frac{2}{5}\beta = -\frac{2}{5}\,40 = \textbf{-16} \text{ current gain} \qquad \textbf{2c}$$

3. $v_{be} = i_b r_i = 1 \text{ k}\Omega i_b$
$v_0 = -\beta i_b(2.5 \text{ k}\Omega)$

$$\frac{v_0}{v_{be}} = -\frac{40 i_b(2.5 \text{ k}\Omega)}{1 \text{ k}\Omega i_b} = -\frac{100 \text{ k}\Omega}{1 \text{ k}\Omega} = \textbf{-100} \text{ voltage gain} \qquad \textbf{3d}$$

4.41 Using Laplace, we have

$$I(S) = \frac{V(S)}{Z(S)} = \frac{(V_1 + V_0)/S}{R + (1/CS)} = \frac{(V_1 + V_0)C}{RCS + 1}$$

$$= \frac{(V_1 + V_0)C}{RC} \frac{1}{S + (1/RC)} = \frac{V_1 + V_0}{R} \frac{V_1 + V_0}{S + (1/RC)}$$

Taking the inverse transform (from tables),

$$i(t) = \frac{V_1 + V_0}{R} \epsilon - \frac{t}{RC}$$

$$= \frac{200 + 150}{150} \epsilon - \frac{t}{2.25(10^{-3})} = 2.33^{-445t}$$

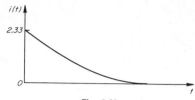

Fig. 4.41

At $t = 0$, $i(t) = 2.33$. At $t = \infty$, $i(t) = 0$. At $t = 2.25$ ms, $i(t) = 2.33\epsilon^{-1}$
$= 2.33/\epsilon = \mathbf{0.86}$ A.   **a**

4.42 No resistance values being furnished, assume a dissipationless parallel or antiresonant circuit for which the angular frequency of antiresonance is merely

$$1/\sqrt{LC} = 2\pi f_0$$

Hence the required capacitance is given by

$$C = \frac{1}{(2\pi f_0)^2 xL} = \frac{1}{(6.28 \times 60 \times 10^3)^2 \times 260 \times 10^{-6}}$$
$$= 0.0271 \times 10^{-6} \text{ F}$$

4.43 *a.*

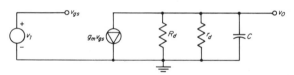

Fig. 4.43a

b. $v_0 = -g_m v_{gs} \dfrac{R(1/jwc)}{R + (1/jwc)}$

where $$R = \frac{R_d r_d}{R_d + r_d} \; 12 \text{ k}\Omega$$

Solving, we have

$$\frac{v_0}{v_{gs}} = -g_m \frac{R}{1 + jwRc}$$

c. $\dfrac{v_0}{v_{gs}} = -g_m R \dfrac{1}{1 + j(w/w_1)}$

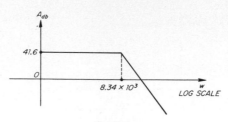

Fig. 4.43c

where $\quad w_1 = \dfrac{1}{Rc} = \dfrac{1}{12 \times 10^3 \times 0.01 \times 10^{-6}} = 8.34 \times 10^3 \text{ rad/s}$

$$g_m R = 10^{-2} \times 12 \times 10^3 = 120$$
$$20 \log 120 = \textbf{41.6 dB}$$

4.44 *a.* $R = 12.6 + 50 = 62.6$ ohms. The general equation for the current is

$$i = I_0 \epsilon^{-(R/L)t}$$
$$e_{ab} = i(50) = 50 I_0 \epsilon^{-(R/L)t}$$
$$20 = 50(9.2)\epsilon^{(R/L)}$$

$$\epsilon^{(R/L)t} = 23 \qquad \frac{R}{L}t = 3.14$$

$$L = \frac{Rt}{3.14} = \frac{(62.6)(2)}{3.14} = \textbf{40 H}$$

b. Induced voltage is

$$e_i = iR = I_0 R \epsilon^{-(R/L)t}$$

$$= \frac{(9.2)(62.6)}{23} = \textbf{25.1 V}$$

c. Since $e_{ab} = 50 I_0 \epsilon^{-(R/L)t}$, max e_{ab} occurs at $t = 0$ or max $e_{ab} = 50(9.2)$ = **460** V.

d. $i = I_0 \epsilon^{-(R/L)t} \qquad$ or $\qquad 0.01 I_0 = I_0 \epsilon^{-(R/L)t}$

$$\frac{R}{L}t = 4.6 \qquad \text{or} \qquad t = 4.6 \frac{L}{R} = \frac{(4.6)(40)}{62.6} = \textbf{2.94 s}$$

4.45 $\quad P_M = 3600 \times R \times \dfrac{K_H}{S} = \dfrac{3600 \times 15 \times \frac{2}{3}}{50} = 720 \text{ W}$

$$P_{\text{system}} = P_M \times (CT \text{ ratio})(PT \text{ ratio}) = 720 \times 20 \times 20$$
$$= \textbf{288 kW} \qquad \textbf{b}$$

4.46 When the bridge shown in Fig. 4.46 is balanced,

a. $Z_A Z_C = Z_B Z_D$ or $Z_C = \dfrac{Z_B Z_D}{Z_A} = \dfrac{Y_A Z_B}{Y_D}$

$$Y_A = \frac{1}{R_A} + j\omega C_A$$

$$Z_B = R_B + \frac{1}{j\omega C_B}$$

$$Y_D = j\omega C_D$$

Then $\quad Z_C = \dfrac{[(1/R_A) + j\omega C_A][R_B + (1/j\omega C_B)]}{j\omega C_D}$

$$= \left(\frac{R_B}{R_A} + j\omega C_A R_B + \frac{1}{j\omega C_B R_A} + \frac{C_A}{C_B}\right)\frac{1}{j\omega C_D}$$

$$= \left(\frac{C_A R_B}{C_D} - \frac{1}{\omega^2 C_B C_D R_A}\right) + \frac{1}{j\omega C_D}\left(\frac{R_B}{R_A} + \frac{C_A}{C_B}\right)$$

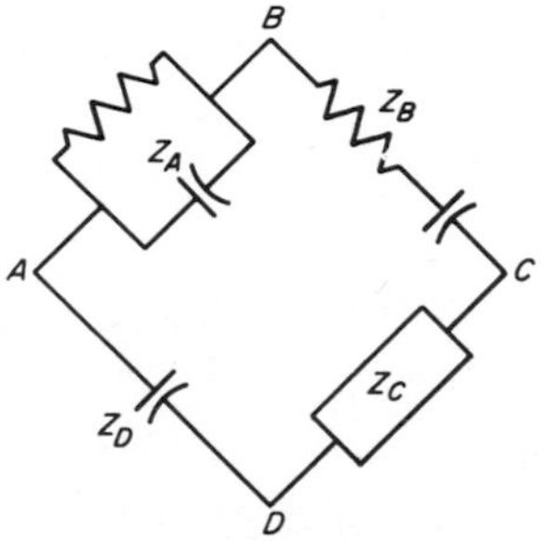

Fig. 4.46

The impedance Z_C to be connected between C and D should be a resistor R_C in series with a capacitor C_C, where

$$R_C = \frac{C_A R_B}{C_D} - \frac{1}{\omega^2 C_B C_D R_A}$$

$$= \frac{0.053 \times 1500}{0.265} - \frac{10^{12}}{25 \times 10^6 \times 0.530 \times 0.265 \times 10^3}$$

$$= 300 - 285 = 15 \text{ ohms}$$

$$C_C = \frac{C_D}{(R_B/R_A) + (C_A/C_B)} = \frac{0.265}{(1500/1000) + (0.053/0.53)}$$

$$= 0.166 \ \mu\text{F}$$

$b.\ \dfrac{E_{BA}}{E_{AC}} = \dfrac{Z_A}{Z_A + Z_B} - \dfrac{Z_D}{Z_C + Z_D}$

$Y_A = \dfrac{1}{10^3} + j5 \times 10^3 \times 0.053 \times 10^{-6} = (1.0 + j0.265)10^{-3}$

$Z_A = \dfrac{1}{Y_A} = \dfrac{10^3}{1.0 + j0.265} = 965\underline{/-14.85°} = 936 - j247$

$Z_A + Z_B = 936 - j247 + 1500 - j372 = 2510\underline{/-14.6°}$

$\dfrac{Z_A}{A_A + Z_B} = \dfrac{965\underline{/-14.85°}}{2510\underline{/-14.6°}} = 0.385\underline{/-0.25°}$

$Z_D = \dfrac{10^6}{j5 \times 10^3 \times 0.27} = 740\underline{/-90°}$

$\dfrac{Z_D}{Z_C + Z_D} = \dfrac{740\underline{/-90°}}{15 + 10^6/(j5 \times 10^3 \times 0.166) - j740}$

$= \dfrac{740\underline{/-90°}}{15 - j(1210 + 740)} = \dfrac{740\underline{/-90°}}{1950\underline{/-90°}} = 0.379\underline{/0°}$

$\dfrac{E_{BA}}{E_{AC}} \doteq 0.385 - 0.379 = 0.006 = 0.6\ \text{percent}$

4.47 $a.$ Overall efficiency at full load equals

$\dfrac{\text{Generator output}}{\text{Motor input}} \times 100 = \dfrac{6000 \times 100}{220 \times 42} = \mathbf{65}\ \text{percent}$

$b.$ Field current $= {}^{220}\!/_{110} = 2\ \text{A}$

Approximate armature current $= \dfrac{\text{stray power}}{\text{voltage}} = \dfrac{500}{220} = 2.27\ \text{A}$

First estimate of line current at no load $= 2.27 + 2 = 4.27\ \text{A}$

For speed determination, assume constant flux, and

$$S = \dfrac{K(V - R_A I - B_{\text{brush}})}{\phi} = K_1 E_{\text{full load}}$$

$E_{\text{full load}} = 220 - (42 - 2)0.5 - 1 = 199\ \text{V}$

$K_1 = 1480/199 = 7.45\ \text{rpm/V}$

$E_{\text{no load}} = 220 - 0.5 \times 2.27 - 1 = 218\ \text{V}$

$S_{\text{no load}} = 7.45 \times 218 = \mathbf{1620}\ \text{rpm}$

$$\text{S.P.}_{\text{no load}} = \frac{1620 \times 500}{1480} = 546 \text{ W}$$

Then the corrected armature current is

$I_{\text{S.P.}} = \frac{546}{220} = 2.48$ A

$R_A I_A^2 = 0.5 \times (2.48)^2 = 3.08$ W

$V_b I_A = 1 \times 2.48 = 2.48$ W

Armature copper plus brush loss $= 3.08 + 2.48 = 5.56$ W

$\Delta I_A = 5.56/220 = 0.025$ A

Total armature current $= 2.48 + 0.025 = 2.5$ A

Line current at no load $= 2.5 + 2 = $ **4.5** A

Power $= 4.5 \times 220 = $ **990** W

c. Try #6 AWG wire, good for 55 A (N.E.C.).

Ohms per 1000 ft $= 0.4028$

R (300 ft) $= 0.3 \times 0.4028 = 0.1208$ ohm

Voltage drop $= 0.1208 \times 42 = 5.07$ V

Percent drop $= 5.07 \times \frac{100}{220} = 2.30$ percent

This is a permissible voltage drop in the line.

Try #8 AWG wire, good for 45 A.

Ohms per 1000 ft $= 0.64$

Voltage drop $= 0.3 \times 0.64 \times 42 = 8.07$ V

Percent drop $= 8.07 \times \frac{100}{220} = 3.67$ percent

Use **#6** AWG wire.

4.48 The model is (neglecting R_a and R_b)

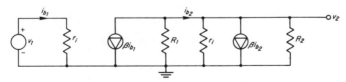

Fig. 4.48 Model A.

$v_1 = i_{b_1}(r_i)$

$i_{b_2} = \beta i_{b_1} \dfrac{R_1}{R_1 + r_i}$

$v_2 = -\beta i_{b_2}(R_2)$

$\quad = -\beta(\beta i_{b_1}) \dfrac{R_1}{R_1 + r_i} R_2 = -\beta^2 i_{b_1} \dfrac{R_1 R_2}{R_1 + r_i}$

$$= -\beta^2 \frac{v_1}{r_i} \frac{R_1 R_2}{R_1 + r_i}$$

Solving for the gain,

$$\frac{v_2}{v_1} = -\beta^2 \frac{R_1 R_2}{r_i(R_1 + r_i)}$$

4.49 The reactances are transformed to per unit values on a common kVA base by the equation

$$Z_{\text{per unit}} = \frac{\text{kVA}_{\text{base}} \times Z_{\text{ohms}}}{\text{kV}^2 \times 1000}$$

Taking 65,000 kVA as the base, the circuit becomes as shown in Fig. 4.49a.

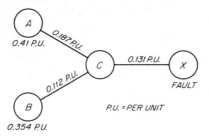

Fig. 4.49a

The network which represents this system is shown in Fig. 4.49b.

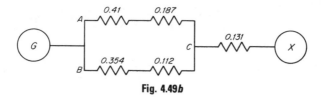

Fig. 4.49b

The equivalent reactance between the generator and the fault is

$$X = \frac{(0.41 + 0.187)(0.354 + 0.112)}{0.41 + 0.187 + 0.354 + 0.112} + 0.131 = 0.392 \text{ per unit}$$

For a 65,000-kVA base, the normal line current at the fault is

$$I = \frac{65,000}{13.2 \sqrt{3}} = 2850 \text{ A}$$

With a three-phase short circuit, the current in each line at the fault is $I = 2850/0.392 = 7270$ A.

For generator A and transmission line AC, each line current is

$$I = \frac{7270(0.41 + 0.187)}{0.41 + 0.187 + 0.354 + 0.112} = 4080 \text{ A}$$

For generator B and transmission line BC, each line current is

$$I = \frac{7270(0.354 + 0.112)}{0.41 + 0.187 + 0.354 + 0.112} = 3180 \text{ A}$$

 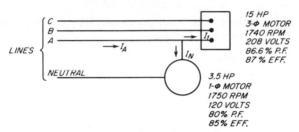

Fig. 4.50

From Fig. 4.50,
For the 15-hp motor:

$$I_L = \frac{\text{power out}}{\sqrt{3} \times E_L \times \cos \theta \times \text{eff.}}$$

$$= \frac{15 \times 746}{\sqrt{3} \times 208 \times 0.866 \times 0.87} = 41.3 \text{ A, } 30° \text{ lag}$$

For the 3.5-hp motor:

$$I_L = \frac{\text{power out}}{E_L \times \cos \theta \times \text{eff.}} = \frac{3.5 \times 746}{115 \times 0.8 \times 85} = 33.4 \text{ A, } 36.8° \text{ lag}$$

$$I_1 = 41.3\underline{/-30°} = 35.8 - j20.7$$
$$I_N = 33.4\underline{/-36.8°} = 26.7 - j20.0$$
$$I_A = I_1 + I_N = 62.5 - j40.7 = |74.6|$$

The line currents are

$$I_A = 74.6 \text{ A}$$
$$I_B = I_C = 41.3 \text{ A}$$
$$I_N = 33.4 \text{ A}$$

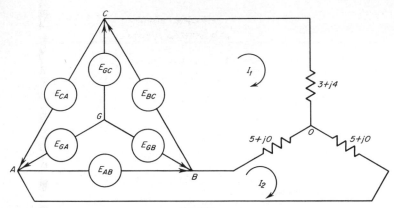

Fig. 4.51a

4.51 The complete system is shown in Fig. 4.51a. It is assumed that the Y-connected supply voltages are equal. Then the ground point G lies at the centroid of the equilateral vector triangle ABC, and since the line voltages are 1000 V, the phase voltages at the supply will be $1000/1.732$ = 577 V and $E_{GC} = 577\underline{/90°}$, $E_{GB} = 577\underline{/-30°}$, and $E_{GA} = 577\underline{/-150°}$.

Considering the load, there are two meshes shown, for which the self-impedances are $Z_{11} = 8 + j4$ and $Z_{22} = 10 + j0$. There is one neutral impedance $Z_{12} = 5 + j0$. There are two unknown mesh currents I_1 and I_2; so two Kirchoff voltage rule equations may be written traversing any paths through the supply:

$$E_{BC} = Z_{11}I_1 - Z_{12}I_2 \tag{1}$$
$$E_{AB} = -Z_{12}I_1 + Z_{22}I_2 \tag{2}$$
$$-1000 \cos 60° + j1000 \sin 60° = (8 + j4)I_1 - (5 + j0)I_2 \tag{1}$$
$$+1000 = -(5 + j0)I_1 + (10 + j0)\mathrm{I}_2 \tag{2}$$

whence

$$-500 + j866.7 = 8I_1 + j4I_1 - 5I_2 \tag{1}$$
$$+1000 = -5I_1 + 10I_2 \tag{2}$$

Solving the second equation for I_2 gives

$$I_2 = 100 + 0.5I_1$$

Putting this in the first equation gives

$$-500 + j866.7 = 8I_1 + j4I_1 - 500 - 2.5I_1$$

and $$(5.5 + j4)I_1 = j866.7$$
$$I_1 = 866.7\underline{/90°}/6.80\underline{/36°} = 127.5\underline{/54°}$$

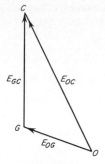

Fig. 4.51*b*

The load phase voltage $= E_{OC} = I_1(3 + j4) = 127.5\underline{/54°} \times 5\underline{/53.2°} = 637.5\underline{/107.2°}$. From the subsidiary vector diagram, Fig. 4.51*b*, and by Kirchhoff's rule, $E_{OC} = E_{OG} + E_{GC}$,

$$E_{OG} = 637.5\underline{/107.2°} - 577\underline{/90°} = -188 + j618 - j577$$
$$= -188 + j41 = 192\underline{/167.7°} \text{ V}$$

4.52 1. For maximum power to the load, allow the collector voltage to reach $2V_{cc}$.

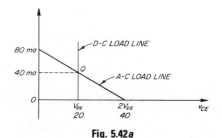

Fig. 5.42*a*

2. $R'_L = \dfrac{V_{cc}}{I_c} = \dfrac{20}{40 \text{ mA}} = 0.5 \times 10^3 = 500$ ohms resistance seen from primary.

$$R_L = \frac{R'_L}{(N_1/N_2)^2} = \frac{500}{25} = \textbf{20 ohms} \qquad \textbf{2a}$$

3. $P_L = E_{rms}I_{rms} = \dfrac{V_{cc}}{\sqrt{2}}\dfrac{I_c}{\sqrt{2}} = \dfrac{20 \times 40 \times 10^{-3}}{2} = \textbf{0.4 W} \qquad \textbf{3b}$

4.53 1. For a given transformer or magnetic circuit $\phi_{max} = E_{rms}/4.44fN$, and since ϕ_{max} was equal in both tests,

$$\frac{E_{60}}{4.44(60)(N)} = \frac{E_{25}}{4.44(25)(N)}$$

$$E_{25} = (2200)(25)/60 = \textbf{917 V}_{rms} \qquad \textbf{1c}$$

2. If the flux or flux density is maintained constant, the core-loss equation may be arranged as follows:

Core loss = hysteresis loss + eddy-current loss
$$CL = K_H f + K_E f^2$$

Dividing by f, the core loss per cycle equals

$$\frac{CL}{f} = K_H + K_E f$$

Substituting the two sets of data and solving the equations simultaneously for K_H and K_E gives

(1) $\qquad\qquad 240/60 = 4 = K_H + 60K_E$
(2) $\qquad\qquad 75/25 = 3 = K_H + 25K_E$
(1) − (2) $\qquad\qquad 1 = 35K_E$
$$K_E = 0.0286 \text{ and } K_H = 2.285$$

Eddy-current loss $= K_E f^2 = (0.0286)(60)^2 = \textbf{103 W} \qquad \textbf{2b}$
Hysteresis loss $= K_H f = (2.285)(60) = \textbf{137} \qquad \textbf{3a}$

4.54 The equivalent current is as shown by Fig. 4.54.

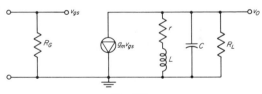

Fig. 4.54

Since the coil Q is high, the parallel resistance of the coil is

$$R_P = (Q_c)^2 r = \left(\frac{wL}{r}\right)^2 10 = \left(\frac{10^6 \times 10^{-3}}{10}\right)^2 10 = 100 \text{ k}\Omega$$

where Q_c = coil Q.

$a.$ $C = \dfrac{1}{w_0^2 L} = \dfrac{1}{(10^6)^2 10^{-3}} = 10^{-9} \text{ F} = \textbf{0.001 } \mu\text{F}$

$b.$ $Q_T = \dfrac{R_T}{w_0 L} = \dfrac{(R_L R_P)/(R_L + R_P)}{w_0 L} = \dfrac{50\ \text{k}\Omega}{10^6 \times 10^{-3}} = 50$

$$BW = \dfrac{w_0}{Q_T} = \dfrac{10^6}{50} = 0.02 \times 10^6 = \mathbf{20R}\ \text{H}$$

$c.$ At resonance the circuit is resistive and equal to

$R_T = 50$ ohms

$v_0 = -g_m v_{gs} R_T$

$G_{am} = \dfrac{v_0}{v_{gs}} = -g_m R_T = -5 \times 10^{-3} \times 50 \times 10^3 = \mathbf{-250}$

4.55 The design formulas for constant-K filters are to be found in all communications handbooks. The low-pass T section has the configuration shown in Fig. 4.55. If the resistive load R is made equal to the characteristic impedance Z_0, then

$$L = \dfrac{R}{\pi f_c} = \dfrac{600}{\pi \times 1000} = 0.191\ \text{H}$$

$$C = \dfrac{1}{\pi f_c R} = \dfrac{1}{\pi \times 1000 \times 600} = 0.53 \times 10^{-6}\ \text{F}$$

Note that each coil used in Fig. 4.55 is to have inductance $L/2$.

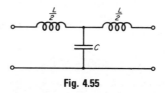

Fig. 4.55

5.01–5.25

Engineering Economics and Business Relations

5.01 The next best thing to a profit is to reduce the amount of loss. At half capacity, a loss of $111,000 − $96,000 (600 @ $160) = $15,000 will occur. Reduction of this loss is an obvious move.

$$\text{Cost of 1,200 items} = \$168,000$$

$$\begin{array}{lll} \text{Less} & 600 \text{ @ } 160 & = & \$\ 96,000 \\ & 600 \text{ @ } 100 & = & \underline{60,000} \\ & & & \$156,000 \end{array}$$

$$\text{Net loss} = \$168,000 − \$156,000 = \$12,000$$

A loss of $12,000 instead of $15,000 is preferable, and therefore the offer should be accepted.

5.02 This break-even type of problem is most easily solved by setting up an algebraic equation. In this case, let x = number of shoes to be manufactured. Then

$$20.00x − (6.00x + 5.50x + 2.50x) = 420,000$$
$$x = \mathbf{70,000} \qquad \mathbf{c}$$

Checking this:

$$\begin{array}{lll} 70,000 \text{ @ } \$20.00 & = & \$1,400,000 \\ 70,000 \text{ @ } \$14.00 & = & \underline{980,000} \end{array}$$

Left to cover fixed charges: $420,000

5.03 Depreciation rate is the rate at which a physical asset lessens in value with the passage of time.

In the straight-line method, it is assumed that the depreciation is equal for each year of the asset's life. A formula for the depreciation per year would be

$$D = \frac{A - B}{n}$$

where D = depreciation per year
    A = first cost
    B = estimated salvage value at end
    n = estimated life of asset in years

In the sinking-fund method of depreciation, one of a series of equal amounts is deposited into a sinking fund at the end of each year of the asset's life. This is usually compounded annually. This amount is equal to the estimated total depreciation times the sinking-fund factor corresponding to the estimated life and the interest rate taken.

$D = (A - B)\text{sf}$, where D, A, and B are as explained above and sf is a sinking-fund factor based on the well-known, equal-payment-series, sinking-fund-factor formula. The payments under the sinking-fund method becomes progressively greater. This method therefore is not as conservative as the straight-line method, particularly if there is any likelihood that an asset may be retired prior to the end of the estimated period of depreciation. The ICC, it would appear, is simply following the custom of imposing conservative practices on a public utility.

5.04 In these straight-line, depreciation-plus-average-interest problems, the most correct method of solution is to consider the first cost (minus salvage if any) divided by n (number of years involved) as capital recovery at 0 percent capital recovery. Depreciation at 0 percent = $(P - L)/n$. With a salvage value,

$$\text{Average interest} = (P - L)\frac{i}{2}\frac{n + 1}{n} + Li$$

Reinforced bridge:

$$\text{Depreciation} = \frac{60,000 + 25,000 - 20,000}{5} = \$13,000$$

Maintenance and inspection   =   1,500
Average interest at 8 percent; $(85,000 - 20,000)$

$$\times \frac{0.08}{2}\frac{6}{5} + 20,000(0.08) = 3120 + 1600 = \quad 4,720$$

Total annual cost   = $\overline{\$19,220}$

New bridge:

$$\text{Depreciation} = \frac{250{,}000 - 50{,}000}{30} \qquad\qquad = \$\ 6{,}666$$

Average interest at 8 percent; $(250{,}000 - 50{,}000)$

$$\times \frac{0.08}{2}\frac{31}{30} + 50{,}000(0.08) = 8266 + 4000 = \$12{,}266$$

Total annual cost $\qquad\qquad\qquad\qquad\qquad = \overline{\$18{,}932}$

Cheaper to replace.

5.05   *a.* Interest on $200{,}000 = 200{,}000(0.10)$ = \$20,000

Interest on present worth of
$200{,}000 = 200{,}000(0.3855)(0.10) =$ $\underline{7{,}710}$

Total annual cost   = \$27,710

   b. Interest on $300{,}000 = 300{,}000(0.10)$ = \$30,000

This being the total annual cost, plan *a* is the more economical.

5.06   Furnishings type *A*:

First cost   = \$10,000

$$\text{Infinite renewals} = 10{,}000\,\frac{0.06903}{0.08} \quad = \quad 8{,}626$$

$$\text{Annual maintenance}\,\frac{100}{0.08} \qquad\quad = \quad 1{,}250$$

$$\text{Present worth of repairs}\,\frac{500(0.17046)}{0.08} = \quad 1{,}065$$

Total capitalized cost   = $\overline{\$20{,}941}$

Furnishings type *B*:

First cost   = \$15,000

$$\text{Infinite renewals} = (15{,}000 - 3000)\,\frac{0.03683}{0.08} = \quad 5{,}524$$

Annual maintenance   =   0

$$\text{Present worth of repairs}\,\frac{200(0.17046)}{0.08} \quad = \quad 426$$

                                    = $\overline{\$20{,}950}$

5.07   The fundamental interest formula for this type of problem involving a systematic reduction of the principal of the debt by repayments of principal and interest uniformly is

$$R = P \frac{i(1 + i)^n}{(1 + i)^n - 1} = P \frac{1}{\text{present worth factor}}$$

where R = end of period payment
P = present sum of money
i = interest rate per period
n = number of periods

$$R = \frac{1000}{3993} = \$250.44$$

5.08　Haskold's valuation formula

$$P = R \frac{1}{i' + i/[(1 + i)^n - 1]}$$
$$= \frac{R}{\text{return interest} + \text{sinking fund factor}}$$

is the most commonly used formula for finding present worth with two interest rates. Its greatest use seems to be in mine-valuation studies.

In this problem

$$R = 500,000 \qquad i = 0.06$$
$$i' = 0.10 \qquad n = 20$$

$$P = 500,000 \frac{1}{0.10 + 0.01842} = \$3,931,306$$

This means that the \$500,000 is actually divided into two parts. One part \$393,130.60 is a 10 percent return on \$3,931,306. The other \$106,869.40 invested each year at 6 percent will yield \$3,931,306 in 20 years, or \$106,869.40 (compound amount factor of 36.786) = \$3,931,306. In other words, the syndicate will receive a 10 percent return each year and in addition will recover the investment sum \$3,931,306 at the end of 20 years.

5.09　Comparing by annual costs:
Method A:

$$
\begin{array}{lr}
\text{Tunnel interest} = \$1,000,000 \times 0.08 & = \$\ \ 80,000 \\
\text{Annual operation and upkeep} & = \ \ \ \ 25,000 \\
\text{Total annual cost} & = \$105,000
\end{array}
$$

Method B:

Ditch interest = 350,000 × 0.08   = $ 28,000
Annual operation and upkeep   = 20,000
Flume capital recovery at 5 percent
  = (200,000 − 30,000)(0.14903) + 30,000(0.08) = 27,735
Annual operation and upkeep   = 40,000
Total annual cost   = $115,735

Note that where annual cost of a structure with perpetual life is involved, interest on the investment is used; i.e., as the number of years approaches infinity the capital-recovery factor approaches the interest rate. Also that where salvage is involved, it is necessary to make the capital-recovery factor equal to the interest earned on the prospective salvage value plus the recovery with interest the difference between first cost and salvage.

5.10 The net return on the sale was $900,000 − $20,000 = $880,000. This problem becomes one of saying that $62,000 times the present worth factor for 15 equal payments plus $1,000,000 times the P.W. factor for a single payment 15 years hence should equal $880,000.

Try 6 percent:

$$P.W. = 62,000 × 9.712 + 1,000,000 × 0.4173 = \$1,019,444$$

Try 7 percent:

$$P.W. = 62,000 × 9.108 + 1,000,000 × 0.3624 = \$927,096$$

Try 8 percent:

$$P.W. = 62,000 × 8.559 + 1,000,000 × 0.3152 = \$845,858$$

By straight-line interpolation, the interest rate is found to be 7.42 percent or, to the nearest 0.1 percent, say, 7.5 percent.

5.11 Holders of simply participating, cumulative preferred stock are paid the stipulated rate of dividend, first for past unpaid dividends, then for current dividends, and after that the common stockholders are paid their dividend up to a rate equal to the preferred. Then both classes of stock participate at equal rates.

1. Earnings for 1968   = $78,500
  Balance preferred share, 1965 dividend = 2,500
          $76,000
  Preferred dividends 1966, 1967, 1968
      = 3 × $5 × 1000 = 15,000
          $61,000

Common dividends 1968
$$= \$5 \times 10,000 = \underline{\quad 50,000 \quad}$$
$$\$11,000$$
Total shares = \$1 \times 11,000 \qquad\quad = \underline{\quad 11,000 \quad}$$
$$0$$

Payment per common share = \$5 + \$1 = **\$6**   **1b**

2. Earnings for 1969 \qquad\qquad = \$77,000
   \$5 \times 1000 (preferred shares) = \underline{\quad 5,000 \quad}
   \qquad\qquad\qquad\qquad\qquad\quad \$72,000
   \$5 \times 10,000 (common shares) = \underline{\quad 50,000 \quad}
   \qquad\qquad\qquad\qquad\qquad\quad \$22,00
   \$2 \times 11,000 (total shares) \quad = \underline{\quad 22,000 \quad}
   \qquad\qquad\qquad\qquad\qquad\qquad\quad 0

Payment per preferred share = \$5 + \$2 = **\$7**   **2c**

3. There are 11,000 shares of voting stock.   **3d**
4. There are no redeemable shares.   **4a**

5.12   1. The normal ordering point is **1800** units.   **1a**
2. The maximum storage space required is for $1800 - 200 + 2000 =$ **3600** units.   **2c**
3. The normal maximum inventory is $1800 - 30 \times 50 + 2000 =$ **2300** units.   **3b**
4. The average inventory is equal to the minimum inventory limit plus one-half the standard order, or $1800 - 30 \times 50 + 2000/2 =$ **1300** units.   **4d**
5. The annual inventory rate of turnover is the yearly total of units used divided by the average inventory, or $15,000/1300 =$ **11.54**.   **5c**

5.13   Reference is made to Alford and Beatty, "Principles of Industrial Management," rev. ed., pp. 324–325, Ronald, 1951.

Data for Inventory Computation

Size of order	Unit pur- chase price	Unit buying expense	Average inventory in units	Value of average inventory
1500	\$2.00	\$0.00500	1050	\$2100.00
2000	1.85	0.00375	1300	2405.00
2500	1.75	0.00300	1550	2712.50
3000	1.70	0.00250	1800	3060.00

1. The value of the average inventory if lots of 2000 units are purchased is **$2,405.00**   **la**

2. The unit buying expense if a lot of 3000 units is purchased is **$0.00250.**   **2d**

3. The cost to store 2000 units for 1 year at the average inventory per unit charge is 2000 × $0.30 = **$600.**   **3c**

4. The inventory carrying charge for an order of 2500 units is $2712.50 × 30 percent = **$813.75.**   **4a**

5.14

Inventory Cost Computations

Units ordered	1500	2000	2500	3000
Purchase cost = 15,000 × unit purchase price	$30,000	$27,750	$26,250	$25,500
Buying expense = 15,000 × unit buying expense	75	56	45	38
Cost of storage = $0.30 × average inventory	315	390	465	540
Inventory carrying charge = 30 percent × estimated value of average inventory	610	722	814	918
Total	$31,000	$28,918	$27,574	$26,996

The economic order size is **3000** units.   **b**

5.15   **1.**

Date	Change	Accumulative valuation
June 1	500 units at $100	$ 50,000
June 12	+500 units at $125	112,500
June 25	−600 units	
	500 at $125	50,000
	100 at $100	40,000
July 6	+500 units at $110	95,000
July 15	−600 units	
	500 at $110	40,000
	100 at $100	30,000
Aug. 3	+500 units at $105	82,500
Aug. 19	−600 units	
	500 at $105	30,000
	100 at $100	20,000

The value of the inventory on August 20 is **$20,000.**   **1d**

2.

Date	Change	Accumulative valuation
June 1	500 units at $100	$ 50,000
June 12	+500 units at $125	112,500
June 25	−600 units	
	500 at $100	62,500
	100 at $125	50,000
July 6	+500 units at $110	105,000
July 15	−600 units	
	400 at $125	55,000
	200 at $110	33,000
Aug. 3	+500 units at $105	85,500
Aug. 19	−600 units	
	300 at $110	52,500
	300 at $105	21,000

The value of the inventory on August 20 is **$21,000.**   **2b**

3.

Date	Change	Weighted average evaluation
June 1	500 units at $100	$500 \times \$100 = \$ 50{,}000$
June 12	+500 units at $125	$1000 \times \$112.5 = 112{,}500$
June 25	−600 units at $112.5	$400 \times \$112.5 = 45{,}000$
July 6	+500 units at $110	$900 \times \$111.111 = 100{,}000$
July 15	−600 units at $111.111	$300 \times \$111.111 = 33{,}333.33$

The value of the inventory on July 31 is **$33,333.33.**   **3a**

5.16 Reference is made to Alford and Beatty, "Principles of Industrial Management," rev. ed., pp. 390–392, Ronald, 1951, who in turn refer to formulas by J. W. Roe.

1.
$$N = \frac{I(A + B + C + 1/H) + LY}{S(1 + T)}$$

where N = number of pieces manufactured per year
I = cost of fixture
A = interest rate
B = fixed-charges rate
C = upkeep rate
H = estimated life of equipment
L = number of lots processed per year
S = estimated unit saving in direct labor cost
T = overhead saving due to direct labor saved
V = gross operating profit per year (less setups and fixed charges)
Y = estimated cost of each setup

$$N = \frac{\$1000(0.06 + 0.06 + 0.13 + 1/4) + \$50}{\$0.05(1 + 0.40)}$$

$$= \frac{1000(0.25 + 0.25) + 50}{0.05 \times 1.4} = \frac{550}{0.07} = \textbf{7857} \text{ units}$$ **1b**

2.
$$H = \frac{I}{NS(1 + T) - LY - I(A + B + C)}$$

$$= \frac{\$1000}{9200(\$0.05)(1.4) - \$50 - \$1000(0.25)}$$

$$= \frac{1000}{644 - 50 - 250} = \frac{1000}{344} = 2.91, \text{ or } \textbf{3} \text{ years}$$ **2c**

3. $V = NS(1 + T) - LY - I(A + B + C + 1/H)$
   $= 10,000(\$0.05)(1 + 0.40) - \$50 - \$1000(0.50)$
   $= \$700 - \$50 - \$500 = \mathbf{\$150}/\text{year} \quad \mathbf{3d}$

4. $I = \dfrac{NS(1 + T) - LY}{(A + B + C + 1/H)}$

   $= \dfrac{12,000(\$0.05)(1 + 0.40) - 4 \times \$50}{0.50}$

   $= \dfrac{\$840 - \$200}{0.50} = \mathbf{\$1280} \quad \mathbf{4c}$

5.17 1. The total current liabilities to the tangible net worth = $10,000/$29,000 = **0.345.**   **1a**

2. The fixed assets to tangible net worth = $25,000/$29,000 = **0.861.**   **2b**

3. The net working capital is the difference between the total current assets and the total current liabilities, or $13,000 − $10,000 = $3000. The ratio of the net sales to net working capital = $30,000/$3000 = **10.**   **3c**

5.18 3000 gal/year = 60 drums/year
Purchase price = 50 × $3.30 = $165/drum

Size of order in drums	Price per drum	Estimated drum-buying expense	Average inventory in drums	Maximum inventory in drums	Estimated value of average inventory
6	$165	$3.500	7	10	$1155
8	165	2.625	8	12	1320
10	165	2.100	9	14	1485
12	165	1.750	10	16	1650
15	165	1.400	11.5	19	1898
20	165	1.050	14	24	2310

	6	8	10	12	15	20
Buying expense = 60 × drum-buying expense	210.0	157.5	126.0	105.0	84.0	63.0
Cost of storage = 50 cents × maximum inventory	5.0	6.0	7.0	8.0	9.5	12.0
Inventory carrying charge = 12 percent × estimated value of average inventory	138.6	158.4	178.2	198.0	227.7	277.2
Total variable expense	353.6	321.9	311.2	311.0	321.2	352.2

Order in 12-drum lots.   c

5.19 Since no data were given, cost of setting up the well-point system will be considered as part of the rental charge. Number of days April through September = 30 + 31 + 30 + 31 + 31 + 30 = 183 days = 26 weeks + 1 day, including 3 legal holidays which fall on weekdays. Also, April 1 and September 30 fall on Sunday.

Wages one shift:
Weekdays (26 × 5) − 3	= 127 days @ $ 72 =	$ 9,144.00
Saturdays	= 26 days @ $108 =	2,808.00
Sundays + 3 legal holidays	= 30 days @ $144 =	4,320.00
		16,272.00
Wages on three shifts	= 3 × $16,272	= $48,816.00
Payroll taxes and insurance @ 13 percent		
	× $48,816	= 6,346.08
Total wages		$55,162.08
Rental charge = 6 × $1000	=	6,000.00
Fuel charge = 183 × $40	=	7,320.00
Wages, rental, and fuel	=	$68,482.08
Overhead and maintenance = 15 percent		
	× $68,482.08	= 10,272.31
		$78,754.39

Assume contractor borrowed $\dfrac{\$78,754}{0.96}$ at start of job

@ 8 percent/year for 6 months, discount in advance:

Interest (apparent @ 4 percent) = $82,035

$$- \$78,754 = \underline{\quad 3,281.00 \quad}$$
$$\$82,035.39$$

Profit and (assume and/or) contingency @ 10

$$\text{percent} = \underline{\quad 8,203.54 \quad}$$

Minimum lump-sum bid   d   = **$90,238.93**

5.20 $5000 = X \times$ capital-recovery factor (equal payment series)

$$= X \frac{0.04(1 + 0.04)^3}{(1 + 0.04)^3 - 1} = X(0.36035)$$

$$X = \$5000/0.36035 = \$13,875$$

$$\$13,875 + \$5000 = Y \frac{(1 + 0.04)^{17} - 1}{0.04}$$

$$\$18,875 = Y(23.698)$$

$$Y = \$18,875/23.698 = \textbf{\$796.48} \qquad \textbf{1a}$$

for **17** equal yearly payments.   **2c**

5.21 Selling price   = $10,000.00

Commission @ 5 percent = $ 500.00

1965 charge   =   100.00

1964 charge = $100(1.07) =   107.00

1963 charge = $107(1.07) =   114.49

1962 charge = $200(1.07)^3 =   245.01

$$\underline{\quad} \$1,066.50 \qquad \underline{\quad 1,066.50 \quad}$$

Net return at time of selling   = $ 8,933.50

Amount invested	Number of years	Product
$5000	6	30,000
100	5	500
100	4	400
($5150)	6	30,900

Weighted amount invested for 6 years = $5150

$$\$5150 (1 + i)^6 = \$8933.50$$

$$(1 + i)^6 = 8933.50/5150 = 1.73466$$

From compound-interest tables

$$(1.08)^6 = 1.587 \qquad (1.09)^6 = 1.772$$

$$\frac{1.73466 - 1.587}{1.772 - 1.587} = \frac{0.14766}{0.185} = 0.798$$

Rate of return = **8.8** percent   **b**

5.22 From 6 percent compound interest tables, use sinking-fund factors (SFF) starting with $n = 20$ and working up through $n = 13$, for uniform annual series payments.

End year	n	Value of mortgage × SFF	= Reduction in mortgage
0			= $100,000
	20	$100,000 × 0.02718	= − 2,718
1			$ 97,282
	19	$ 97,282 × 0.02962	= − 2,881
2			$ 94,401
	18	$ 94,401 × 0.03236	= − 3,054
3			$ 91,347
	17	$ 91,347 × 0.03544	= − 3,237
4			$ 88,110
	16	$ 88,110 × 0.03895	= − 3,432
5			$ 84,678
	15	$ 84,678 × 0.04296	= − 3,638
6			$ 81,040
	14	$ 81,040 × 0.04758	= − 3,856
7			$ 77,184
	13	$ 77,184 × 0.05296	= − 4,087
8			$ 73,097

Capital-recovery factor = 0.08718 for 20 equal payments of $8718 for 6 percent interest and principal reduction.

Capital-recovery factor = 0.08883, and 0.08883 × $73,097 = $6493.20 for 30 equal payments for 8 percent interest and principal reduction. Reduction of yearly payment = $8718 − $6493.20 = **$2,224.80.**   **c**

5.23 The compound amount of a fund A for a series of 20 $100 yearly payments at 10 percent at the end of the twentieth year (from interest tables).

$$\text{Fund } A = \$100 \, \frac{(1 + 0.10)^{20} - 1}{0.10}$$

$$= \$100 \times 57.275 = \$5727.5$$

but   Fund $A - 0.10 \times$ fund $A = 0.90 \times$ fund $A = \$5727.5$
and   Fund $A = 5727.5/0.90 = \$6363.8$

Let ratio of actual fund needed to fund $A = X$. Then

$$\$100X + 0.10(\$6363.8X) = \$2000$$
$$(100 + 636.4)X = 2000$$
$$X = 2000/736.4 = 2.716$$

$\$636.8 \times 2.716 = \mathbf{\$17{,}284} = $ max cost of sprinkler system   **a**

Check:

Loss of 10 percent interest/year on $\$17{,}284 = 1728.4$
and $0.9 \times 17{,}284 = \$15{,}556$

$(\$2000 - \$1728.4) \times 57.275 = \$271.6 \times 57.275 = \$15{,}556$

5.24 $\$500 - X \times$ capital-recovery factor (equal payment series)

$$= X \frac{0.04 (1 + 0.04)^8}{(1 + 0.04)^8 - 1}$$

$$= X(0.14853)$$

$$X = \$500/0.14853 = \$3366.32$$

$$\text{Present worth} = \frac{\$500 + \$3366.32}{(1 + 0.04)^{11}}$$

$$= \$3866.32 \times 0.6496$$

$$= \mathbf{\$2{,}511.56} = \text{single payment required} \qquad \mathbf{d}$$

5.25

End year	Cost	Trade-in value	Net	Cost per year
1	$3,000	$1,800	$1,200	$1,200
	+ 100	− 200		
2	3,100	1,600	1,500	750
	+ 200	− 200		
3	3,300	1,400	1,900	633
	+ 300	− 200		
4	3,600	1,200	2,400	600
	+ 400	− 200		
5	4,000	1,000	3,000	600
	+ 500	− 200		
6	4,500	800	3,700	617

Trade in car at end of fifth year.   **c**

6.01–6.55

Mechanical Engineering

6.01 c_v = specific heat at constant volume = 0.1714 for air
R_{air} = 53.3 p_2 = 155 psia (1068.7 kPa abs)
t_1 = 100°F (37.78°C) Q_k = 50 Btu (52.8 kJ)
p_1 = 13.75 psia (94.81 kPa abs) k = 1.4
V_1 = 1 ft³ (0.02832 m³)

1. $p_1 V_1^k = p_2 V_2^k$

$$(13.75 \times 144)(1)^k = (155 \times 144)V_2^k$$

$$V_2^{1.4} = \frac{(13.75)(1)}{155} = 0.0887$$

$$V_2 = 0.1772 \text{ ft}^3 \ (0.0050 \text{ m}^3)$$

$$\text{Compression ratio} = r_k = \frac{V_1}{V_2} = \frac{1}{0.1722} = \textbf{5.642} \qquad \textbf{1a}$$

2. Percent clearance $= \dfrac{V_2}{V_1 - V_2} = \dfrac{0.1722}{1 - 0.1772}$

$$= \textbf{21.5 percent} \qquad \textbf{2b}$$

3. Q_A = area under 2 − 3 on Ts diagram
   $= wc_v(T_3 - T_2)$
   T_3 = temperature after combustion
   p_3 = pressure after combustion

$$w = \frac{p_1}{RT_1} = \frac{13.75 \times 144}{(53.3)(100 + 460)} = 0.0663$$

$$T_2 = \frac{p_2 V_2}{wR} = \frac{(155 \times 144)(0.1772)}{(0.0663)(53.3)}$$

or $\qquad T_2 = 1119°\text{R}$

$$Q_A = 50 = (0.0663)(0.1714)(T_3 - 1119)$$

or $\qquad T_3 = \textbf{5518}°\text{R} \ (\textbf{3066 K}) \qquad \textbf{3c}$

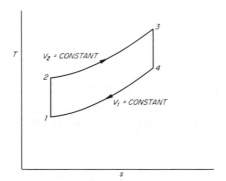

Fig. 6.01a

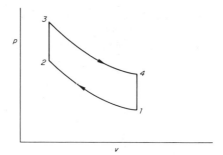

Fig. 6.01b

4. $C_2 = p_3 = \dfrac{wRT_3}{V_3} \qquad V_3 = V_2$

$$= \frac{(0.0663)(53.3)(5518)}{0.1772} = 110,000 \text{ psf}$$

$$= \textbf{764.2} \text{ psia } (\textbf{5269} \text{ kPa abs}) \qquad \textbf{4d}$$

5. Net work per cycle = area enclosed on Ts diagram

$$W = Q_A - Q_R = Q_A - wc_v(T_4 - T_1)$$

To find T_4:

$$p_3 V_3^k = p_4 V_4^k$$
$$(764.2 \times 144)(0.1772)^{1.4} = p_4(1)^{1.4} \quad \text{or} \quad p_4 = 9759 \text{ psf } (467.3 \text{ kPa})$$

$$p_4 V_4 = wRT_4$$
$$(9759)(1) = (0.0663)(53.3)T_4 \quad T_4 = 2762°R$$

$$
\begin{aligned}
W &= 50 - (0.0663)(0.1714)(2762 - 560) \\
&= 24.98 \text{ Btu } (26.38 \text{ kJ}) \\
&= (24.93)(778) = \mathbf{19{,}432} \text{ ft·lb } (26{,}346 \text{ N·m}) \qquad \mathbf{5b}
\end{aligned}
$$

6. Heat rejected = $Q_R = wc_v(T_4 - T_1)$
$$
\begin{aligned}
&= (0.0663)(0.1714)(2762 - 560) \\
&= \mathbf{25.02} \text{ Btu } (\mathbf{26.40} \text{ kJ}) \qquad \mathbf{6a}
\end{aligned}
$$

7. m.e.p. = net work per $\dfrac{\text{cycle}}{\text{stroke}}$ length

$$= \frac{W}{V_1 - V_2} = \frac{19{,}432}{1 - 0.1772} = 23{,}617 \text{ psf}$$

$$= \frac{23{,}617}{144} = \mathbf{164} \text{ psi } (\mathbf{1131} \text{ kPa}) \qquad \mathbf{7c}$$

6.02 $q = 10$ Btu/(h)(s ft^2) [31.52 W/(m^2)(h)] of outside area of steel pipe. (*Reference:* Kreith, "Principles of Heat Transfer," 3d ed., Intext, 1973.)

$$q = \frac{t_1 - t_0}{\dfrac{r_{oa}}{h_s r_{is}} + \dfrac{r_{oa}}{k_s} \ln\left(\dfrac{r_{os}}{r_{is}}\right) + \dfrac{r_{oa}}{k_a} \ln\left(\dfrac{r_{oa}}{r_{os}}\right) + \dfrac{1}{h_o}}$$

where r_{is} = inside radius of pipe
      r_{oa} = outside radius of asbestos

$$10 =$$
$$\frac{300 - 70}{\dfrac{r_{oa}}{(20)(2.4485)} + \dfrac{r_{oa}}{(30)(12)} \ln\left(\dfrac{3.3125}{2.4485}\right) + \dfrac{r_{aa}}{(0.06)(12)} \ln\left(\dfrac{r_{oa}}{3.3125}\right) + \dfrac{1}{6}}$$

$$1 = \frac{23}{\dfrac{r_{oa}}{48.97} + \dfrac{r_{oa}}{360}(1.1977 - 0.8955) + \dfrac{r_{oa}}{0.720}(\ln r_{oa} - 1.1977) + \dfrac{1}{6}}$$

$$23 = + 0.0204 r_{oa} + 0.0008 r_{oa} + 1.389 r_{oa}(\ln r_{oa} - 1.1977) + 0.1666$$

$$1.389(\ln r_{oa}) - 1.642 = \frac{22.833}{r_{oa}}$$

By trial: $r_{oa} = 12$ in (30.48 cm).

$$1.389(2.485) - 1.642 = 22.833/12$$
$$3.41 - 1.64 = 1.77 < 1.90$$

Try $r_{oa} = 3.31 + 9 = 12.31$ in (31.27 cm).

$$1.389(2.510) - 1.64 = 22.833/12.31$$
$$3.49 - 1.64 = 1.85 = 1.85$$

Use **9**-in (**22.86**-cm)-thick pipe covering $[r_{oa} = 12.31$ in (31.27 cm)]   **d**

6.03 *a*. Figure 6.03*a* is a line diagram of the apparatus showing the flow of steam and condensate in the system.

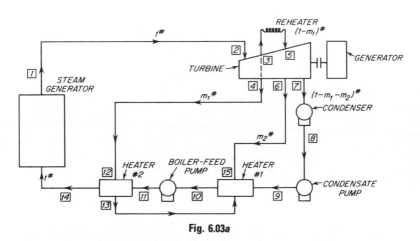

Fig. 6.03*a*

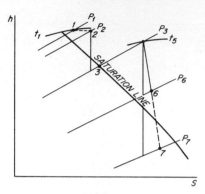

Fig. 6.03b

b. The *hs* diagram is shown in Fig. 6.03b. All values taken from a Mollier chart or steam tables are as follows:

Point	p	t	h
1	700	744.1	1371.1
2	655	740.0	1371.1 (turbine)
3	109	334.1	1188.7
4	109	334.1	1188.7
5	109	700.0	1378.4
6	15	270.0	1177.6
7	1	101.7	973.0
8	1	101.7	69.7
9	15	101.7	69.7 (approx)
10	15	213.0	181.1
11	700	214.0	183.2
12	109	334.1	1188.7
13	109	334.1	305.0
14	700	324.1	296.8
15	15	270.0	1177.6

c. Temperature leaving boiler: $p_1 = 700$ psia (4826.5 kPa abs), $h_i = 1371.1$ Btu/lb (3191.9 kJ/kg), so $t_1 = $ **744.1°F (395.6°C)**.

d. Temperature of feedwater leaving the last heater:

$$t_{14} = t_{13} - 10° = 334.1° - 10° = \textbf{324.1°F (162.3°C)}$$

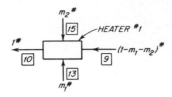

Fig. 6.03e

e. The heat balance around heater no. 1 is shown in Fig. 6.03*e.*

$$m_2 h_{15} + (1 - m_1 - m_2)h_9 + m_1 h_{13} = h_{10}$$
$$m_2(1177.6) + (1 - 0.1262 - m_2)(69.7) + (0.1262)(305) = 181.1$$
$$m_2 = \textbf{0.074} \text{ lb/lb (kg/kg) throttle steam}$$

f. The *hs* diagram is shown in Fig. 6.03*f.*

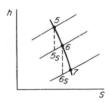

$$\frac{h_5 - h_6}{h_5 - h_{5,s}} = 0.80$$

$$h_6 = h_5 - 0.80(h_5 - h_{5,s})$$
$$= 1378.4 - 0.80(1378.4 - 1171)$$
$$= 1212 \text{ Btu/lb (2821.5 kJ/kg)}$$

Fig. 6.03f

$$\frac{h_6 - h_7}{h_6 - h_{6,s}} = 0.80$$

$$h_7 = h_6 - 0.80(h_6 - h_{6,s})$$
$$= 1212 - 0.80(1212 - 1026) = 1063 \text{ Btu/lb (2474.7 kJ/kg)}$$

For $p_7 = 1$ psia and $h_7 = 1063$ Btu/lb (from the Mollier chart), there is 3.2 percent moisture. The quality of the steam at exhaust = **96.8** percent.

g. From the Mollier diagram using h_2 as h for the turbine entrance condition,

$$h_{2,s} = \text{enthalpy of exhaust} = \textbf{900} \text{ Btu/lb (\textbf{2095.2} kJ/kg)}$$

6.04   Assume cement plaster: $k_1 = k_4 = 8.0$

$$Q = \frac{A(t_1 - t_2)}{1/h_1 + L_1/K_1 + L_2/K_2 + L_3/K_3 + L_4/K_4 + 1/h_2}$$

$$= \frac{1 \times 40}{1/1.65 + 0.5/8 + 1.61 + 2/0.30 + 0.5/8 + 1/1.65}$$

$$= \frac{40}{0.606 + 0.063 + 1.61 + 6.67 + 0.063 + 0.606}$$
$$= 40/9.62 = \textbf{4.16} \text{ Btu/(ft}^2)(\text{h}) \ (\textbf{13.11 W/m}^2) \qquad \textbf{a}$$

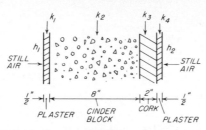

Fig. 6.04

where $t_1 - t_2 = 40°F\ (4.44°C)$

A = area = 1 ft² (0.0929 m²)

L = thickness of material, in (cm)

K = thermal conductivity, Btu·in/(h)(ft²)(°F) [0.144 W/(m²)(°C)]

H = surface coefficient of heat transmission, Btu/(h)(ft²)(°F) [5.67 W/(m²)(°C)]

(From "Heating, Ventilating, and Air Conditioning Guide," ASHVE, 1950.)

For 8-in (20.32-cm) cinder block, $L_2/K_2 = 1.61$.

6.05 See answer 6.04.

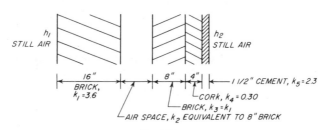

Fig. 6.05

$$Q = \frac{A(t_1 - t_2)}{1/h_1 + L_1/K_1 + L_2/K_2 + L_3/K_3 + L_4/K_4 + L_5/K_5 + 1/h_2}$$
$$= \frac{1(70 - 30)}{1/1.65 + 16/3.6 + 8/3.6 + 8/3.6 + 4/0.3 + 1.5/2.3 + 1/1.65}$$

$$= 40/24.1 = 1.66 \text{ Btu/ft}^2 \text{ per hour}$$
$$= \mathbf{39.8} \text{ Btu/(ft}^2)(°F)[\mathbf{126} \text{ W/(m}^2)(°C)] \text{ per day} \qquad \mathbf{b}$$

6.06

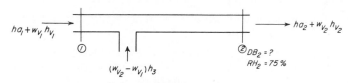

Fig. 6.06

Where RH = relative humidity
DB = dry-bulb temperature, °F
ha = enthalpy of dry air, Btu/lb dry air
hf = enthalpy of water spray = Btu/lb water
w_v = lb vapor/lb dry air
v = ft^3/lb dry air
w_a = lb dry air/ft^3
Evaluating the entering energy to the leaving energy for 1 lb (0.4536 kg) dry air,

$$h_{a1} + w_{v1}h_{v1} + (w_{v2} - w_{v1})h_3 = h_{a2} + w_{v2}h_{v2} \qquad (1)$$

a. $h_{a1} + w_{v1}h_{v1} = 23.7$ Btu/lb dry air* (55.17 kJ/kg)

$$w_{v1} = 44 \text{ gr vapor/lb dry air*}$$
$$= 44/7000 = 0.0063 \text{ lb vapor/lb dry air}$$
$$\therefore 23.7 + (w_{v2} - 0.0063)18.1 = h_{a2} + w_{v2}h_{v2} \qquad (2)$$

A trial-and-error solution is applicable. Assume DB_2. Use chart to find $h_{a2} + w_{v2}h_{v2}$ and w_{v2} based on this assumption. Apply Eq. (2) to check validity of assumption. As an aid in choosing a trial DB_2, note that $(w_{v2} - 0.0063)18.1$ is very small compared to the other quantities in Eq. (2). $\therefore h_{a,2} + w_{v2}h_{v2}$ is approximately equal to 23.7 and RH_2 = 75 percent (given).

First trial: $DB_2 = 61°F$, $RH_2 = 75$ percent, gives

$$h_{a2} + w_{v2}h_{v2} = 24 \text{ Btu/lb dry air*}$$
and $\qquad w_{v2} = 60/7000 = 0.0086$ lb vapor/lb dry air*

Check: $\qquad 23.7 + (0.0083 - 0.0063)(18.1) = 23.74$, or 24

*Using psychrometric chart.

Second trial: Assume $DB_2 = 60.5°F$ (15.85°C):

$$\therefore h_{a2} + w_{v2}h_{v2} = 23.8 \text{ Btu/lb dry air*}$$

$$w_{v2} = 59/7000 = 0.0084 \text{ lb vapor/lb dry air*}$$

Check: $23.7 + (0.0084 - 0.0063)(18.1) = 23.74$, or 23.8

This is close enough. Therefore $v_1 = 13.48$ ft^3/lb dry air* and $w_{a1} = 1/13.48 = 0.0743$ lb dry air/ft^3.

Weight of incoming air $= (1800)(0.0743) = 134$ lb/min
$\therefore$ Water required $= 134(w_{v2} - w_{v1}) = 134(0.0084 - 0.0063)$
$$= \textbf{0.282} \text{ lb } (\textbf{0.128} \text{ kg})/\text{min}$$

$$DB_2 = \textbf{60.5}°F \text{ and } WB_2 = \textbf{55.8}°F \text{ } (\textbf{13.22}°C)$$

b. Equation (2) becomes

$$23.7 + (w_{v2} - 0.0063)1150 = h_{a2} + w_{v2}h_{o2}$$

First trial: As in (*a*), assume $DB_2 = 71°F$ (21.7°C).

$$\therefore (h_{a2} + w_{v2}h_{v2}) = 30.4 \text{ Btu/lb dry air*}$$
$$w_{v2} = 85/7000 = 0.01214 \text{ lb steam/lb dry air*}$$

Check: $23.7 + (0.0121 - 0.0063)(1150) = 30.4$

$$\text{Steam required} = \frac{\text{lb air}}{\text{min}} \frac{\text{lb steam}}{\text{lb air}} = \text{lb/min}$$

$$134(w_{v2} - w_{v1}) = 134 \times 0.0058 = \textbf{0.785} \text{ lb } (\textbf{0.356} \text{ kg})$$
$$\text{steam/min}$$

$$DB_2 = \textbf{71}°F \quad \text{and} \quad WB_2 = \textbf{65.5}°F \text{ } (\textbf{18.61}°C)$$

6.07 See answer 6.06 for symbol notation.

$$W_{v,B} = w_{v,A} - (w_{v2} - w_{v1}) = \text{lb departing water/lb dry air}$$

$$\frac{\text{Dry air, lb}}{\text{Water, lb}} = \frac{1}{w_{v,A}} = \frac{h_{f,A} - h_{f,B}}{0.24(t_2 - t_1) + w_{v2}(h_{v2} - h_{f,B}) - w_{v1}(h_{v1} - h_{f,B})}$$

$$= \frac{93.91 - 48.0}{0.24(115 - 85) + w_{v2}(1111.6 - 48) - w_{v1}(1098.8 - 48)}$$

$$\text{Eq. (1)} = \frac{45.9}{7.2 + w_{v2}(1063.6) - w_{v1}(1050.8)}$$

$$P_{v1} = RH_1 \times P_{\text{sat}} = 0.47 \times 0.596 = 0.280 \text{ psi}$$

and

$$P_{v2} = 1.471 \text{ psi}$$

*Using psychrometric chart.

$$w_{v1} = 0.622 \frac{P_{v1}}{P_a - P_{v1}} = 0.622 \frac{0.280}{14.7 - 0.280}$$

$$= 0.0121 \text{ lb vapor/lb dry air}$$

$$w_{v2} = 0.622 \frac{1.471}{14.7 - 1.47} = 0.0693 \text{ lb vapor/lb dry air}$$

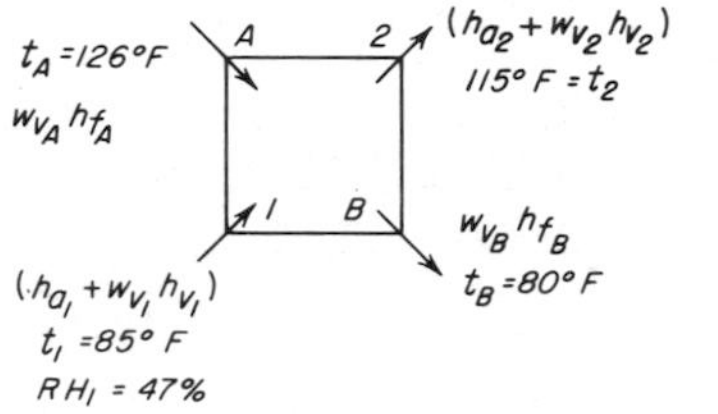

Fig. 6.07 Energy diagram for a cooling tower where weight of dry air is 1 lb. h_{v2} = h_{g2} since air leaves in a saturated condition; P_v = vapor pressure, psi; P_{sat} = saturated vapor pressure, psi.

1. Eq. (1) = $\dfrac{45.9}{7.2 + 0.0693(1063.6) - 0.0121(1050.8)}$

   $$= 0.674 \text{ lb dry air/lb water}$$

   $$V = \frac{WRT}{P} = \frac{0.674 \times 53.3(85 + 460)}{(14.7 - 0.28) \times 144}$$

   $$= \textbf{9.43} \text{ ft}^3 \text{ d/lb } (\textbf{0.588} \text{ m}^3\text{/kg}) \qquad \textbf{1c}$$

   $$= 9.43 \text{ ft}^3 \text{ wet air/lb water (by Dalton's law)}$$

   $$\frac{\text{Wet air, lb}}{\text{Water, lb}} = \frac{w_{v1}}{w_{v,A}} + \frac{1}{w_A} = 0.0121 \times 0.674 + 0.674 = 0.682$$

2. $\dfrac{\text{Air, ft}^3\text{/min}}{\text{Air, ft}^3\text{/lb water}} = \dfrac{2000}{9.43} = \textbf{212} \text{ lb } (\textbf{96.16} \text{ kg})\text{/min} \qquad \textbf{2d}$

6.08 Outside air = 7.5 ft^3/min per person (not smoking)
1. Total outside air = 7.5 × 1800
   $$= \textbf{13,500} \text{ ft}^3 \ (\textbf{382.3} \text{ m}^3)\text{/min} \qquad \textbf{1a}$$
2. Sensible heat loss per person = 225 Btu/h

   Sensible heat loss of audience = 225 × 1800 = 405,000 Btu/h

   Solar heat loss = 120,000 Btu/h

   Total heat loss = Q_s = 525,000 Btu/h

   $$Q_s = W_s(0.24)(t_i - t_e)$$

where t_i = temperature in auditorium, °F
t_e = temperature of conditioned air
W_s = conditioned air per hour, lb

$$525,000 = W_s \times 0.24(78 - 65)$$
$$W_s = 168,000 \text{ lb air/h}$$

From psychrometric chart, air at 65°F and at an estimated relative humidity = 70 percent shows specific volume = 13.4 ft³/lb dry air.

$$\text{Air circulated} = \frac{\text{Air, lb/h}}{60 \text{ min}} \frac{\text{ft}^3}{\text{lb dry air}} = \frac{168,000}{60} \times 13.4$$

$$= \textbf{37,400} \text{ ft}^3 \text{ (\textbf{1059.2} m}^3)/\text{min} \qquad \textbf{2b}$$

3. From "Heating, Ventilating, and Air Conditioning Guide," chap. 6, fig. 7, curve D, ASHVE, 1950.

165 Btu/h = latent heat loss per person
Latent heat loss of audience = 165 × 1800 = 296,000 Btu/h
Moisture per person = 1100 gr
Moisture from audience, gr = 1100 × 1800 = 1,980,000 gr/h

$$\frac{\text{Moisture/h}}{\text{Dry air circulated/h, lb}} = \frac{1,980,000}{168,000}$$

$$= 11.8 \text{ gr moisture/lb dry air}$$

Moisture content of inside air at 78 to 67°F = 82 gr/lb dry air (from chart).

$$\text{Moisture content of entering air} = 82.0 - 11.8$$
$$= 70.2 \text{ gr/lb dry air}$$

Air with 70.2 gr moisture/lb dry air and 65°F has a wet-bulb temperature = **60.2**°F (**15.67**°C)   **3c**. Specific volume = 13.43 ft³/lb dry air is sufficiently close to 13.40 ft³/lb dry air used in part *b*.

6.09 1. Sensible heat load = Q_S = $W \times C_P(t_a - t_b)$

$$W = \frac{Q_S}{C_P(t_a - t_b)} = \frac{450,000}{0.24(75 - 65)} = \textbf{187,500} \text{ lb (\textbf{85.05} Mg)/h} \qquad \textbf{1b}$$

2. $\dfrac{\text{Moisture, gr}}{\text{Dry air, lb}} = \dfrac{\text{gr moisture/h}}{\text{lb dry air/h}} = \dfrac{1,200,000}{187,500}$

$$= 6.40 \text{ gr/lb dry air to be removed}$$

$$\frac{\text{Moisture, gr}}{\text{Supply air, lb}} = 78 - 6.40 = 71.6 \text{ gr/lb air}$$

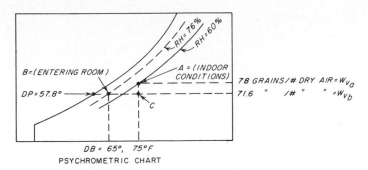

Fig. 6.09

From Fig. 6.09, $DB = 65°F$ at 71.6 gr/lb air and dew point = **57.8**°F (**14.33**°C)   **2c** and $RH = $ **76** percent   **2h**.

3. $\dfrac{1,200,000}{7000} = $ **171.2** lb (**77.66** kg) moisture/h   **3a**

Heat load picked up in auditorium

$$= \frac{\text{lb moisture}}{\text{h}} \; \frac{\text{Btu}}{\text{lb moisture}} \quad \text{(from psychometric chart)}$$

$$= \frac{\text{lb moisture}}{\text{h}} \, h_{fg} = 171.2 \times 1051 = 180,000 \text{ Btu/h}$$

6.10 h_f at 1 psia = 69.70 Btu/lb

$w_{cw} = $ lb circulating water/h

Heat given up by steam = heat picked up by water

$(1 \times 10^6)(1090 - 69.70) = w_{cw}(95 - 85)$

$w_{cw} = (1 \times 10^6)(1020.3)/10 = 102 \times 10^6$ lb/h

$= (102 \times 10^6)/(3600 \times 62.4)$ 454 ft³/s

Flow area $= Q/V = {}^{454}\!/_6 = 75.8$ ft²

Inside tube area $= \pi(0.875^2)/(4 \times 144) = 0.00417$ ft²

No. of tubes/pass $= 75.8/0.00417 = $ **18,200** tubes   **1b**

$q = vA_0\Delta t_m$ \quad where $\Delta t_m = $ log mean temperature difference

$$= \frac{\theta_{max} - \theta_{min}}{\log_e (\theta_{max}/\theta_{min})}$$

$$= \frac{16.74 - 6.74}{\log_e (16.74/6.74)} = \frac{10}{0.912} = 11°F$$

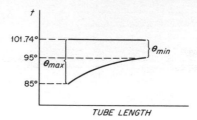

Fig. 6.10

where Fig. 6.10 shows $\theta_{max} = 101.74 - 85 = 16.74°F$ and
$\theta_{min} = 101.74 - 95 = 6.74°F$

$$(1 \times 10^6)(1090 - 69.70) = 480 \times A_0 \times 11$$
$$A_0 = (1 \times 10^6 \times 1020.3)/(480 \times 11) = 193,000 \text{ ft}^2$$
$$= \pi \times d_0 \times L \text{ tubes/pass} \times 2 \text{ passes}$$

where $d_0 = \frac{1}{12}$ ft.

$$L = 193,000 \times 12/3.14 \times 18,200 \times 2 = \mathbf{20.3} \text{ ft } (\mathbf{6.187} \text{ m}) \qquad \mathbf{2c}$$

6.11 For $P_1 = 1$ psig $= 15.75$ psia (101.35 kPa abs) at 12 percent moisture, $h_1 = 1037$ Btu/lb (2,412,1 kJ/kg).

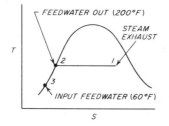

Fig. 6.11

Neglecting the effect of subcooling, which is nil in this case,
For $t_2 = 200°F$, $\qquad h_2 = 168$ Btu/lb
For $t_3 = 60°F$, $\qquad h_3 = 28.1$ Btu/lb

Heat absorbed by water = heat given up by steam
$$1000(h_2 - h_3) = w_s \text{ lb } (h_1 - h_2)$$
$$1000(168 - 28) = w_s \text{ lb } (1037 - 168)$$
$$w_s = 1000 \times 140/869 = \mathbf{161} \text{ lb } (\mathbf{73.03} \text{ kg}) \qquad \mathbf{a}$$

6.12 $c_{P,A}$ = specific heat of air (constant pressure)
$c_{P,G}$ = specific heat of flue gas (constant pressure)

$$\Delta t_m = \text{log mean temperature difference} = \frac{\Delta t_1 - \Delta t_2}{\log_e (\Delta t_1/\Delta t_2)}$$

$$= \frac{(t_{G1} - t_{A1}) - (t_{G2} - t_{A2})}{\log_e (t_{G1} - t_{A1})/(t_{G2} - t_{A2})} = \frac{325 - 246}{\log_e {}^{325}\!/_{246}} = \frac{79}{\log_e 1.321}$$

$$= \frac{79}{0.278} = \textbf{284}°\text{F } (\textbf{140}°\text{C}) \quad \textbf{1b}$$

$$Q = W_A c_{P,A}(t_{A2} - t_{A1}) = 357{,}000 \times 0.24 \times 399$$
$$= \textbf{34.2} \times \textbf{10}^6 \text{ Btu/h } (\textbf{100.2} \times \textbf{10}^5 \text{ W}) \quad \textbf{2c}$$

$$Q = W_G c_{P,G}(t_{G2} - t_{G1}) \quad \text{or} \quad W_G = \frac{Q}{c_{P,G}(t_{G2} - t_{G2})}$$

$$W_G = \frac{34.2 \times 10^6}{0.24 \times 320} = \textbf{447,000} \text{ lb } (\textbf{202,759} \text{ kg}) \text{ flue gas} \quad \textbf{3d}$$

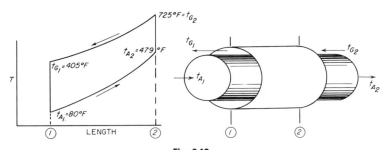

Fig. 6.12

Note: This specific heat of the flue gas is equal to the weighted average of the specific heats of its constituents. A typical flue-gas analysis is

$$
\begin{aligned}
CO_2 &= 0.15 \times 0.202 = 0.0303 \\
N_2 &= 0.80 \times 0.248 = 0.1983 \\
O_2 &= \underline{0.05} \times 0.219 = \underline{0.0110} \\
&\quad 1.00 \qquad\qquad\quad 0.2396, \text{ or } 0.24
\end{aligned}
$$

6.13   Let V_1 = steam velocity leaving nozzle, ft/s
V_{1r} = relative velocity entering blade, ft/s
V_2 = absolute exit velocity, ft/s
V_{2r} = relative velocity leaving blade, ft/s
Δh = available energy, Btu/lb
K = velocity coefficient
V_b = blade velocity, ft/s

β = blade-entrance angle, degrees
F = force/(lb steam)(s)
w = lb steam/s

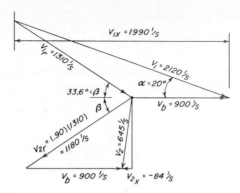

Fig. 6.13

a. $V_1 = 223.8\ k\ \sqrt{\Delta h} = 223.8 \times 0.95 \times \sqrt{100}$
  $= $ **2120** ft/s (**646.2** m/s)
b. **1310** ft/s (**399.3** m/s)
c. **1180** ft/s (**359.7** m/s)
d. **645** ft/s (**196.6** m/s)
e. $\beta = $ **33.6°**

f. $F = \dfrac{V_{1,x} - V_{2x}}{g} = \dfrac{1990 - (-84)}{32.2} = \dfrac{\textbf{64.4 lb (29.2 kg)}}{\text{lb(kg) steam/s}}$

g. Blade hp $= \dfrac{W(V_{1x} - V_{2x})V_b}{g \times 550} = \dfrac{1800}{3600} \times \dfrac{64.4 \times 900}{550}$

  $= $ **52.6** hp (**39.22** kW)

h. Shaft hp = blade hp − friction loss hp − windage loss hp
  $= 52.6 - 6 - 2 = $ **44.6** hp (**33.26** kW)

i. Losses converted into enthalpy of steam
  (1) Nozzle loss $= (1 - k^2)100$
  $= [1 - (0.95)^2]\ 100 = $ 9.75 Btu/lb

  (2) Blade friction loss $= \dfrac{V_{1,r} - V_{2r}}{2gJ}$

  $= \dfrac{(1310)^2 - (1180)^2}{64.4 \times 778} = $ 6.52 Btu/lb

(3) Residual velocity loss

$$= \frac{V_2^2}{2gJ} = \frac{(645)^2}{64.4 \times 778} = 8.30 \text{ Btu/lb}$$

(4) Friction between steam and wheel

$$= \frac{\text{hp} \times 550}{wJ} = \frac{6 \times 550 \times 2}{778} = 8.48 \text{ Btu/lb}$$

$$\text{total of (1) to (4)} = \overline{33.05 \text{ Btu/lb}}$$

$$\text{Enthalpy of exhaust steam} = 1200 - 100 + 33$$
$$= \textbf{1133} \text{ Btu/lb} (\textbf{3637.6} \text{ kJ/kg})$$

6.14   Load factor = (average power)/(peak power)

Efficiency of steam-generating unit = 80 percent

Heating value of Pittsburgh bituminous coal = HV
$$= 13{,}400 \text{ Btu/lb}$$

Average steam rate = 11 lb steam/kWh

Average power = (load factor)(peak power per turbine)
$$= 0.75 \times 10{,}000 = 7500\text{-kW output per turbine}$$

$$\text{Efficiency} = \frac{w_n(h_3 - h_1)}{w_f(HV)}$$

Obtain h_3 and h_1 from tables and charts

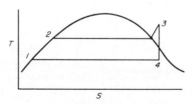

Fig. 6.14

1. w_n = lb steam supplied/h = (steam rate)(kW output)
$$= 11 \times 7500 = 82{,}500 \text{ lb steam/h}$$
w_f = weight of fuel fired

$$= \frac{82{,}500(1318 - 69.1)}{0.80 \times 13{,}400} = 9620 \text{ lb/(h)(turbine)}$$

$$\text{Tons coal/day for 4 units} = \frac{4 \times 9620 \times 24}{2000}$$

$$= \mathbf{461}\ (\mathbf{418.2}\ \text{Mg}) \qquad \mathbf{1a}$$

2. Assume 10°F rise in cooling water and no subcooling.

$$Q_R = \text{heat rejected by four turbines} = 4w_n(h_4 - h_1)$$
$$= W_W(\Delta T) = 4 \times 82,500(921 - 69.1)$$
$$= 281,000,000\ \text{Btu/h}$$

$$W_W = \frac{Q_R}{\Delta T} = \frac{281,000,000}{10} = 28,100,000\ \text{lb/h}$$

$$\text{Gal. water/min} = \frac{28,100,000 \times 7.48}{60 \times 62.4}$$

$$= \mathbf{56,200}\ (\mathbf{212.7}\ \text{kl/min}) \qquad \mathbf{2b}$$

6.15 $P_1 = 20$ psia $h_1 = 606.2$ Btu/lb
$P_2 = 190$ psia $h_2 = 753$ Btu/lb
$t_3 = 80°F$ $h_3 = h_4 = 132$ Btu/lb
$S_1 = S_2 = 1.3700$

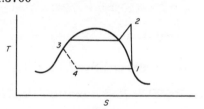

Fig. 6.15

1a. $W = h_2 - h_1 = 753 - 606.2$
$\qquad\qquad\quad = \mathbf{146.8}$ Btu/lb ($\mathbf{341.6}$ kJ/kg)   **1a**
 b. $Q_R = h_2 - h_3 = 753 - 132$
$\qquad\qquad\quad = \mathbf{621}$ Btu/lb ($\mathbf{1445.7}$ kJ/kg)   **1d**
 c. $Q_A = $ refrigeration $= h_1 - h_4 = 606.2 - 132$
$\qquad\quad = \mathbf{474.2}$ Btu/lb ($\mathbf{1103.9}$ kJ/kg)   **1c**

2. cop $= \dfrac{QA}{W} = \dfrac{474.2}{146.8} = \mathbf{3.23}$ **2b**

3. Circulation $= \dfrac{200\ \text{Btu/(ton)(min)}}{474.2\ \text{Btu/lb}}$

$\qquad\qquad = \mathbf{0.422}$ lb/ton ($\mathbf{0.211}$ kg/Mg) per min of
$\qquad\qquad\quad$ refrigeration   **3d**

6.16
$$\frac{W_{bo}}{W_f} = 12.77$$

$$\text{Boiler efficiency} = \frac{W_{bo}(h_2 - h_1)}{W_f \times q_H}$$

From chart,

$$h_2(200 \text{ psia } 87°\text{F superheat}) = 1251.5 \text{ Btu/lb}$$
$$h_1(93°\text{F}) = 61 \text{ Btu/lb}$$

$$qh = \frac{12.77(1251.5 - 61)}{0.828}$$

$$= \textbf{18,400} \text{ Btu/lb } (\textbf{42,835.2} \text{ kJ/kg}) \qquad \textbf{a}$$

6.17   Overall boiler efficiency $= \dfrac{W_{bo}(h_2 - h_1)}{W_f \times q_H}$

where W_{bo} = lb steam delivered by boiler/h
$\quad\;\; h_2$ = enthalpy of delivered steam
$\quad\;\; h_1$ = enthalpy of water entering boiler
$\quad\;\; W_f$ = lb fuel fired/h
$\quad\;\; q_H$ = higher heating value of fuel/lb

$$W_{bo} = \frac{885,000}{4} = 221,250 \text{ lb steam/h}$$

$$h_2(\text{at 400 psi, 700°F}) = 1362.7 \text{ Btu/lb} \qquad h_1(280) = 249.1$$
$$W_f = \frac{221,250(1362.7 - 249.1)}{0.825 \times 13,850} = 21,500 \text{ lb/h}$$

$$= 21,500/2000 = \textbf{10.75} \text{ short tons } (\textbf{9752.4} \text{ kg)/h} \qquad \textbf{b}$$

6.18   1. A boiler hp is equal to the power necessary to evaporate 34.5 lb water/h from saturated water at 212°F to saturated steam at 212°F.

$$1 - \text{boiler hp} = \textbf{33,475} \text{ Btu } (\textbf{9805} \text{ W})/\text{h} \qquad \textbf{1c}$$

2. $1 - $ mechanical hp $= 33,000$ ft·lb/min 1 Btu $= 778$ ft·lb
$$= 33,000/778 = 42.4 \text{ Btu/min}$$

$$\therefore 1 - \text{boiler hp} = \frac{33,475/60}{42.4} = \textbf{13.15} \text{ mechanical hp} \qquad \textbf{2d}$$

3. The factor of evaporation is the ratio that changes the actual evaporation rate (lb/h) to the rate that would result if the boiler were operated from and at 212°F.

Equiv. evaporation = (actual evaporation)(factor of evaporation)
= 4000 × 1.06 = 4240 lb/h

$$\text{Boiler hp} = \frac{(\text{evaporation})(\text{latent heat @ 212°F})}{33{,}475}$$

$$= \frac{4240 \times 970}{33{,}475} = \textbf{123} \text{ boiler hp } (\textbf{1207} \text{ kW}) \qquad \textbf{3a}$$

6.19 Horsepower input $= \dfrac{0.0001573 \times \text{ft}^3/\text{min} \times \text{static pressure}}{\text{static efficiency}}$

$$= \frac{0.0001573 \times (60{,}000/60) \times 2}{0.40}$$

$$= 0.786 \text{ hp}$$

Use **1**-hp (**0.746**-kW) motor   **b**.

6.20 Velocity of air $= \dfrac{Q}{A} = \dfrac{9300 \text{ ft}^3/\text{min}}{(\pi/4)(3)^2 \text{ ft}^2}$

$$= 1317 \text{ ft/min or } 21.95 \text{ ft/s}$$

$$V = \sqrt{2g\frac{Dh_v}{12d}}$$

and $$h_v = \frac{12dV^2}{2gD}$$

where D = 62.1 lb/ft³ H₂O at gage temperature
  h_v = velocity pressure, inH₂O
  d = density of air flowing
  V = ft/s
  h_t = total pressure difference created by fan
  h_s = static pressure

$$PV = WRT \quad \text{and} \quad d = \frac{W}{V}$$

$$\therefore d = \frac{P}{RT} = \frac{144 \times 14.1}{53.3(460 + 83)} = 0.0703 \text{ lb/ft}^3$$

where $P = \dfrac{28.75}{29.9} \, 14.7 = 14.1 \text{ psi}$

$$h_v = \frac{12 \times 0.0703 \times (21.95)^2}{2 \times 32.2 \times 62.1} = 0.102 \text{ inH}_2\text{O}$$

$$h_t = h_s + h_v = 0.85 + 0.102 = 0.952 \text{ inH}_2\text{O}$$

$$\text{Air hp} = \frac{Qh_tD}{12 \times 33,000} = \frac{9300 \times 0.952 \times 62.1}{12 \times 33,000} = 1.39 \text{ hp}$$

$$\text{Fan mechanical efficiency} = \frac{\text{air hp}}{\text{input}} = \frac{1.39}{3.55} \times 100$$

$$= \textbf{39.2} \text{ percent} \qquad \textbf{1c}$$

$$\text{Static efficiency} = \text{mechanical efficiency} \left(\frac{h_s}{h_t}\right)$$

$$= 39.2 \frac{0.850}{0.952} = \textbf{35} \text{ percent} \qquad \textbf{2d}$$

6.21 $Q = A \times V = (2 \times 3) \times 20 = 120 \text{ ft}^3/\text{min}$
For standard air conditions

$$V = 0.33 \text{ fps}$$

$$= 66.76 \sqrt{h_v} \quad \text{and} \quad h_v = \left(\frac{0.33}{66.67}\right)^2 = 0.000025 \text{ inH}_2\text{O}$$

$$h_t = h_s + h_v = 3.00 \text{ inH}_2\text{O}$$

$$\text{Air hp} = \frac{Qh_t}{6355} = \frac{120 \times 3.00}{6355} = \textbf{0.0566} \text{ hp} (\textbf{0.0422} \text{ kW}) \qquad \textbf{a}$$

6.22 Horsepower input $= \dfrac{Qh_t}{6355 \times \text{eff}} = \dfrac{(300,000/60)(2)}{6355 \times 0.40}$

$$= 3.934$$

Use **4**-hp (**2.98** kW)   **d**.

6.23 $N =$ standard commercial ton
cop = coefficient of performance
1-ton capacity = 12,000 Btu (3514.8 W)/h
20-ton capacity = 240,000 Btu (70,296 W)/h
1. $Q = W \times C_p(t_2 - t_1)$

$$W = \frac{240,000}{0.24(90 - 70)} = \textbf{50,000} \text{ lb} (\textbf{22,680} \text{ kg}) \qquad \textbf{1b}$$

2. Assuming cop = 4,

$$\text{hp} = \frac{(4.71)N}{\text{cop}} = \frac{4.71 \times 20}{4} = \textbf{23.5} \text{ hp} (\textbf{17.52} \text{ kW}) \qquad \textbf{2c}$$

6.24 Refer to ammonia tables or charts for property values.

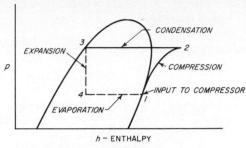

Fig. 6.24

1. Vol efficiency $= \dfrac{\text{vol ammonia delivered/min}}{\text{piston displacement, rpm}}$

Vol ammonia delivered/min to compressor
$$= 2 \times \text{vol efficiency} \times \text{piston displacement, rpm}$$
$$= 2 \times 0.82 \left\{ 1 \left[\frac{\pi}{4} (1)^2 \right] \right\} (150) = 193 \text{ ft}^3/\text{min}$$

For ammonia at 31.16 psi (from tables)—saturated,
$$v_1 = 8.92 \text{ ft}^3/\text{lb}$$

Weight to be circulated $= \dfrac{193}{8.92} = 21.60$ lb/min

Refrigeration effect $= h_1 - h_3 = 612 - 143$
$$= 469 \text{ Btu/lb } (1091.8 \text{ kJ/kg})$$

Capacity $= \dfrac{(\text{lb/min})(\text{Btu/lb})}{200} = \dfrac{(21.6)(469)}{200}$
$$= \textbf{50.65} \text{ short tons } (\textbf{45.95} \text{ Mg}) \quad \textbf{1d}$$

2. $\dfrac{\text{Btu extracted}}{\text{Pounds ice at } 16°\text{F}} = c_{p,W}(t_2 - 32) + \text{heat of fusion}$
$$+ c_{p,I}(32 - 16)$$
$$= 1(80 - 32) + 144 + 0.5(32 - 16)$$
$$= 48 + 144 + 8 = 200 \text{ Btu/lb ice}$$
$$= 400,000 \text{ Btu/ton ice}$$

$$\frac{\text{Ice, tons}}{24 \text{ h}} = \frac{\text{capacity} \times 288{,}000 \text{ Btu}/24 \text{ h}}{\text{Btu extracted/ton ice}} = \frac{50.5 \times 288{,}000}{400{,}000}$$

$$= \textbf{36.4} \text{ tons (}\textbf{33.02} \text{ Mg) ice} \qquad \textbf{2a}$$

3. Ideal hp $= \dfrac{778W(h_2 - h_1)}{33{,}000}$ (for isentropic expansion)

$$= \frac{778 \times 21.6(724 - 612)}{33{,}000} = 57.0 \text{ hp}$$

Brake hp $= \dfrac{\text{ideal hp}}{\text{overall efficiency}} = \dfrac{57.0}{0.80} = \textbf{71.2}$ hp (**53.09** kW)   **3b**

6.25 P_D = diametral pitch
   N = number of teeth
   D = pitch diameter
   T_V = train value
   w = rpm

$$P_D = \frac{N}{D} \qquad N_A = P_D D_A = 12 \times 3 = 36 \text{ teeth}$$

$$N_C = N_A + 2N_B = 36 + 2 \times 18 = 72 \text{ teeth}$$

$$\frac{W_A - W_D}{W_C - W_D} = T_V = \frac{N_C}{N_A} = \frac{72}{36} = -2$$

Case I: Gear A clutched in, gear C fixed.

$$\frac{W_A - W_{D(I)}}{0 - W_{D(I)}} = -2 \qquad W_A - W_{D(I)} = 2W_{D(I)}, \; W_{D(I)} = \frac{W_A}{3} \qquad (1)$$

Case II: Gear A fixed, gear C clutched in.

$$\frac{0 - W_{D(II)}}{W_C - W_{D(II)}} = -2 \qquad W_{D(II)} = 2W_C - 2W_{D(II)}, \; W_{D(II)} = \frac{2W_C}{3} \qquad (2)$$

(For same input speed, $W_A = W_C$)

$$\therefore \frac{W_{D(II)}}{W_{D(I)}} = \frac{\frac{2}{3}W_C}{\frac{1}{3}W_A} = \frac{2W_C}{1W_C} = \textbf{2} \qquad \textbf{c}$$

6.26 1. Displacement vol = stroke $\times$ area

$$= 6.00 \times \frac{\pi}{4} \times (5.375)^2 = 136 \text{ in}^3$$

Initial volume = displacement volume + clearance volume
$$= 136 + 23.4 = 159.4 \text{ in}^3$$

Compression ratio $= \dfrac{\text{initial volume}}{\text{clear volume}} = \dfrac{159.4}{23.4} = \mathbf{6.82 : 1}$ **1d**

2. Brake hp $= \dfrac{Tn}{63,000} = \dfrac{(12 \times 2100) \times 2500}{63,000} = 1000 \text{ hp}$

$b_{\text{mep}} = \dfrac{33,000 \times \text{bhp}}{LAN} = \dfrac{33,000 \times 1000}{0.5[(\pi/4)(5.375)^2] \times 17,500}$

$$= \mathbf{166} \text{ psig } (\mathbf{1.145} \text{ MPa gage}) \quad \mathbf{2a}$$

where T = torque, in·lb
$\quad n$ = rpm
$\quad L$ = stroke, ft
$\quad A$ = piston area, in^2

N = power cycles/min $= \dfrac{14 \times 2500}{2}$

3. Brake hp/lb $= 1000/1450 = \mathbf{0.689} \; (\mathbf{1.133} \text{ kW/kg}) \quad \mathbf{3b}$

6.27 *a.* Mean torque $T_m = \dfrac{\text{max torque} \times \text{min torque}}{2}$

$$= \dfrac{20,000 + (-6000)}{2}$$

$$= 7000 \text{ ft·lb } (9492 \text{ N·m})$$

Variable torque $T_v = \dfrac{\text{max } T - \text{min } T}{2}$

$$= \dfrac{20,000 - (-6000)}{2}$$

$$= 13,000 \text{ ft·lb } (17,628 \text{ N·m})$$

Mean component of stress $S_m = \dfrac{16 T_m}{\pi D^3} = \dfrac{16 \times 7000}{\pi D^3}$

$$= \dfrac{35,600 \text{ psi}}{D^3}$$

Variable component of stress $S_v = \dfrac{16 T_r}{\pi D^3} = \dfrac{16 \times 13,000}{\pi D^3}$

$$= \dfrac{66,200 \text{ psi}}{D^3}$$

$$S_n(\text{reversed torsion}) = S_{t,\text{ult}} \frac{0.50}{2} = \frac{97,000}{4}$$
$$= 24,250 \text{ psi (167.20 MPa)}$$

$$S_{s,yp} = 0.55 \, S_{t,yp} = 0.55 \times 58,000 = 31,900 \text{ psi (219.95 MPa)}$$

$$\frac{1}{\text{Factor of safety}} = \frac{S_v}{S_n} + \frac{S_m}{S_y}$$

$$\frac{1}{2} = \frac{66,200}{24,250 \, D^3} + \frac{35,600}{31,900 \, D^3} = \frac{2.73}{D^3} + \frac{1.12}{D^3} = \frac{3.85}{D^3}$$

$$D^3 = 7.70 \qquad D = \mathbf{1.975} \text{ in (}\mathbf{5.017} \text{ cm)}$$

$$b. \quad \frac{S_{s,yp}}{\text{Factor of safety}} = \frac{16T}{\pi D^3} = \frac{31,900}{2} = \frac{16 \times 20,000}{\pi D^3}$$

$$D^3 = 6.39 \qquad D = \mathbf{1.855} \text{ in (}\mathbf{4.712} \text{ cm)}$$

c. Working stresses which have been determined from the ultimate or yield-point values of a material with a factor of safety will give safe and reliable results only for static loading. Failures in machine parts subjected to variable loading occur at stresses considerably below yield point. Because of this "fatigue" phenomenon, the diameter calculated on the basis of a static torque loading in part b would not stand up under the variable loading acting on the shaft in part a.

6.28 For Corliss engine, diagram factor = 0.85.

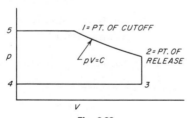

Fig. 6.28

$$\text{Conventional diagram} \quad \frac{V_1}{V_2} = \frac{3}{8} = \frac{1}{r_e} \qquad \text{where } r_e = 2.67$$

$$\text{Head end piston area} \quad = \frac{\pi}{4} (28)^2 = 615.0 \text{ in}^2$$

$$\text{Piston rod area} \quad = \frac{\pi}{4} (5)^2 = 19.6 \text{ in}^2$$

$$\text{Crank end piston area} \quad = \overline{595.4 \text{ in}^2}$$

1. p_{mc} = mean efficiency pressure (conventional)

$$= P_1 \frac{1 + \log_e r_e}{r_e} - P_3 = 150 \frac{1 + \log_e 2.67}{2.67} - P_3$$

$$= 56.3(1 + 0.981) - 15 = \textbf{96.8} \text{ psia (\textbf{667.4} kPa abs)} \qquad \textbf{1a}$$

P_{ma} = mean effective pressure (actual)

= diagram factor $\times P_{mc}$

= $0.85 \times 96.8 = \textbf{82.3}$ psia (**567.5** kPa abs)   **1c**

2. $i$hp head end $= \dfrac{P_{ma}LA_H N}{33,000}$

$$= \frac{82.3 \times \, ^{54}\!/_{12} \times 615 \times 60}{33,000} = 413$$

$i$hp crank end $= \dfrac{P_{ma}LA_C N}{33,000}$

$$= \frac{82.3 \times \, ^{54}\!/_{12} \times 595.4 \times 60}{33,000} = 401$$

Indicated hp = $\overline{814}$

Assuming 90° mechanical efficiency,

hp delivered at crank shaft = (mechanical efficiency)($i$hp)

= 0.90×814

= **733** hp (**546.6** kW)   **2c**

3. hp delivered to mill = (733)(0.90)(0.90)(0.85)

= **505** hp (**376.6** kW)   **3d**

6.29 Assuming uniform wear which is proportional to product of velocity and pressure,

$$T = \pi \mu P_{max} r_i (r_0^2 - r_i^2)$$

with one side effective and twice with two sides effective

where T = torque, in·lb

μ = coefficient of friction

P_{max} = maximum pressure

r_i = inside radius

r_0 = outside radius = $1.25 r_i$

N = total axial pressure

$\therefore T = 2\pi \times 0.3 \times 12 \times r_i^3 [(1.25)^2 - (1)^2] = 6516$

$$r_i^3 = \frac{6516}{22.7(1.562 - 1000)} = 512$$

$$r_i = \textbf{8.00} \text{ in } (\textbf{20.32} \text{ cm}) \qquad \text{and} \qquad r_0 = \textbf{10.00} \text{ in } (\textbf{25.4} \text{ cm})$$

$$N = \frac{T}{\mu(r_0 + r_i)} = \frac{6516}{0.3(10 + 8)} = \textbf{1205} \text{ lb } (\textbf{5360} \text{ N})$$

6.30 Indicated work/rev $= \dfrac{\text{output hp} \times 33{,}000}{N \times \text{efficiency}}$

$$= \frac{20 \times 33{,}000}{1800 \times 0.80} = 458 \text{ ft·lb/rev}$$

Let N = mean rpm
   N_2 = maximum rpm
   N_1 = minimum rpm

   $\omega = \dfrac{2\pi N}{60}$, rad/s

   c_f = coefficient of fluctuation
   J = mass moment of inertia, lb·s²/ft
   $\bar{r}$ = radius to concentrated mass, in

Excess energy delivered to flywheel = change in flywheel KE
$$= \Delta \text{KE} = (0.20)(458)$$

$$\Delta \text{KE} = 91.6 \text{ ft·lb} = \frac{1}{2} J(\omega_2^2 - \omega_1^2)$$

$$= \frac{1}{2} \frac{W}{g} \left(\frac{\bar{r}}{12}\right)^2 \left[\left(\frac{2\pi N_2}{60}\right)^2 - \left(\frac{2\pi N_1}{60}\right)^2 \right]$$

$$= \frac{W}{2g} \left(\frac{\bar{r}\pi}{360}\right)^2 (N_2^2 - N_1^2)$$

since $N = \dfrac{N_2 + N_1}{2}$ and $c_f = \dfrac{N_2 - N_1}{N}$,

$$N_2 + N_1 = 2N$$
$$\underline{N_2 - N_1 = c_f N}$$
$$N_2^2 - N_1^2 = 2c_f N^2$$

$$\Delta \text{KE} = 91.6 = \frac{W}{2g} \left(\frac{\bar{r}\pi}{360}\right)^2 2c_f N^2 = \frac{50}{64.4} \left(\frac{7\pi}{360}\right)^2 2c_f (1800)^2$$

$$c_f = 0.00488$$

$$N_2 + N_1 = 2N = 3600 \text{ rpm}$$
$$N_2 - N_1 = c_f N = \quad\quad \textbf{8.8} \text{ rpm} = \text{total speed variation} \quad \textbf{b}$$
$$2N_2 \quad\quad\quad = 3608.8 \text{ rpm}$$
$$N_2 \quad\quad\quad\quad = \textbf{1804.4} \text{ rpm}$$
$$N_1 \quad\quad\quad\quad = \textbf{1795.6} \text{ rpm}$$

6.31 W = indicated work
$\quad\quad t$ = hours
$\quad\quad T$ = torque, in·lb
$\quad\quad Q_A$ = head added
$\quad\quad e$ = thermal efficiency
$\quad\quad w_f$ = fuel consumption, lb/h
$\quad\quad L$ = stroke, ft
$\quad\quad q_L$ = lower heating value, Btu/lb
$\quad\quad A$ = area, in^2
$\quad\quad P_m$ = mean effective pressure, psi
$\quad\quad N$ = explosions/min

Indicated power = $P_m LAN = 110 \times 1 \times \dfrac{121\pi}{4} \times \dfrac{300}{2}$

$$= 1,570,000 \frac{\text{ft·lb}}{\text{min}} = \frac{1,570,000}{33,000} = 47.6 \text{ hp}$$

$$W/\text{h} = \frac{1,570,000 \times 60}{778} = 121,000 \text{ Btu } (127,776 \text{ kJ})/\text{h}$$

$$Q_A = q_L w_f = 18,000 \times 24 = 433,000 \text{ Btu } (457,248 \text{ kJ})/\text{h}$$

1. $e = \dfrac{W/\text{h}}{Q_A} = \dfrac{121,000}{433,000} \times 100 = \textbf{28} \text{ percent} \quad \textbf{1c}$

2. Mechanical efficiency = brake hp/indicated hp

$$\text{Brake hp} = \frac{T \times \text{rpm}}{63,000} = \frac{12 \times 700 \times 300}{63,000} = 40 \text{ hp}$$

$$\text{Mechanical efficiency} = \frac{40 \times 100}{47.6} = \textbf{84.2} \text{ percent} \quad \textbf{2d}$$

3. Specific gravity of fuel = 0.75 and there are 7.5 gal/ft^3, so

$$\frac{62.4 \times 0.75}{7.5} = 6.24 \text{ lb/gal,}$$

and
$$\frac{24 \text{ lb/h}}{6.24 \text{ lb/gal}} = 3.84 \text{ gal/h}$$

$$\text{Fuel cost/h} = 3.84 \times \$1.30 = \$4.99$$

$$\frac{\text{Fuel cost/h}}{\text{hp}} = \frac{\$4.99}{40} = \textbf{\$0.125}/\text{hp} \ (\textbf{\$0.167}/\text{kW}) \qquad \textbf{3a}$$

6.32 Dimensions are shown in Fig. 6.32 of the question.

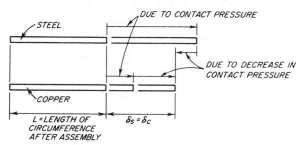

Fig. 6.32

$$\text{Stress due to centrifugal force} = \frac{w}{g}\, \omega^2 r^2$$

where w = specific weight
ω = rad/s
r = mean radius of thin cylinder

$$\Delta S_s' (\text{due to centrifugal force}) = \frac{w_s}{g}\left(\frac{6.28N}{60}\right)^2 \left(\frac{d + h_s}{2}\right)^2$$

$$= \frac{490}{1728 \times 386} \times (0.0523N)^2(6 + 0.125)^2$$

$$= \frac{1}{1360}(0.320N)^2 = \frac{N^2}{(115.2)^2}$$

$$\Delta S_c' (\text{due to centrifugal force}) = \frac{w_c}{g}\left(\frac{6.28N}{60}\right)^2 \left(\frac{d - h_c}{2}\right)^2$$

$$= \frac{558}{1728 \times 386} \times (0.0523N)^2(6 - 0.063)^2$$

$$= \frac{1}{1248}(0.310N)^2 = \frac{N^2}{(114.0)^2}$$

$\Delta S_s''$ (due to decrease in contact pressure) $= -Xd/2h_s$

$\Delta S_c''$ (due to decrease in contact pressure) $= +Xd/2h_c$

where $X =$ change in contact pressure.

Figure 6.32 with this answer shows the increments of change of length due to the forces acting.

Total change in stress in the steel $= \Delta S_s = \Delta S_s' + \Delta S_s''$

Total change in stress in the copper $= \Delta S_c = \Delta S_c' + \Delta S_c''$

$$\delta_c = L\epsilon_c = L \Delta S_c/E_c = 2000L/E_c$$
$$\epsilon_c = \Delta S_c/E_c = 2000/E_c$$
$$\delta_s = L\epsilon_s = L \Delta S_s/E_s = \delta_c = 2000L/E_c$$
$$\epsilon_s = \Delta S_s/E_s = 2000/E_c$$

$$\frac{2000}{E_c} = \frac{\Delta S_c}{E_c} = \frac{1}{E_c}\left[\left(\frac{N}{114}\right)^2 + \frac{Xd}{2h_c}\right] \qquad (1)$$

$$h_c(1) = 2000h_c = \left[\left(\frac{N}{114}\right)^2 h_c + \frac{Xd}{2}\right] \qquad (1a)$$

$$\frac{2000}{E_c} = \frac{\Delta S_s}{E_s} = \frac{1}{E_s}\left[\left(\frac{N}{115.2}\right)^2 - \frac{Xd}{2h_s}\right] \qquad (2)$$

$$h_s(2) = 2000\left(\frac{E_s}{E_c}\right) h_s = \left[\left(\frac{N}{115.2}\right)^2 h_s - \frac{Xd}{2}\right] \qquad (2a)$$

$(1a) + (2a) = 2000 [h_c + (^{15}/_8)h_s] = 2000(0.0625 + 0.2345)$

$$= \left(\frac{N}{114}\right)^2 \left(\frac{1}{1.02} \times 0.125 + 0.0625\right)$$

$N = 1140 \sqrt{(20 \times 0.297)/0.185} = 1140 \sqrt{32.1}$

$= 1140 \times 5.67 = $ **6450** rpm   **a**

6.33 $A_A = 2 \times 0.785 \times 1^2 = 1.570 \text{ in}^2$

$A_B = 0.785(2.50^2 - 1.50^2) = 3.14 \text{ in}^2 = 2A_A$

After the screw has been lowered, *subsequent* deformations of tube B and rods A are equal, and $\delta_A = \delta_B$.

For static equilibrium

$$\Delta P_A + \Delta P_B = 0 \qquad \text{and} \qquad \Delta P_A = -\Delta P_B$$
$$\Delta S_A A_A = - \Delta S_B A_B \qquad \text{and} \qquad \Delta S_A = -2\Delta S_B$$

$\Delta S_B =$ final stress $-$ initial stress (after screw has been lowered)

$= [(-10,000) - (-5000)] = -5000 \text{ psi}$

$$\Delta S_A L_A/E_A + L_A \alpha_A \Delta t = \Delta S_B L_B/E_B + L_B \alpha_B \Delta t$$

where the first term on each side of the equation represents constraint, and the second term free temperature expansion. Substituting $-2\Delta S_B$ for ΔS_A in the last equation, and by combining terms,

$$\Delta t = \frac{\Delta S_B(L_B/E_B + 2L_A/E_A)}{L_A\alpha_A - L_B\alpha_B}$$

$$= \frac{-5000[(23/14.5) + (2 \times 30/30)](1/10^6)}{[(30 \times 6.5) - (23 \times 10.5)](1/10^6)}$$

$$= \frac{-5000(1.582 + 2)}{195 - 241.5} = \frac{-5000 \times 3.582}{-46.5} = +384°F$$

Final temperature $= 70 + 384 = $ **454**°F (**234.4**°C)   **b**

6.34 The torque-time relationship for load and input is shown in Fig. 6.34.

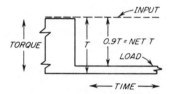

Fig. 6.34

Let W = flywheel weight = 3 tons = 6000 lb

k = radius of gyration = 5 ft

α = angular acceleration

T = torque input, ft·lb

n_1 = initial rpm

n_2 = final rpm

$\Delta\theta$ = revolutions after reduction of load = 5 rev

$T_{net} = J\alpha = 0.9T$

$\alpha = 0.9T/J$

$J = Wk^2/g = 6000 \times 5^2/32.2 = 4670$ ft·lb·s^2

$T = 63,000 \times hp/n = 63,000 \times {}^{80}\!/_{150} = 33,600$ in·lb

$= 2800$ ft·lb

$\alpha = 0.9 \times 2800/4670 = 0.539$ rad/s$^2 = 309$ rpm^2

For constant angular acceleration

$$n_2^2 - n_1^2 = 2\alpha \, \Delta\theta$$

$n_2 = \sqrt{n_1^2 + 2\alpha \, \Delta\theta} = \sqrt{150^2 + 2 \times 309 \times 5} = 160$ rpm

Change in speed $= n_2 - n_1 = 160 - 150 = $ **10** rpm   **c**

6.35 T_1 = torque in
   T_2 = torque out
   n_1 = rpm in
   n_2 = rpm out, and n_1/n_2 = 4.11
   r = radius of wheel, ft
   F = tractive force
   v = velocity, fpm

1. hp out = efficiency × hp in

$$T_2 n_2 = \text{efficiency} \times T_1 \times n_1$$

$$T_2 = \frac{n_1}{n_2} \times T_1 \times \text{efficiency} = 4.11 \times 230 \times 0.97 = 915 \text{ ft·lb}$$

$$F = \frac{T_2}{r} = \frac{915}{15/12} = \frac{915}{1.25}$$

$$= \textbf{732} \text{ lb } (\textbf{3256} \text{ N}) \text{ at both rear wheels}\qquad\textbf{1d}$$

2. $\text{hp} = \dfrac{F_v}{33,000} = \dfrac{732 \times (40 \times 5280/60)}{33,000} = \dfrac{732 \times 3520}{33,000}$

$$= \textbf{78.1} \text{ hp } (\textbf{58.24} \text{ kW})\qquad\textbf{2a}$$

6.36 ihp = indicated horsepower per cylinder
   bhp = brake horsepower
   P_m = mean effective pressure, psi
   n = no. of cylinders
   L = stroke, ft
   A = piston area, psi
   D = diameter, in
   N = explosions/min

$$\frac{12 \times L(\text{ft})}{D(\text{in})} = 1.4 \qquad \text{and} \qquad D^2 = 73.5L^2$$

$$A = \frac{\pi D^2}{4} = 57.6L^2$$

$$\text{ihp} = \frac{\text{bhp}}{\text{mechanical efficiency} \times n} = \frac{30}{0.85 \times 6} = 5.88$$

$$\text{ihp} = \frac{P_m LAN}{33,000} = 5.88 = \frac{90 \times L \times 57.6L^2 \times (1800/2)}{33,000}$$

$$L^3 = 0.0416 \qquad L = \textbf{0.346} \text{ ft or } \textbf{4.15} \text{ in } (\textbf{10.54} \text{ cm})\qquad\textbf{1b}$$

$$D = \frac{4.15}{1.4} = \textbf{2.97} \text{ in } (\textbf{7.54} \text{ cm})\qquad\textbf{2c}$$

6.37 Let E_{ss} = shearing modulus of steel

E_{sa} = shearing modulus of aluminum

I_{ps} = polar moment of inertia of steel rod

I_{pa} = polar moment of inertia of aluminum tube

L = length of shaft = 5 ft = 60 in

T_s = torque carried by steel rod

T_a = torque carried by aluminum tube

The angle of twist and the dimensions are shown in Fig. 6.37a, and the

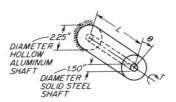

DIAMETER
HOLLOW
ALUMINUM
SHAFT —1.50″
DIAMETER
SOLID STEEL
SHAFT

Fig. 6.37a

shear-stress variation is shown in Fig. 6.37b. The two basic equations are:

$$\theta = \theta_a = \theta_s \qquad (1)$$

$$T = T_a + T_s \qquad (2)$$

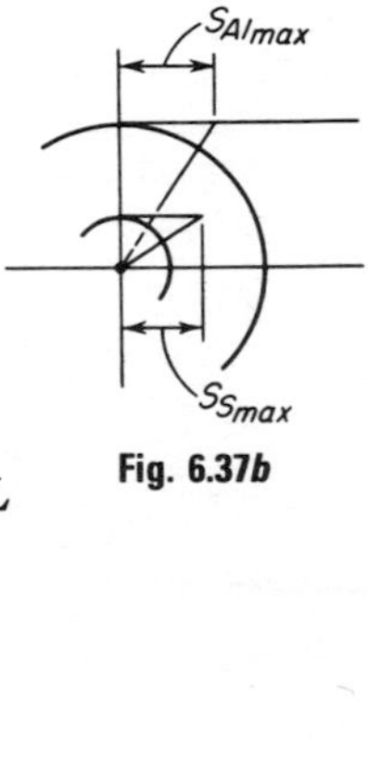

$$I_{ps} = \frac{\pi}{32}(1.5)^4 = 0.497 \text{ in}^4$$

$$I_{pa} = \frac{\pi}{32}[(2.25)^4 - (1.5)^4] = 2.01 \text{ in}^4$$

Fig. 6.37b

1. $\theta_a = T_aL/I_{pa}E_{sa}$ and $\theta_s = T_sL/I_{ps}E_{ss}$

$T = \theta_aI_{pa}E_{sa}/L + \theta_sI_{ps}E_{ss}/L = \theta(I_{pa}E_{sa} + I_{ps}E_{ss})/L$

$\theta = TL/(I_{pa}E_{sa} + I_{ps}E_{ss})$

$$= \frac{(15{,}000)(60)}{(2.01)(3.5 \times 10^6) + (0.497)(12 \times 10^6)}$$

$$= \mathbf{0.0693} \text{ rad} \qquad \mathbf{1a}$$

2. $S_{a(max)} = T_a\dfrac{2.25/2}{I_{pa}} = \dfrac{\theta_aI_{pa}E_{sa}}{L}\dfrac{2.25/2}{I_{pa}} = \dfrac{\theta E_{sa}(2.25)}{2L}$

$$= \frac{(0.0693)(3.5 \times 10^6)(2.25)}{(2)(60)} = \mathbf{4540} \text{ psi } (\mathbf{31.3} \text{ MPa}) \qquad \mathbf{2b}$$

$$S_{s(max)} = \frac{\theta E_{ss}(1.5)}{2L}$$

$$= \frac{(0.0693)(12 \times 10^6)(1.5)}{120} = \textbf{10,400} \text{ psi } (\textbf{71.71 MPa}) \quad \textbf{2d}$$

6.38 N = rpm
T = number of teeth
x = number of gear pairs
F = force, lb
v = head velocity, fpm

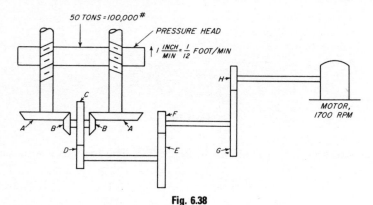

Fig. 6.38

½-in travel of head/1 revolution of A
1 in/min travel of head = 2 rpm of A

$$N_A = 2 \text{ rpm} \quad \text{and} \quad N_H = 1700 \text{ rpm}$$

and

Reduction per train = ⅙

$$\text{Train valve} = T_V = \frac{N_A}{N_H} = \frac{2}{1700} = \frac{1}{850} = \left(\frac{1}{y}\right)^x$$

when $x = 4$, as in sketch, $y = 5.4 < 6$ as required

$$T_V = \frac{T_H}{T_G} \times \frac{T_F}{T_E} \times \frac{T_D}{T_C} \times \frac{T_B}{T_A}$$

and assume y for spur-gear pairs to be 5, 5, and 6, and T multiples of 12.

a. $T_V = \dfrac{1}{850} = \dfrac{12}{60} \times \dfrac{12}{60} \times \dfrac{12}{72} \times \dfrac{T_B}{T_A}$

$$\frac{T_B}{T_A} = \frac{150}{850} = \frac{15}{85} = \frac{1}{5.67}$$

Make T_B and T_A bevel gears, with teeth 15 and 85.

$b.$ $\mathrm{hp} = \dfrac{F_v}{\text{efficiency} \times 33,000} = \dfrac{100,000 \times \frac{1}{12}}{0.40 \times 33,000}$

$$= 0.632 \text{ hp } (0.471 \text{ kW})$$

Use a ¾-hp (0.559-kW) motor.

6.39 V = pitch line velocity
        F_t = transmitted load
        P_D = diametral pitch
        n = number of teeth
        b = face width
        I = increment load
        S_n = endurance limit
        Y = form factor
       F_w = wear load
        N = rpm

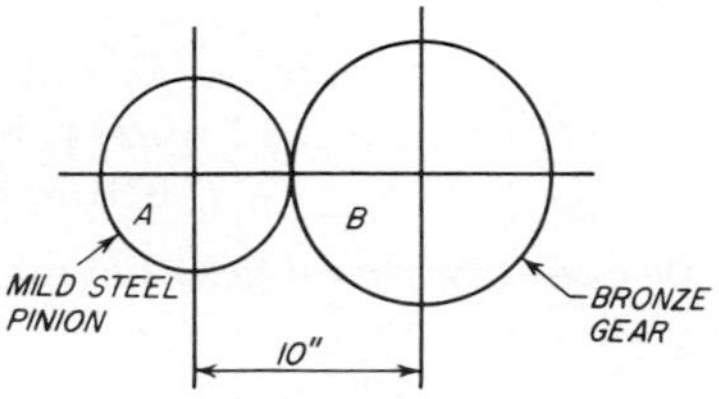

Fig. 6.39

$$\frac{R_A}{R_B} = \frac{\omega_B}{\omega_A} = \frac{2}{3} \text{ (given)}$$

$$R_A = \tfrac{2}{3} R_B$$
$$R_A + R_B = 10 \text{ in (given)}$$
$$\tfrac{2}{3} R_B + \tfrac{3}{3} R_B = \tfrac{5}{3} R_B = 10$$
$$R_B = 6 \text{ in} \quad \text{and} \quad R_A = 4 \text{ in}$$
$$D_B = 12 \text{ in} \quad \text{and} \quad D_A = 8 \text{ in}$$
$$V = 2\pi N_A R_A = 2\pi \times 150 \times 4 = 3770 \text{ in/min}$$
$$= 314 \text{ ft (95.71 m)/min}$$

$$\mathrm{hp} = \frac{F_t V}{33,000} \qquad F_t = \frac{33,000 \times 8}{314} = 841 \text{ lb (3741 N)}$$

All references to tables and pages are in Faires, "Design of Machine Elements," Macmillan, 1941.

   e = maximum permissible error = 0.0045 (p. 216)

   Use first-class commercial gears ($P_D > \frac{1}{2}$) (p. 216)

   Assume $P_D = 6$ and expected error = 0.002 in. (p. 216)

Recommended proportions for cut teeth:

$$\frac{9.5}{P_D} < b < \frac{12.5}{P_D} \qquad \text{or} \qquad 1.58 < b < 2.08$$

Use $b = $ **2** in (**5.08** cm).

Find dynamic load F_D. For $14\frac{1}{2}°$ full depth, $k = 0.107e$:

$$C = \text{deformation factor} = \frac{kEgEp}{Eg + Ep}$$

$$= \frac{0.107 \times 0.002 \times 15 \times 10^6 \times 30 \times 10^6}{(15 + 30)10^6} = 2140$$

$$F_D = \frac{0.05V(bC + F_t)}{0.05V + (bC + F_t)^{0.5}} + F_t = I + F_t$$

$$bC + F_t = 2 \times 2140 + 841 = 5121$$
$$I = 910 \text{ (p. 214) and 920 (from the equation)}$$
$$F_D = 910 + 841 = \mathbf{1751} \text{ lb } (\mathbf{7788} \text{ N})$$

Design is based on gear having smallest product of S_nY, where

$$F_s = \frac{S_n Y_b}{P_D}$$

For gear (phosphor bronze):

$$T = P_D D = 6 \times 12 = 72 \text{ teeth} \qquad \text{and} \qquad Y = 0.360 \text{ (p. 204)}$$
$$S_n Y = 24,000 \times 0.360 = \mathbf{8640}$$

For pinion (SAE 1035):

$$T = P_D D = 6 \times 8 = 48 \text{ teeth} \qquad \text{and} \qquad Y = 0.345 \text{ (p. 204)}$$

$$S_n Y = \frac{S_{ultimate}}{2} \times Y = \frac{87,000}{2} \times 0.345 = \mathbf{15,000}$$

$\therefore$ bronze gear is weaker

$$F_s = \frac{24,000 \times 2 \times 0.360}{6} = 2880 \text{ lb}$$

$$\text{Margin of safety} = \frac{F_s}{F_d} - 1 = \frac{2880}{1751} - 1 = 0.643$$

$\therefore$ Suitable for speed reducers, pumps, mining machinery, etc. Check for wear:

$$Q = \frac{2D_B}{(D_A + D_B)} = \frac{2 \times 12}{(8 + 12)} = 1.2$$

Use SAE 1035 steel, tempering temperature $= 600°F$
Brinell No. 240 (Tables XXX and XXXIII)
By interpolation, $K = 126$.

$$F_W = D_A bKQ = 8 \times 2 \times 126 \times 1.2 = 2420 \text{ lb}$$
$$F_W > F_D \qquad 2420 \text{ lb} > 1751 \text{ lb}$$

$\therefore b = 2$ in and $P_D = 6$ is suitable. b could be decreased somewhat to obtain a more economical design.

Recalculate for $b = 1\frac{7}{8}$ in or $b = 1\frac{3}{4}$ in

6.40   Assume:
1. The wrench moves in horizontal plane.
2. No energy is absorbed by the bolted connection.
3. The design stress $= 40,000$ psi (275.8 MPa).
4. No frictional losses.

Let Δ = max deflection
  $\bar{F}$ = force causing Δ
  W = 5 lb (2.268 kg)
  v = 12 fps = 144 in (366 cm)/s
  L = 18 in (45.72 cm)
  d = wrench diameter
  g = 386 in (980.44 cm)/s^2
  $\bar{M}$ = moment caused by $\bar{F}$

Figure 6.40 shows the wrench at the moment the 5-lb force strikes the handle at point (1) and the deflection to point (2) of the handle due to

Fig. 6.40

the blow. The total energy at position (1) = the total energy at position (2), or

$$PE_1 + KE_1 = PE_2 + KE_2$$
$$0 + KE_1 = PE_2 + 0$$
$$KE_1 = PE_2$$
$$= \frac{Wv^2}{2g} = \frac{\bar{F}\Delta}{2} = \frac{\bar{M}^2L}{6EI_{xx}} = \frac{\bar{S}_t^2 I_{xx} L}{c^2 6E} = \frac{\bar{S}_t^2 L}{6E}\frac{\pi d^2}{16}$$

where $\Delta = \bar{F}L^3/3EI$
  $\bar{F} = \bar{M}/L$
  $I = \pi d^4/64$
  $c = d/2$
  $\bar{S}_t = \bar{M}c/I_{xx}$, or $\bar{M} = (\bar{S}_t I_{xx})/c$

$$d = \frac{4v}{\bar{S}_t}\sqrt{\frac{3WE}{\pi gL}} = \frac{(4)(144)}{40,000}\sqrt{\frac{(3)(5)(30\times10^6)}{\pi(386)(18)}} = 2.07 \text{ in}$$

Use **2**-in (**5.08**-cm) diameter wrench.   **c**

6.41 1. $v_1 = 500$ ft^3/min, and $c = 3$ percent. Values are schematically shown in Fig. 6.41.

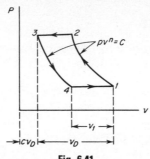

Fig. 6.41

$$
\begin{aligned}
\text{Volumetric efficiency} &= 1 + c - c(p_2/p_1)^{1/n} \\
&= 1 + 0.03 - 0.03 \times (105/14.7)^{1/1.34} \\
&= 1.03 - 0.03 \times (7.15)^{0.745} \\
&= 1.03 - 0.03 \times 4.32 = 0.90 = 90 \text{ percent} \\
&= \text{capacity of compressor/displacement}
\end{aligned}
$$

Displacement $= v_D = 500/0.90 = \mathbf{555}$ ft^3 **(15.72** m^3)/min   **1b**

2. $p_1 = 23.8$ inHg $= 23.8 \times 0.491 = 11.7$ psia, with $t_1 = 70°$F; since $v_2/v_1 = r_k$ is constant, but $p_1 v_1^n = p_2 v_2^n$ and $v_2/v_1 = (p_1/p_2)^{1/n} = $ constant, it means that the volumetric efficiency must be constant along with n and v_D.

Capacity in cubic feet per minute free air (measured at 23.8 inHg at 70°F) **would be the same.** Since $r_k = c$, and $n = c$,

$$
r_k = (p_2/p_1)^{1/n} = (p_2'/p_1')^{1/n}
$$
$$
105/14.7 = p_2'/11.7
$$

and

$$
p_2' = \frac{11.7 \times 105}{14.7} = \mathbf{83.5} \text{ psia } \mathbf{(575.7} \text{ kPa abs)}   \mathbf{2a}
$$

3. **Increase the rpm.**

6.42 1. One gallon $= 8.34$ lb (3.785 kg) water.

Work done by pump $= 8.34 \times 500 \times 90 = 375,000$ ft·lb/min

With an efficiency of 60 percent,

Input to pump $= 375,000/0.60 = 625,000$ ft·lb/min

or   Motor output $= 625,000/33,000 = $ **18.93** (**14.12** kW)   **1c**

2. Let N = rpm, H = head, and Q = capacity, gal/min.

$$\frac{N_1^2}{N_2^2} = \frac{H_1}{H_2}$$

$$N_2 = \left(\frac{H_2}{H_1} N_1^2\right)^{1/2} = \left[\frac{120}{90}(100)^2\right]^{1/2} = \mathbf{116} \text{ rpm} \qquad \mathbf{2d}$$

$$\frac{Q_1}{Q_2} = \frac{N_1}{N_2}$$

$$Q_2 = \frac{Q_1 N_2}{N_1} = 500 \times {}^{116}\!/_{100}$$

$$= \mathbf{580} \text{ gal } (\mathbf{2195} \text{ liters})/\text{min} \qquad \mathbf{2f}$$

3. Let P = power in horsepower.

$$\frac{P_1}{P_2} = \frac{N_1^3}{N_2^3}$$

$$P_2 = P_1 \frac{N_2^3}{N_1^3} = 18.93 \times \frac{116^3}{100^3} = 29.6 \text{ hp}$$

A 25-hp motor would **not be adequate.**

6.43 $F = 50$ lb (in horizontal plane and at right angles to bar)

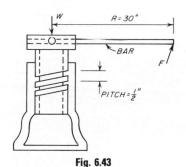

Fig. 6.43

$$\frac{\text{Work (out)}}{\text{Revolutions}} = \text{efficiency} \times \frac{\text{work (in)}}{\text{revolutions}}$$

$$W \times \text{pitch} = \text{efficiency} \times F \times 2\pi R$$
$$W \times \frac{1}{2} = 0.80 \times 50 \times 2\pi \times 30$$
$$W = \mathbf{15,100} \text{ lb } (\mathbf{6849} \text{ kg}) \qquad \mathbf{d}$$

6.44 $R = 12 + \frac{1}{2} \times \frac{7}{8} = 12.438$ in $= 1.0365$ ft (31.59 cm). Since lead of worm = two pitch distances, 40 revolutions of worm = one revolution of drum.

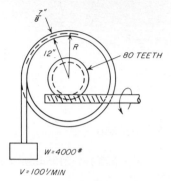

Fig. 6.44

$$\text{hp in} = \frac{\text{hp out}}{\text{efficiency}}$$

$$= \frac{W \times V}{\text{efficiency} \times 33,000} = \frac{4000 \times 100}{0.60 \times 33,000}$$

$$= \textbf{20.2} \text{ hp } (\textbf{15.06} \text{ kW}) \qquad \textbf{1a}$$

$$\text{rpm drum} = \frac{V}{2\pi R} = \frac{100}{6.28 \times 1.0365} = 15.33 \text{ rpm}$$

$$\text{rpm worm} = 40 \times 15.33 = 613 \text{ rpm}$$

$$T, \text{in} = \frac{63,000 \times \text{hp}}{\text{rpm worm}} = \frac{63,000 \times 20.2}{613} = 2075 \text{ in} \cdot \text{lb}$$

$$T, \text{ft} = \textbf{173} \text{ ft} \cdot \text{lb } (\textbf{234.6} \text{ N} \cdot \text{m}) \qquad \textbf{2b}$$

6.45 $P = pA = 205 \times \dfrac{\pi}{4} \times (1.25)^2 = 252$ lb

N = effective coils = total coils $- 2 = 9.5 - 2 = 7.5$ coils
D = diameter of coil
d = diameter of wire $\qquad$ and $\qquad y$ = deflection, in

$$C = \frac{D}{d} = \frac{4.5 - 0.5}{0.5} = 8$$

$$y = \frac{8\,PC^3N}{E_s d} = \frac{8 \times 252 \times 8^3 \times 7.5}{11.5 \times 10^6 \times 0.5} = 1.345 \text{ in}$$

Initial compressed length = free length − y
$$= 8.00 \text{ in} - 1.345 \text{ in}$$
$$= \textbf{6.66} \text{ in } (\textbf{16.92} \text{ cm}) \quad \textbf{1c}$$

Solid length $= dN + 2d = 0.5 \times 7.5 + 2 \times 0.5 = 4.75$ in

Deflection from free length to solid length $= y_c$
$$= 8.00 - 4.75 = 3.25 \text{ in}$$

$$P_c = \frac{y_c}{y} \times P = \frac{3.25 \text{ in}}{1.345 \text{ in}} \times 252 \text{ lb} = 609 \text{ lb}$$

$$k = 1.18$$

(See curves of stress factors versus spring index, fig. 322, Faires, "Design of Machine Elements," Macmillan, 1941.)

$$S_s = \frac{k \times 2.55 P_c D}{d^3} = 1.18 \times 2.55 \times 609 \times 4 \times 8$$

$$= \textbf{58,800} \text{ psi } (\textbf{405} \text{ MPa}) \quad \textbf{2d}$$

6.46 N = number of teeth, w = angular velocity. Gear 1 is input gear, gear 4 is output gear.

$$\frac{w_4}{w_1} = \frac{1131}{2000} = \frac{N_1 N_3}{N_2 N_4}$$

Factoring, we have

$$1131 = 3 \times 377 = 3 \times 13 \times 29 = 39 \times 29$$

Let $N_1 = \textbf{39}$ and $N_3 = 29$ **1a**

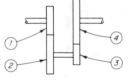

Fig. 6.46

$$N_1 + N_2 = N_3 + N_4$$
$$39 + N_2 = 29 + N_4$$
or $$N_4 = N_2 + 10$$

Also

$$\frac{1131}{2000} = \frac{N_1 N_3}{N_2 N_4} = \frac{39 \times 29}{N_2 N_4} \quad \text{or} \quad N_2 N_4 = 2000$$

or $$N_2(N_2 + 10) = 2000$$
$$N_2^2 + 10 N_2 - 2000 = 0$$

$$N_2 = -5 \pm \sqrt{25 + 2000} = -5 + 45 = 40$$
and $$N_4 = N_2 + 10 = 40 + 10 = \textbf{50} \text{ teeth} \quad \textbf{2d}$$

6.47 Let the smaller gear be 1 and the larger gear be 2. Let

$$\Sigma = \text{angle between shafts}$$
$$\psi = \text{helix angle}$$

p_n = normal circular pitch
N = number of teeth
w = angular velocity
c = center-to-center distance

$$\Sigma = \psi_1 + \psi_2$$
$$90° = 35° + \psi_2 \quad \text{and} \quad \psi_2 = 55°$$

$$\text{Speed ratio} = \frac{w_2}{w_1} = \frac{1}{2} = \frac{N_1}{N_2} = \frac{24}{N_2} \quad \text{or} \quad N_2 = 48 \text{ teeth}$$

$$c = \frac{p_n}{2\pi}\left(\frac{N_1}{\cos\psi_1} + \frac{N_2}{\cos\psi_2}\right) = \frac{0.785}{2\pi}\left(\frac{24}{\cos 35°} + \frac{48}{\cos 55°}\right)$$

$$= \textbf{14.125} \text{ in } (\textbf{35.876} \text{ cm}) \qquad \textbf{b}$$

where $\cos 35° = 0.819$ and $\cos 55° = 0.573$.

6.48 Call points of contact on bodies 1 and 2, B and P, respectively.
1, 2. Find angular velocity of body 1:

$$\mathbf{V}_P = \mathbf{V}_B + \mathbf{V}_{P/B}$$
$$V_P = w_2(\overline{O_2P}) = (10)(2) = 20 \text{ in/s}$$

Solve graphically or by geometry.

$$V_B = 0.5V_P\sqrt{2} = 14.142 \text{ in/s}$$
$$V_{P/B} = 0.5V_P + 0.866V_P = 1.366V_P = 27.32 \text{ in/s}$$

$$w_1 = \frac{V_B}{\overline{O_1B}} = \frac{0.5(V_P)\sqrt{2}}{1\sqrt{2}} = \textbf{10} \text{ rad/s} \qquad \textbf{1d; Clockwise} \qquad \textbf{2a}$$

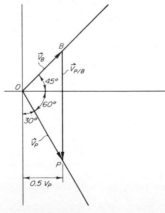

Fig. 6.48a   Velocity polygon.

3, 4. Find angular acceleration of body 1:

$$\mathbf{A}_P = \mathbf{A}_B + \mathbf{A}_{P/B} + 2\mathbf{w}_1\mathbf{V}_{P/B}$$
$$= \mathbf{A}_B^N + \mathbf{A}_B^T + \mathbf{A}_{P/B}^N + \mathbf{A}_{P/B}^T + 2\mathbf{w}_1\mathbf{V}_{P/B}$$

$$A_P = w_2^2(\overline{O_2P}) = (10)^2(2) = 200.0 \text{ in/s}^2$$

$$A_B^N = w_1^2(\overline{O_1B}) = (10)^2\sqrt{2} = 141.4 \text{ in/s}^2$$

$$A_{P/B}^N = \frac{V_{P/B}^2}{R_1} = \frac{\overline{27.32^2}}{(1)} = 746.4 \text{ in/s}^2$$

$$2w_1V_{P/B} = (2)(10)(27.32) = 546.4 \text{ in/s}^2$$

By geometry $A_B^T = 669.2 \text{ in/s}^2$.

$$\alpha_1 = \frac{A_B^T}{(O_1B)} = \frac{669.2}{\sqrt{2}}$$

$$= \mathbf{473.2} \text{ rad/s}^2 \quad \mathbf{3c; \text{ Counterclockwise}} \quad \mathbf{4b}$$

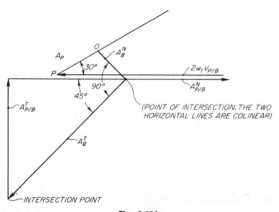

Fig. 6.48*b*

6.49 Break unbalance effects into horizontal and vertical components. Find vertical components:

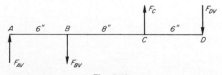

Fig. 6.49*a*

At B: $\quad\quad\quad F_B = 4 \quad F_{BV} = F_B \sin 30° = 2.000$

$$F_{BH} = F_B \cos 30° = 3.464$$

$$\Sigma M_A = 0 \quad 20F_{DV} + 6F_{BV} - 14F_C = 0$$

or $\quad\quad 20F_{DV} + (6)(2.00) - (14)(2.00) = 0$

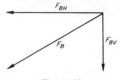

Fig. 6.49b

therefore $F_{DV} = 0.800$ oz·in.

$$\Sigma M_D = 0 \quad 20F_{AV} + (6)(2) - (14)(2) = 0$$

$$F_{AV} = \mathbf{0.800} \text{ oz·in}$$

Find horizontal components:

Fig. 6.49c

$$\Sigma M_A = 0 \quad 20F_{DH} - 6F_{BH} = 0$$

or $\quad\quad 20F_{DH} - (6)(3.464) = 0 \quad F_{DH} = 1.039$ oz·in

$$\Sigma M_D = 0 \quad 20F_{AH} - 14F_{BH} = 0$$

or

$$F_{AH} = \frac{(14)(3.464)}{20} = \mathbf{2.425} \text{ oz·in}$$

At left end:

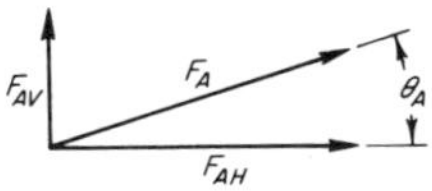

Fig. 6.49d

$$F_A = \sqrt{F_{AV}^2 + F_{AH}^2} = \sqrt{(0.800)^2 + (2.425)^2} = \mathbf{2.554} \text{ oz·in}$$

$$\theta_A = \tan^{-1}\frac{0.800}{2.425} = \mathbf{18.3°}$$

At right end:

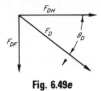

Fig. 6.49e

$$F_D = \sqrt{F_{DV}^2 + F_{DH}^2} = \sqrt{(0.800)^2 + (1.039)^2} = \textbf{1.311} \text{ oz·in}$$

$$\theta_D = \tan^{-1} \frac{0.800}{1.039} = \textbf{37.6}°$$

6.50

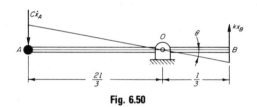

Fig. 6.50

$$J_0 = m \left(\frac{2l}{3}\right)^2 \qquad x_B = \frac{l\theta}{3} \qquad \dot{x}_A = \frac{2}{3} l\dot{\theta}$$

where J_0 = polar moment of inertia
　　　m = total mass

$$\Sigma T_0 = J_0 \theta$$

$$-(Kx_B) \left(\frac{l}{3}\right) - C\dot{x}_A \left(\frac{2l}{3}\right) = J_0 \ddot{\theta}$$

$$-\frac{Kl^2}{9} \theta - \frac{4Cl^2}{9} \dot{\theta} = \frac{4ml^2}{9} \ddot{\theta}$$

$$\ddot{\theta} + \frac{C}{m} \dot{\theta} + \frac{K}{4m} \theta = 0$$

Damped natural frequency $= w_{nd} = \sqrt{\frac{K}{4m} - \left(\frac{c}{2m}\right)^2}$

for critical damping $w_{nd} = 0$. Therefore,

$$\left(\frac{C_c}{2m}\right)^2 = \frac{K}{4m}$$
$$C_c = \sqrt{Km}$$

6.51 $C_1 = S_y(4 - e)b = (35,000)(4 - e)6 = 210,000(4 - e)$

$C_2 = \dfrac{S_y}{2} eb = \dfrac{35,000}{2} (e)6 = 105,000e$

$M = 240,000 \times 12 = 2,880,000 \text{ lb·in}$
$T_1 = C_1 \quad \text{and} \quad T_2 = C_2$

External moment is balanced by summation of internal moments:

$$M = 2\left(2 + \frac{e}{2}\right)C_1 + 2\left(\frac{2}{3}e\right)C_2 = (4 + e)C_1 + 1.33C_2e$$

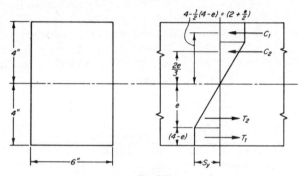

Fig. 6.51

or $2,880,000 = (4 + e)(210,000)(4 - e) + 1.333(105,000)e^2$
$70e^2 = 480 \qquad e^2 = 6.857 \qquad e = 2.619 \text{ in } (6.65 \text{ cm})$
Distance from surface $= 4 - e = 4 - 2.619$
$$= \textbf{1.381 in (3.51 cm)} \qquad \textbf{a}$$

6.52 1. The discharge $Q = \dfrac{\text{gal/min}}{448}$

$$= \frac{4000}{448} = 8.93 \text{ ft}^3 \ (0.253 \text{ m}^3)/\text{s}$$

$$h_f = f\frac{L}{D}\frac{V^2}{2g} \qquad (1)$$

$$Q = AV = \frac{\pi}{4} D^2 V \quad \text{or} \quad V^2 = \frac{16Q^2}{\pi^2 D^4} \quad (2)$$

where f = friction factor. If h_f = head loss, ft, then substituting Eq. (2) into (1), we have

$$h_f = f \frac{L}{D} \frac{Q^2}{2g(D^2\pi/4)^2}$$

$$D^5 = \frac{8LQ^2}{h_f g \pi^2} f = \frac{(8)(10,000)(8.93)^2}{(75)(32.2)\pi^2} f = 267.0f \quad (3)$$

The Reynolds number Re is

$$\text{Re} = \frac{VD}{\nu} = \frac{4Q}{\pi\nu} \frac{1}{D} = \frac{(4)(8.93)}{\pi(0.0001)D} = \frac{113,800}{D} \quad (4)$$

where ν = kinematic viscosity, f²/s.

From Moody diagram, roughness height ϵ = 0.00015 ft. Solve by trial and error. For first trial, let f = 0.02, get D = 1.393 from Eq. (3), and then Re = 81,400 from Eq. (4) and ϵ/D = 0.00011. Using Re and ϵ/D in Moody diagram gives f = 0.019. Repeat above procedure using f = 0.019, resulting in the following values: D = 1.382, Re = 82,300, and f = 0.019. The trial value of f checks out. Therefore D = (1.382)(12) = **16.6** in (**42.16** cm)   **1b.**

2. $\dfrac{V_1^2}{2g} + \dfrac{p_1}{\gamma} + z_1 = \dfrac{V_2^2}{2g} + \dfrac{p_2}{\gamma} + z_1 + h_f \quad V_1 = V_2$

$$\frac{(40)(144)}{(0.85)(62.4)} + 200 = \frac{p_2(144)}{(0.85)(62.4)} + 50 + 75$$

$$p_2 = \textbf{67.5} \text{ psi } (\textbf{465.4 kPa}) \quad \textbf{2d}$$

6.53   Principles of dynamic similitude will be applied.

$$\text{Inertia force} = Ma = \frac{\rho L^3 L}{T^2}$$

where ρ = density. Since $V = L/T$,

$$Ma = \rho L^2 V^2$$

$$\text{Viscous force} = \tau A = \mu \left(\frac{dV}{dy}\right) L^2$$

where μ = viscosity
   τ = shear stress
   V = velocity

$$\text{Viscous force} = \mu \left(\frac{V}{L}\right) L^2 = \mu V L$$

$$\frac{\text{Inertia force}}{\text{Viscous force}} = \frac{\rho L^2 V^2}{\mu V L} = \frac{\rho L V}{\mu}$$

$\mu/\rho = \nu =$ kinematic viscosity, ft²/s. The Reynolds number Re $= LV/\nu$.

$$\text{Re(model)} = \text{Re(prototype)}$$

$$\frac{L_m V_m}{\nu_m} = \frac{L_p V_p}{\nu_p}$$

$\nu_m = 1.57 \times 10^{-4}$ ft²/s, $\nu_p = 1.05 \times 10^{-5}$ ft²/s; from temperature versus kinematic viscosity curve,

$$L_m = L_p \frac{V_p}{V_m} \frac{\nu_m}{\nu_p} = \frac{6}{12} \frac{60}{200} \frac{1.57 \times 10^{-4}}{1.05 \times 10^{-5}}$$

$$= \mathbf{2.24} \text{ ft } (\mathbf{68.28} \text{ cm}) \qquad \mathbf{b}$$

6.54 Index of curvature $= c_1 = \dfrac{2R}{h} = \dfrac{(2)(5)}{4} = 2.500$

$$\text{Area of cross section} = A = 4 \text{ in}^2$$

$$\text{Eccentricity} = e = \frac{hc_1}{2(3c_1^2 - 0.8)} = \frac{(4)(2.5)}{2[(3)(2.5)^2 - 0.8]} = 0.2785$$

$$h_1 = \frac{h}{2} - e = \frac{4}{2} - 0.2785 = 1.7214$$

$$h_2 = \frac{h}{2} + e = \frac{4}{2} + 0.2785 = 2.2785$$

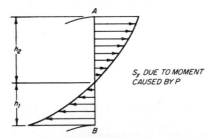

Fig. 6.54a

S_y *due to moment caused by P:*

At B: $S_B = \dfrac{PRh_1}{Aea} = \dfrac{(1000)(5)(1.7214)}{(4)(0.2785)(3)} = 2575.4$-psi compression

At A: $S_A = \dfrac{PRh_2}{Aec} = \dfrac{(1000)(5)(2.2785)}{(4)(0.2785)(7)} = 1461.0$-psi tension

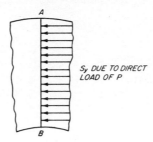

Fig. 6.54b

S_y *due to direct load of P:*

$$S_A = S_B = \frac{P}{A} = \frac{1000}{4} = 250\text{-psi compression}$$

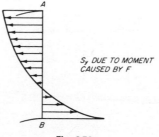

Fig. 6.54c

S_y *due to moment caused by F:*

At B: $S_B = \dfrac{FRh_1}{Aea} = \dfrac{F(5)(1.7214)}{(4)(0.2785)(3)} = 2.5754F$ tension

At A: $S_A = \dfrac{FRh_2}{Aec} = \dfrac{F(5)(2.2785)}{(4)(0.2785)(7)} = 1.4609F$ compression

Total magnitude stress at A = total magnitude stress at B
$$1461.0 - 250.0 - 1.4609F = -(-2575.4 - 250 + 2.5754F)$$
or $\quad\quad\quad\quad\quad\quad\quad\quad F = \mathbf{1448}$ lb $(\mathbf{6441}$ N$)$ **d**

6.55 For E6010,

$$\text{Tension yield point} = S_{ty} = 50{,}000 \text{ psi}$$
$$\text{Shear yield point} = S_{sy} = 25{,}000 \text{ psi}$$

$$\text{Allowable shear stress} = S_{sA} = \frac{S_{sy}}{2} = 12{,}500 \text{ psi}$$

Point B most critically stressed.

$$\text{Left throat area} = 6 \times 2 \times \tfrac{1}{4} \times 0.707 = 2.121 \text{ in}^2$$
$$\text{Right throat area} = 2 \times \tfrac{3}{8} \times 4 \times 0.707 = 2.121 \text{ in}^2$$
$$\text{Total area} = 4.242 \text{ in}^2$$

Find centroid by taking moments about left end.

$$\bar{x} = \frac{\Sigma Ax}{\Sigma A} = \frac{(2.121)(3) + (2.121)(12)}{4.242} = 7.50 \text{ in}$$

Find area moment of inertia about centroid.

$$J = \Sigma A \left(\frac{l^2}{12} + a^2 \right) = 2.121 \left(\frac{6^2}{12} + 4.5^2 \right) + 2.121 \left(\frac{4^2}{12} + 4.5^2 \right)$$
$$= 95.1 \text{ in}^4$$

where l = length of weld and a = distance from centroid to weld center.

$$S_{sP} = \frac{P}{\Sigma A} = \frac{P}{4.242} = 0.236P$$

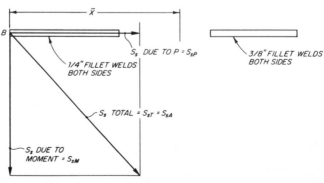

Fig. 6.55

$$S_{sM} = \frac{Tr}{J} = \frac{(5P)(7.5)}{95.1} = 0.394P$$

$$S_{sT} = \sqrt{S_{sP}^2 + S_{sM}^2}$$

$$12{,}500 = \sqrt{(0.236P)^2 + (0.394P)^2} = 0.460P$$

$$P = \mathbf{27{,}200} \text{ lb } (\mathbf{121.0} \text{ kN}) \qquad \mathbf{b}$$

SECTION SEVEN

7.01–7.75

Structural Engineering

7.01 Assume punched hole = $\frac{13}{16}$ in diam.

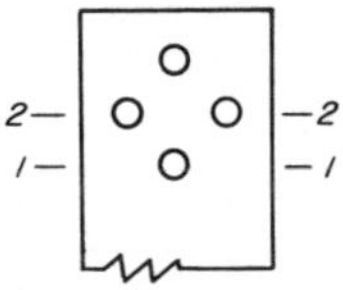

Fig. 7.01

Area net sec. 1-1 $= \frac{1}{4} \times (5 - \frac{13}{16})$
$= 1.04$ in^2

Area net sec. 2-2 $= \frac{1}{4} \times (5 - 2 \times \frac{13}{16})$
$= 0.84$ in^2

Hanger load from gross sec. $= 5 \times \frac{1}{4} \times 16{,}000$
= **20,000** lb (**88,960** N)

Hanger load from net sec. 1-1 $= 1.04 \times 16{,}000 =$ **16.650** lb

Hanger load from net sec. 2-2 $= \frac{4}{3} \times 0.84 \times 16{,}000$
$= 17{,}950$ lb

Four rivets in shear $= 4 \times 0.442 \times 12{,}000 = 21{,}000$ lb

Four rivets in bearing $= 4 \times \frac{1}{4} \times 0.75 \times 24{,}000 = 18{,}000$ lb

Joint weakest in **tension in net sec. 1-1** **1c**

$$\text{Efficiency of joint} = \frac{\text{smallest allow. load}}{\text{allow. load in gross section}}$$

$= (16{,}650/20{,}000) \times 100 =$ **83.3** percent **2b**

7.02 Area gross sec. $= \frac{3}{8} \times 9 = 3.37$ in^2

Area net sec. 1-1 (2 holes) $= \frac{3}{8} \times (9 - 2 \times \frac{3}{4})$
$= 2.81$ in^2

Area net sec. 2-2 (4 holes) = $\frac{3}{8} \times (9 - 4 \times \frac{3}{4})$
$$= 2.25 \text{ in}^2$$

Plate tens. from gross sec. = $3.37 \times 20,000 = 67,500$ lb

Plate tens. from net sec. 1-1 = $2.81 \times 20,000$
$$= \mathbf{56,200} \text{ lb } \mathbf{(249,978} \text{ N)} \qquad \mathbf{1a}$$

Plate tens. from net sec. 2-2 = $8/6 \times 2.25 \times 20,000$
$$= 60,000 \text{ lb}$$

Eight rivets in shear = $8 \times 0.442 \times 16,000$
$$= 56,500 \text{ lb}$$

Eight rivets in bearing = $8 \times 3/8 \times 3/4 \times 32,000$
$$= 72,000 \text{ lb}$$

Max. permissible tens. in plate = 56,200 lb

Efficiency of joint = $(56,200/67,500) \times 100$
$$= \mathbf{83.3} \text{ percent} \qquad \mathbf{2b}$$

7.03 p = three rivet diams. = $3 \times 3/4 + 2.25$ in

$2p$ = 4.50 in

Investigate on basis of 4.5-in length of main plate

Area gross plate = $4.5 \times 1/2 = 2.25 \text{ in}^2 (14.52 \text{ cm}^2)$

Area net sec. 1-1 (1 hole) = $1/2 (4.5 - 13/16) = 1.84 \text{ in}^2$

Area net sec. 2-2 (2 holes) = $1/2 (4.5 - 2 \times 13/16)$
$$= 1.44 \text{ in}^2 (9.29 \text{ cm}^2)$$

Tens. in plate from gross sec. = $2.25 \times 16,000$
$$= 36,000 \text{ lb} (160,128 \text{ N})$$

Tens. in plate from net sec. 1-1 = $1.84 \times 16,000$
$$= 29,500 \text{ lb} (131,216 \text{ N})$$

Tens. in plate from net sec. 2-2 = $3/2 \times 1.44 \times 16,000$
$$= 34,600 \text{ lb} (153,901 \text{ N})$$

Three rivets in shear = $6 \times 0.442 \times 10,000 = 26,500$ lb

Three rivets in bearing = $3 \times 1/2 \times 3/4 \times 20,000 = 22,500$ lb

Allow load/in seam = 22,500/4.5 = **5000** lb **(22,240** N)   **1b**

Working efficiency = $(22,500/36,000) \times 100$
$$= \mathbf{62.5} \text{ percent} \qquad \mathbf{2c}$$

7.04   Tens. main plate must carry

$$= \frac{\text{diam. tank, in}}{2} \times \text{psi} = (60/2) \times 500 = 15,000 \text{ lb/in}$$

$$= 180,000 \text{ lb/ft}$$

Min thickness of plate probably

$$\frac{12/11 \times 15,000}{20,000} = 0.818 \text{ in, or say } 13/16 \text{ in } (20.64 \text{ mm}) \qquad \mathbf{2b}$$

Assuming 7/8-in rivets used, pitch for inside row rivets = $3 \times 7/8+$ =
$21/8+$ = **3** in (**76.2** mm)　　**1a**
Shear 11 rivets = $22 \times 0.601 \times 15,000 = 198,500$ lb
Bearing 11 rivets (double shear) = $11 \times 7/8 \times 13/16 \times 30,000$
$$= 235,000 \text{ lb } (1,045,280 \text{ N})$$
Tens. in plate from gross sec. = $12 \times 13/16 \times 20,000$
$$= 195,000 \text{ lb } (867,360 \text{ N})$$
Tens. in plate from sec. 1-1 = $11/11 \times (12 - 0.94) \times 13/16 \times 20,000$
$$= 178,000 \text{ lb } (791,744 \text{ N})$$
Tens. in plate from sec. 2-2 = $11/10 (12 - 1.88) \times 13/16 \times 20,000$
$$= 181,000 \text{ lb } (805,088 \text{ N})$$
Tens. in plate from sec. 3-3 = $11/8 \times (12 - 3.76) \times 13/16 \times 20,000$
$$= 184,000 \text{ lb } (818,432 \text{ N})$$
Assume splice plate = 9/16 in thick
Tens. in splice plate at sec. 4-4
$$= 11/11 \times (12 - 3.76) \times 2 \times 9/16 \times 20,000$$
$$= 185,000 \text{ lb } (822,880 \text{ N})$$
Bearing 11 rivets (single shear) = $11 \times 7/8 \times 2 \times 9/16 \times 24,000$
$$= 260,000 \text{ lb } (1,156,480 \text{ N})$$
Design satisfactory: p = **3.0** in (**76.2** mm)　　**1a**
$$t_{\text{main plate}} = \textbf{13/16} \text{ in } (\textbf{20.64} \text{ mm})　　\textbf{2b}$$
$$t_{\text{splice plate}} = \textbf{9/16} \text{ in } (\textbf{14.29} \text{ mm})　　\textbf{3c}$$
Eff. = $(178,000/195,00) \times 100$ = **91.2** percent　　**4d**

7.05　Long. tens. = $\dfrac{\text{diam. tank, in}}{4}$ psi

$$= 60/4 \times 500 = 7500 \text{ lb/in}$$

Shear five rivets = $9 \times 0.601 \times 15,000 = 81,100$ lb
(360,732 N) (four double shear, one single shear)

$81,000 \div 7500 = 10.8$ in. Use pitch = 5.25 in (**133.35** mm)　　**1c**

Actual tens./10.5 in = $10.5 \times 7500 = 78,750$ lb
Tens. gross sec. = $10.5 \times 13/16 \times 20,000 = 171,000$ lb
Tens. in two cover plates at net sec. (inner row rivets) = $(10.5 - 1.87)$
$\times t \times 20,000 = 78,750$ lb (350,280 N)

$$t = \frac{78,750}{8.63 \times 20,000} = 0.457 \text{ in}$$

This is split $5/9 \times 0.457 = 0.25$ in min (to inside plate) and $4/9 \times 0.457$
$= 0.20$ in min (to outside plate).
Use two splice plates **7/16** in (**11.11** mm) thick to take care of bearing of

five rivets = $9 \times 7/16 \times 7/8 \times 24{,}000 = 82{,}700$ lb.   **2d**

Eff. splice$(81{,}100/171{,}000) \times 100 = $ **47.4** percent

low because thickness of main plate was dictated by hoop tension, not by longitudinal tension.   **3a**

7.06   Eleven rivets; eight double shear, three single shear, eight bearing on 3/4-in (19.05-mm) plate, three on 1/2-in (12.70-mm) plate.

Single shear one rivet = $0.994 \times 8800 = 8750$ lb per rivet
Bearing one rivet on 3/4-in plate
$$= 1.125 \times 3/4 \times 19{,}000 = 16{,}060 \text{ lb per rivet.}$$
Bearing one rivet on 1/2-in plate = $1.125 \times 1/2 \times 19{,}000$
$$= 10{,}700 \text{ lb per rivet}$$
Allow. shear in 11 rivets = $19 \times 8750 = 166{,}250$ lb (739,480 N)
Allow. bearing in 11 rivets = $3 \times 10{,}700 + 8 \times 16{,}060$
$$= 32{,}100 + 128{,}480 = 160{,}580 \text{ lb (714,260}$$
$$\text{N)}$$
Tens. gross sec. main plate = $16.5 \times 3/4 \times 11{,}000$
$$= 136{,}000 \text{ lb (604,928 N)}$$
Tens. in main plate from sec. 1-1 = $(16.5 - 1.19) \times 3/4 \times 11{,}000$
$$= 126{,}200 \text{ lb (561,338 N)}$$
Tens. in main plate from sec. 2-2 = $19/18 \times (16.5 - 2.38) \times 3/4 \times$
$$11{,}000$$
$$= 123{,}000 \text{ lb (547,104 N)}$$
Tens. in main plate from sec. 3-3 = $19/16 \times (16.5 - 4.76) \times 3/4 \times$
$$11{,}000 = 115{,}000 \text{ lb (511,520 N)}$$
Tens. in inner cover plates sec. 4-4 = $19/11 \times 1/2(16.5 - 4.76) \times 11{,}000$
$$= 111{,}530 \text{ lb } (\textbf{critical})   \textbf{1d}$$
Eff. splice = $(111{,}530/136{,}000) \times 100 = $ **82** percent   **2c**

7.07   Tens. in tank shell/in = $(D/2) \times \text{psi} = 72/2 \times 125 = 4500$ lb (20,116 N). Assume rivet pitch = 3 in (76.2 mm) and river diam. = 7/8 in (22.23 mm), hole = 15/16 (23.81 mm).

$$\text{Min thickness main plate} = \frac{1}{3 - 0.94} \frac{3 \times 4500}{16{,}000} = 0.41 \text{ in}$$

Use **7/16**-in (**11.11**-mm) plate, two rows of rivets.   **1b**

Shear one rivet = $0.601 \times 12{,}000 = 7200$ lb, two rivets
$$= \textbf{14,400} \text{ lb } (\textbf{64,051 N})   \textbf{2a}$$
Bearing one rivet = $7/8 \times 0.437 \times 24{,}000 = 9200$ lb
Tens. gross sec. of plate = $7/16 \times 3 \times 16{,}000$
$$= 21{,}000 \text{ lb}$$

Tens. sec. 1-1 of plate = 7/16 × (3 − 0.94) × 16,000
= 14,420 lb (64,140 N)
Eff. splice = (14,400/21,000) × 100 = **68.5** percent   **3d**

7.08 Slot 3-in (76.2-mm) leg of angle with 0.50-in (12.70-mm) slot just
inside 4-in (101.6-mm) leg for length of 6.75 in (171.5 mm).

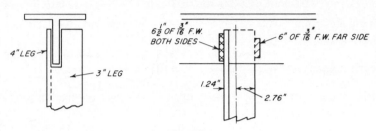

Fig. 7.08

Because of rounded edge of angle, fillet weld should not exceed 3/16 in
(4.76 mm) in size (AISC).

Throat depth = 0.707 × 3/16 = 0.1325 in (3.37 mm)
Allow. load/in fillet weld = 0.1325 × 15,000
= 1990 lb/in (348.49 N/mm)
Tensile strength 4 × 3 × 1/4 angle = 1.69 × 20,000
= 33,800 lb

No. in fillet weld needed = 33,800/1990
= **17.0** in (**431.8** mm).   **1c**

Take moments about back of 3-in leg of angle

$$1.24 \times 33,800 = 4 \times L_R \times 1990$$

$$L_R = \frac{1.24 \times 33,800}{4 \times 1990} = 5.27 \text{ in (133.9 mm)}$$

Use one length = 5.5 in + 0.50 in = 6 in (152.4 mm). Take moments
about edge of 4-in leg of angle

$$2.76 \times 33,800 \times L_L \times 2 \times 4 \times 1990$$

$$L_L = \frac{2.76 \times 33,800}{8 \times 1990} = \textbf{5.86} \text{ in (\textbf{148.9} mm)} \quad \textbf{2d}$$

Check:
5.27
5.86
5.86
___
16.99 in (17.0 in)

Use two lengths = 6 in + 0.50 in = 6.50 in (165.1 mm).

7.09 Slot stem of T with 0.75-in (19.05-mm) slot 10.5 in (266.4 mm) long just beneath flange.

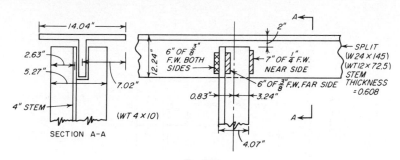

Fig. 7.09

WT 4 × 10:
Area = 2.94 in² (1897 mm²)
Flange thickness = 0.378 in (9.60 mm)
Stem thickness = 0.248 in (6.30 mm)
Tensile strength T (WT 4 × 10) = 2.94 × 20,000
$$= 58{,}800 \text{ lb (261.54 kN)}$$
Take moments about back of flange (WT 4 × 10)

$$0.83 \times 58{,}800 = 4 \times 2000 \times L_R$$

$$L_R = \frac{0.83 \times 58{,}800}{4 \times 2000} = \textbf{6.10} \text{ in (\textbf{154.9} mm)} \qquad \textbf{1c}$$

Use 6.5 in + 0.5 in = 7 in (177.3 mm).
Take moments about edge of stem (WT 4 × 10)

$$3.24 \times 58{,}800 = (2 \times 4 + 1 \times 3.62) \times 3000 \times L_L$$

$$L_L = \frac{3.24 \times 58{,}800}{11.62 \times 3000} = \textbf{5.47} \text{ in (\textbf{138.9} mm)} \qquad \textbf{2b}$$

Use three lengths of 5.5 in + 0.5 in or 6 in (152.4 mm). *Check:* 6.5 × 2000 + 3 × 5.5 × 3000 = 62,500 lb (278 kN) > 58,800 lb (262 kN)

7.10 Let f = torsional force 1 in (25.4 mm) from rivet group center of gravity.

Torsion M (about cg of rivet group) = 0

$$= +20P - 10 \times 3f \times 3$$

or

$$3f = \frac{20P}{30} = 0.667P$$

Lower right-hand rivet = most stressed rivet

$$\sqrt{(0.667P)^2 + (0.833P)^2} = 1.075P$$
$$1.075P = 9500 \qquad \text{given}$$

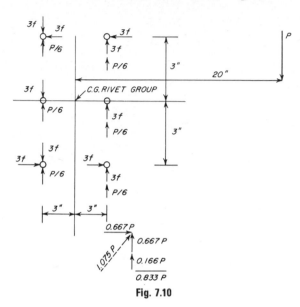

Fig. 7.10

Then $P = 9500/1.075 = \textbf{8850}$ lb (**39,365** N)   **1b**

$$A = \text{area of bolt}$$

$$A \times 16,000 = 9500, A = \frac{9500}{16,000} = 0.593 \text{ in}^2$$

Use **7/8**-in (**22.23**-mm) diam. bolt ($A = 0.601$)   **2c**
Thickness of plate for bearing $= t_b$

$$t_b \times \text{⅞} \times 32,000 = 9500 \qquad t_b = 9500/28,000 = 0.34 \text{ in}$$

Thickness of plate for tension (bending) $= t_t$

$$I_{\text{(net of plate)}} = t_t \times 9^3 \times \text{1/12} - 2 \times (1 \times t_t) \times 3^2 = 42.75t_t$$

$$f = \frac{Mc}{I}, \quad 20,000 = \frac{17 \times 8850 \times 4.5}{42.75t_t}$$

$$t_t = \frac{17 \times 8850 \times 4.5}{42.75 \times 20,000} = 0.79 \text{ in}$$

Use **13/16**-in (**20.64**-mm) plate.   **3d**

7.11   Let f be the force in a rivet 1 in (25.4 mm) from the center of the rotation. Then torsion in the rivets of one plate (about center of gravity of rivet group)

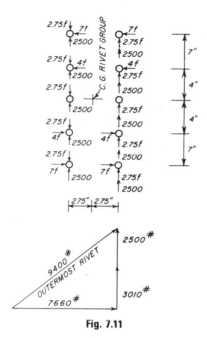

Fig. 7.11

$$10 \times 2.75 \times 2.75f = 76f$$
$$4 \times 4 \times 4f = 64f$$
$$4 \times 7 \times 7f = \underline{196f}$$
$$ 336f$$

$$M = 0$$
$$ = 2 \times 336f - 14.75 \times 50,000$$
$$f = 1094 \text{ lb}$$
$$7f = 7660 \text{ lb}$$
$$2.75f = 3010 \text{ lb } (13,388 \text{ N})$$

A 1-in rivet in single shear and bearing on a 1/2-in plate may carry 11,780 lb (52,397 N). O.R. carries **9400** lb (**41,811** N).   **1d**

$I_{(net)}$ of plates $= 1/12 \times 1 \times 17^3 - 2 \times 1 \times 1 \times (4^2 + 7^2)$

$$f = \frac{Mc}{I} = \frac{12 \times 50,000 \times 8.5}{280} = \textbf{18,200} \text{ psi } (\textbf{125.49} \text{ MPa}) \quad \textbf{2c}$$

7.12   Let f = shearing value of the weld 1 in (25.4 mm) from center of gravity of welds.

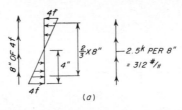

Fig. 7.12a   Shearing value of welds along one edge.

M of welds about cg of welds

$$= 4 \times [4 \times 8 \times 4f + 4 \times (4f/2) \times (\tfrac{2}{3} \times 8)]$$
$$\doteq 512f + 171f = 683f$$
$$683f = 12 \times 10,000, f = 176 \text{ lb/in}, 4f = 704 \text{ lb/in}$$

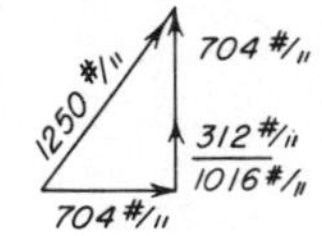

Max weld force 1250 lb/in (219 N/mm). Allowable for ¼-in (6.35-mm) fillet weld = 2000 lb/in (350 N/mm).

$$I = \tfrac{1}{12} \times \tfrac{1}{2} \times 8 \times 8 \times 8 = 21.3 \text{ in}^4$$

$$f = \frac{Mc}{I} = \frac{8 \times 10,000 \times 4}{21.3} = 15,000 \text{ psi } (103.43 \text{ MPa})$$

A better use of welds is not to weld into corners.
If only 5 in (127 mm) of weld is used on each of the four sides

$$4[4 \times 5 \times 4f + 2.5 \times (2.5f/2) \times (\tfrac{2}{3} \times 5)]$$
$$= 320f + 42f = 362f$$
$$362f = 12 \times 10,000, \quad f = 331,$$
$$2.5f = 824, \quad 4f = 1324 \text{ lb/in}$$

$$\frac{10,000}{4 \times 5} = 500 \text{ lb/in } (87.6 \text{ N/mm})$$

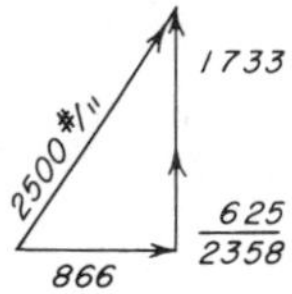

If four 5/16-in (7.94-mm) fillet welds are used, each 4 in (101.6 mm) long, then

$$4[4 \times 4 \times 4f + 2 \times (2f/2) \times (\tfrac{2}{3} \times 4)] = 256f + 21f = 277f$$
$$277f = 12 \times 10,000, \quad f = 433, \quad 2f = 866, \quad 4f = 1733 \text{ lb/in}$$

$$\frac{10,000}{4 \times 4} = 625 \text{ lb/in } (109.5 \text{ N/mm})$$

Fig. 7.12b

Fig. 7.12c

Fig. 7.12d

2500 lb/in (438 N/mm), OK for 5/16-in (7.94-mm) fillet weld.

7.13 Gage from back of 4-in (101.6-mm) leg to rivets in 3½-in (88.9-mm) leg = 2 in (50.8 mm).

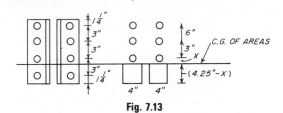

Fig. 7.13

Each 7/8-in (22.23-mm) rivet good for 9020 lb (40,121 N) in shear and bearing on 3/8-in (9.53-mm) leg. Area of two 7/8-in (22.23-mm) rivets = 1.2 in² (774.2 mm²). Assume that center of gravity of areas lies x distance below the second rivet from the bottom. Taking statical moments about the center-of-gravity line

$$\frac{8(4.25 - x)^2}{2} = 1.2[x + (x + 3) + (x + 6)] = 1.2(3x + 9)$$

$$(4.25 - x)^2 = \frac{1.2}{4}(3x + 9)$$

$$18.1 - 8.5x + x^2 = 0.9x + 2.7$$

$$(x^2 - 9.4x + 15.4) + 6.7 = 6.7$$

$$x - 4.7 = \pm 2.59, x = +4.7 - 2.59 = 2.11 \text{ in } (53.6 \text{ mm})$$

Let f be the intensity of tension or compression 1 in (25.4 mm) from the center-of-gravity line

$$1.2(2.11^2 f + 5.11^2 f + 8.11^2 f) + \tfrac{8}{3} \times 2.14^3 f = 115.8f + 26.2f$$
$$= 142.0f$$
$$= 8 \times 9020 \times 2$$
$$= \text{double shear in rivets}$$
$$\text{through web of beam}$$
$$\times 2 \text{ in eccentricity.}$$

$$f = 1016 \text{ psi/in}$$
$$\text{Tension top rivet} = 0.601 \times 1016 \times 8.11$$
$$= 4950 \text{ lb } (\mathbf{22{,}018} \text{ N}) \qquad \mathbf{b}$$

7.14 Similar to question 7.13, assume that center of gravity lies x distance below third rivet above the bottom. M about center-of-gravity line is

$$\frac{8(7.25 - x)^2}{2} = 1.2[x + (x + 3) + (x + 6) + (x + 9)$$

$$+ (x + 12) + (x + 15) + (x + 18) + (x + 21)]$$

$$(7.25 - x)^2 = \frac{1.2}{4}(8x + 84)$$

$$52.6 - 14.5x + x^2 = 2.4x + 25.2$$

$$(x^2 - 16.9x + 27.4) + 44 = 44$$

$$(x - 8.45)^2 = 44$$

$$x - 8.45 = \pm 6.63$$

$$x = +8.45 - 6.63 = 1.82 \text{ in } (46.23 \text{ mm})$$

Let f be the intensity of tension or compression 1 in (25.4 mm) from the center-of-gravity line.

$$1.2 \times 1.82^2 f = 1.2f \times 3.3$$
$$1.2 \times 4.82^2 f = 1.2f \times 23.2$$
$$1.2 \times 7.82^2 f = 1.2f \times 61.2$$
$$1.2 \times 10.82^2 f = 1.2f \times 117.3$$
$$1.2 \times 13.82^2 f = 1.2f \times 191.0$$
$$1.2 \times 16.82^2 f = 1.2f \times 283.0$$
$$1.2 \times 19.82^2 f = 1.2f \times 393.0$$
$$1.2 \times 22.82^2 f = 1.2f \times 522.0$$

$$1.2f \times 1596 = 1915f$$

$$8 \times (5.43)^3 \times f/3 = 427f$$

$$20 \times 9020 \times 2 = 2342f$$

$$f = 154.0 \text{ lb/(in}^2)(\text{in})$$

$$\text{Tens. top rivets} = 0.601 \times 154.0 \times 22.82$$
$$= \textbf{2110} \text{ lb } (\textbf{9385} \text{ N}) \qquad \textbf{a}$$

7.15 Side dimension of fillet weld must be 5/8 in (15.88 mm) or less.

$$\text{Throat dimension of } \frac{1}{2}\text{-in (12.7-mm) fillet weld} = 0.707 \times \frac{1}{2}$$
$$= 0.354 \text{ in } (8.99 \text{ mm})$$

½-in fillet weld can carry $0.354 \times 13,000 = \textbf{4600}$ lb/in (**806** N/mm) in shear.   **b**

$4 \times 12 \times 4600 = 221,000$ lb (983,008 N), which is greater than 200,000 lb (889,600 N).

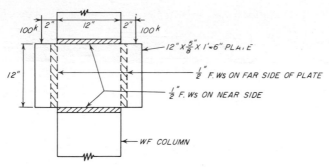

Fig. 7.15

7.16   Gross I of girder:

I of four angles about cg of angles = 4×7.5 = 30 in⁴
plus Ad^2 for angles = $4 \times 5.86(21.25 - 1.03)^2$ = 9,600 in⁴
I of two cover plates = $2 \times 14 \times \frac{1}{2} \times 21.50^2$ = 6,470 in⁴
I of web plate = $\frac{1}{12} \times \frac{3}{8} \times 42^3$ = 2,320 in⁴

$$I = 18,420 \text{ in}^4$$

$$f = \frac{Mc}{I} = 18,000 = \frac{\frac{1}{8} \times W \times 30 \times 30 \times 12 \times 21.75}{18,420}$$

$W =$ **11,292** lb/ft (**164,784** N/m)
    = total uniform load per foot (0.305 m) of girder   **2d**

Center of gravity of flange is 1.05 in below back of cover plate

Plate: $14 \times \frac{1}{2}$ = 7.00 in² × ¼ in = 1.75 in³
Angles: 2×5.86 = $\underline{11.72} \times 1.53$ = $\underline{17.92}$
Flange: = 18.72 (1.05) = 19.67

$$sb = \frac{VQ}{I} = \frac{75,000 \times 18.72 \times 20.70}{18,420} = 1580 \text{ lb/in}$$

One ¾-in rivet may carry $2 \times 0.4418 \times 15,000 = 13,250$ lb in double shear and 11,300 lb in double bearing on ⅜-in web maximum rivet pitch for one line of rivets = 11,300/1580 = 7.15 in. Use **6** in (**152.4** mm). (Max usually allowed by specifications for this condition.)   **1c**

7.17 Try four angles 6 × 6 × ⅞ in placed 36½ in back to back of angles.
Gross I of girder:

I of four angles = 4 × 31.9	= 127 in^4
Ad^2 for angles = 4 × 9.73(18.25 − 1.82)^2	= 10,508 in^4
I of web = 1/12 × 3/8 × 36^3	= 1,458 in^4
I gross	= 12,093 in^4
I of holes = 2 × 17/8 × 13/16(18.25 − 3.50)^2	= 750 in^4
I net	= 11,343 in^4
Weight four angles = 4 × 33 lb	= 132 lb/ft
Weight web plate = 490/144 × 36 × 3/8	= 46 lb/ft
Weight girder	= 178 lb/ft
Dead-load moment = 1/8 × 178 × 50 × 50 × 12	= 666,000 in·lb
Live-load moment = 1/4 × 70,000 × 50 × 12	= 10,500,000 in·lb
	M = 11,166,000 in·lb

$$f = \frac{Mc}{I} = \frac{11{,}166{,}000 \times 18.25}{11{,}343} = 17{,}950 \text{ psi } (123.77 \text{ MPa})$$

$$V = 70{,}000 + 25 \times 178 = 74{,}450 \text{ lb}$$

$$sb = \frac{VQ}{I} = \frac{74{,}450 \times 19.46 \times 16.43}{11{,}343} = 2085 \text{ lb/in } (365.1 \text{ N/mm})$$

One 3/4-in rivet in bearing = 3/4 × 3/8 × 40,000 = 11,250 lb. Max pitch of 3/4-in rivets = 11,250/2085 = 5.40 in (137.2 mm).

7.18 Trial design by flange-area method.
Let girder be 1/15 by 60 ft or 4 ft in depth.

Shear at end = 5000 × 30 = 150,000 lb
Web area needed at 13,500 psi = 150,000/13,500
= 11.1 in^2

Use web plate 48 × 3/8 = 18.0 in^2
Assume lateral support of girder sufficient to permit bending stress of 18,000 psi in compression.

Moment = flange area × 18,000 × 45 in (couple arm)
= 1/8 × 5000 × 60 × 60 × 12
= 27,000,000 in·lb

$$\text{Flange area} = \frac{27{,}000{,}000}{18{,}000 \times 45} = 33.33 \text{ in}^2$$

1/8 area of web = flange area	= 2.25 in^2
	31.08 in^2
Area of two angles 8 × 6 × 3/4	= 19.88 in^2
	11.20 in^2

Area of one cover plate $18 \times \frac{5}{8}$ = **11.25** in²

Check design by gross-moment-of-inertia method.

Distance back to back of 8-in legs of angles	= 48½ in
I of two angles about horizontal cg of angles	= 68 in⁴
Plus area of two angles × distance² to N.A. of web	
$19.88(48.50/2 - 1.56)^2$	= 10,250 in⁴
I of cover plate = $18 \times \frac{5}{8} \times (48.50/2 + \frac{1}{2} \times \frac{5}{8})^2$ =	6,775 in⁴
I of top flange	= 17,093 in⁴
I of bottom flange	= 17,093 in⁴
I of web plate = $\frac{1}{12} \times \frac{3}{8} \times 48^3$	= 3,456 in⁴
I total	= 37,642 in⁴

$$f = \frac{Mc}{I} = \frac{27,000,000 \text{ in·lb} \times 24.875 \text{ in}}{37,642 \text{ in}^4} = 17,850 \text{ psi } (123.08 \text{ MPa})$$

7.19 Center of gravity of T (above bottom of W)   18.16 in

W: 28.22 in² ×   9.08 in = 256 in³ + 0.40 in

Channel: 9.90 in² × 17.77 in = 176 in³   18.56 in

38.12 in² (11.32 in) = 432 in³ − 0.79 in

17.77 in

− 11.32 in

6.45 in

I of T (about cg):

Channel:	I about channel cg	=	8.2 in⁴
	$A \times d^2 = 9.90 \times 6.45^2$	=	412.0 in⁴
W:	I about W cg	=	1674.7 in⁴
	$A \times d^2 = 28.22 \times 2.24^2$	=	141.5 in⁴
		I =	2236.4 in⁴

Moment:

$$S = \frac{I}{c} \text{ (for top flange)}$$

$$= 2236.4/7.24 \qquad\qquad = 309 \text{ in}^3$$

$$M = f \times S = 17,050 \text{ psi} \times 309 \text{ in}^3 = 5,260,000 \text{ in·lb}$$

$$S = \frac{I}{c} \text{ (for bottom flange)}$$

$$= 2236.4/11.32 \qquad\qquad = 197 \text{ in}^3$$

$$M = f \times S = 20,000 \text{ psi} \times 197 \text{ in}^3 = 3,940,000 \text{ in·lb (use)}$$

$$\frac{P \times 30}{4} \times 12 + \frac{1}{8} \times 129.9 \times 30 \times 30 \times 12 = 3{,}940{,}000 \text{ lb}$$

$$90P = 3{,}940{,}000 - 175{,}500 = 3{,}764{,}500$$

$$P = \textbf{41,900} \text{ lb}$$

Shear:

Two ⅝-in rivets in single shear $= 2 \times 0.307 \text{ in}^2 \times 15{,}000 \text{ psi}$
$= 9200 \text{ lb}$

Horizontal shear/in beam (at end) may not exceed 9200 lb/6 in = 1534 lb/in $= sb = VQ/I$.

$$V = \frac{sb \times I}{Q} = \frac{1534 \times 2236.4}{9.90 \times 6.45} = 53{,}600 \text{ lb} = \frac{P}{2} + 15 \text{ ft} \times 129.9 \text{ ft} \cdot \text{lb}$$

$$P = 2(53{,}600 - 1950) = \textbf{103,300} \text{ lb} > \textbf{41,900} \text{ lb}$$

$$\text{Deflection} = \frac{1}{360} \times 30 \times 12 = \frac{P \times (30 \times 12)^3}{48 \times 30{,}000{,}000 \times 2236.4}$$

$$P = \frac{4 \times 1{,}000{,}000 \times 2236.4}{30 \times 30 \times 12 \times 12} = \textbf{69,000} \text{ lb} > \textbf{41,900} \text{ lb}$$

Permissible $P = \textbf{41,900}$ lb (**186,371** N)

7.20 (Refer to answer 1.616.)

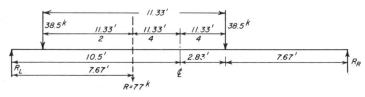

Fig. 7.20

Max live-load moment

$$= \frac{7.67 \text{ ft}}{21 \text{ ft}} \times 77 \text{ kips} \times 7.67 \text{ ft} \qquad\qquad = 215 \text{ ft} \cdot \text{kips}$$

Dead-load moment at centerline
$= \frac{1}{8} \times 100.7 \text{ ft/lb} \times 21 \text{ ft} \times 21 \text{ ft} \qquad = \underline{ 5.5 \text{ ft} \cdot \text{kips}}$
Total moment $= (299.0 \text{ kN} \cdot \text{m}) \qquad\qquad\qquad = 220.5 \text{ ft} \cdot \text{kips}$
Max end live-load shear

$$= 38.5 \text{ kips} + \frac{21 \text{ ft} - 11.33 \text{ ft}}{21 \text{ ft}} \times 38.5 \text{ kips} \quad = 56.2 \text{ kips}$$

Dead-load shear = 10.5 ft × 100.7 lb   = 1.06 kips

Total shear   = 57.26 kips

Center of gravity of T (above bottom of W)   24.280 in

W: 23.54 in² × 12 in   = 282.5 in²   − 0.70 in

Channel: $\dfrac{6.03\ \text{in}^2 \times 23.58\ \text{in} = 142.0\ \text{in}^3}{29.57\ \text{in}^2\ (14.35\ \text{in})\quad = 424.5}$ $\begin{array}{r}23.58\ \text{in}\\ -14.35\ \text{in}\end{array}$

I of T [about horizontal line through cg] $\begin{array}{r}9.23\ \text{in}\\ 14.35\ \text{in}\\ -12.00\ \text{in}\end{array}$

Channel: Ad^2 = 6.03 in² × 9.23 in × 9.23 in $\begin{array}{r}2.35\ \text{in}\\ = \quad 514\ \text{in}^4\end{array}$

      I about its cg   =   4 in⁴

W: Ad^2 = 23.54 × 2.35 × 2.35   =   130 in⁴

    I about its cg   = 2,230 in⁴

    I total   = 2,878 in⁴

$$S = \frac{I}{c} = \frac{2878}{14.35} = 201 \text{ in}^3 \text{ (sec. modulus for bottom fiber)}$$

$$S = \frac{2878}{9.93} = 290 \text{ in}^3 \text{ (sec. modulus for top fiber)}$$

$$f(\text{bottom}) = \frac{Mc}{I} = \frac{220{,}500 \text{ ft·lb} \times 12 \text{ in/ft} \times 14.35 \text{ in}}{2878 \text{ in}^4}$$

$$= \textbf{13,170} \text{ psi } (\textbf{90.81 MPa}) \quad \textbf{1b}$$

$$f(\text{top}) = \frac{220{,}500 \times 12 \times 9.93}{2878} = \textbf{9130} \text{ psi } (\textbf{62.95 MPa}) \qquad \textbf{1a}$$

$$f(\text{allow.}) \text{ for } \frac{L}{b} = 21$$
$$= 18{,}000 \text{ psi } (124.11 \text{ MPa}) \text{ in compress. (also tens.)}$$

Bridge crane runway girders are subjected to lateral loads = to 10 percent of suspended load on crane plus an increase of vertical live loads due to impact. Full data not given in question, but fiber stresses will increase due to impact plus bending about vertical axis. Above bending stresses do not appear to be unreasonable. Horizontal shear per running inch of beam at plane between channel and the W top flange is:

$$sb = \frac{VQ}{I} = \frac{57{,}260 \text{ lb} \times 6.03 \text{ in}^2 \times 9.23 \text{ in}}{2878 \text{ in}^4}$$

$$= 1108 \text{ lb/in } (194.03 \text{ N/mm})$$

Single shear val. of two ¾-in rivets = 2 × 0.4418 in² × 15,000 psi
$$= 13{,}250 \text{ lb } (58{,}936 \text{ N})$$

$$\frac{13{,}250}{1108} = 12 \text{ in } (304.8 \text{ mm})$$

Use **6**-in (**152.4**-mm) maximum rivet pitch.   **2c**

7.21 $R_L = R_R = 3/8wL = 3/8 \times 10 \times 12 = 45$ kips (200.2 kN)
 $R_C = 2 \times 5/8wL = 5/4 \times 10 \times 12 = 150$ kips (667.2 kN)
M at R_C (no deflection of R_C) $= 45$ kips $\times 12$ ft $- 10$ kips/ft $\times$
12 ft $\times 6$ ft $= -180$ ft·kips (244.1 kN·m)

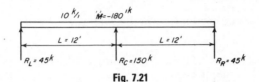

Fig. 7.21

1. Remove R_C and under 10 kips/ft load on 24-ft simple span

$$\Delta \text{ centerline} = \frac{5}{384} \times \frac{240,000 \text{ lb} \times 24 \text{ ft} \times 24 \text{ ft} \times 24 \text{ ft} \times 1728}{30,000,000 \text{ psi} \times 1373 \text{ in}^4}$$

$$= 1.81 \text{ in (45.97 mm)}$$

2. R_C must push the load up 1.81 in for no deflection or 1.31 in for a 0.5-in deflection

$$R_C \text{ for a 0.5-in deflection} = \frac{1.31}{1.81}\, 150 \text{ kips} = 108.5 \text{ kips (482.6 kN)}$$

and $\qquad R_L = R_R = 120 - \dfrac{108.5}{2} = 120 - 54.25$
$$= 65.75 \text{ kips (292.5 kN)}$$

M at R_C (0.5-in deflection of R_C)
$$= 65.75 \text{ kips} \times 12 \text{ ft} - 10 \text{ kips/ft} \times 12 \text{ ft} \times 6 \text{ ft}$$
$$= +69 \text{ ft·kips (93.6 kN·m)}$$

3. Change in bending moment at $R_C = 180 + 69 = 249$ ft·kips. This may be found by $(65.75 - 45) \times 12 = 249$ ft·kips.
Alternate: Change in bending moment at $\text{R}_C = \Delta$ bending moment$_C =$
$\dfrac{\Delta_{R,C}}{1000} \times \dfrac{2 \times 12}{4}$ where $\Delta_{R,C}$ is found from the deflection at C

$$\Delta_C = 0.5 \text{ in} = \frac{1}{48} \times \frac{\Delta_{R,C}(24)^3 \times 1728}{30,000,000 \times 1378}$$

$$\frac{\Delta_{R,C}}{1000} = 41.5 \text{ kips (184.6 kN)}$$

Δ bending moment$_C = \dfrac{41.5}{2} \times 12 = \mathbf{249.0}$ ft·kips (337.6 kN·m)   **d**

7.22 Determine reactions:

$$M_{R,L} = 0 \qquad 6R_R - 4 \times 2400 - 2 \times 4800 = 0$$
$$R_R = 3200 \text{ lb (14,233 N)}$$
$$M_R = 0 \qquad 6R_L - 4 \times 4800 - 2 \times 2400 = 0$$
$$R_L = 4000 \text{ lb (17,792 N)}$$

Check reactions:

$$F_v = 0 \qquad 3200 + 4000 - 4800 - 2400 = 0 \qquad \text{OK}$$

$$I_x = \frac{bh^3}{12} = \frac{1 \times 6^3}{12} = 18 \text{ in}^4 \qquad Q = A'\bar{y} = (1 \times 1.5)2.25$$
$$= 3.375 \text{ in}^3$$

$$S_{xy} = \frac{VQ}{I_x b} = \frac{4000 \times 3.375}{18 \times 1} = 750 \text{ psi (5.17 MPa)}$$

$$M_{a-a} = 1.5 \times 4000 = 6000 \text{ ft·lb (8.14 kN·m)}$$

$$S_x = \frac{Mc}{I_x} = \frac{72,000 \times 1.5}{18} = 6000 \text{ psi (41.37 MPa)}$$

Max normal stress:

$$S_1 = \frac{S_x}{2} + \sqrt{\left(\frac{S_x}{2}\right)^2 + S_{xy}^2} = \frac{6000}{2} + \sqrt{\left(\frac{6000}{2}\right)^2 + (750)^2}$$
$$= \textbf{6090}\text{-psi (\textbf{41.99}-MPa) \textbf{tension}} \qquad \textbf{1d} \text{ and } \textbf{2a}$$

Min normal stress:

$$S_2 = \frac{S_x}{2} - \sqrt{\left(\frac{S_x}{2}\right)^2 + S_{xy}^2} = \frac{6000}{2} - \sqrt{\left(\frac{6000}{2}\right)^2 + (750)^2}$$
$$= \textbf{90}\text{-psi (\textbf{621}-kPa) \textbf{compression}} \qquad \textbf{3a} \text{ and } \textbf{4b}$$

Max shearing stress:

$$S_S = \frac{S_1 - S_2}{2} = \frac{6090 - (-90)}{2} = \textbf{3090} \text{ psi (\textbf{21.31} MPa)} \qquad \textbf{5c}$$

$$\tan 2\theta_n = -\frac{S_{xy}}{(S_x - S_y)/2} = -\frac{750}{6000/2} = -0.25$$

$$2\theta_n = 166° \text{ and } 346° \text{ or } \theta_n = 83° \text{ and } 173°$$

7.23 1. Beam *EF* carries a moment of 499.7 ft·kips (677.6 kN·m) and end shear = 20.7 kips (92.1 kN). Use A36 (248.2 MPa) structural steel.

$$\text{Sec. modulus required} = S \text{ in}^3 = \frac{M \text{ in} \cdot \text{lb}}{f \text{ psi}}$$

$$= \frac{499,700 \times 12}{22,000}$$

$$= 273 \text{ in}^3 \ (f = 0.6F_y = 22,000)$$

Use W24 $\times$ 110, $S = 274.4$ in³. Use two connection angles 4 $\times$ 3½ $\times$ ⁵⁄₁₆ and four ¾-in A307 (96.5-MPa) bolts per angle.

2. Beam CD carries a moment of 251 ft·kips (340.4 kN·m) and end shear = 25.1 kips (111.6 kN).

$$S = \frac{251,000 \times 12}{22,000} = 137 \text{ in}^3$$

Use W 24 $\times$ 68, $S = 153.1$ in³.
Use same end connections as for (1).

3. Beam AB carries a moment of 931 ft·kips (1262.4 kN·m) and end shear = 39.4 kips (175.3 kN).

$$S = \frac{931,000 \times 12}{22,000} = 508 \text{ in}^3$$

Use W 36 $\times$ 160, $S = 541.0$ in³. Use two connection angles 4 $\times$ 3½ $\times$ 5/16 and six ¾-in A307 (96.5-MPa) bolts per angle.

7.24   (The following formulas for deflections of simple beams due to two concentrated loads at the third points are given in many texts and handbooks.)

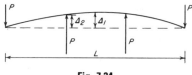

Fig. 7.24

First, neglect centerline load of 4P

$$\Delta_1(\text{at centerline}) \uparrow \ = \frac{P \times L/3}{24EI} \left(3L^2 - 4\frac{L^2}{9} \right)$$

$$= \frac{PL}{3 \times 9 \times 24EI} (27L^2 - 4L^2) = \frac{23PL^3}{648EI}$$

$$\Delta_2(\text{at third point}) \uparrow \; = \frac{P \times L/3}{6EI}\left(3L\frac{L}{3} - 3\frac{L^2}{9} - \frac{L^2}{9}\right)$$

$$= \frac{4PL}{4 \times 3 \times 9 \times 6EI}(9L^2 - 3L^2 - L^2) = \frac{20PL^3}{648EI}$$

Deflection of centerline point above third points $= \dfrac{3PL^3}{648EI}$

Second, neglect end loads of P

$$\Delta_3(\text{at centerline}) \downarrow \; = \frac{Fx^3}{48EI} = \frac{4P \times L^3/27}{48EI}$$

$$= \frac{2PL^3}{648EI}$$

$= $ deflection of centerline point below third points

Resultant deflection at centerline due to all P loads is $PL^3/648EI$ *upward.*

7.25 Solve by use of conjugate-beam method, requiring no handbook formulas.

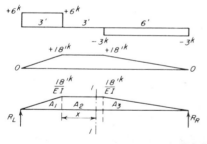

Normal shear diagram.

Normal moment diagram.

Normal moment diagram divided by EI becomes the loading for conjugate beam.

Fig. 7.25

$$R_L \text{ for conjugate beam, } M \text{ about } R_R$$

Area no.	Area	Arm (from R_R)	Area × arm = M
A_1	$= \dfrac{18}{EI} \times 3 \times \frac{1}{2} = \dfrac{27}{EI}$	$\times (9 + \frac{1}{3} \times 3) =$	$\dfrac{270}{EI}$
A_2	$= \dfrac{18}{EI} \times 3 \qquad = \dfrac{54}{EI}$	$\times (6 + \frac{1}{2} \times 3) =$	$\dfrac{405}{EI}$
A_3	$= \dfrac{18}{EI} \times 6 \times \frac{1}{2} = \dfrac{54}{EI}$	$\times (\frac{2}{3} \times 6) \qquad =$	$\dfrac{216}{EI}$
Σ areas $=$	$\dfrac{135}{EI}$	$\Sigma M = $	$\dfrac{891}{EI}$

$$\text{and } R_L = \frac{891}{12EI} = \frac{74.25}{EI} \left(\frac{\text{ft}^2 \text{ kips}}{EI}\right)$$

$$R_R \text{ for conjugate beam, } M \text{ about } R_L$$

Area	×	Arm	=	M
$A_1 = \dfrac{27}{EI}$	$\times$	$(\frac{2}{3} \times 3)$	$=$	$\dfrac{54}{EI}$
$A_2 = \dfrac{54}{EI}$	$\times$	$(3 + \frac{1}{2} \times 3)$	$=$	$\dfrac{243}{EI}$
$A_3 = \dfrac{54}{EI}$	$\times$	$(6 + \frac{1}{3} \times 6)$	$=$	$\dfrac{432}{EI}$

$$\Sigma M = \frac{729}{EI} \text{ and } R_R = \frac{729}{12EI} = \frac{60.75}{EI}.$$

$$R_L + R_R = \frac{74.25}{EI} + \frac{60.75}{EI} = \frac{135}{EI}$$

Find where shear for conjugate-beam loading passes through zero.

$$R_L - A_1 = \frac{74.25}{EI} - \frac{27}{EI} = \frac{47.25}{EI}$$

which is less than $54/EI$ or A_2.

Therefore shear passes through zero x distance to right of A_1 area. So

$$\frac{18x}{EI} = \frac{47.25}{EI} \text{ and } x = \frac{47.25}{18} = 2.625 \text{ ft (0.800 m)}$$

and deflection is maximum at section 1-1, 5.625 ft to right of R_L.

$$\Delta \text{ max} = \text{moment conjugate-beam loading about section 1-1}$$

$$= +R_L \times 5.625 - A_1 \times (1 + 2.625) - \frac{47.25}{EI} \frac{2.625}{2}$$

$$= \frac{74.25}{EI} 5.625 - \frac{27}{EI} 3.625 - \frac{47.25}{EI} 1.312$$

$$= (+417.65 - 97.87 - 62.01) \frac{1}{EI} = \frac{257.77}{EI} \left(\frac{\text{ft}^3 \text{ kips}}{EI} \right)$$

7.26   For simplicity, divide this problem into two parts and solve by the conjugate-beam method.

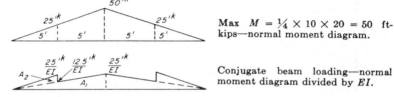

Max $M = \frac{1}{4} \times 10 \times 20 = 50$ ft-kips—normal moment diagram.

Conjugate beam loading—normal moment diagram divided by EI.

Fig. 7.26a

1. *Concentrated load:*

$$A_1 = \frac{1}{2} 10 \frac{25}{EI} = \frac{125}{EI} \left. \vphantom{\frac{12.5}{EI}} \right\} \text{ the same for}$$
$$A_2 = \frac{1}{2} 5 \frac{12.5}{EI} = \frac{31.25}{EI} \left. \vphantom{\frac{12.5}{EI}} \right\} \begin{array}{l} \text{right or left} \\ \text{half of beam} \end{array}$$

$$R_R = R_L = A_1 + A_2 = \frac{156.25}{EI}$$

$$M \text{ at centerline} = \frac{156.25}{EI} 10 - \frac{125}{EI} \frac{10}{3} - \frac{31.25}{EI} \left(5 + \frac{5}{3} \right)$$

$$= (+1562.5 - 416.7 - 208.3) \frac{1}{EI}$$

$$= \frac{937.5}{EI} \left(\frac{\text{ft}^2 \text{ kips}}{EI} \right)$$

2. *Uniform load:*

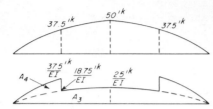

Max $M = \frac{1}{8} \times 1 \times 20 \times 20 = 50$ ft-kips—normal moment diagram is a parabola and M at quarter point = $\frac{3}{4}M$ at center line.

Conjugate beam loading – normal moment diagram divided by EI

Fig. 7.26b

$$A_3 = \frac{2}{3}\frac{25}{EI}\,10 = \frac{166.67}{EI}$$ area under parabola from vertex outward = two-thirds area enclosing rectangle

$$A_4 = \frac{1}{2}\,5\,\frac{18.75}{EI} = \frac{46.87}{EI}$$ it is close enough to treat area as triangle

$$R_R = R_L = A_3 + A_4 = \frac{213.54}{EI}$$

M at centerline $= +\dfrac{213.54 \times 10}{EI} - \dfrac{46.875}{EI}\left(5 + \dfrac{5}{3}\right) - \dfrac{166.67}{EI}\dfrac{3}{8}\,10$

$$= (+2135.4 - 312.5 - 625)\frac{1}{EI}$$

$$= \frac{1197.9}{EI}\left(\frac{ft^3\ kips}{EI}\right)$$

Total Δ at centerline $= (937.5 + 1197.9)\dfrac{1}{EI}$

$$= \frac{2135.4}{EI}\left(\frac{ft^3\ kips}{EI}\right)$$

(To understand the conversion of Δ to inches, assume I of beam $= 344$ in⁴.)

$$\Delta\ max = \frac{2{,}135{,}400 \times 1728}{344 \times 30{,}000{,}000} = 0.357\ in\ (9.1\ mm)$$

(The exact way in getting and dealing with A_4 would have shown in the fourth figure to the right of the decimal by about -1.)

7.27 Take statical moment from back of plate to find center of gravity.

Member	Area	×	Arm	=	Product
Plate	4.88	×	$\frac{3}{16}$	=	0.915
2 Channels	6.72	×	$4\frac{3}{8}$	=	29.400
	11.60 in²	×	(2.61 in)	=	30.315 in³

Center of gravity lies 2.61 in (66.3 mm) from back of plate and on axis of symmetry.

$$I_{(1-1)}$$

Plates $\frac{1}{12} \times 13 \times \frac{3}{8} \times \frac{3}{8} \times \frac{3}{8} = \quad 0.06 \text{ in}^4$

$+4.88 \times (2.61 - 0.19)^2 \qquad = \quad 28.80$

2 channels $2 \times 32.3 \qquad\qquad = \quad 64.60$

$+6.72 \times (4.37 - 2.61)^2 \qquad = \quad \underline{20.82}$

$I \text{ (total)} \qquad\qquad\qquad = \quad 114.28 \text{ in}^4$

$$r^2_{(1-1)} = \frac{I}{A} = \frac{114.28}{11.60} \qquad = \quad 9.85 \text{ in}^2$$

$$\frac{L^2}{r^2} = \frac{(20 \times 12)^2}{9.85} \qquad = 5840$$

Allow. stress $= 17,000 - 0.485 \times 5840 = 14,170$ psi (97.90 MPa)

Allow. load $= A_s f_s = 11.60 \times 14,170 = \mathbf{164,300}$ lb **(730.8 kN)** **d**

$$I_{(2-2)}$$

Plate $\frac{1}{12} \times \frac{3}{8} \times 13 \times 13 \times 13 \quad = \quad 68.7 \text{ in}^4$

2 channels $2 \times 1.3 \qquad\qquad\qquad = \quad 2.6$

$+6.72 \times 4.58^2 \qquad\qquad\qquad = \quad \underline{141.0}$

$\qquad\qquad\qquad\qquad\qquad\qquad\qquad 212.3 \text{ in}^4$

$$r^2_{(2-2)} = \frac{212.3}{11.60} \qquad\qquad = 18.3 > 9.85$$

Use $r^2_{(1-1)}$ only.

7.28 15 channel 50:

$$A = 14.64 \text{ in}^2 \qquad x = 0.80 \text{ in},$$
$$I_1 = 401.4 \text{ in}^4 \qquad I_2 = 11.2 \text{ in}^4$$

Center-of-gravity distance from top of cover plate = 5.42 in (137.7 mm)

Member		Area	×	Arm	=	Product
Plate 22 × ¾	=	16.50	×	⅜	=	6.18
2 Channels	=	29.28	×	8.25	=	242.00
		45.78 in²	×	(5.42 in)	=	248.2 in³

$I_{(1-1)}$
Plate = ¹⁄₁₂ × 22 × ¾ × ¾ × ¾ =   0.8
+ 16.50 (5.42 − 0.37)²   = 420.2
2 channels = 2 × 401.4   = 802.8
+ 29.28 (8.25 − 5.42)²   = 234.2
$I_{(1-1)}$ = 1458.0 in⁴

$$r^2_{(1-1)} = \frac{I}{A} = \frac{1458.0}{45.78} \qquad = \quad 31.8 \text{ in}^2$$

$I_{(2-2)}$ (axis of symmetry)
Plate = ¹⁄₁₂ × ¾ × 22 × 22 × 22 = 665.0
2 channels = 2 × 11.2   =   22.4
+ 29.28 × (6.8)²   = 1358.0
$I_{(2-2)}$ = 2045.4 in⁴

$$r^2_{(2-2)} = \frac{2045.4}{45.78} \qquad = 44.7 \text{ in}^2 > 31.8$$

Permissible fiber stress $= 15,000 - \dfrac{(25 \times 12)^2}{4 \times 31.8} = 14,292$ psi

Allow. load $= 45.78 \times 14,292 =$ **655,000** lb (**2913** kN)   **c**

7.29 A of four angles   = 9.92
      A of 8 × ⅜ plate   = 3.00
                             12.92 in²
      Plate = ¹⁄₁₂ × 8 × ⅜ × ⅜ × ⅜ = 0.04
      Four angles = 4 × 4   = 16.00
      + 9.92 (0.19 + 1.28)²   = 21.40
      $I_{(1-1)}$ = 37.44 in⁴

$$r^2_{(1-1)} = \frac{I}{A} = \frac{37.44}{12.92} = 2.90 \text{ in}^2, \; r_{(1-1)} = 1.70 \text{ in}$$

$$\text{Use } K = 1, \frac{Kl}{r} = \frac{1 \times 15 \times 12}{1.70} = 106 < 120 < C_c$$

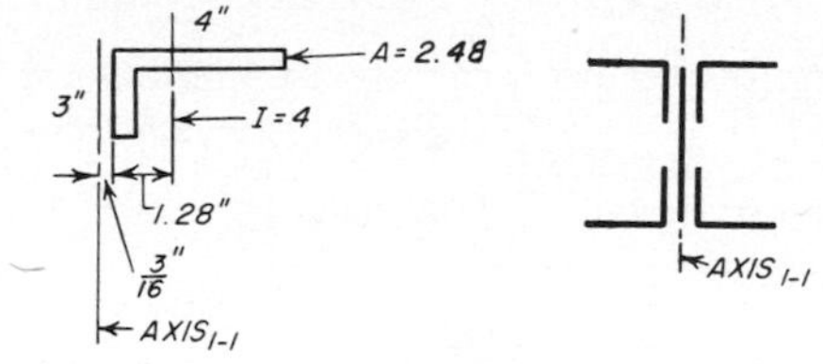

Fig. 7.29

$$C_c^2 = \frac{2\pi^2 E}{F_y} = \frac{2\pi^2 29{,}000{,}000}{36{,}000}, C_c = 126$$

$$\text{F.S.} = \frac{5}{3} + \frac{3}{8}\left(\frac{Kl/r}{C_c}\right) - \left(\frac{Kl/r}{2C_c}\right)^3 = 1.909$$

$$F_a = \left[1 - \frac{1}{2}\left(\frac{Kl/r}{C_c}\right)^2\right]F_y = 12{,}200 \text{ psi (84.12 MPa)}$$

Allow. load $= 12.92 \times 12{,}200 = $ **157,600** lb (**701** kN)   **b**

7.30   Find A, I, and r of section.

$$
\begin{array}{lll}
A = 84.37 + 2 \times 22.00 \times 2 = & 172.37 \text{ in}^2 \\
I_{(1-1)}: \text{W } 14 \times 287 & = & 3912.1 \text{ in}^4 \\
\text{Two } 22 \times 2 \text{ in plates} \\
2 \times \frac{1}{12} \times 22 \times 2 \times 2 \times 2 & = & 29.4 \\
2 \times 44 \times 9.40 \times 9.40 & = & \underline{7770.0} \\
& & 11{,}711.5 \text{ in}^4 \\
I_{(2-2)}: \text{W } 14 \times 287 & = & 1466.5 \\
\text{Plates } 2 \times \frac{1}{12} \times 2 \times 22^3 & = & \underline{3550.0} \\
& & 5016.5 \text{ in}^4
\end{array}
$$

$$r^2 = \frac{I}{A} = \frac{5016.5}{172.37} = 29.1 \text{ in}^2, r = 5.4 \text{ in}$$

$$K = 1, \frac{Kl}{r} = 1 \times 20 \times 12/5.4 = 44.5 < 120 < C_c = 126$$

F.S. $= 1.797$, $F_a = 18{,}820$ psi, $F_b = 22{,}000$ psi, $F'_e = 75.44$ psi (Table 2, Appendix A, AISC specifications)

$$f_a = \frac{P}{A} = \frac{1{,}250{,}000}{172.37} = 7250 \text{ psi (49.99 MPa)}$$

$$f_b = \frac{Pec}{I} = \frac{1{,}250{,}000 \times 8.40 \times 10.40}{11{,}711.5} = 9325 \text{ psi (64.30 MPa)}$$

Assume $C_m = 1$. When $f_a/F_a \geq 0.15$, AISC column formula is

$$\frac{f_a}{F_a} + \frac{C_m f_b}{[1 - (f_a/F_e')]F_b} \leq 1 \qquad \begin{array}{l} \text{AISC 1.6.1} \\ \text{(1.6-1a)} \end{array}$$

Check: $f_a/F_a = 7250/18{,}820 = 0.385 > 0.15$.
Column formula becomes

$$0.385 + \frac{9325}{[1 - (7250/75{,}440)]22{,}000} = 0.385 + 0.469$$

$$= 0.854 < 1$$

Assumed section is OK.

7.31 Check slenderness ratios about axes x and y.

$$\frac{K_x L}{r_x} = \frac{1.80(15)(12)}{6.42} = 50.47$$

$$\frac{K_y L}{r_y} = \frac{0.80(15)(12)}{4.01} = 35.91$$

Slenderness ratio about the x axis controls

$$C_c = \sqrt{\frac{2\pi^2 E}{F_y}} = \sqrt{\frac{2(3.14)^2(29{,}000)}{36}} = 126.1$$

Find F_a. From above, $K_x L/r_x < C_c$, therefore,

$$F_a = \frac{\left[1 - \frac{(K_x L/r_x)^2}{2C_c^2}\right] F_y}{\dfrac{5}{3} + \dfrac{3(K_x L/r_x)}{8C_c} - \dfrac{(K_x L/r_x)^3}{8C_c^3}} \qquad \begin{array}{l} \text{AISC 1.5.1.3.1} \\ \text{(1.5-1)} \end{array}$$

$$\frac{(K_x L/r_x)^2}{2C_c^2} = \frac{(50.47)^2}{2(126.1)^2} = 0.08$$

$$\frac{3(K_x L/r_x)}{8C_c} = \frac{3(50.47)}{8(126.1)} = 0.15$$

$$\frac{(K_x L/r_x)^3}{8C_c^3} = \frac{(50.47)^3}{8(126.1)^3} = 0.01$$

$$F_a = \frac{(1.00 - 0.08)(36)}{1.66 + 0.15 - 0.01} = 18.30 \text{ ksi}(126.2 \text{ MPa})$$

Unsupported length of compression flange $= \dfrac{76b_f}{\sqrt{F_y}} = \dfrac{76(15.6)}{\sqrt{36}}$

$$= 197.60 \text{ in or } 16.47 \text{ ft } (5.02 \text{ m}) > 15 \text{ ft } 0 \text{ in } (4.57 \text{ m}).$$

Therefore, $F_b = 0.66F_y = 0.66(36) = 24 \text{ ksi } (165.5 \text{ MPa})$

$$f_a = \frac{P}{A} = \frac{500}{49.1} = 10.18 \text{ ksi } (70.2 \text{ MPa})$$

$$f_{b_x} = \frac{M}{S_x} = \frac{220 \times 12}{267} = 9.89 \text{ ksi } (68.2 \text{ MPa})$$

$$\frac{f_a}{F_a} = \frac{10.18}{18.30} = 0.56 > 0.15$$

Therefore, the following interaction formula has to be satisfied:

$$\frac{f_a}{F_a} + \frac{C_{mx}f_{bx}}{[1 - (f_a/F_{ex})]F_{bx}} \le 1 \qquad \begin{array}{l} \text{AISC 1.6.1} \\ (1.6\text{-}1a) \end{array}$$

Since column is subject to sidesway, $C_m = 0.85$

$$F_e' = \frac{12\pi^2 E}{23 \left(\dfrac{KL_b}{r_b}\right)^2} = \frac{12 \times 3.14^2 \times 29 \times 10^3}{23 \left(\dfrac{1.80 \times 15 \times 12}{6.42}\right)^2}$$

$$= 58.56 \text{ ksi } (403.8 \text{ MPa}) \qquad \text{AISC 1.6.1}$$

$$C_{m_x}f_{b_x} = 0.85 \times 9.89 = 8.42$$

$$\left(1 - \frac{f_a}{F_e'}\right)F_b = \left(1 - \frac{10.18}{58.56}\right)24 = 19.82$$

Substituting the above values in the interaction formula,

$$0.56 + \frac{8.42}{19.82} = 0.98 < 1.00$$

Therefore the column is **adequate**.

7.32 A $\frac{7}{8}$-in (22.22-mm) rivet in double shear and in double bearing on a $\frac{1}{2}$-in (12.7-mm) plate is good for 18,040-lb (80.24-kN) A36 (248.2-MPa) steel. Four rivets required at each end. Assume two angles $8 \times 4 \times 1$ in ($203.2 \times 101.6 \times 25.4$ mm) with 4-in (101.6-mm) legs vertical and stitch riveted.

Weight $= 74.8$ lb/ft (111.3 kg/m)
$$A = 22.0 \text{ in}^2$$
$$I = 23.3 \text{ in}^4$$
$$x = 1.05 \text{ in}$$
$$r = 1.03 \text{ in}$$

Rivet gage 4-in leg $= 2.50$ in (6.35 mm)

$$\frac{f}{r} = \frac{10 \times 12}{1.03} = 116.5 < 200 \qquad \text{OK}$$

Distance of cg for net section from back of 8-in (203.2-mm) legs.

$$y \times A_{(\text{net})} = y[A_{(\text{gross})} - A_{(\text{holes})}] = A_{(\text{gross})} \times \text{arm} - A_{(\text{holes})} \times \text{arm}$$

$$y(22.00 - 2.00) = 22.00 \times 1.05 - 2.00 \times 2.50$$

$$= \frac{23.10 - 5.00}{20.00} = \frac{18.10}{20.00}$$

$$= 0.905 \text{ in} \ (22.99 \text{ mm})$$

$$I_{(\text{net section})} = I_{(\text{gross section})} - I_{(\text{hole})} - A_{(\text{net})} \times d^2$$
$$= 23.30 - 2.00(2.50 - 1.05)^2 - 20.00(1.050 - 0.905)^2$$
$$= 23.30 - 2.00(1.45)^2 - 20.00(0.145)^2$$
$$= 23.30 - 4.20 - 0.42 = 18.68 \text{ in}^4$$

Tension stress at end of member caused by

1. Direct stress $= \dfrac{P}{A} = \dfrac{65,000}{20.00}$ $\qquad = \quad 3250$ psi

2. Eccentric bending $= \dfrac{Pec}{I}$

$$= \frac{65,000}{18.68}(2.50 - 0.905)(4.00 - 0.905)$$

$$= 3480 \times 1.595 \times 3.095 \qquad\qquad = 17,200 \text{ psi}$$

3. Negative bending at end due to weight of member

$$= \frac{Mc}{I} = \frac{1}{12} wl^2 \frac{c}{I}$$

$$= \tfrac{1}{12} \times 74.8 \times 10 \times 10 \times 12 \times \frac{3.095}{18.68} \qquad = \quad \underline{1240 \text{ psi}}$$

$$= \mathbf{21{,}690} \text{ psi}$$
$$(\mathbf{149.6} \text{ MPa}) \qquad \mathbf{c}$$

7.33   Assume weight of the member with rivets and lacing = 112 lb/ft (166.7 kg/m).

Moment due to dead weight = $\frac{1}{12} \times 112 \times 25 \times 25 \times 12 = 70,000$ in·lb (7910 Nm).

Given two 15-in channels at 33.9 lb/ft with four holes out of web for $\frac{7}{8}$-in rivets.

$$\frac{l}{r} = \frac{25 \times 12}{5.62} = 53.4 \qquad \text{OK for primary tension member}$$

$$\begin{aligned}
\text{Area (gross) of 1 channel} &= 9.90 \text{ in}^2 \\
\text{Area of holes} = 4 \times 1 \times 0.400 &= \underline{1.60 \text{ in}^2} \\
\text{Area (net)} &= 8.30 \text{ in}^2
\end{aligned}$$

$$f = \frac{P}{A_{(n)}} + \frac{Mc}{I_{(g)}} = \frac{350,000}{2 \times 8.30} + \frac{70,000 \times 7.5}{2 \times 312.5}$$

$$= 21,100 + 840 = \mathbf{21,940} \text{ psi } (\mathbf{151.3} \text{ MPa}) \qquad \mathbf{b}$$

$\therefore$ assumed section OK

7.34   Try an ST 8 WF 22.5 lb/ft.

$A = 6.62$ in^2

$d = 8.06$ in

$I = 37.8$ in^4

$r = 2.39$ in and 1.52 in

$y = 1.87$ in

Negative moment due to weight

$= \frac{1}{12} w l^2 = \frac{1}{12} \times 22.5 \times 8 \times 8 \times 12 = \quad 1440$ in·lb

Negative moment due to purlin

$$= \frac{PL}{8} = \frac{4000}{8} \times 8 \times 12 \qquad\qquad = \underline{48,000 \text{ in·lb}}$$

$$M_T \qquad\qquad\qquad\qquad\qquad\qquad\qquad\qquad = \overline{49,400 \text{ in·lb}}$$

$$\frac{Kl}{r} = \frac{1 \times 8 \times 12}{1.52} = 63.1, \text{ F.S.} = 1.838$$

$$C_m = 0.85, F'_e = 37,500 \text{ psi } (258.6 \text{ MPa})$$

(Table 2, Appendix A, AISC specifications)

$$F_a = 17,130 \text{ psi } (118.1 \text{ MPa})$$

$$F_b = 22,000 \text{ psi } (151.7 \text{ MPa})$$

$$f_a = \frac{P}{A} = \frac{60,000}{6.62} = 9070 \text{ psi } (62.5 \text{ MPa})$$

$$f_b = \frac{Mc}{I} = \frac{49{,}440}{37.8} \times (8.06 - 1.87) = 1308 \times 6.19$$

$$= 8090 \text{ psi (55.8 MPa)}$$

Check: $f_a/F_a = 9070/17{,}130 = 0.529 > 0.15$.

$$\frac{f_a}{F_a} + \frac{C_m f_b}{[1 - (f_a/F_e')]F_b} = 0.529 + \frac{0.85 \times 8090}{[1 - (9070/37{,}500)]22{,}000}$$

$$= 0.529 + 0.413 = 0.942 < 1$$

Assumed section OK.

7.35 Choose two angles $3\frac{1}{2} \times 3\frac{1}{2} \times \frac{3}{8}$.
$A = 4.96 \text{ in}^2$
$I = 5.8 \text{ in}^4$
$y = 1.01 \text{ in}$
$w = 17 \text{ lb/ft}$

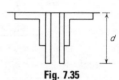

Fig. 7.35

Normal rivet gage for $3\frac{1}{2}$-in leg is 2 in from back of angle. Design plates to have cg of plate and angle at rivet gage. Use $\frac{3}{8}$-in (9.53-mm) plates. For one angle and one plate: statical moment from back of L.

$$2.48 \times 1.01 + \frac{3}{8} \times d \times \frac{d}{2} = (\frac{3}{8}d + 2.48) \times 2$$

$$2.50 + 0.1875d^2 = 0.75d + 4.96$$

$$+d^2 = 4d + 13.12$$

$$d^2 - 4d + (4) = +13.12 + (4) = +17.12$$

$$d - 2 = \pm\sqrt{17.12} = \pm 4.14 \text{ in}$$

$$d = +4.14 + 2 = 6.14 \text{ in}$$

Try two plates $6 \times \frac{3}{8}$ in (weight = 16.0 lb)

$$A = 2 \times \frac{3}{8} \times 6 + 4.96 = 4.50 + 4.96 = 9.46 \text{ in}^2$$

I (for two angles and two plates)
(weight = $16.0 + 17.0 = 33$lb/ft)

Two angles	=	5.80 in^4
$+4.96 \times 0.99^2$	=	4.86
Two plates $= \frac{1}{12} \times \frac{3}{4} \times 6^3$	=	13.50
$+4.50 \times 1^2$	=	4.50
		$\overline{28.60 \text{ in}^4}$

$$r^2 = \frac{I}{A} = \frac{28.60}{9.46} = 3.02 \text{ in}^2, \text{ and } \frac{Kl}{r} = \frac{1 \times 8 \times 12}{1.74} = 55.4$$

Kl/r in the other plane is 33.1, as the unbraced length is 4×12.
$F_a = 17,860$ psi (123.1 MPa), $C_m = 0.85$, $F'_e = 48,670$ psi (335.6 MPa)
(Table 2, Appendix A, AISC specifications)

$$F_b = 22,000 \text{ psi (151.7 MPa)}$$

$$f_a = \frac{P}{A} = \frac{75,000}{9.46} = 7920 \text{ psi (54.6 MPa)}$$

$$f_b = \frac{Mc}{I} = \frac{(48,000 + \frac{1}{12} \times 33 \times 8 \times 8 \times 12) \times 4}{28.60}$$

$$= \frac{48,000 + 2110}{7.15} = 7000 \text{ psi (48.3 MPa)}$$

Check: $f_a/F_a = 7920/17,860 = 0.443 > 0.15$

$$\frac{f_a}{F_a} + \frac{C_m f_b}{[1 - (f_a/F'_e)]F_b} = 0.443 + \frac{0.85 \times 7000}{[1 - (7920/48,670)]22,000}$$

$$= 0.443 + 0.323 = 0.766 < 1$$

7.36 Vertical load is divided into two components:

$$\text{Normal (to roof surface)} = \frac{2}{\sqrt{5}} \times 160 = 143 \text{ lb/ft} = w_n$$

$$\text{Parallel (to roof surface)} = \frac{1}{\sqrt{5}} \times 160 = 71.5 \text{ lb/ft} = w_p$$

Bending moment about xx axis
$$= \frac{1}{8}w_n l^2 = \frac{1}{8}(143 + 90) \times 20 \times 20 \times 12 = 140,000 \text{ in·lb}$$
Bending moment about yy axis

$$= \frac{1}{10}w_p \left(\frac{l}{3}\right)^2 = \frac{1}{10} \times 71.5 \times \frac{20 \times 20 \times 12}{3 \times 3} = 3820 \text{ in·lb}$$

Assume 7-in I beam 15.3 lb/ft
$$S_{xx} = 10.4 \text{ in}^3$$
$$S_{yy} = 1.5 \text{ in}^3$$

$$f = \frac{M_x}{S_x} + \frac{M_y}{S_y} = \frac{140,000}{10.4} + \frac{3820}{1.5} = 13,460 + 2540$$

$$= 16,000 \text{ psi (110.3 MPa)} < 20,000 \text{ psi (137.9 MPa)}$$

Assumed section OK.

7.37 Assume beam weight of 140 lb/ft (208.3 kg/m). Maximum moment induced in beam is

$$\text{Max } M = \frac{5.14 \times 24^2}{8} = 370.08 \text{ ft·kips (501.9 kN·m)}$$

Assume allow. stress $F_b = 19.5$ ksi (134.5 MPa)

$$\text{Required } S_x = \frac{370.08 \times 12}{19.5} = 227.74 \text{ in}^3$$

Try a W 14 × 142, $S_x = 227$ in³, $r_T = 4.33$ in, $b_f = 15.5$ in. (Compression flange, see AISC specifications 1.5.1.4.1e)

$$\frac{76b_f}{\sqrt{36}} = 12.67b_f, \quad \frac{12.67 \times 15.5}{12} = 16.37 \text{ ft} < 24 \text{ ft (7.32 m)}$$

Design the beam without lateral supports.

$$\frac{L}{r_T} = \frac{24 \times 12}{4.33} = 66.51$$

In this case $C_b = 1$, see AISC specifications 1.5.1.4.6a

$$\sqrt{\frac{102 \times 10^3 \times 1}{36}} = 53.25 \leq \frac{L}{r_T} \leq \sqrt{\frac{510 \times 10^3 \times 1}{36}} = 119 \qquad (1.5\text{-}6a)$$

From the above it can be seen that L/r_T is between the above calculated square roots, therefore the allowable stress F_b is

$$F_b = \left[\frac{2}{3} - \frac{(L/r_T)^2 F_y}{1530 \times 10^3 C_b} \right] F_y \qquad (1.5\text{-}6a)$$

$$= \left[0.67 - \frac{(66.51)^2 \times 36}{1530 \times 10^3 \times 1} \right] 36 = 20.37 \text{ ksi (140.5 MPa)}$$

$$\text{Actual } f_b = \frac{M}{S_x} = \frac{370.08 \times 12}{227} = 19.56 \text{ ksi (134.9 MPa)} < 20.37 \text{ ksi}$$

W 14 × 142 is OK.   **1b**

7.38 Try a bolt of 1½-in diameter.
Diameter at root of thread = 1.283 in
Area at root of thread   = 1.294 in²

$$S = \frac{P}{A} = \frac{20,000}{1.294} = 15,500 \text{ psi (106.9 MPa)}$$

$$S_t = \frac{Tc}{J} = \frac{6000 \times d/2}{\pi d^4/32} = \frac{6000 \times 16}{\pi d^3} = \frac{30,600}{(1.283)^3}$$

$$= 14,600 \text{ psi (100.7 MPa)}$$

$$\text{Max normal stress} = \frac{1}{2}(S) + \sqrt{\left(\frac{S}{2}\right)^2 + (S_t)^2}$$

$$= \frac{15,500}{2} + \sqrt{(7750)^2 + (14,600)^2}$$

$$= 7750 + 16,500$$

$$= \textbf{24,250} \text{ psi } (\textbf{167.2} \text{ MPa}) \qquad \textbf{b}$$

1½-in bolt OK.

7.39
$$\frac{E_s}{E_w} = \frac{30,000,000}{1,500,000} = 20$$

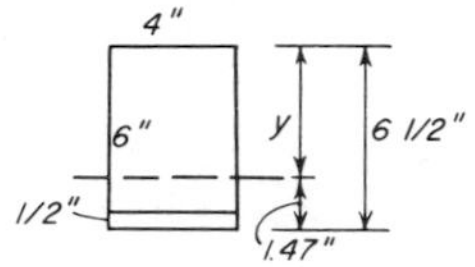

$$
\begin{array}{rll}
A \times \text{arm} & = \text{product} \\
4 \times 6 = 24 \times 3 & = 72 \\
\frac{1}{2} \times (4 \times 20) = \underline{40} \times 6\frac{1}{4} & = \underline{250} \\
64 \ (5.03) & = 322
\end{array}
$$

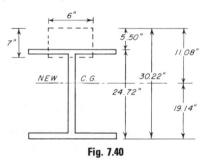

Fig. 7.39

$$\frac{f_t}{600} = \frac{1.47}{5.03}, f_t = 600 \times \frac{1.47}{5.03} = 175.1 \qquad \text{equivalent wood}$$

Stress in steel, when wood is stressed to 600 psi in compression, is $20 \times 175.1 = \textbf{3502}$ psi (**24.15** MPa) tension.   **a**

7.40

Fig. 7.40

$$f'_c = 3000 \text{ psi}$$
$$f_c = 1350 \text{ psi}$$
$$\frac{E_s}{E_c} = \frac{30,000,000}{3,000,000} = n = 10$$
$$I_w = 5110.3 \text{ in}^4$$
$$A = 47.04 \text{ in}^2$$

Assume concrete slab to be equivalent billet of steel $= \dfrac{5 \times 12}{n} = \dfrac{60}{10} =$ 6 in wide. Find new cg (of composite section) from bottom of W.

Member	Area	×	Arm	=	Product
W	47.04	×	12.36	=	582 in³
Steel block	42.00	×	26.72	=	1124 in³
	89.04	×	(19.14 in)	=	1706 in³

I of composite section:

Steel block $\frac{1}{12} \times 6 \times 7 \times 7 \times 7$ $\qquad = \quad 171.5$ in⁴
$+42(26.72 - 19.14)^2 = 42 \times 7.58^2 \qquad = 2410.0$
W $\qquad\qquad\qquad\qquad\qquad\qquad\qquad\quad = 5110.3$
$+47.04(19.14 - 12.36)^2 = 47.04 \times 6.78^2 = \underline{2160.2}$
$\qquad\qquad\qquad\qquad\qquad\qquad\qquad\qquad\quad 9852$ in⁴

Wt concrete/ft $= \dfrac{7 \times 60}{144} \times 150 \qquad\qquad = 438$ lb/ft

Wt W/ft $\qquad\qquad\qquad\qquad\qquad\qquad = \underline{160 \text{ lb/ft}}$

$Wt_T \qquad\qquad\qquad\qquad\qquad\qquad = 598$ lb/ft or say 600

Dead moment $= \frac{1}{8} \times 600 \times 40 \times 40 \times 12 = 1,440,000$ in·lb

Fiber stresses in W $= \dfrac{Mc}{I} = \dfrac{1,440,000 \times 12.36}{5110} = 3480$ psi

Fiber stress in bottom of steel beam due to concentrated load may be $20,000 - 3480 = 16,520$ psi and fiber stress in top of concrete may be $\dfrac{1}{10} \times \dfrac{11.08}{19.14} \times 16,520 = 956$ psi (6.59 MPa).

$$\text{Live-load moment} = \dfrac{PL}{4} = \dfrac{P \times 40 \times 12}{4} = 120P$$

$$f = \dfrac{Mc}{I} = 956 \times 10 = \dfrac{120P \times 11.08}{9852}$$

$$P = \dfrac{9852 \times 79.6}{11.08} = \mathbf{70{,}815} \text{ lb } (\mathbf{315.0} \text{ kN}) \qquad \mathbf{c}$$

7.41 $sb = \dfrac{VQ}{I} = \dfrac{33{,}000 \times 42 \times 7.58}{9852}$

$\qquad\qquad = 1068 \text{ lb/in } (187.0 \text{ kN/m})$

Two ⅞-in rivets in single shear $= 2 \times 0.601 \times 15{,}000 = 18{,}040 \text{ lb } (80.24$ kN).

End spacing of Z bars is $18{,}040/1068 = 16.9$ in.

Use **17**-in (**431.8** mm) spacing, max.   **d**

7.42   Total horizontal force at wall $= (540 \times 18)/2 = 4860 \text{ lb/ft } (70.92$ kN/m).

$$\text{Max moment} = 4860 \times \frac{18}{3} \times 12$$

$$= 350{,}000 \text{ in}\cdot\text{lb } (39.55 \text{ kN}\cdot\text{m})$$

$\text{Max } p = 0.75 \left(0.85 \times 0.85 \times \dfrac{3.5}{50} \times \dfrac{87}{87 + 50} \right)$ 　　1971 ACI 8.6 (8-1)

$\qquad\qquad\qquad\qquad\qquad\qquad\qquad\qquad\qquad$ 10.3.2

$\qquad = 0.024$ 　　　　　　　　　　　　　　　　7.13

Use steel ratio of

$$p = 0.18 \frac{f'_c}{f_y} = 0.18 \times \frac{3.5}{50} = 0.013$$

Resisting moment of a section is

$$M_u = \phi p f_y b d^2 \left(1 - 0.59 p \frac{f_y}{f'_c} \right) \qquad 1963 \text{ ACI (16-1)}$$

or

$$d^2 = \frac{350{,}000}{0.90 \times 0.013 \times 50{,}000 \times 12 \left(1 - \dfrac{0.59 \times 0.013 \times 50{,}000}{3500} \right)}$$

or $d = 7.48$ in.

Required d for shear:

$$v_c = 2 \sqrt{f'_c} = 2 \sqrt{3500} = 118.32 \text{ psi} \qquad 1971 \text{ ACI } 11.4.1$$

$$d = \frac{4860}{118.32 \times 12} = 3.42 \text{ in}$$

Bending controls. Use $d = 7.5$ in (190.5 mm). $A_S = pbd = 0.013 \times 12$

$\times 7.5 = 1.17$ in²/ft. Use #9 at 10 in (28.65 mm at 254 mm) = **1.2** in²/ft
(**774** mm²/0.304 m).   **1b**
ACI requires 3-in (76.2-mm) cover. Diameter of #9 bar is 1.13 in. There-
fore, minimum thickness of wall is $h = 7.5 + 3 + (1.13/2) = 11.07$ in.
Use $h =$ **11** in (**279.4** mm).   **2c**
Check distribution of steel: $d_c = 3.57$ in (90.68 mm).

$$z = 0.40 \times 50 \sqrt[3]{3.57 \frac{2 \times 3.57 \times 12}{1.2}}$$

$= 127 < 145$ kips/in (25.39 MN/m)   OK

1971 ACI 10.6.3 and 10.6.4
(10-2)

7.43 Deflection considerations require minimum slab thickness h equal
to

$$\text{Min } h = \frac{1}{20}\left(0.4 + \frac{f_y}{100,000}\right) \qquad \text{ACI 9.5.1}$$

$$= \frac{10 \times 12}{20}\left(0.4 + \frac{50,000}{1,000,000}\right) = 5.4 \text{ in (137.2 mm)}$$

Assume 6-in (152.4-mm) slab.

Slab weight = $\frac{6}{12}(0.150) = 0.075$ kips/ft
Total load $w_u = 1.4D + 1.7L = 1.4(0.075) + 1.7(0.400)$
$= 0.785$ kips/ft (11.46 kN/m)

Maximum moment induced in slab:

$$\text{Max } M_u = \frac{w_u l^2}{8} = \frac{0.785(10)^2}{8} = 9.81 \text{ ft·kips (13.30 kN·m)}$$

To ensure that tensile yielding controls the strength of the section,

$$\text{Max } p = 0.75 p_b \qquad \text{ACI 10.3.2}$$

$$= (0.75)(0.85)\beta_1 \left(\frac{f'_c}{f_y}\right)\left(\frac{87,000}{87,000 + f_y}\right) \qquad \text{ACI 8.1}$$

$$= (0.75)(0.85)(0.85)\left(\frac{3}{50}\right)\left(\frac{87}{87 + 50}\right) = 0.021$$

Try #5 bars (15.9-mm diam.) at 5¾ in, $A_s = 0.65$ in²

$$d = 6 - \left(0.75 + \frac{0.625}{2}\right) = 4.94 \text{ in}$$

Supplied tensile reinforcement is

$$p = \frac{0.65}{12 \times 4.94} = 0.011 < 0.021 \quad \text{OK}$$

To find service moment of slab, first find a.

$$a = \frac{A_s f_y}{0.85 f'_c b} = \frac{0.65 \times 50}{0.85 \times 3.00 \times 12}$$

$$= 1.062 \text{ in} \quad \begin{array}{l} \text{1971 ACI 10.2.7} \\ \text{1963 ACI (16-1)} \end{array}$$

$$M_u = \phi A_s f_y \left(d - \frac{a}{2} \right) \quad \text{ACI (16-1)}$$

$$= 0.90 \times 0.65 \times 50 \left(4.94 - \frac{1.062}{2} \right) \frac{1}{12}$$

$$= 10.75 \text{ ft·kips} \ (14.58 \text{ kN·m}) \quad \text{OK}$$

Check shear:

$$\text{Max } V_u = 0.785 \times \frac{10}{2} = 3.93 \text{ kip} \ (17.48 \text{ kN})$$

$$v_u = \frac{V_u}{\phi b d} = \frac{3930}{0.85 \times 12 \times 4.94}$$

$$= 77.88 \text{ psi} \ (537.0 \text{ kPa}) \quad \text{1971 ACI (11-3)}$$

Maximum shear stress carried by concrete is

$$v_c = 2\sqrt{f'_c} = 2\sqrt{3000} = 109 \text{ psi} \ (751.6 \text{ kPa}) \quad \text{ACI 11.4.1}$$

Since $v_u < v_c$, no stirrups are necessary (ACI 11.1.1).
Check distribution of tensile reinforcement (ACI 10.6.3).

$$d_c = \text{cover} + \frac{1}{2} \text{ bar diam.} = 0.75 + \frac{1}{2} \times 0.625 = 1.063$$

$$\text{No. bars} = \frac{12}{5.75} = 2.087 \text{ per ft}$$

$$A = \frac{2 d_c b}{\text{no. bars}} = \frac{2 \times 1.063 \times 12}{2.087} = 12.22 \text{ in}^2/\text{ft}$$

$$z = f_s \sqrt[3]{d_c A} = 0.60 \times 50 \sqrt[3]{1.063 \times 12.22}$$

$$= 70.50 < 175 \text{ kips/in} \ (31.14 \text{ MN/m}) \quad \text{1971 ACI (10-2)}$$

7.44 Centroid of tensile reinforcement is located at distance of 3.74 in (95.0 mm) from bottom of beam. Therefore, $d = 40 - 3.74 = 36.26$ in (921 mm).

Check maximum tensile reinforcement requirements.

$$A_s = 2 \times 0.79 + 2 \times 1.00 + 4 \times 1.27 = 8.66 \text{ in}^2$$

$$\text{Actual steel ratio} = \frac{A_s}{bd} = \frac{8.66}{18 \times 36.26} = 0.0133$$

Max allowed $p = 0.75\, p_b$ ACI 10.3.2

$$= 0.75(0.85)\beta_1 \left(\frac{f'_c}{f_y} \right) \left(\frac{87,000}{87,000 + f_y} \right)$$

$$= 0.75 \times 0.85 \times 0.85 \left(\frac{3}{40} \right) \left(\frac{87}{137} \right)$$

$$= 0.0278 \quad \text{OK}$$

Find a.

$$a = \frac{A_s f_y}{0.85 f'_c b} = \frac{8.66 \times 40}{0.85 \times 3 \times 18} = 7.55 \text{ in}$$

Usable moment is equal to

$$M_u = \phi A_s f_y \left(d - \frac{a}{2} \right) = 0.90 \times 8.66 \times 40 \left(36.26 - \frac{7.55}{2} \right)\frac{1}{12}$$

$$= 843.96 \text{ ft·kips}$$

$$\text{Dead load of beam} = 0.145 \times \frac{18 \times 40}{144} = 0.725 \text{ kips/ft}$$

$$\text{Dead-load moment} = \frac{0.725 \times 15^2}{8} = 20.39 \text{ ft·kips}$$

$$\text{Live-load moment} = \frac{P_u L}{4} = \frac{P_u(15)}{4}$$

$$\frac{P_u(15)}{4} = \text{usable net } M_u = 843.96 - 20.39 \times 1.4$$

$$= 815.41 \text{ ft·kips (1106 kN·m)}$$

$$P_u = 217.44 \text{ kips (967.2 kN)}$$

or dividing by live-load safety factor,

$$P = \frac{217.44}{1.7} = 127.91 \text{ kips (568.9 kN)}$$

If no stirrups are to be used, then

$$\text{Max } v_u = 0.5v_c \qquad \text{ACI } 11.1.1(d)$$

$$v_c = 2\sqrt{f_c'} = 2\sqrt{3000} = 109.6 \text{ psi (756.7 kPa)} \qquad \text{ACI } 11.4.1$$

$$\text{Max } V_u = \phi v_u bd = 0.85 \times 0.5 \times 109.6 \times 18 \times 36.26\,\frac{1}{1000}$$

$$= 30.2 \text{ kips (134.3 kN)}$$

$$\text{Dead load } V = \frac{0.725 \times 15}{2} = 5.4 \text{ kips}$$

$$\text{Dead load } V_u = 5.4 \times 1.4 = 7.6 \text{ kips (33.8 kN)}$$

$$P_u = 2(30.2 - 7.6)$$

$$= 45.2 \text{ kips (201.0 kN)}$$

Assuming there are no stirrups and dividing by live-load safety factor,

$$P = \frac{45.2}{1.7} = 26.6 \text{ kips (118.3 kN)}$$

Check for deflection. According to the problem, the allowable deflection is

$$\frac{L}{800} = \frac{15 \times 12}{800} = 0.225 \text{ in (5.7 mm)}$$

or $\quad \Delta = \dfrac{PL^3}{48EI} = 0.225$

$$E = 57,000\sqrt{f_c'} = 57,000\sqrt{3000}$$

$$= 3.12 \times 10^6 \text{ psi} \qquad \text{ACI } 8.3.1$$

$$I = \left(\frac{M_{cr}}{M_a}\right)^3 (I_g) + \left[1 - \left(\frac{M_{cr}}{M_a}\right)^3\right] I_{cr} \qquad \begin{array}{l}\text{ACI } 9.5.2.2 \\ (9\text{-}4)\end{array}$$

$$f_r = 7.5\sqrt{f_c'} = 7.5 \times 54.8 = 411 \text{ psi (2.83 MPa)} \qquad \begin{array}{l}\text{ACI } 9.5.2.2 \\ (9\text{-}5)\end{array}$$

$$I_g = \frac{18 \times 40^3}{12} = 96,000 \text{ in}^4$$

$$M_{cr} = \frac{411 \times 96,000}{20} = 1,972,800 \text{ in·lb (222.9 kN·m)} \qquad \begin{array}{l}\text{ACI } 9.5.2.2 \\ (9\text{-}5)\end{array}$$

To find I_{cr}, find moment of inertia of transformed section about its centroidal axis.

$$n = \frac{29 \times 10^6}{3.12 \times 10^6} = 9$$

$$nA_s = 9 \times 8.66 = 77.94$$

Distance of the centroid from the top is

$$y = \frac{(18 \times 7.55)3.78 + (77.94)36.26}{18 \times 7.55 + 77.94} = 15.62 \text{ in (396.7 mm)}$$

Moments of inertia are

Compression area:

$$\frac{1}{12} \times 18 \times 7.55^3 + 18 \times 7.55 \left(15.62 - \frac{7.55}{2}\right)^2 = 19,712 \text{ in}^4$$

Transformed area: $77.94(36.26 - 15.62)^2 = 33,203 \text{ in}^4$

$$I_{cr} = 19,712 + 33,203 = 52,915 \text{ in}^4$$

$$M_a = \frac{843.96}{0.90} = 937.73 \text{ ft} \cdot \text{kips (1272 kN} \cdot \text{m)}$$

$$\left(\frac{M_{cr}}{M_a}\right)^3 = \left(\frac{1,972,800}{937.73 \times 12,000}\right)^3 = (0.175)^3 = 0.005$$

$$I = 0.005 \times 96,000 + (1 - 0.005)52,915 = 53,100 \text{ in}^4$$

Therefore,

$$\frac{P(15 \times 12)^3}{48 \times 53,100 \times 3.12 \times 10^3} = 0.225$$

or
$$P = 306.5 \text{ kips (1363.3 kN)}$$

From the above it can be seen that shear controls. Maximum live-load $P = \mathbf{26.6}$ kips (**118.3** kN).

7.45 Total uniformly distributed ultimate load on the beam is

$$w_u = 1.4 \times 0.6 + 1.7 \times 1.2 = 2.88 \text{ kips/ft (42.03 kN/m)}$$

$$\text{Max } M = \frac{2.88 \times 20^2}{8} = 144.00 \text{ ft} \cdot \text{kips (196.3 kN} \cdot \text{m)}$$

Max steel ratio $p = 0.75p_b$

$$= 0.75 \times 0.85\beta_1 \left(\frac{f'_c}{f_y}\right) \left(\frac{87,000}{87,000 + f_y}\right)$$

$$= 0.75 \times 0.85 \times 0.85 \left(\frac{4}{50}\right) \left(\frac{87}{87 + 50}\right)$$

$$= 0.0275$$

Min ratio $p = \dfrac{200}{f_y} = \dfrac{200}{50,000} = 0.004$ ACI 10.5.1

Any steel ratio between the above limits is acceptable. The steel ratio that yields the most economical section is

$$p = 0.18\frac{f'_c}{f_y} = 0.18 \times \frac{4}{50} = 0.0144$$

To determine the values of b and d of the section, use the equation

$$bd^2 = \frac{M_u}{\phi p f_y[1 - (0.59pf_y/f'_c)]}$$

$$= \frac{144.0 \times 12}{0.90 \times 0.0144 \times 50[1 - 0.59 \times 0.0144 \times (50/4)]}$$

$$= 2983 \text{ in}^3$$

Assume $b = d/2$. This value yields an economical section. $d^3 = 5966$ or $d = 18.14$ in (460.8 mm) and $b = 9.07$ in (230.4 mm).

Deflection considerations require that minimum thickness h of beam be

$$\text{Min } h = \frac{L}{16}\left(0.4 + \frac{f_y}{100,000}\right)   \text{ACI 9.5.1}$$

$$\text{Min } h = \frac{20 \times 12}{16}\left(0.4 + \frac{50}{100}\right) = 13.5 \text{ in}$$

Use $b = 10$ in (254 mm) and $d = 17.5$ in (444.6 mm). Then $A_s = pbd = 0.0144 \times 10 \times 17.5 = 2.52$ in.^{2}

Use three #9 (28.7 mm diam.) bars. Check to see if the width of the beam can accommodate the bars.

Required $b = 2 \times 1.5 + 3 \times 1.128 + 2 \times 1.81 = 10$ in

Check for shear,

$$v_u = \frac{V_u}{\phi bd} = \frac{0.5 \times 2.88 \times 20 \times 1000}{0.85 \times 10 \times 17.5} = 193.61 \text{ psi (1.33 MPa)}$$

Shear stress at d distance from face of support is

$$v_u = \frac{(10 - 1.46) \times 193.61}{10} = 165.34 \text{ psi } (1140 \text{ kPa}) \qquad \text{ACI } 11.2.2$$

Shear stress carried by concrete is

$$v_c = 2\sqrt{f_c'} = 2\sqrt{4000} = 126.5 \text{ psi } (873 \text{ kPa}) \qquad \text{ACI } 11.4.1$$

Stirrups are required (ACI 11.1.1). Use #2 U bars placed vertically. Spacing required is

$$s = \frac{A_v f_y}{(v_u - v_c)b} \qquad \text{ACI } 11.6.1$$

$$= \frac{2 \times 0.05 \times 50,000}{(165.34 - 126.4)10} = 12.84 \text{ in} > \frac{d}{2} = 8.75 \text{ in}$$

Place stirrups 8½ in on centers. See ACI 11.1.4(b). Place stirrups to a distance x from centerline of beam. At this distance v_u becomes equal to $\frac{1}{2}v_c = 63.25$ psi. From the shear diagram,

$$x = \frac{(10 - 1.46)63.25}{165.34} = 3.27 \text{ ft } (1.00 \text{ m})$$

Since f_y does not exceed 40,000 psi, it is not necessary to check distribution of the tensile reinforcement (ACI 10.6.3).

7.46 Assume clear span of $l_n = 18$ ft 0 in (5.49 m):

$$\text{Negative moment at interior support } M = -\frac{wl_n^2}{10}$$

$$\text{Shear } V = 1.15\frac{wl_n}{2}$$

$$\text{Positive moment in end span } M = \frac{wl_n}{14} \qquad \text{ACI } 8.4.2$$

$$\text{Positive moment in center span } M = \frac{wl_n}{16}$$

$$w_u = 1.4 \times 0.6 + 1.7 \times 1.2 = 2.88 \text{ kips/ft } (42.03 \text{ kN/m})$$

$$\text{Max moment in slab } M_u = \frac{2.88 \times 18^2}{10}$$

$$= 93.31 \text{ ft} \cdot \text{kips}$$

Max allowable steel ratio $p = 0.75p_b$

$$= 0.75 \times 0.85\beta_1 \left(\frac{f_c'}{f_y}\right) \left(\frac{87}{87 + 50}\right)$$

$$= 0.75 \times 0.85 \times 0.85 \left(\frac{4}{50}\right) \left(\frac{87}{87 + 50}\right)$$

$$= 0.0275$$

$$\text{Min } p = \frac{200}{f_y} = \frac{200}{50,000} = 0.004$$

An economical steel ratio $p = 0.18 \, (f_c'/f_y) = 0.18 \, (4/50) = 0.0144$. Using this ratio it is

$$bd^2 = \frac{M_u}{\phi p f_y \left(1 - 0.59p \dfrac{f_y}{f_c'}\right)}$$

$$= \frac{93.31 \times 12}{0.90 \times 0.0144 \times 50[1 - 0.59 \times 0.0144 \times (50/4)]}$$

$$= 1934 \text{ in}^3$$

Assume $b = 0.50d$, then $d^3 = 3868$ or $d = 15.7$ in.
Deflection considerations allow

$$\text{Min } h = \frac{l}{16} \left(0.4 + \frac{f_y}{100,000}\right) = \frac{20 \times 12}{16} \left(0.4 + \frac{50,000}{100,000}\right)$$

$$= 13.5 \text{ in (342.9 mm)} \qquad \text{ACI 9.5.1}$$

Use $b = 8$ in (203.2 mm) and $d = 16$ in (406.4 mm).
Then over interior supports

$$A_s = 0.0144 \times 8 \times 16 = 1.84 \text{ in}^2, \text{ three \#7 (22.22-mm) bars}$$

Positive steel in center span

$A_s = {}^{10}\!/_{16} \times 1.84 = 1.15 \text{ in}^2$, two #6 (19.05-mm) bars, one #5 (15.88-mm) bar

Positive steel in end span

$$= {}^{10}\!/_{14} \times 1.84 = 1.31 \text{ in}^2, \text{ three \#6 (19.05-mm diam.) bars}$$

Check to see whether three #7 (22.22-mm) bars can fit in the beam.

$$\text{Required } b = 2 \times 1.5 + 3 \times 0.875 + 2 \times 1.1875 = 8 \text{ in}$$

Check for shear:

$$V_u = 1.15 \left(\frac{w_u l_n}{2}\right) = 1.15 \left(\frac{2.88 \times 18}{2}\right) = 29.81 \text{ kips (132.6 kN)}$$

$$v_u = \frac{29{,}810}{0.85 \times 8 \times 16} = 273.99$$

$$v_c = 2 \sqrt{f_c'} = 2 \sqrt{4000} = 126.5 \text{ psi (872.2 kPa)}$$

U stirrups are required (ACI 11.2.2). Use #3 (9.5 mm diam.).

$$\text{Required spacing } s = \frac{A_v f_y}{(v_u - v_c)^b} \qquad \text{ACI 11.6.1}$$

Shear stress v_u at distance d from face of support (see ACI 11.2.2) is

$$v_u = \frac{9.0 - 1.33}{9} \, 273.99 = 233.50 \text{ psi (1.61 MPa)}$$

$$s = \frac{2 \times 0.11 \times 50{,}000}{(233.50 - 126.5)8} = 12.85 \text{ in} > 8 \text{ in}$$

Place stirrups to a distance x from centerline of beam equal to

$$x = \frac{(9.0 - 1.33)63.25}{233.50} = 2.08 \text{ ft (0.63 m)}$$

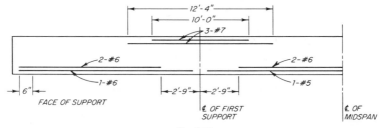

Fig. 7.46

7.47 By 1973 ACI specifications: $n = 9, f_s = 18{,}000$.

$$8y \times \frac{y}{2} + 2 \times (18 - 1) \times 0.60 \times (y - 2)$$

$$= 2 \times 9 \times 1.27 \times (16 - y)$$

$4y^2 + 20.4 \times (y - 2) = 22.86 \times (16 - y)$

$y^2 + 5.1y - 10.2 = 91.44 - 5.71 y$

$y^2 + 10.81y + (5.405)^2 = 101.64 + (5.405)^2 = 130.8$

$y = 11.42 - 5.40 = 6.02 \text{ in (152.9 mm)}$

Moment of inertia:

$$\begin{aligned}
\text{Compression area} &= \tfrac{1}{3} \times 8 \times 6.02^3 &= 580 \\
\text{Compression steel} &= 20.4 \times 4.02^2 &= 330 \\
\text{Tension steel} &= 22.86 \times 9.98^2 &= \underline{2280} \\
& & 3190 \text{ in}^4
\end{aligned}$$

If $f_s = 20{,}000$ psi, $f_c = \dfrac{6.02 \times 20{,}000}{9.98 \times 9} = 1340$ psi

$$f = \frac{Mc}{I} \text{ or } M = \frac{fI}{c} = \frac{1340 \times 3190}{6.02} = 710{,}066 \text{ in}\cdot\text{lb } (80.24 \text{ kN}\cdot\text{m})$$

This is $> 12 \times 50{,}000$ or $600{,}000$ in·lb (67.78 kN·m). Therefore, beam OK to carry load.

7.48 Total ultimate moment $M_u = 1.4 \times 75 + 1.7 \times 120$
$$= 309 \text{ ft}\cdot\text{kips } (419.0 \text{ kN}\cdot\text{m})$$
Maximum steel ratio for singly reinforced beam is

$$p_{\max} = 0.75p_b = 0.75 \times 0.85 \times 0.80 \left(\frac{5}{50}\right)\left(\frac{87}{87 + 50}\right) = 0.0324$$

$$\text{Max } A_s = 0.0324 \times 12 \times 15.5 = 6.02 \text{ in}^2$$

$$\frac{a}{2} = \frac{A_s f_y}{1.70 f_c' b} = \frac{6.02 \times 50}{1.70 \times 5 \times 12} = 2.95$$

$$M_u = \phi A_s f_y \left(d - \frac{a}{2}\right)$$

$$= 0.90 \times 6.02 \times 50(15.5 - 2.95)/12$$

$$= 284 \text{ ft}\cdot\text{kips } (385.1 \text{ kN}\cdot\text{m}) < 309 \text{ ft}\cdot\text{kips}$$

A doubly reinforced beam is needed and M_u above becomes M_{u1} and $M_u = 309$ ft·kips (419.0 kN·m).

$$M_{u2} = M_u - M_{u1} = 309 - 284 = 25 \text{ ft}\cdot\text{kips } (33.9 \text{ kN}\cdot\text{m})$$

$$A_s' = \frac{M_{u2}}{\phi f_y(d - d')} = \frac{25 \times 12}{0.90 \times 50(15.5 - 2.5)} = 0.513 \text{ in}^2$$

$$A_{s1} = 6.02$$

$$\frac{A_{s2} = 0.513}{A_s = 6.533}$$

Use three #9 (28.7 mm) bars $= 3.00$ in^2 in upper row of tension steel

One #9 bar + two #10 (32.3 mm) bars = 3.54 in^2 in lower row of tension steel

$$\overline{6.54 \text{ in}^2 \text{ total}}$$

Assume compressive steel stress $f'_s = f_y$ at failure, but to ensure failure by tensile yielding use

$$A'_s = \frac{0.513}{0.75} = 0.69 \text{ in}^2, \text{ one #8 (2.54 mm) bar} = 0.79 \text{ in}^2$$

Use $A_{s1} = 6.54 - 0.79 = 5.75$ and $A_{s2} = 0.79$ in^2

$$p = \frac{6.54}{12 \times 15.5} = 0.0352$$

$$p' = \frac{0.79}{186} = 0.0042$$

$$(p - p') = \frac{5.75}{186} = 0.0310$$

$$(p - p')_{max} = 0.75p_b = 0.0324$$

$$(p - p')_{min} = 0.85 \times 0.80 \left(\frac{5}{50}\right) \left(\frac{2.5}{15.5}\right) \left(\frac{87}{37}\right) = 0.0258$$

$$(p - p')_{used} = 0.0310 < 0.0324 \text{ max} > 0.0258 \text{ min} \quad \text{OK}$$

Recheck resisting moments.

$$M_{u2} = \phi A'_s f_y (d - d') = \frac{0.90 \times 0.79 \times 50 \times 13}{12}$$

$$= 38.4 \text{ ft} \cdot \text{kips} (52.1 \text{ kN} \cdot \text{m})$$

$$\frac{a}{2} = \frac{(A_s - A'_s)f_y}{1.70 f'_c b} = \frac{5.75 \times 50}{1.70 \times 5 \times 12} = 2.82 \text{ in}$$

$$M_{u1} = \frac{0.90 \times 5.75 \times 50(15.5 - 2.82)}{12}$$

$$= 274 \text{ ft} \cdot \text{kips} (371.5 \text{ kN} \cdot \text{m})$$

$$M_u = 38.4 + 274 = 312.4 \text{ ft} \cdot \text{kips} (423.6 \text{ kN} \cdot \text{m})$$
$$> 309 \text{ ft} \cdot \text{kips} (419.0 \text{ kN} \cdot \text{m}) \quad \text{OK}$$

Check distribution of tensile reinforcement.

$$d_c = 1.50 + 1.27/2 = 2.13 \text{ in}$$
$$A = (1.50 + 2.13)12/3 = 14.5$$
$$z = 0.60 \times 50 \sqrt[3]{2.13 \times 14.5}$$
$$= 94 < 145 \text{ kips/in} (25.39 \text{ MN/m}) \quad \text{OK}$$

7.49 New total load is

$$U = 1.4 \times 0.6 + 1.7(1.2 + 0.6) = 3.9 \text{ kips/ft (56.91 kN/m)}$$

New ultimate moment $M_u = \dfrac{3.90}{2.88} \, 144 = 195.00 \text{ ft·kips (264.4 kN·m)}$

$$\text{Max } p = 0.75 \times 0.85 \times 0.85 \times \frac{4}{50} \times \frac{87}{137} = 0.0275$$

$$\text{Min } p = \frac{200}{50,000} = 0.004$$

Try maximum reinforcement to see if compressive reinforcement is required.

$$\text{Max } A_s = 0.0275 \times 10 \times 17.5 = 4.81 \text{ in}^2$$

Maximum usable moment beam can carry is

$$M_u = \phi A_s f_y \left(d - \frac{a}{2} \right) = 0.90 \times 4.81 \times 50 \left(17.5 - \frac{a}{2} \right) \frac{1}{12}$$

$$= 18.04 \left(17.5 - \frac{a}{2} \right)$$

where $a = (4.81)(50)/(0.85)(4)(10) = 7.07$ in. Therefore,

$$M_u = 18.04 \left(17.5 - \frac{7.07}{2} \right) = 251.93 \text{ ft·kips (341.6 kN·m)}$$

Compressive reinforcement is not necessary. For new reinforcement, try three #10 (32.3-mm diam.) bars.

Required $b = 2(1.5) + 3(1.27) + 2(1.595) = 10$ in　　OK

$$A_s = 3.81 \text{ in}^2$$

Then

$$p = \frac{3.81}{10 \times 17.5} = 0.022 < 0.0275 \qquad \text{OK}$$

$$a = \frac{3.81 \times 50}{0.85 \times 4 \times 10} = 5.60$$

$$M_u = 0.90 \times 3.81 \times 50 \left(17.50 - \frac{5.60}{2} \right) \frac{1}{12}$$

$$= 210.02 \text{ ft·kips (284.8 kN·m)} < 195.0 \text{ ft·kips} \qquad \text{OK}$$

Use three #10 (32.3-mm diam.) bars. Check shear

$$V_u = 0.5 \ \times 3900 \times 20 = 39,000 \text{ lb (173.5 kN)}$$

$$v_u = \frac{39,000}{0.85 \times 10 \times 17.5} = 262.18 \text{ psi (1.81 MPa)} \qquad \text{ACI (11-3)}$$

The concrete can carry

$$v_c = 2\sqrt{4000} = 126.5 \text{ psi (872 kPa)} \qquad \text{ACI 11.4.1}$$

Stirrups are required (see ACI 11.1.1).

Use #3 (9.5 mm diam.) U bars. Shear stress d distance from the face of the support is

$$v_u = \frac{10.0 - 1.46}{10.0} \times 262.18 = 223.90 \text{ psi (1.54 MPa)}$$

$$\text{Required spacing } s = \frac{2(0.11)50,000}{(223.90 - 126.5)10} = 11.31 > \frac{d}{2}$$

$$= 8.75 \text{ in (222.2 mm)}$$

Use #3 (9.5 mm diam.) U stirrups 8.5 in (215.9 mm) centers to a distance x from centerline of beam

$$x = \frac{10.0 - 1.46}{223.90} \, 63.25 = 2.42 \text{ ft (0.74 m)}$$

Check distribution of tensile reinforcement (ACI 10.6.3.).

$$d_c = \text{cover} + \tfrac{1}{2} \text{ bar diam.} + \text{stirrup diam.}$$

$$= 1.5 + \tfrac{1}{2} \times 1.27 + 0.38 = 2.51 \text{ in}$$

$$A = \frac{2d_c b}{\text{no. bars}} = \frac{2 \times 2.51 \times 10}{3} = 16.73$$

$$z = 0.60 \times 50 \sqrt[3]{2.51 \times 16.73}$$

$$= 104.5 < 145 \text{ kips/in (25.39 MN/m)} \qquad \text{OK}$$

7.50 Determine the effective width of the beam:

$$\frac{L}{4} = \frac{20 \times 12}{4} = 60 \text{ in (1524 mm)}$$

$$16h_f + b_w = 16 \times 4 + 12 = 76 \text{ in}$$

$$\text{Beam spacing} = 45 \text{ in (1143 mm)}$$

The latter controls, therefore $b_f = 45$ in, $d = 24 - 2 = 22$ in.

Determine whether the beam has a real T-beam action. Assume $a = h_f = 4$ in. Then

$$A_s = \frac{M_u}{\phi f_y \left(d - \dfrac{a}{2}\right)} = \frac{985 \times 12}{0.90 \times 60(22 - 2)} = 10.94 \text{ in}^2$$

$$a = \frac{10.94 \times 60}{0.85 \times 4 \times 45} = 4.29 \text{ in}$$

Since a is larger than $h_f = 4$ in, the beam acts like a T beam. Tensile reinforcement due to compression in flanges is

$$A_{s1} = \frac{0.85 f_c'(b_f - b_w)h_f}{f_y} = \frac{0.85 \times 4(45 - 12)4}{60} = 7.48 \text{ in}^2$$

Corresponding moment is

$$M_1 = \phi A_{s1} f_y \left(d - \frac{h_f}{2} \right) = 0.90 \times 7.48 \times 60(22 - 2)\frac{1}{12}$$

$$= 673.20 \text{ ft} \cdot \text{kips} (912.8 \text{ kN} \cdot \text{m})$$

$$M_2 = M - M_1 = 985.00 - 673.20 = 311.80 \text{ ft} \cdot \text{kips} (422.8 \text{ kN} \cdot \text{m})$$

For the rectangular part of the beam, assume $p = 0.0138$ or A_{s2} = 3.65 in^2

$$a = \frac{3.65 \times 60}{0.85 \times 4 \times 12} = 5.37 \text{ in}$$

$$M_u = 0.90 \times 3.65 \times 60 \left(22 - \frac{5.37}{2} \right) \frac{1}{12}$$

$$= 317.25 \text{ ft} \cdot \text{kips} (430.2 \text{ kN} \cdot \text{m})$$

Total moment M_u
$$= 673.20 + 317.25$$
$$= 990.45 \ (1343 \text{ kN} \cdot \text{m}) > 985 \text{ ft} \cdot \text{kips} (1336 \text{ kN} \cdot \text{m}) \qquad \text{OK}$$

Total tensile reinforcement is

$$A_s = A_{s1} + A_{s2} = 7.48 + 3.65 = 11.13 \text{ in}^2 (7181 \text{ mm}^2)$$

Check to see if the beam fails by yielding of tensile reinforcement. Balanced reinforcement of the T beam is

$$p = p_b + p_f$$

where
$$p_b = 0.85(0.85) \left(\frac{4}{60} \right) \left(\frac{87}{87 + 60} \right) = 0.0285$$

$$p_f = \frac{A_{s1}}{b_w d} = 0.0283$$

$$\text{Max } p = 0.75(0.0285 + 0.0283) = 0.0426$$

$$\text{Actual } p = \frac{11.13}{12 \times 22} = 0.0421 < 0.0426 \qquad \text{OK}$$

7.51 Weight/ft $= (2 \times 2 + \frac{1}{2} \times 4) \times 145 = 870$ lb/ft

$\begin{aligned}
M_D &= \frac{1}{8} \times 870 \times 40 \times 40 \times 12 &= 2{,}080{,}000 \text{ in·lb} \\
M_L &= 172{,}000 \times 12 &= 2{,}064{,}000 \text{ in·lb} \\
M_I &= 52{,}000 \times 12 &= \underline{624{,}000} \text{ in·lb} \\
M_T &= (538.8 \text{ KN·m}) &= 4{,}768{,}000 \text{ in·lb}
\end{aligned}$

Assuming steel and concrete is stressed to maximum allowable,

$\begin{aligned}
k &= 0.377 \\
j &= 0.874 \\
kd &= 0.377 \times 21.5 \text{ in} = 8.10 \text{ in (205.7 mm)} \\
&21.50 \text{ in} - 8.10 \text{ in} = 13.40 \text{ in (340.4 mm)}
\end{aligned}$

350 + 1000 = 1350

6"

2.10"

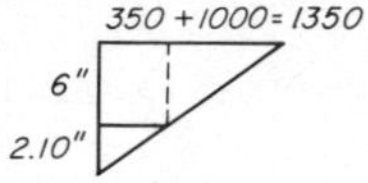

Fig. 7.51

$$\left(\frac{8.10 - 6}{8.10}\right) 1350 = 350 \text{ psi}$$

$$350 \times 6 = 2100 \times 3 = 6{,}300$$

$$\frac{1000}{2} 6 = \frac{3000 \times 2}{5100} \frac{= 6{,}000}{(2.41) = 12{,}300}$$

Compressive force centers 2.41 in (61.2 mm) from top of slab. Use six #11 bars ($A_{s1} = 9.36$ in²) in bottom row of steel.

Moment carried by #11 (35.81-mm diam.) bars
$$= 20{,}000 \times 9.36 \times 19.09 = 3{,}580{,}000 \text{ in·lb (404.5 kN·m)}$$

Use second row of steel 3 in above #11 (35.81-mm) bars.

$$M = 4{,}768{,}000 - 3{,}580{,}000 = 1{,}188{,}000$$

$$= \frac{10.40}{13.40} \times 20{,}000 \times 16.09 A_{s2}$$

$$A_{s2} = \frac{1{,}188{,}000 \times 13.40}{16.09 \times 20{,}000 \times 10.40} = 4.75 \text{ in}^2$$

Use six #8 (25.4-mm diam.) bars. $A_{s2} = 4.76$ in²

Cg of $A_s = 3 \times 4.76/(4.76 + 9.36) = 1.01$ in above A_{s1}
M beam can carry in concrete $= 72 \times 5100 \times 18.08$
$$= 6{,}640{,}000 \text{ in·lb}$$
and no A_s' needed. [$b = 72$ in $< l/4 < (24 + 2 \times 8 \times 6)$]
$72 \times 6(y - 3) = 4.76 \times 9 \times (18.5 - y) + 9.36 \times 9 \times (21.5 - y)$
$(y - 3) = 1.83 - 0.099y + 4.20 - 0.195y$
$1.294y = 9.03$, and $y = 6.98$ in

$I = \frac{1}{12} \times 72 \times 6 \times 6 \times 6 \quad = \quad 1,296$
$+ \; 72 \times 6 \times 3.98^2 \qquad\qquad = \quad 6,850$
$+ \; 47.6 \times 11.52^2 \qquad\qquad = \quad 5,680$
$+ \; 93.6 \times 14.52^2 \qquad\qquad = \quad \underline{17,820}$
$\qquad\qquad\qquad\qquad\qquad\qquad 31,646 \text{ in}^4$

$$f_c = \frac{4,768,000}{31,646} \times 6.98 = 1035 \text{ psi (7.14 MPa)}$$

$$f_s = \frac{4,768,00}{31,646} \times 14.52 \times 9 = 19,700 \text{ psi (135.8 MPa)}$$

$$v = \frac{V}{b'd} = \frac{40,000}{24 \times 21.5} = 77.5 \text{ (534.4 kPa)}$$

> 60 psi (413.7 kPa) (no stirrups)

$v'bs < 2A_v f_v$, and $s = 21.5/2 = 10.75$ in
Min stirrup steel $= 2A_v = 0.0015bs$
$A_v = 0.0015 \times 24 \times 10.75/2 = 0.194 < 0.20 \text{ in}^2$

Use #4 (12.7-mm diam.) U stirrups.

$17.5 \times 24 \times 10.75 = 4520 < 2 \times 0.20 \times 20,000 = 8000$ lb

First stirrups 5 in (127 mm) from ends, then on 10-in (254-mm) centers.

7.52
$\qquad\qquad f_c = 1800, f_s = 20,000$
$\qquad\qquad n = 8, \, k = 0.419, j = 0.860$
$\qquad\qquad R = 324$
b for T beam $= l/4 = 16 \times 12/4$
$\qquad\qquad = 48$ in $< b' + 2 \times (8 \times 4.5) < 96$ in
Assume dead weight of T beam $= 600$ lb/ft
$+M = \frac{8}{10} \times (\frac{1}{8} \times 8600 \times 16 \times 16 \times 12)$
$\qquad = 2,640,000 \text{ in} \cdot \text{lb (298.3 kN} \cdot \text{m)}$
$-M = \frac{1}{4} \times (2,640,000) = 660,000 \text{ in} \cdot \text{lb (74.6 kN} \cdot \text{m)}$
$M = Rbd^2 = 324 \times 48 \times d^2 = 2,640,000 \text{ in} \cdot \text{lb}$
$d^2 = 169, d = 13.0$, use 14 in; $kd = 5.87$ in

As neutral axis lies below slab, A_s' is required.

$$d - kd = 14 \text{ in} - 5.87 \text{ in} = 8.13 \text{ in}$$

$jd = 14$ in $- 1.96$ in
$\qquad\qquad = 12.04$ in $\qquad$ to bottom row tension steel, approx

$$A_s = \frac{M}{f_s jd} = \frac{2,640,000}{20,000 \times 12.04 \text{ in}}$$

$$= 10.96 \text{ in}^2 \qquad \text{if all } A_s \text{ in 1 row}$$

$b' = 24$ in (610 mm) for six #11 (35.81 mm) bars in one row.
Use six #11 bars, $A_{s1} = 9.36$ in bottom row.

$$+M_1 = 9.36 \times 20,000 \times 12.04 = 2,255,000 \text{ in·lb}$$
$$+M_2 = 2,640,000 - 2,255,000 = 385,000 \text{ in·lb}$$

Place A_{s2} 3 in (76.2 mm) above A_{s1}

$$M_2 = A_{s2} \times \frac{5.13}{8.13} \times 20,000 \times 9.04 = 385,000 \text{ in·lb}$$

$$A_{s2} = \frac{385,000}{114,000} = 3.38 \text{ in}^2.$$

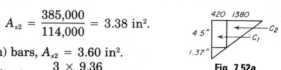

Fig. 7.52a

Use six #7 (22.22-mm) bars, $A_{s2} = 3.60 \text{ in}^2$.

cg of steel (below A_{s2} line) $= \dfrac{3 \times 9.36}{9.36 + 3.60}$

$$= 2.17 \text{ in or } 0.83 \text{ in above } A_{s1}$$

Center of gravity of compression forces (from top of beam)

	Force	Arm	Moment
$c_1 = 4.5 \times 420 =$	1890 lb	$\times 2.25$	$= 4253$
$c_2 = 4.5 \times 1380/2 =$	3105 lb	$\times 1.5$	$= 4657$
	4995 lb	(1.79)	$= 8910$

New $jd = 14.00 - 0.83 - 1.79 = 11.38 \text{ in} > 11.35 \text{ in}$
$M_{\text{(concrete)}} = 4995 \times 48 \times 11.38 = 2,730,000 \text{ in·lb}$

$M_T = 2,640,000$ in·lb. No A'_s needed.
Check kd:

$48 \times 4.5 \times (kd - 2.25)$
$$= 9.36 \times 8 \times (14 - kd) + 3.60 \times 8 \times (11 - kd)$$
$$kd = 5.80 \text{ in} < 5.87, \text{ but } f_c \text{ and } f_s \text{ OK.}$$

At end of beam for negative moment

$$A_s \text{ (at top)} = \frac{660,000}{20,000 \times 0.860 \times 14} = 2.74 \text{ in}^2$$

Six #7 (22.22-mm) bars bent up from $A_{s2} = 3.60 \text{ in}^2$

$M_{\text{(concrete stem)}} = Rb'd^2 = 324 \times 24 \times 14 \times 14$
$$= 1,525,000 \text{ in·lb} > 660,000 \text{ in·lb}$$
V [14 in $(= d)$ from end] $= 8 \times 8600 \times (96 - 14)/96$
$$= 58,700 \text{ lb } (261 \text{ kN})$$

$$u = \frac{V}{\Sigma_o jd} = \frac{58,700}{6 \times 4.43 \times 11.38} = 194 < \frac{4.8\sqrt{4,000}}{1.41}$$

$$= 215 \text{ psi } (14.82 \text{ kPa})$$

$$v \text{ (14 in from end)} = \frac{58,700}{24 \times 14} = 175 > 60 \text{ and } < 190 \text{ psi}$$

Min stirrup area $= 2A_v = 0.0015\ bs$

$A_v = 0.0015 \times 24 \times \tfrac{1}{2} = 0.127$ or #4 (12.7 mm) U stirrup

$$\text{Max } v' \text{ for \#4 U stirrup} = \frac{2 \times 0.20 \times 20,000}{24 \times 7} = 47 \text{ psi}$$

$$v'bs = (175 - 60) \times 24 \times 6 = 16,560 = 2A_v \times 20,000$$

$A_v = 0.413$ or #6 (19.05-mm) U stirrups at end on 6-in centers

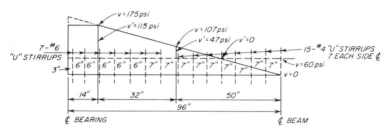

SHEAR DIAGRAM AND STIRRUP ARRANGEMENT

Fig. 7.52*b*

7.53 Total tensile reinforcement $A_s = 6 \times 1.00 = 6.00 \text{ in}^2$. Check whether a extends in web.

$$a = \frac{A_s f_y}{0.85 b_f d} = \frac{6 \times 60}{0.85 \times 25 \times 24} = 4.24 > 4.00 \text{ in}$$

A T-beam analysis is required. Check to see whether the beam will fail by yielding of tensile reinforcement. Tensile reinforcement due to compression in flanges is

$$A_{s1} = \frac{0.85 f_c'(b_f - b_w)h_f}{f_y} = \frac{0.85 \times 4(25 - 10)4}{60} = 3.40 \text{ in}^2$$

$$A_{s2} = 6.00 - 3.40 = 2.60 \text{ in}^2$$

$$p_w = \frac{A_s}{b_w d} = \frac{6}{10 \times 24} = 0.025$$

$$p_f = \frac{A_{s1}}{b_w d} = \frac{3.40}{10 \times 24} = 0.0142$$

$$p_b = (0.85)(0.85)\left(\frac{4}{60}\right)\left(\frac{87}{147}\right) = 0.0285$$

$$\text{Max } p = 0.75(p_b + p_f) = 0.75(0.0285 + 0.0142)$$
$$= 0.0320 > 0.025 \qquad \text{OK}$$

Find capacity.
Moment due to flanges:

$$\frac{M_1}{\phi} = A_{s1}f_y\left(d - \frac{h_f}{2}\right) = 3.4 \times 60(24 - 2)\frac{1}{12} = 374.00 \text{ ft·kips}$$

Moment due to web:

$$a = \frac{2.6 \times 60}{0.85 \times 4 \times 10} = 4.59 \text{ in}$$

$$\frac{M_2}{\phi} = A_{s2}f_y\left(d - \frac{a}{2}\right) = 2.6 \times 60\left(24 - \frac{4.59}{2}\right)\frac{1}{12}$$
$$= 282.17 \text{ ft·kips}$$

Capacity of beam:

$$M = \phi\left(\frac{M_1}{\phi} + \frac{M_2}{\phi}\right) = 0.90(374.00 + 282.17)$$
$$= \mathbf{590.55} \text{ ft·kips } (\mathbf{800.8} \text{ kN·m}) \qquad \mathbf{d}$$

7.54 $P_u = 1.4 \times 266,000 + 1.7 \times 530,000$
$$= 372,400 + 901,000 = 1,273,400 \text{ lb } (5.66 \text{ MN})$$

Here compression controls. Use $e = 0.05 D$ minimum.
Try $D = 21$ in (533.4 mm), $e = 1.05$ in (26.7 mm).

$$D_s = 21 - 2 \times \left(1.5 + 0.375 + \frac{1.693}{2}\right)$$

$$= 15.56 \text{ in } (395.2 \text{ mm})$$

$A_{st} = $ eight #14S (43.2-mm diam.) bars $= 8 \times 2.25 = 18.00 \text{ in}^2$
$A_g = 346 \text{ in}^2$, $p_t = A_{st}/A_g = 18.00/346 = 5.2\% < 8\%$

Clear spacing of bars $= (3.14 \times 15.56/8) - 1.69 = 4.41$ in (112 mm)

$$P_u = \phi\left(\frac{A_{st} \times f_y}{\dfrac{3e}{D_s} + 1} + \frac{A_g \times f_c'}{\dfrac{9.6 \times D \times e}{(0.8D + 0.67D_s)^2} + 1.18}\right)$$

$$= 0.75 \left(\frac{18 \times 60,000}{\dfrac{3.15}{15.56} + 1} + \frac{346 \times 4000}{\dfrac{9.6 \times 21 \times 1.05}{(16.8 + 10.4)^2} + 1.18} \right)$$

$$= 0.75 \left(\frac{1,080,000}{1.202} + \frac{1,384,000}{0.286 + 1.18} \right)$$

$$= 0.75 \times (896,000 + 944,000) \text{ lb}$$

$$= \text{kips} = 1380 \text{ kips } (6.14 \text{ MN}) > 1273 \text{ kips}$$

Spiral steel:

$$p_s = 0.45 \left(\frac{A_g}{A_c} - 1 \right) \frac{f'_c}{f_y} = 0.45 \left(\frac{21^2}{18^2} - 1 \right) \frac{4000}{60,000}$$

$$= \frac{0.45}{15} (1.360 - 1) = 0.0108$$

Try #⅜-in (9.5-mm diam.) spiral, $a_s = 0.11$ in²

$$g = \frac{4a_s}{p_s D_c} = \frac{4 \times 0.11}{0.0108 \times 18} = 2.26 \text{ in } (57.4 \text{ mm})$$

Use $g = 2.25 < 2.26 < 18/6 =$ or < 3-in clear max

7.55 $P_u = 1273$ kips as in answer 7.54.
Here compression controls. Use $e = 0.05t$ min.
Try $t = 19$ in (482.6 mm), $e = 0.95$ in, $D_s = 13.31$ in

$A_{st} = $ six #14S (43.2-mm diam.) + two #11 (35.8-mm diam.) bars
$= 6 \times 2.25 + 2 \times 1.56 = 16.62$ in²

$A_g = 361$ in², $p_t = A_{st}/A_g = 16.62/361 = 4.61\% < 8\%$
Clear spacing of bars $= (3.14 \times 13.31/8) - 1.69 = 3.54$ in

$$P_u = \phi \left(\frac{A_{st} f_y}{\dfrac{3e}{D_s} + 1} + \frac{A_s f'_c}{\dfrac{12te}{(t + 0.67D_s)^2} + 1.18} \right)$$

$$= 0.75 \left(\frac{16.62 \times 60,000}{\dfrac{3 \times 0.95}{13.31} + 1} + \frac{361 \times 40,000}{\dfrac{12 \times 19 \times 0.95}{(19 + 8.91)^2} + 1.18} \right)$$

$$= 0.75 \times (821,000 + 991,000) \text{ 1b}$$

$$= 1359 \text{ kips } (6.04 \text{ MN}) > 1273 \text{ kips } (5.66 \text{ MN})$$

For spiral design:

$$p_s = 0.45 \left(\frac{19 \times 19}{3.14 \times 8.3 \times 8.3} - 1 \right) \frac{4000}{60,000} = \frac{0.45}{20} (1.67 - 1)$$

$$= 0.0201$$

Try #4 (12.7-mm diam.) spiral, $a_s = 0.20$ in².

$$g = \frac{4a_s}{p_s D_c} = \frac{4 \times 0.20}{0.0201 \times 16.62} = 2.40 \text{ in } (61.0 \text{ mm})$$

Use $g = 2.25 < 2.40 < {}^{16}\!/_6 < 3$ in (76.2 mm) clear max.

7.56 Total axial load and moment are

$$P_u = 1.4 \times 210 + 1.7 \times 140 = 532.00 \text{ kips } (2.37 \text{ MN})$$

$$\frac{P_u}{\phi} = \frac{532.00}{0.7} = 760.00 \text{ kips } (3.38 \text{ MN})$$

$$M_u = 1.4 \times 63 + 1.7 \times 42 = 159.60 \text{ ft·kips } (216.4 \text{ kN·m})$$

$$\frac{M_u}{\phi} = \frac{159.60}{0.7} = 228.00 \text{ ft·kips } (309.2 \text{ kN·m})$$

Actual eccentricity is

$$e = \frac{228 \times 12}{760.00} = 3.60 \text{ in } (91.4 \text{ mm})$$

Check to see if there is a slenderness effect.

$$r = 0.3 \times 20 = 6.00 \qquad \text{ACI } 10.11.2$$

Since the column is braced against sidesway, $k = 1$ (ACI 10.11.3).

$$\frac{kl_u}{r} = \frac{1.0 \times 8.0 \times 12}{6} = 16.00$$

$$34 - 12 \left(\frac{M_1}{M_2} \right) = 34 - 12 \left(\frac{-1}{2.5} \right) = 38.8$$

Therefore,

$$\frac{kl_u}{r} < 38.8 \qquad \text{ACI } 10.11.4$$

No slenderness effect. For longitudinal reinforcement assume six #9 (28.7 mm diam.) bars. Effective depth $d = 20.00 - 2.50 = 17.50$ in. Find the balanced eccentricity of the section.

$$\frac{P_{u_b}}{\phi} = 0.85\beta_1 f_c' bd \left(\frac{87}{87 + f_y}\right)$$

$$= 0.85 \times 0.85 \times 4 \times 14 \times 17.5 \left(\frac{87}{87 + 60}\right)$$

$$= 419.05 \text{ kips } (1.86 \text{ MN})$$

$$a_b = \beta_1 d \left(\frac{87}{147}\right) = 0.85 \times 17.5 \times \frac{87}{147} = 8.80 \text{ in}$$

$$\frac{P_{u_b} e_b'}{\phi} = 0.85 f_c' a_b b \left(d - \frac{a_b}{2}\right) + A_s' f_y'(d - d')$$

$$419.05 e_b' = 0.85 \times 4 \times 8.80 \times 14 \left(17.5 - \frac{8.8}{2}\right)$$

$$+ 3.0 \times 60(17.5 - 2.5)$$

$$e_b' = 19.5 \text{ in}$$

Then $e_b = 19.5 - 7.5 = 12.0$ in $= e$. Therefore compression controls. Proceed by successive approximations. Assume $a = 14.5$, then $c = 14.5/0.85 = 17.06$.

Stress in tension steel is

$$f_s' = 87 \left(\frac{d - c}{c}\right) = 87 \left(\frac{17.5 - 17.06}{17.06}\right) = 2.24 \text{ ksi } (15.4 \text{ MPa})$$

$$e' = 3.60 + 7.50 = 11.10 \text{ in}$$

$$\frac{P_u e'}{\phi} = 0.85 f_c' ab \left(d - \frac{a}{2}\right) + A_s' f_y'(d - d')$$

$$\frac{P_u 11.10}{\phi} = 0.85 \times 4 \times 14.5 \times 14 \left(17.5 - \frac{14.5}{2}\right) + 3 \times 60(17.5 - 2.5)$$

$$\frac{P_u}{\phi} = 880.59 \text{ kips } (3.92 \text{ MN})$$

Find a using the equation

$$P_u = 0.85 f_c' ab + A_s' f_y' - A_s f_s$$

$$880.59 = 0.85 \times 4 \times a \times 14 + 3.0 \times 60 - 3.0 \times 2.24$$

or $$a = 14.18$$

close enough to the assumed value.

For $$a = 14.8 \frac{P_u}{\phi} = 880.59 \text{ kips}$$

$$\frac{M_u}{\phi} = 880.59 \times \frac{3.60}{12}$$

$$= 264.18 \text{ ft·kips (358.2 kN·m)}$$

Usable load $P_u = 0.70 \times 880.59 = 616.41$ kips > 532 kips
Usable moment $M_u = 0.70 \times 264.18 = 184.93$ kips > 160 kips

Use #3 (9.5-mm diam.) ties (see ACI 7.12.3). Spacing of ties

$$16 \times 1.13 = 18.08 \text{ in}$$
$$48 \times 0.38 = 18.24 \text{ in}$$
$$\text{Least dimension} = 14 \text{ in (355.6 mm)}$$

Use #3 ties 14 in on centers.

7.57 Assume $d' = 2.5$ in. Then $d = 22 - 2.5 = 19.5$ in. Determine e_b:

$$P_u = 0.85 f_c' \beta_1 \left(\frac{87}{87 + f_y} \right)$$

$$= 0.85 \times 4 \times 14 \times 19.5 \times 0.85 \left(\frac{87}{147} \right) = 466.94 \text{ kips (2.08 MN)}$$

$$a_b = \frac{466.94}{0.85 \times 4 \times 14} = 9.81 \text{ in}$$

$$\frac{P_{u_b} e_b'}{\phi} = 0.85 f_c' a_b b \left(d - \frac{a}{2} \right) + A_s' f_y'(d - d')$$

$$\frac{466.94 e_b'}{0.70} = 0.85 \times 4 \times 9.81 \times 14 \left(19.5 - \frac{9.81}{2} \right)$$
$$+ 2 \times 1.27 \times 60(19.5 - 2.5)$$

$$e_b' = 14.10 \text{ in}$$

$$e_b = 14.10 - 8.55 = 5.55 < 16 \text{ in}$$

Therefore tension controls.

$$p = \frac{A_s}{bd} = \frac{2 \times 1.27}{14 \times 19.5} = 0.0093$$

$$e' = e + 8.5 = 16 + 8.5 = 24.5$$

$$m = \frac{f_y}{0.85 f_c'} = \frac{60}{0.85 \times 4} = 17.65$$

$$\left(\frac{e'}{d} - 1 \right)^2 = \left(\frac{24.5}{19.5} - 1 \right)^2 = (0.256)^2 = 0.066$$

$$2pm \left(1 - \frac{d'}{d}\right) = 2 \times 0.0093 \times 17.65 \left(1 - \frac{2.5}{19.5}\right) = 0.286$$

$$\frac{P_u}{\phi} = 0.85 f'_c bd \left[- \left(\frac{e'}{d} - 1\right)\right.$$

$$\left. + \sqrt{\left(\frac{e'}{d} - 1\right)^2 + 2pm \left(1 - \frac{d'}{d}\right)}\right]$$

$$= 0.85 \times 4 \times 14 \times 19.5 \left[-0.256 + \sqrt{0.066 + 0.286}\right]$$

$$= 312.8 \text{ kips}$$

Usable load $P = 0.7 \times 312.8 = \textbf{218.96}$ kips (**974** kN)   **b**

7.58 Ultimate loads:

$$\frac{P_u}{\phi} = \frac{1}{0.7} (1.4 \times 350 + 1.7 \times 153) = 1071.43 \text{ kips (4.77 MN)}$$

At the top:

$$\frac{M_2}{\phi} = \frac{1}{0.7} (1.4 \times 50 + 1.7 \times 50) = 221.43 \text{ ft·kips (300.3 kN·m)}$$

At the bottom:

$$\frac{M_1}{\phi} = \frac{1}{0.7} (1.4 \times 20 + 1.7 \times 15) = 76.43 \text{ ft·kips (103.6 kN·m)}$$

Slenderness effect is

$$r = 0.3 \times 24 = 7.2$$

Since the column is not braced against sidesway, $k > 1$ (ACI 10.11.3). Since it bends in double curvature, $k = 2$.

$$\frac{kl_u}{r} = \frac{2 \times 17 \times 12}{7.2} = 56.66 > 22$$

the slenderness effect has to be considered (ACI 10.11.4).

$$I_g = \frac{1}{12} \times 24 \times 24^3 = 27,648 \text{ in}^4$$

Distance of centroid of reinforcement to centroidal axis of section is

$$\frac{24}{2} - 2.14 = 9.86 \text{ in}$$

where $2.14 = d'$.

Total $A_s = 8 \times 1.27 = 10.16 \text{ in}^2$
then $I_{se} = 10.16(9.86)^2 = 987.75 \text{ in}^4$
          $E_s = 29 \times 10^6 \text{ psi}$
          $E_c = 57{,}000 \sqrt{f'_c} = 4.03 \times 10^6 \text{ psi}$ ACI 8.3.1

Top moment M_2 governs.

$$\beta_d = \frac{50 \times 1.4}{50(1.4 + 1.7)} = \frac{70}{155} = 0.45$$

$$EI = \frac{(E_c I_g / 5) + (E_s I_{se})}{1 + \beta_d}$$

$$= \frac{4.03 \times 2.76 \times 10^{10} \times 0.2 + 29 \times 9.88 \times 10^8}{1.45}$$
          ACI 10.11.5
          (10-7)

or $EI = 3.5 \times 10^{10}$.

Then

$$P_{cr} = \frac{\pi^2 EI}{(kl_u)^2} = \frac{9.86 \times 3.5 \times 10^{10}}{(2 \times 17 \times 12)^2 \times 10^3}$$

$$= 2073.00 \text{ kips (9.22 MN)}$$
    ACI 10.11.5
    (10-6)

Column is not braced against sidesway, therefore $c_m = 1$.

$$\delta = \frac{c_m}{1 - (P_u/\phi P_c)} = \frac{1}{1 - (1071.43/2073.00)}$$

$$= 2.08$$
    ACI 10.11.5
    (10-5)

Magnified moment $M_c = \delta M_2 = 2.08 \times 221.43 = 460 \text{ ft·kips}$

Design for an axial load of $1071.43/\phi$ kips and a moment of $460/\phi$ ft·kips. For these values

$$e = \frac{460 \times 12}{1071.43} = 5.16 \text{ in}$$

Find e_b:

$$a_b = \frac{1071.43}{0.85 \times 5 \times 24} = 10.5 \text{ in}$$

$$1071.43 e'_b = 0.85 \times 5 \times 10.5 \times 24 \left(21.86 - \frac{10.5}{2}\right)$$

$$+ 5.08 \times 60(21.86 - 2.14)$$

$$e'_b = 22.21 \text{ in}$$
$$e_b = 22.21 - 9.86 = 12.35 \text{ in}$$

$e < e_b$, therefore compression controls.

Proceed using successive approximations. Assume $a = 16.00$ in.

$$c = \frac{16.00}{0.80} = 20.00$$

$$f_s = 87\left(\frac{21.86 - 20.00}{20.00}\right) = 8.1 \text{ ksi}$$

$$e' = 5.16 + 9.86 = 15.02 \text{ in}$$

$$15.02\,\frac{P_u}{\phi} = 0.85 \times 5 \times 16.00 \times 24\left(21.86 - \frac{16.00}{2}\right)$$
$$+ \, 5.08 \times 60(21.86 - 2.14)$$

$$\frac{P_u}{\phi} = 1905 \text{ kips}$$

$$1905 = 0.85 \times 5 \times 24a + 5.08(60 - 8.1)$$

or
$$a = 16.1 \text{ in}$$

close enough to the assumed value.

For $a = 16.1$, $P_u/\phi = 1905 > 1071.43$. Column is adequate.

7.59

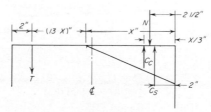

Fig. 7.59

Take $\Sigma M_N = 0$

Force	Area ×	Intensity	× Arm ×	Factor (to simplify products)	= Moment
$+C_c$	$24x$	$f_c/2$	$(+)\left(\dfrac{x}{3} - 2.5\right)$	$x/4f_c$	$= +x^2(x - 7.5)$
$+C_s$	$(18 - 1)3$	$\dfrac{x - 2}{x}f_c$	$(-)(0.5)$	$x/4f_c$	$= -6.375(x - 2)$
$-T$	9×3	$\dfrac{13 - x}{x}f_c$	$(+)(10.5)$	$x/4f_c$	$= -70.875(13 - x)$

$$\Sigma M_N = 0 = +x^2(x - 7.5) - 6.375(x - 2) + 70.875(x - 13)$$
$$x^3 - 7.5x^2 + 64.5x - 908.56 = 0$$

Solve for x by trial.

$$x = 10.00, +1000 - 750 + 645 - 908.56 = -13.56$$
$$x = 10.06, +1018.1 - 759 + 647.7 - 908.6 = -1.8$$
$$x = 10.07, +1021.1 - 760.5 + 649.5 - 908.6 = +1.5$$

Use $x = 10.06$ in $> \dfrac{15}{4}$ in $= 3.75$ in min

$$+C_c = 24 \times 10.06 \times f_c/2 \quad = +120.8f_c$$

$$+C_s = 17 \times 3 \times \frac{8.06}{10.06} \times f_c = +40.9f_c$$

$$\overline{+161.7f_c}$$

$$-T = 9 \times 3 \times \frac{2.94}{10.06} \times f_c = -7.9f_c$$

$$\overline{+153.8f_c}$$

$$\Sigma V = +C_c + C_s - T - N = 0$$
$$N = 153.8f_c = 150,000$$

$f_c =$ **975** psi (**6.72** MPa) > 750 psi $(0.25f'_c) < 1125$ psi $(0.315\ f'_c)$ **1b**

$$f_s(\text{tens.}) = 9\,\frac{2.94}{10.06}\,975 = \textbf{2.57}\ \text{ksi}\ (\textbf{17.72}\ \text{MPa}) < 20\ \text{ksi} \qquad \textbf{2a}$$

$$f_s(\text{compress.}) = 18 \times 975 \times \frac{8.06}{10.06}$$

$$= \textbf{14.1}\ \text{ksi}\ (\textbf{97.22}\ \text{MPa}) < 20\ \text{ksi} \qquad \textbf{2c}$$

7.60 Assume a W 12 × 65 (A36 steel).
Area $= A_r = 19.11$ in^2, $d = 12.12$ in, $b = 12$ in.

Diameter of enclosing circle for W = 17.1 in. Use spiral OD = 22 in and ID = 21.25 in.

$$A_{col} = \frac{\pi}{4}(25)^2 = 490 \text{ in}^2$$

$$D_s = 25 - 3 - 0.75 - 1.41 = 19.84 \text{ in (18.43 in clear inside)}$$

Use ten #11 (35.8-mm diam.) bars, $A_s = 15.6 \text{ in}^2$, $p = 15.6/490 = 0.0317$.

$$\text{Clear spacing} = (3.14 \times 19.84/10) - 1.41 = 4.82$$
$$A_{conc} = A_{col} - A_{st} - A_r = 490 - 15.6 - 19.11 = 455 \text{ in}^2$$

Ultimate axial load that the column may carry is

$$\frac{P_u}{\phi} = 0.85 f_c' A_{conc} + A_r f_{ry} + A_s f_y$$

$$= 0.85 \times 3.0 \times 455 + 19.11 \times 36 + 15.6 \times 40$$

$$= 1160.3 + 687.6 + 624.0 = 2471.9 \text{ kips}$$

Working stress load is $P_u(C/\phi)$, and C/ϕ may be taken to be 0.40.

$$P = 0.40 \times 2471.9 = 988.8 \text{ kips} < 1000 \text{ kips}$$

If W 12 × 65 is changed to **W 12 × 72**, $P = 1016$ kips.   OK
Spiral steel:

$$p_s = 0.45 \left(\frac{A_g}{A_c} - 1\right) \frac{f_c'}{f_y}$$

$$= 0.45 \left(\frac{490}{380} - 1\right) \frac{3000}{60,000} = \frac{0.45}{20}(1.296 - 1) = 0.00652$$

Try #3 in (9.5-mm diam.) spiral, $a_s = 0.11$

$$g = \frac{4a_s}{p_s D_c} = \frac{4 \times 0.11}{0.00652 \times 22} = 3.07 \text{ in}$$

Use $g = 3$ in (76.2 mm)(max) < 22/6.

7.61 Area and moment of inertia of W 12 × 65 are $A_t = 19.1 \text{ in}^2$ and $L_t = 533.0 \text{ in}^4$. Assume an 18 × 18 in concrete encasement.

$$A_{conc} = 18 \times 18 - 19.1 = 304.9 \text{ in}^2$$

Ultimate axial load the above section can carry is

$$\frac{P_u}{\phi} = 0.85 f_c' A_{conc} + A_s f_y$$

Assume a concrete with $f'_c = 3$ ksi and A36 structural steel.

$$\frac{P_u}{\phi} = 0.85 \times 3 \times 304.9 + 19.1 \times 36 = 777.5 + 687.6$$

$$= 1465.1 \text{ kips}$$

Working stress axial load is $P = CP_u/\phi$, where $C/\phi = 0.40$.

$$P = 0.40(1465.1) = \mathbf{586} \text{ kips } (\mathbf{2.61} \text{ MN}) \quad \mathbf{d}$$

Check to see if there is slenderness effect (ACI 10.11.4 10.15.3).

$$I_g = \frac{18 \times 18^3}{12} - 533 = 8215 \text{ in}^4$$

$$E_c = 57,000 \sqrt{3000} = 3.12 \times 10^3 \text{ ksi}$$

$$r = \sqrt{\frac{3.12 \times 10^3 \times 8215/5 + 29 \times 10^3 \times 533}{3.12 \times 10^3 \times 304.9/5 + 29 \times 10^3 \times 19.1}}$$

$$= 5.26 \qquad (10\text{-}10)$$

Assume $k = 1$, $kl_u/r = 1 \times 20 \times 12/5.26 = 45.63 < 100$. No slenderness effect.

7.62 Assume a 12-in extra strong pipe, with $f_y = 36$ ksi. OD = 12.75 in, ID = 11.75 in, thickness of wall = 0.5 in, $A = 19.2$ in^2, $I = 362$ in.^{4}
ACI requires that thickness of pipe is larger than

$$h \sqrt{\frac{f_y}{8E_s}} \quad \text{or} \quad 12.75 \sqrt{\frac{36}{8 \times 29,000}} = 0.159 < 0.500 \text{ in}$$

Assume a concrete with $f'_c = 3$ ksi.

$$A_c = \frac{\pi}{4} (11.75)^2 = 108.38 \text{ in}^2$$

Ultimate axial load is

$$\frac{P_u}{\phi} = 0.85 f'_c A_c + f_y A_s = 0.85 \times 3 \times 108.38 + 36 \times 19.2$$

$$= 967.57 \text{ kips}$$

Working stress load is

$$P = 0.40(967.57) = 387.03 \text{ kips } (1.72 \text{ MN}) > 350 \text{ kips } (1.56 \text{ MN})$$

Check to see if there is a slenderness effect.

$$E_c = 57,000 \sqrt{3000} = 3.12 \times 10^3 \text{ ksi}$$

$$I_g = 0.785 \left(\frac{11.75}{2} \right)^4 = 935.19 \text{ in}^4$$

$$r = \sqrt{\frac{3.12 \times 10^3 \times 935.19/5 + 29 \times 10^3 \times 362}{3.12 \times 10^3 \times 108.38/5 + 29 \times 10^3 \times 19.2}}$$

$$= 4.21 \qquad (10\text{-}10)$$

Assume $k = 1$, $kl_u/r = 1 \times 20 \times 12/4.21 = 57.01 < 100$. No slenderness effect.

7.63 Find the area of the footing required to satisfy the soil conditions. Total dead and live load is

$$P = 350 + 125 = 475 \text{ kips (2.11 MN)}$$

Assume a footing 2 ft 0 in deep. Then the weight of the soil above footing and the weight of the footing are

$$\text{Soil weight} = 2.5 \times 0.120 = 0.300 \text{ kips/ft}^2$$
$$\text{Footing weight} = 2.00 \times 0.150 = 0.300 \text{ kips/ft}^2$$
$$\text{Total} = 0.600 \text{ kips/ft}^2 \text{ (28.72 kPa)}$$

$$\text{Net bearing soil capacity} = 2.75 \times 2 - 0.600$$
$$= 4.9 \text{ kips/ft}^2 \text{ (234.6 kPa)}$$
$$\text{Required area of footing} = \frac{475}{4.90} = 96.94 \text{ ft}^2$$

Use a footing: 9 ft 10 in × 9 ft 10 in (3 × 3 m) ($A = 96.63 \text{ ft}^2$). Design the footing.

$$P_u = 1.4 \times 350 + 1.7 \times 125 = 702.5 \text{ kips}$$

Soil pressure is

$$w_u = \frac{702.5}{96.63} = 7.27 \text{ kips/ft}^2$$

Punching shear usually controls, so

$$v_c = 4 \sqrt{f_c'} = 4 \sqrt{4000} = 253.00 \text{ psi} \qquad \text{ACI 11.10.3}$$

Critical section is at $0.5d$ from the face of the column (see ACI 11.10.2). Perimeter of the critical section is

$$b_p = 4(^{20}/_{12} + d) = 6.67 + 4d$$
$$V_u = P_u - w_u(^{20}/_{12} + d)^2 = 702.5 - 7.27(1.67 + d)^2$$

$$v_u = \frac{V_u}{\phi b_p d} = \frac{702.5 - 7.27(1.67 + d)^2}{0.85(6.67 + 4d)d}$$

$$v_c = \frac{253.0 \times 144}{100} = 36.36 \text{ kips/ft}^2$$

Above equations yield $d = 1.56$ ft or $d = 18.72$ in. Use $d = 19$ in (482.6 mm) to center of top steel for a total depth of footing of 24 in (610 mm) to provide coverage of 3 in (76.2 mm) for two layers of reinforcing steel. Check beam shear. Critical section at d from face of column.

$$v_c = 2\sqrt{f_c'} = 2\sqrt{4000} = 126.50 \text{ psi}$$

Distance of the critical section from the edge of the footing is

$$l = \frac{118 - 20}{2} - 19 = 30 \text{ in or 2.5 ft}$$

$$V_u = 2.5 \times 7.27 = 18.18 \text{ kips/ft of width}$$

$$v_u = \frac{V_u}{\phi bd} = \frac{18,180}{0.85 \times 12 \times 19}$$

$$= 93.81 \text{ psi} < 126.5 \qquad \text{OK} \qquad \text{ACI 11.1.1}(a)$$

Determine flexural steel. The critical section at face of wall [see ACI 15.4.2(a)].

$$\frac{M_u}{\phi} = \frac{1}{2} \times \frac{7.27 \times 4.08^2}{0.90} = 67.23 \text{ ft·kips}$$

$$\text{Min steel ratio } p = \frac{200}{f_y} = \frac{200}{50,000} = 0.004$$

Use minimum steel ratio $A_s = 12 \times 19 \times 0.004 = 0.912 \text{ in}^2$/ft of width. Total steel $A_s = 0.912 \times 9.83 = 8.96 \text{ in}^2$. Use nine #9 (28.7 mm diam.) bars both ways. Development length required is

$$l_d = \frac{0.04 A_b f_y}{\sqrt{4000}} = \frac{0.04 \times 1.0 \times 50,000}{63.25} = 31.62 \text{ in} \qquad \text{ACI 12.5}$$

but not less than $0.0004 d_b f_y = 0.004 \times 1.13 \times 50,000 = 22.6$ in.

$$\text{Development provided} = (9.83 - 4.08 - 0.25)12$$
$$= 66.00 \text{ in (1676 mm)} \qquad \text{OK}$$

7.64 Use soil pressure from column loads $= 5500$ psf (263.3 kPa).

$$\text{Length of footing} = \frac{700,000}{9 \times 5500}$$

$$L = 14.17 \text{ ft (4.32 m)}$$

$$M_{xy} = 9 \times \frac{5.91^2}{2} \times 5500 = 862{,}000 \text{ ft·lb (1169 kN·m)}$$

$$M_{PQ} = 14.17 \times \frac{3.33^2}{2} \times 5500 = 432{,}000 \text{ ft·lb (587 kN·m)}$$

$$d^2_{(xy)} = \frac{M}{Rb} = \frac{862{,}000}{223 \times 9} = 429, \; d = 20.7 \text{ in}$$

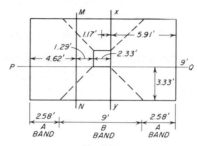

Fig. 7.64

Use $d = 31.5$ in, and overall depth $= 34.5$ in (876 mm) with weight
under 500 psf (23.9 kPa):

$$V = \left(2.58 \times 9 + \frac{9 + 2.58 + 2.33}{2} \times 2.04\right) \times 5500$$

$$= 127{,}800 + 77{,}500 = 205{,}300 \text{ lb (913 kN)}$$

$$v = \frac{V}{bd} = \frac{205{,}300}{59.5 \times 31.5} = 109.5 \text{ psi} < 110 \text{ psi (no stirrups)}$$

Long steel for 9-ft width:

$$A_s = \frac{M}{f_s jd} = \frac{862{,}000 \times 0.85 \times 12}{20{,}000 \times 0.874 \times 31.5} = 15.92 \text{ in}^2$$

Use 16 #9 (28.7 mm diam.) bars, $A_s = 16.0$ in^2.

$$u = \frac{V}{\Sigma_0 jd} = \frac{205{,}300}{56.8 \times 0.874 \times 31.5} = 131.5 \text{ psi} < 233 \text{ psi}$$

Short steel for 14-ft 2-in length

$$A_s = \frac{0.85 \times 432{,}000 \times 12}{20{,}000 \times 0.874 \times 31.5} = 7.98 \text{ in}^2$$

$$A_s \text{ for } B \text{ band} = 7.98 \times \frac{2}{s+1} \qquad \text{where } s = \frac{14.17}{9} = 1.575$$

$$= 7.98 \times \frac{2}{2.575} = 6.20 \text{ in}^2$$

Use 11 #7 (22.2-mm) bars on 10-in centers, $A_s = 6.60 \text{ in}^2$.

$$a = \frac{77,500}{30.24 \times 0.874 \times 31.5} = 93 \text{ psi} < 301 \text{ psi}$$

$$A_s \text{ for } A \text{ bands} = \frac{7.98 - 6.20}{2} = \frac{1.78}{2} = 0.89 \text{ in}^2$$

Use three #6 (19.1-mm diam.) bars on 10-in centers in each 31-in end A band. $A_s = 1.32$ in.

7.65

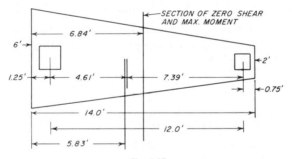

Fig. 7.65

cg distance of column loads from centerline 120-kips load

$$= \frac{12 \times 75}{120 + 75} = \frac{900}{195} = 4.61 \text{ ft } (1.41 \text{ m})$$

Use 3500 psf (167.6 kPa) as working presure on underside of concrete slab from column loads.

$$\text{Area footing} = \frac{195,000}{3500} = 56 \text{ ft}^2 \ (5.20 \text{ m}^2)$$

Trapezoidal footing 14 ft (4.27 m) long, width 2 to 6 ft (0.61 to 1.83 m), and A 56 ft².

cg of footing from 6 ft end = 5.83 ft

$2 \times 14 \quad = 28 \times 7$ ft $\quad = 196$

$2 \times \dfrac{14 \times 2}{2} = 28 \times 4.67$ ft $= 130.7$

$\overline{56} \ (5.83 \text{ ft}) \quad = \overline{326.7}$

Place centerline of 15×15 column 15 in from 6-ft end of footing.
Place centerline of 12×12 column 9 in from 2-ft end of footing.
cg of column loads and cg of upward pressure coincide within 0.03 ft.
Say OK.
Find distance from left end to point of zero shear

$$6 \times 3500 = 21 \text{ kips/ft at 6-ft end}$$

Upward pressure varies from 21 to 7 kips in 14 ft or by 1 kip/ft

$$+21x - 1x^2/2 - 120 = 0$$
$$+42x - x^2 - 240 = 0$$
$$x^2 - 42x + (21)^2 = -240 + (21)^2 = +201$$
$$x - 21 = \pm\sqrt{201} = \pm 14.16$$
$$x = +21 - 14.16 = 6.\overset{.}{8}4 \text{ ft} = \text{point of max moment}$$

$$M_{(max)} = (21 - 6.84) \text{ kips} \times \frac{6.84^2}{2} + 6.84 \text{ kips} \times \frac{6.84}{2}$$

$$\times \tfrac{2}{3} \times 6.84 - 120 \times 5.59 \text{ ft}$$

$$= 331.3 + 106.7 - 671$$

$$= 233 \text{ ft·kips } (316 \text{ kN·m})$$

Width 8.5 in inward from face of 15×15 column

$$= 6 - \frac{2.58 \times 4}{14} = 6 - 0.74 = 5.26 \text{ ft } (1.60 \text{ m}) \text{ at point } A$$

Width 8.5 in inward from face of 12×12 column

$$= 2 + \frac{1.96 \times 4}{14} = 2 + 0.56 = 2.56 \text{ ft } (0.78 \text{ m}) \text{ at point } B$$

Width at point of max moment $= 6 - \dfrac{6.84 \times 4}{14}$

$$= 6 - 1.95 = 4.05 \text{ ft } (1.23 \text{ m})$$

Shear at point $A = 21 \times 2.58 - \dfrac{2.58^2 \times 1}{2} - 120$

$$= 54.2 - 3.33 - 120$$

$$= 69.13 \text{ kips } (307.5 \text{ kN})$$

Shear at point $B = 7 \times 1.96 + 1 \times \dfrac{1.96^2}{2} - 75$

$$= 13.7 + 1.92 - 75 = 59.38 \text{ kips } (264.1 \text{ kN})$$

$$k = \dfrac{1350}{1350 + 20{,}000/9} = 0.377, j = 0.874$$

$$R = \dfrac{1350}{2} \times 0.874 \times 0.377 = 223$$

$$v = \dfrac{V}{bd} \text{ and } d = \dfrac{V}{vb} = \dfrac{59{,}380}{110 \times 12 \times 2.56}$$

$$= 17.5 \text{ in} + 3 \text{ in}$$

Depth footing $= 20.5$ in overall

Wt footing $= 248$ psf $< 500 \qquad \therefore$ OK

Moment concrete could take $= M_c = Rbd^2$
$$= 223 \times 4.05 \times 17.5 \times 17.5$$
$$= 275 \text{ ft·kips} > 233 \text{ ft·kips} \qquad \therefore \text{ OK}$$

$$A_s = \dfrac{M}{f_s jd} = \dfrac{233{,}000 \times 12}{20{,}000 \times 0.874 \times 17.5} = 9.14 \text{ in}^2$$

Use six #11 (35.8-mm diam.) bars, $A = 9.36$ in².

$$\text{Max lateral bending} = 3500 \times \dfrac{2.38^2}{2} \times 12$$

$$= 119{,}000 \text{ in·lb}$$

$$A_s = \dfrac{M}{f_s jd} = \dfrac{119{,}000}{20{,}000 \times 0.874 \times 17.5} = 0.388 \text{ in}^2/\text{ft}$$

Use #4 (12.7-mm diam.) bars 6 in on centers, $A_s = 0.40$ in²/ft.
For longitudinal bottom steel, and transverse steel top and bottom, use #4 bars at suitable spacings.

7.66 6 to 12 triangle $= 1, 2, \sqrt{5}$

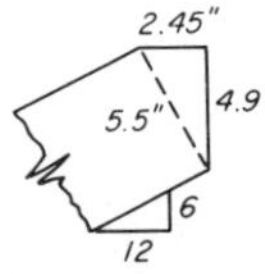

$$\cos^2 \theta = \dfrac{(2)^2}{5} = 0.80, \sin^2 \theta = \dfrac{(1)^2}{5} = 0.20$$

Fig. 7.66

$$N_1 = \dfrac{1200 \times 380}{1200 \times 0.20 + 380 \times 0.80}$$

$$= \dfrac{456{,}000}{240 + 304} = \dfrac{456{,}000}{544} = 838 \text{ psi } (5.78 \text{ MPa})$$

$$N_2 = \frac{1200 \times 380}{1200 \times 0.80 + 380 \times 0.20} = \frac{456{,}000}{960 + 76}$$

$$= \frac{456{,}000}{1036} = 440 \text{ psi } (3.03 \text{ MPa})$$

Dressed size of timbers $= 5.5 \times 5.5$ in

4-in diam. washer has 11 in^2 bearing area.

Washer bearing $= 4500/11 = 409$ psi < 440 psi $\therefore$ OK

$$6 \times 6 \text{ bearing against } 6 \times 6 = \frac{9000}{4.9 \times (5.5 - 1)}$$

$$= \frac{9000}{4.9 \times 4.5}$$

$$= 408 \text{ psi} < 838 \text{ psi} \quad \therefore \text{ OK}$$

$$2 \times 2.45 \text{ in} = 4.90 \text{ in} > 4 \text{ in (diam. washer)}$$

7.67 4×6 bearing on plane of notch in $6 \times 6 = \dfrac{5000}{3.5 \times 5.5}$

$= \mathbf{260}$ psi $(\mathbf{1.79}$ MPa$) < 838$ psi (see answer 7.66) $\therefore$ OK   **a**

7.68   Let N be the load per ft^2.
For bearing:

$$V = 1.33 \times 14 \times N \times \tfrac{1}{2} = 9.33N$$

$$s_b = \frac{V}{bl} = \frac{9.33N}{2.62 \times 6} = 380$$

$$N = 641 \text{ psf } (30.69 \text{ kPa})$$

For shear:

$$s_s = \frac{3}{2}\frac{V}{A} = \frac{3 \times 9.33N}{2 \times 2.62 \times 11.5} = 120$$

$$N = 259 \text{ psf } (12.40 \text{ kPa})$$

For bending:

$$f = \frac{M}{S} = \frac{\tfrac{1}{8}(1.33N) \times 14 \times 14 \times 12}{\tfrac{1}{6} \times 2.62 \times 11.5 \times 11.5} = 1600$$

$$N = 236 \text{ psf } (11.30 \text{ kPa})$$

For deflection:

$$\frac{\text{Span}}{360} = \frac{5wl^4}{384EI} = \frac{14 \times 12}{360} = \frac{5(1.33N) \times 14^4 \times 12^3}{384 \times 1,600,000 \times 2.62 \times 11.5^3/12}$$

$N = \mathbf{216}$ psf (**10.34** kPa)   **d**

This is the largest loading to satisfy all four conditions.

7.69 From answer 7.68, deflection is most critical.

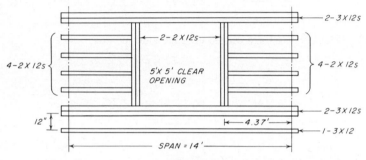

Fig. 7.69

$$\Delta \text{ for one } 3 \times 12 = \frac{5 \times 250 \times 14^4 \times 1728}{384 \times 1,600,000 \times 333} = \frac{5wl^4}{384EI}$$

$$= 0.406 \text{ in (10.3 mm)} < \text{allow.}$$

$$= \frac{14 \times 12}{360} = 0.467 \text{ in (11.9 mm)}$$

Δ for two concentrated loads on one 3×12

$$F = \text{concentrated load} = 2 \times 250 \times \frac{4.37}{2} = 1092 \text{ lb (4857 N)}$$

$$a = 4.37 \text{ ft}$$
$$l = 14 \text{ ft}$$

$$\Delta = \frac{Fa(3l^2 - 4a^2)}{24EI}$$

$$= \frac{1092 \times 4.37(3 \times 14 \times 14 - 4 \times 4.37 \times 4.37)1728}{24 \times 1,600,000 \times 333}$$

$$= 0.330 \text{ in (8.4 mm)} < 0.467 \text{ in } (0.406 + 0.330)/2 = 0.368 \text{ in}$$

The two 3 × 12s will have a Δ = 0.368 in < 0.467 in.

$V = 7 \times 250 + 1092 = 2842$ (9.35 mm)

$$V = \frac{3V}{2A} = \frac{3 \times 2842}{2 \times 5.25 \times 11.5} = 71 \text{ psi (490 kPa)} < 120 \qquad \therefore \text{ OK}$$

$$f = \frac{(\frac{1}{8} \times 250 \times 14 \times 14 \times 12) + (1092 \times 4.37 \times 12)}{\frac{1}{6} \times 5.25 \times 11.5 \times 11.5}$$

$$= \frac{73{,}500 + 57{,}500}{115.7} = 1132 \text{ psi (7.81 MPa)} < 1600 \qquad \therefore \text{ OK}$$

7.70 If four 2 × 8s are placed on edge side by side,

$$I = 4 \times \frac{1}{12} \times 1.62 \times 7.5^3 = 228 \text{ in}^4 \text{ (9490 cm}^4\text{)}$$

If placed in the form of an *I* or hollow box 10.75 in deep,

$$I = \frac{1}{12} \times 7.5 \times 10.75^3 - \frac{1}{12}(7.5 - 2 \times 1.62) \times 7.5^3$$

$$= 777 - 149 = \mathbf{628} \text{ in}^4 \text{ (}\mathbf{26{,}139} \text{ cm}^4\text{)} \qquad \mathbf{c}$$

7.71 One 2 × 8, $A = 1.62 \times 7.5 = 12.15$ in^2.

Q = statical moment = $12.15 \times 4.56 = 55.4$ in^3

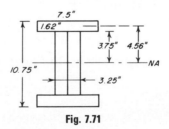

Fig. 7.71

If placed in the form of an *I*,

$$v = \frac{VQ}{bI} \text{ or } \frac{V}{v} = \frac{bI}{Q} = \frac{3.25 \times 628}{55.4} = 36.8 \text{ in}^2 \text{ at joint}$$

$$v = \frac{3V}{2A} \text{ or } \frac{V}{v} = \frac{2A}{3} = \frac{2}{3} \, 3.25 \times 10.75 = 23.2 \text{ in}^2 \text{ at N.A.}$$

If placed side by side on edge,

$$v = \frac{3V}{2A} \text{ or } \frac{V}{v} = \frac{2}{3} \times 4 \times 12.15 = \mathbf{32.4} \text{ in}^2 \text{ (}\mathbf{209.0} \text{ cm}^2\text{)} \qquad \mathbf{b}$$

7.72 Dense southern yellow pine has the following properties:

$$E = 1,600,000 \text{ psi (11.03 GPA)}$$

$$\frac{p}{A} \text{ for } \frac{L}{d} < 10 = 1285 \text{ psi (8.86 MPa)} = f_c$$

and

$$\frac{p}{A} \text{ for } \frac{L}{d} > 10 = 1285C$$

$$k = \frac{\pi}{2}\sqrt{\frac{E}{4f_c}} = \frac{3.14}{2}\sqrt{\frac{1,600,000}{4 \times 1285}} = 27.8 \text{ or } 28$$

$$C = \left[1 - \frac{1}{3}\left(\frac{L}{kd}\right)^4\right]$$

$$\frac{L}{d} = \frac{6 \times 12}{5.5} = 13.1$$

$$C = \left[1 - \frac{1}{3}\left(\frac{13.1}{28}\right)^4\right] = 0.98$$

$$CP/A = 1285 \times 0.98 = 1260 \text{ psi (8.69 MPa)}$$

$$N = 1260 \times (5.5 \times 9.5) = 65.8 \text{ kips (292.7 kN)} > 60 \text{ kips}$$

Use **6 × 10 (152.4 × 254** mm).   **c**

7.73 Try *a* 6 × 6 (152.4 × 152.4 mm)

$$\frac{L}{d} = \frac{12 \times 12}{5.5} = 26.2$$

$$C = \left[1 - \frac{1}{3}\left(\frac{26.2}{28}\right)^4\right] = 0.75$$

$$\frac{CP}{A} = 1285 \times 0.75 = 963 \text{ psi (6.64 MPa)}$$

$$N = 963 \times (5.5 \times 5.5) = 29,100 \text{ lb (129.4 kN)} > 27,500 \text{ lb}$$

$\therefore$ **6 × 6** in (**152 × 152**) OK.   **c**

7.74 Try four 2 × 10s (51 × 254 mm). Box is 1.62 + 9.5 = 11.12 by 11.12 in (282.4 by 282.4 mm) outside

$$\frac{L}{d} = \frac{12 \times 12}{11.12} = 13$$

$$C = [1 - \frac{1}{3}(\tfrac{13}{28})^4] = 0.98$$

$$\frac{CP}{A} = 1285 \times 0.98 = 1260 \text{ psi } (8.69 \text{ MPa})$$

$$N = 4 \times (1.62 \times 9.5) \times 1260 = 77,600 \text{ lb} > 75,000 \text{ lb } (345.2 \text{ kN})$$

$\therefore$ four **2 × 10s** OK.   **c**

7.75   4 × 8 is 3.62 × 7.5 in dressed size, $A = 27.2 \text{ in}^2$
        2 × 4 is 1.62 × 3.62 dressed size, $A = 5.89 \text{ in}^2$
$$\begin{array}{r} 5.89 \\ \hline 38.98 \text{ in}^2 \end{array}$$

Least dimension is $(1.62 + 1.62 + 3.62) = 6.86 \text{ in} < 7.5 \text{ in}$

$$\frac{L}{d} = \frac{10 \times 12}{6.86} = 17.5$$

$$C = \left[1 - \frac{1}{3} \left(\frac{17.5}{28} \right)^4 \right] = 0.949$$

$$\frac{CP}{A} = 1285 \times 0.949 = 1220 \text{ psi}$$

$$N = 38.98 \times 1220 = \textbf{47,500} \text{ lb } (\textbf{211.3} \text{ kN}) \qquad \textbf{c}$$

SYMBOLS
AND
NOTATION

$''$	=	inch (in); second (sec of angle)
$''2$	=	square inch (in^2)
$''3$	=	cubic inch (in^3)
$''4$	=	inches 4th (in^4)
$'$	=	foot (ft); minute (min of angle)
$'2$	=	square foot (ft^2)
$'3$	=	cubic foot (ft^3)
$'4$	=	feet 4th (ft^4)
I	=	moment of inertia ($''4$ or $'4$)
#	=	pound (lb)
k	=	kip (1000-pound unit)
T	=	ton (2000-pound unit)
$''$#	=	inch-pound, (in·lb)
$'$#	=	foot-pound, (ft·lb)
$'$k	=	foot-kip, kip-foot (ft·kip)
psi	=	pounds per square inch, #/$''2$
psia	=	pounds per square inch absolute, #a/$''2$
psig	=	pounds per square inch gage, #g/$''2$
psf	=	pounds per square foot, #/$'2$
pci	=	pounds per cubic inch, #/$''3$
pcf	=	pounds per cubic foot, #/$'3$
ksf	=	kips per square foot, k/$'2$
ksi	=	kips per square inch, k/$''2$
mph	=	miles per hour

fps = feet per second
cps = cycles per second
π = ratio of circumference to diameter of a circle, or 3.1416
g = gram
gr = grain
kg = kilogram or 1000 grams
liter = 0.001 cubic meter
° = degree of temperature or of angle
°F = degree Fahrenheit
°R = degree Rankin [°F (absolute)]
°C = degree Celsius
K = kelvin [°C (absolute)]
hp = horsepower [33,000 ft·lb/min or 550 ft·lb/s; 745.7 watts (W)]
dB = decibel
Re = Reynolds number
A = ampere
C = coulomb
V = volt
W = watt
kW = 1000 watts
cd = candela
lm = lumen
lx = lux
Hg = mercury
J = joule
N = Newton
Pa = pascal
/ = per
rad = radian
sr = steradian (solid angle)
a = acceleration, feet per second per second [ft/(s·s), ft/s^2, or fps/s]
g = acceleration due to gravity (32.2 ft/s^2)
v = velocity, units per second (ft/s)
v_0 = initial velocity
M = mass or w/g, weight/gravity
F = force or ma (mass × acceleration)
f = coefficient of friction; frequency

REFERENCE
TABLES

PREFIXES FOR SI UNITS

10^{12}	= T	= tera
10^9	= G	= giga
10^6	= M	= mega
10^3	= k	= kilo
10^2	= h	= hecto
10	= da	= deka
10^{-1}	= d	= deci
10^{-2}	= c	= centi
10^{-3}	= m	= milli
10^{-6}	= μ	= micro
10^{-9}	= n	= nano
10^{-12}	= p	= pico

CONVERSION FACTORS

(E = exact)

Multiply	By	To obtain
	Length	
inch	25.4 E	millimeter (mm)
inch	2.54 E	centimeter (cm)
inch	0.0254 E	meter (m)
foot	0.3048 E	meter (m)
yard	0.9144 E	meter (m)
meter	3.281	foot (ft)
mile (statute)	1.6093	kilometer (km)
kilometer	0.6214	mile (statute)
rod	5.0292 E	meter (m)
surveyor chain (66 ft)	20.1168	meter (m)

CONVERSION FACTORS (*Continued*)

Multiply	By	To obtain
	Area	
square inch	6.452	square centimeter (cm^2)
square foot	0.0929	square meter (m^2)
square yard	0.8361	square meter (m^2)
square meter	10.764	square feet (ft^2)
square rod	25.293	square meter (m^2)
acre	4046.9	square meter (m^2)
acre	0.4047	hectare (ha)
hectare	2.471	acre
hectare	10,000	square meter (m^2)
	Volume (capacity)	
cubic inch	16.387	cubic centimeter (cm^3)
cubic foot	0.02832	cubic meter (m^3)
cubic yard	0.7646	cubic meter (m^3)
cubic meter	35.32	cubic foot (ft^3)
liter	1000	cubic centimeter (cm^3)
liter	0.001	cubic meter (m^3)
gallon (U.S.)	3.785	liter
gallon (U.S.)	0.003785	cubic meter (m^3)
gallon (U.S.)	0.8327	Imperial gallon
ounce (U.S.)	29.57	cubic centimeter (cm^3)
	Quadruple	
inch to 4th power	41.623	centimeter to 4th power (cm^4)
foot to 4th power	0.00863	meter to 4th power (m^4)
meter to 4th power	115.884	foot to 4th power (ft^4)
	Force	
dyne	0.00001	newton (N)
pound-force	0.4536	kilogram-force (kgf)
pound-force	4.448	newton (N)
pound-force/foot	14.593	newton/meter (N/m)
pound-force/inch	1.7512	newton/centimeter (N/cm)
pound-force	4.448	newton (N)
kip	453.6	kilogram-force (kgf)
kip	4448	newton (N)
kilogram-force	9.806	newton (N)
ton-force (2000 lb)	907.2	kilogram-force (kgf)
ton (2000 lb)	8896	newton (N)

ton (2000 lb)/ft	29,186	newton/meter (N/m)
ton-force (metric)	1000 E	kilogram-force (kgf)
ton (metric)	9806	newton (N)

Pressure or stress (force per area)

dyne/square centimeter	0.10	pascal (Pa)
pascal	1.00	newton/square meter (N/m²)
kilogram-force/square meter	9.807	newton/square meter (N/m²)
kip/square inch (ksi)	70.31	kilogram-force/square centimeter (kgf/cm²)
pound-force/square foot	4.882	kilogram-force/square meter (kgf/m²)
pound-force/square foot	47.88	newton/square meter (N/m²)
pound-force/square inch (psi)	0.07031	kilogram-force/square centimeter (kgf/cm²)
pound-force/square inch (psi)	6895	newton/square meter (N/m²)
pound-force/square inch (psi)	6895	pascal (Pa)
mmHg	0.133	kilopascal (kPa)
atmosphere	101.3	kilopascal (kPa)

Bending moment or torque

inch-pound-force	0.01152	meter-kilogram-force (m·kgf)
inch-pound-force	0.1130	newton-meter (N·m)
foot-pound-force	0.1383	meter-kilogram-force (m·kgf)
foot-pound-force	1.3558	newton-meter (N·m)
meter-kilogram-force	9.807	newton-meter (N·m)
foot-ton (2000 lb)-force	276.5	meter-kilogram-force (m·kgf)
foot-ton (2000 lb)-force	2712	newton-meter (N·m)

Energy

watthour	3600	joule (J)
calorie	4.1868	joule (J)
Btu	1055	joule (J)
Btu/pound	2326	joule/kilogram (J/kg)
Btu/(ft²)(°F)(h)	20.44	kJ/(m²)(°C)(h)
Btu/(ft²)(°F)(h)	5.6745	W/(m²)(°C)

Power

Btu/(ft²)(h)	3.1525	watt/meter² (W/m²)
Btu/(ft²)(day)	75.66	watt/meter² (W/m²)
Btu/hour	0.2929	watt (W)
horsepower (boiler)	9809.5	watt (W)
horsepower (electric)	745.7	watt (W)
horsepower (33,000 ft·lb/min)	42.40	Btu/min

CONVERSION FACTORS (*Continued*)

Multiply	By	To obtain
	Mass	
grain	0.0648	gram (g)
grain/gallon	0.0171	gram/liter (g/liter)
ounce-mass (avdp)	28.35	gram (g)
pound-mass (avdp)	0.4536	kilogram (kg)
ton (metric)	1000 E	kilogram (kg)
ton (2000 lb)	907.2	kilogram (kg)
ton (metric)	1.1	ton (2000 lb)
	Mass per volume	
pound-mass/cubic foot	16.017	kilogram/cubic meter (kg/m^3)
pound-mass-cubic yard	0.5933	kilogram/cubic meter (kg/m^3)
pound-mass/gallon (U.S.)	119.8	kilogram/cubic meter (kg/m^3)
pound-mass/gallon (Imperial)	99.78	kilogram/cubic meter (kg/m^3)
	Temperature	
°R	1/1.8	K
°F	1/1.8	°C
°F	1.0	°R
°C	1.8	°F
°C	1.0	K

Temp. normal: $t_F = 1.8t_C + 32$; $t_C = (t_F - 32)/1.8$.
Temp. absolute: $t_R = (t_F + 459.67)$; $t_K = (t_F + 459.67)/1.8 = t_C + 273.15$.

REINFORCING BARS FOR CONCRETE CONSTRUCTION

Bar no.	Nominal size, in	Diameter		Area	
		in	mm	in^2	mm^2
3	³⁄₈	0.375	9.53	0.11	71.0
4	¹⁄₂	0.500	12.70	0.20	129.0
5	⁵⁄₈	0.625	15.88	0.31	200.0
6	³⁄₄	0.750	19.05	0.44	283.9
7	⁷⁄₈	0.875	22.22	0.60	387.1
8	1	1.000	25.40	0.79	509.7
9	1¹⁄₈	1.128	28.65	1.00	645.2
10	1¹⁄₄	1.270	32.26	1.27	819.4
11	1³⁄₈	1.410	35.81	1.50	967.8
14S	1³⁄₄	1.693	43.20	2.25	1451.7
18S	2¹⁄₄	2.257	57.33	4.00	2580.8

STUDY TEXTS

BASIC FUNDAMENTALS

CHEMISTRY, MATERIALS SCIENCE, AND NUCLEONICS

Anderson, J. C., and K. D. Leauer, "Materials Science," 2d ed., Van Nostrand-Reinhold, 1975.
Barrow, G., "Physical Chemistry," 4th ed., McGraw-Hill, 1979.
Brady, G. S., and H. Clauser, "Materials Handbook," 11th ed., McGraw-Hill, 1977.
Brick, R., R. Gordon, and A. Phillips, "Structure and Property of Alloys," 3d ed., McGraw-Hill, 1965.
Bush, H. D., "Atomic Nuclear Physics," Prentice-Hall, 1962.
DiBennedetto, A., "Structure and Properties of Materials," McGraw-Hill, 1967.
Edelglass, Stephen M., "Engineering Material Science," Ronald, 1966.
Hagelberg, P., "Physics: An Introduction for Students of Science and Engineering," Prentice-Hall, 1973.
Keyser, Carl A., "Materials Science in Engineering, 3d ed., Merrill, 1980.
Quagliano, J. V., and L. M. Vallarino, "Chemistry," 3d ed., Prentice-Hall, 1969.
Sienko, M., and R. Plane, "Chemistry," 5th ed., McGraw-Hill, 1975.

ELECTRICITY

Dawes, C., "Industrial Electricity," vol. II, 3d ed., McGraw-Hill, 1960.
Durney, C., and C. Johnson, "Introduction to Modern Electromagnetics," McGraw-Hill, 1969.
Foecke, H. A., "Introduction to Electrical Engineering Science," Prentice-Hall, 1961.
Grob, B., "Basic Electronics," 4th ed., McGraw-Hill, 1977.
——, and M. Kiver, "Applications of Electronics," 2d ed., McGraw-Hill, 1966.
Leach, D. P., "Basic Electric Circuits," 2d ed., Wiley, 1976.
Zbar, P., "Basic Electricity," 5th ed., McGraw-Hill, 1983.

HYDRAULICS AND FLUID MECHANICS

Allen, T., Jr., and L. Ditsworth, "Fluid Mechanics," McGraw-Hill, 1972.
Dake, J. M. K., "Essentials of Engineering Hydraulics," Wiley, 1972.
King, H., and E. Brater, "Handbook of Hydraulics," 6th ed., McGraw-Hill, 1976.
Shames, I., "Mechanics of Fluids," McGraw-Hill, 1962.

CHEMICAL ENGINEERING

Balzhiser, R. E., M. R. Samuels, and J. D. Eliassen, "Chemical Engineering Thermodynamics," Prentice-Hall, 1972.
Coulson, J. M., and J. F. Richardson, "Chemical Engineering," 3d ed., Pergamon, 1979.
Harriott, P., "Process Control," McGraw-Hill, 1964.
Hougen, P. A., K. M. Watson, and R. A. Ragatz, "Chemical Process Principles," 2d ed., Wiley, 1954.
Levenspiel, O., "Chemical Reaction Engineering," Wiley, 2d ed., Wiley, 1972.
McCabe, W. L., and J. C. Smith, "Unit Operations of Chemical Engineering," 2d ed., McGraw-Hill, 1967.
Meissner, H. P., "Processes and Systems in Industrial Chemistry," Prentice-Hall, 1971.
Newman, J. S., "Electrochemical Systems," Prentice-Hall, 1973.
Perry, R. H., and C. Chilton, "Chemical Engineer's Handbook," 5th ed., McGraw-Hill, 1973.
Peters, M., and K. Timmerhaus, "Plant Design and Economics for Chemical Engineers," 3d ed., McGraw-Hill, 1980.
Prausuitz, J. M., "Molecular Thermodynamics of Fluid Phase-Equilibria," Prentice-Hall, 1969.
Riegel, E. R., "Industrial Chemistry," Reinhold, 1962.
Shreve, R., "Chemical Process Industries," 4th ed., McGraw-Hill, 1977.
Treybal, R. E., "Liquid Extraction," 2d ed., McGraw-Hill, 1963.
Treybal, R. E., "Mass-Transfer Operations," 3d ed., McGraw-Hill, 1979.
Vilbrandt, F. C., "Chemical Engineering Plant Design," 4th ed., McGraw-Hill, 1959.
Williams, D. J., "Polymer Science and Engineering," Prentice-Hall, 1971.

CIVIL ENGINEERING

Abbett, R. W. (ed.), "American Civil Engineering Practice," 3 vols., Wiley, 1956.
Babbitt, H. E., J. J. Doland, and J. L. Cleasby, "Water Supply Engineering," 6th ed., McGraw-Hill, 1962.
Bowles, J. E., "Foundation Analysis and Design," 3d ed., McGraw-Hill, 1982.
Chellis, R., "Pile Foundations," 2d ed., McGraw-Hill, 1961.
Chow, V. T., "Handbook of Applied Hydrology," McGraw-Hill, 1964.
Clark, J. W., "Water Supply and Pollution Control," 3d ed., International Textbook, 1977.
Drew, D. R., "Traffic Flow Theory and Control," McGraw-Hill, 1968.
Eckenfelder, W. W., Jr., "Industrial Water Pollution Control," McGraw-Hill, 1966.
Foster, N., "Construction Estimates From Take-off to Bid," 2d ed., McGraw-Hill, 1972.
Hall, W. A., and J. A. Dracup, "Water Resources Systems Engineering," McGraw-Hill, 1970.
Hennes, R. G., and M. Ekse, "Fundamentals of Transportation Engineering," 2d ed., McGraw-Hill, 1969.
Horonjeff, R., "Planning and Design of Airports," 2d ed., McGraw-Hill, 1975.
King, H., and E. Brater, "Handbook of Hydraulics," 6th ed., McGraw-Hill, 1976.

Legault, A. R., "Highway and Airport Engineering," Prentice-Hall, 1960.
Leonards, G. A., "Foundation Engineering," McGraw-Hill, 1962.
Linsley, R. K., and J. B. Franzini, "Water Resources Engineering," 3d ed., McGraw-Hill, 1979.
Merritt, F. S., "Standard Handbook for Civil Engineers," 2d ed., McGraw-Hill, 1976.
Metcalf and Eddy, Inc., "Wastewater Engineering: Collection, Treatment, Disposal," McGraw-Hill, 1978.
Nemerow, N. L., "Industrial Waste Treatment," Addison-Wesley, 1963.
Oglesby, C. H., and L. I. Hewes, "Highway Engineering," 4th ed., Wiley, 1982.
Peurifoy, R. L., "Construction Planning, Equipment and Methods," 3d ed., McGraw-Hill, 1979.
Pignataro, L. J., "Traffic Engineering: Theory and Practice," Prentice-Hall, 1973.
Rich, L., "Environmental Systems Engineering," McGraw-Hill, 1973.
Scott, R. F., and J. J. Schoustra, "Soil: Mechanics and Engineering," McGraw-Hill, 1968.
Woods, K. "Highway Engineering Handbook," McGraw-Hill, 1960.

ELECTRICAL AND ELECTRONIC ENGINEERING

Balakrishnan, A. V., "Communication Theory," McGraw-Hill, 1968.
Blake, L. V., "Transmission Lines and Waveguides," Wiley, 1969.
Brenner, E., and M. Javid, "Analysis of Electric Circuits," 2d ed., McGraw-Hill, 1967.
Buban, P., and M. Schmitt, "Understanding Electricity and Electronics," 3d ed., McGraw-Hill, 1974.
Carlson, B., "Communication Systems," McGraw-Hill, 1968.
Chirlian, P. M., "Basic Network Theory," McGraw-Hill, 1969.
Fink, D., and J. Carroll, "Standard Handbook for Electrical Engineers," 11th ed., McGraw-Hill, 1978.
Fitzgerald, A. E., and D. H. Higginbotham, "Electrical and Electronic Engineering Fundamentals," McGraw-Hill, 1964.
Gehmlich, D. K., and S. B. Hammond, "Electromechanical Systems," McGraw-Hill, 1967.
Greenwood, A., "Electrical Transients in Power Systems," Wiley, 1971.
Leach, D. P., "Basic Electric Circuits," 2d ed., Wiley, 1976.
Lloyd, T. C., "Electric Motors and Their Applications," Wiley, 1969.
Matick, R., "Transmission Lines for Digital and Communication Networks," McGraw-Hill, 1969.
Murdock, J. B., "Network Theory," McGraw-Hill, 1970.
Stott, G. S., and G. Birchall, "Electrical Engineering Principles," McGraw-Hill, 1969.
Sunde, E. D., "Communication Systems Engineering Theory," Wiley, 1969.
Taub, H., and D. L. Schilling, "Principles of Communication Systems," McGraw-Hill, 1971.
Watt, J., and F. Stetka, "NFPA Handbook of the National Electrical Code," 3d ed., McGraw-Hill, 1972.
Weedy, B. M., "Electric Power Systems," 3d ed., Wiley, 1979.

ENGINEERING ECONOMICS AND BUSINESS RELATIONS

Barish, N. N., "Economic Analysis for Engineering and Managerial Decision Making," McGraw-Hill, 1962.
Canada, J. R., "Intermediate Economic Analysis for Management and Engineering," Prentice-Hall, 1971.
Grant, E. L., and W. G. Ireson, "Principles of Engineering Economy," 5th ed., Ronald, 1970.
Jelen, F. C., "Cost and Optimization Engineering," McGraw-Hill, 1970.
Riggs, J. L., "Economic Decision Models for Engineers and Managers," McGraw-Hill, 1968.
Thuesen, H. G., W. J. Forbrycky, and G. J. Thuesen, "Engineering Economy," 5th ed., Prentice-Hall, 1977.

MECHANICAL ENGINEERING

Baumeister, T., and L. Marks, "Standard Handbook for Mechanical Engineers," 8th ed., McGraw-Hill, 1978.
Black, P. H., and O. E. Adams, Jr., "Machine Design" 3d ed., McGraw-Hill, 1968.
Carrier Air Conditioning Co., "Handbook of Air Conditioning System Design," McGraw-Hill, 1965.
Emerick, R., "Heating Handbook," McGraw-Hill, 1964.
Gebhart, B., "Heat Transfer," 2d ed., McGraw-Hill, 1970.
Greenwood, D., "Mechanical Power Transmission," McGraw-Hill, 1962.
Harris, N., "Modern Air Conditioning Practice," 2d ed., McGraw-Hill, 1983.
Holman, J. P., "Thermodynamics," 3d ed., McGraw-Hill, 1980.
Lichty, L. C., "Combustion Engine Processes," 7th ed., McGraw-Hill, 1967.
Phelan, R. M., "Fundamentals of Mechanical Design," 3d ed., McGraw-Hill, 1970.
Power Magazine, "Plant Energy Systems," McGraw-Hill, 1967.
——, "Power Generation Systems," McGraw-Hill, 1967.
Reynolds, W. C., and H. C. Perkins, "Engineering Thermodynamics," 2d ed., McGraw-Hill, 1977.
Rothbart, H., "Mechanical Design and Systems Handbook," McGraw-Hill, 1964.
Stoecker, W. F., "Design of Thermal Systems," 2d ed., McGraw-Hill, 1980.
Woodruff, E., and H. Lammers, "Steam-Plant Operation," 4th ed., McGraw-Hill, 1976.

STRUCTURAL ENGINEERING

Bowles, J. E., "Foundation Analysis and Design," 3d ed., McGraw-Hill, 1982.
Coull, A., and A. Dyke, "Fundamentals of Structural Theory," McGraw-Hill, 1972.
Gaylord, E. H., and C. N. Gaylord, "Structural Engineering Handbook," 2d ed., McGraw-Hill, 1979.
Johnson, S., and T. Kavanagh, "The Design of Foundations for Buildings," McGraw-Hill, 1968.
Khachaturian, N., and G. Gurfinkle, "Prestressed Concrete," McGraw-Hill, 1969.

LaLonde, W., and M. Janes, "Concrete Engineering Handbook," McGraw-Hill, 1962.

Laursen, H. I., "Structural Analysis," 2d ed., McGraw-Hill, 1977.

Merritt, F., "Structural Steel Designers' Handbook," McGraw-Hill, 1972.

Norris, C. H., and J. B. Wilbur, "Elementary Structural Analysis," 3d ed., McGraw-Hill, 1976.

Shermer, C. L., "Design in Structural Steel," Ronald, 1972.

Tomlinson, M. J. "Foundation Design and Construction," 4th ed., Wiley, 1980.

Winter, G., and A. H. Nilson, "Design of Concrete Structures," 9th ed., McGraw-Hill, 1979.

"Building Code Requirements for Reinforced Concrete," latest ed., American Concrete Institute.

"Code for Arc and Gas Welding in Building Construction," latest ed., American Welding Society. (Brochures)

"Specifications for Railway Bridges," latest ed., American Railway Engineering Association.

"Standard Specifications for Highway Bridges," latest ed., American Association of State Highway Officials.